W0106595

Akira Azuma

The Biokinetics
of Flying
and Swimming

With 273 Figures

Springer-Verlag
Tokyo Berlin Heidelberg New York London
Paris Hong Kong Barcelona Budapest

AKIRA AZUMA
Professor Emeritus of the University of Tokyo
37-3 Miyako-cho,
Saiwai-ku, Kawasaki,
210 Japan

On the front cover: Flying squid, *see* p. 54.

ISBN-13: 978-4-431-68212-7   e-ISBN-13: 978-4-431-68210-3
DOI: 10.1007/978-4-431-68210-3

Printed on acid-free paper

Library of Congress Cataloging-in-Publication Data. Azuma, Akira, 1927– . The blokinetics of
flying and swimming/Akira Azuma. p. cm. Includes bibliographical references and index. ISBN-
13: 978-4-431-68212-7 1. Flying—Physiological aspects. 2. Swimming—Physiological aspects.
3. Biomechanics. I. Title. QP310.F5A98 1992. 574.1'8—
dc20. 92-590

© Springer-Verlag Tokyo 1992
Softcover reprint of the hardcover 1st edition 1992

This work is subject to copyright. All rights are reserved, whether the whole or part of the material
is concerned, specifically the rights of translation, reprinting, reuse of illustrations, recitation,
broadcasting, reproduction on microfilms or in other ways, and storage in data banks.
The use of registered names, trademarks, etc. in this publication does not imply, even in the
absence of a specific statement, that such names are exempt from the relevant protective laws and
regulations and therefore free for general use.

Typesetting: Asco Trade Typesetting Ltd., Hong Kong

Preface

Animals and plants range in body size from small single-celled creatures of the order of microns and picograms, such as bacteria, to huge mammals that may, like the whales, reach tens of meters in length and several hundred tons in weight. Some live in water and some in air, and each selects the mode of locomotion it finds best suited to its particular environment and ecology.

Animals have bodies with a density that is nearly the same as that of water, presumably because they lived in water in the early stages of their evolution. Regardless of size, therefore, they can stay afloat in water without great effort to support their weight. This means that in locomotion they can concentrate on propelling themselves against the resistive and reactive forces caused by their motion relative to the water. The situation is very different, however, for animals that move in the air, since the buoyant force provided by air is, at only 1/820 that of water, substantially negligible. Thus, unless they are very tiny, the first thing these animals must do is to generate an aerodynamic force to support themselves. This necessity puts severe restrictions on the size and weight of airborne animals, and on the modes of locomotion they can adopt. On the other hand, it is to the advantage of airborne animals that they are required to overcome much less resistance than their aquatic cousins.

The same fluid environment feels stickier to a small animal than to a large one: below a certain size the effect of the fluid's viscosity becomes greater than the effect of its inertia. To microorganisms, water is a highly resistive fluid, and swimming in it must feel to them as swimming in honey or warm pitch would feel to us. In contrast, whales experience water to be less sticky than we do and propel themselves using inertial force.

Almost all swimming animals, both fish and mammals, are considerably longer in the direction in which they swim than in the directions at right angles to it. In other words, they are slender. Their modes of swimming can be broadly divided into snaking and fanning. The first mode is used by creatures with long flexible bodies and is characterized by series of traveling waves sent from head to tail. In microscopic organisms the amplitude of the waves is constant along the body length, while in larger creatures it usually increases to the rear. The second mode is used by swimmers that have a relatively inflexible anterior body section and a generally flat, flexible posterior section. Because of this configuration, undulations sent rearward along the body attain a significant amplitude only in the posterior section, rarely reaching more than a quarter- to a half-wavelength. In some swimmers, the fanning motion is limited to the caudal fin. Other swimming methods include the ciliary propulsion used by some minute animals, jetting, whipping, and paddling.

Powered flying is generally confined to animals in the middle of the size spectrum and is performed by using laterally extended wings, rather than a slender body. Small animals fly exclusively by beating wings located near the center of gravity, whereas large animals intermittently discontinue beating and glide with the wings extended. As size decreases, the beating frequency increases and the flight becomes maneuverable with a swift motility and violent beating. As the way of beating changes between the feathered wings of birds and the membranous wings of insects, the effects of air viscosity on the generation of aerodynamic force become more pronounced.

In this book, the ways in which living creatures disperse themselves will be examined. Myriad forms of life have continued to extend their habitat all over the globe, frequently sacrificing the individual in order to enhance the chances for the survival of the species. This can be seen even today in some species of dragonflies, for example. These insects are known to depart from their habitat and even to set off on transoceanic journeys, apparently with the sole purpose of disseminating the species into new regions. This book will investigate the dispersal modes, within the framework of flying and swimming mechanics, of various plants and animals.

Through study of the locomotion of living creatures, the author has come to the conclusion that every creature has the body shape and uses the mode of locomotion that is most economical for the way of life it has developed through natural selection, in response to severe environmental conditions. Thus, one purpose of this book is to shed light on the physical relationships among habitat, form, way of life, and mode of movement in living creatures. The ways in which creatures move about will be introduced and analyzed in depth, both mechanically and mathematically.

It is also a purpose of this book to introduce the results of theoretical and empirical research carried out by various scientists over the years. These will be taken up in the sections to which, from the author's viewpoint, they are most closely related.

Although this study was undertaken to explain the biomechanical aspects of locomotion, it focuses primarily on the mechanical or kinetic principles that animals use for moving from one place to another and gives only secondary attention to the physiological principles of the locomotor organs and neuromuscular processes that enable animals to utilize their muscular energy. From a mechanical viewpoint, the external shape of an animal is a very important factor in locomotion as well as the topological interconnection of the individual body organs viewed in an anatomical sense. Therefore, the author generally confined himself to a consideration of the physical forces and the external body forms involved in the movements of the subjects. Each of the two main parts of this book, Flying Dynamics and Swimming Dynamics,

is written from the viewpoint of mechanics, specifically fluid dynamics and flight dynamics, rather than from the viewpoint of physiology and ecology. Thus the chapters and sections are organized according to mechanical, not biological, principles.

However, the author hopes that the book will also prove a useful reference for biologists who need occasionally to use mechanical analyses for a full understanding of the behavior of animals and the mechanical functions of body parts in relation to their forms and modes of locomotion.

Acknowledgments. The author is grateful to his former graduate students for their contributions: Dr. Keiji Kawachi (Professor of the University of Tokyo) and Dr. Kenichi Nasu (Associate Professor of Ryukyu University) developed the local momentum theory (LMT) and extended the LMT to the local circulation method (LCM), and Dr. Yoshinori Okuno (research staff of the National Aerospace Laboratory) and Dr. Shigeru Sunada (graduate student of the University of Tokyo) assisted with the calculations.

The author is greatly indebted to Mr. Kunio Yasuda (Nihon University), Ms. Eiko Shimizu, Ms. Chizuko Amano, and Ms. Takako Inoue for gathering materials, arranging references and typing the original manuscripts; and to Mr. Kunio Yasuda who contributed greatly to the illustrated drawings.

In writing of this book the author is also greatly indebted to his cousin, Mr. Shogoro Azuma, the president of Sanwa Kogyo Co. Ltd., and Mr. Tadaki Kawada, the president of Kawada Industries Inc., for providing support and heartfelt encouragement.

A special acknowledgement is made to Springer-Verlag Tokyo for their invaluable assistance in the editing and publishing of this book, and to Mr. Richard Foster of IEC for his efforts on behalf of the author's English style.

Finally, the author wishes to express his thanks to his family for their patience and support during a long period of hard work and irregular hours.

AKIRA AZUMA

Contents

Symbols

$A_{.j}$	added masses (tensor of i, j = 1–6)	(B_0, B_1)	amplitudes of bending moment defined in Table 4.2-6
$Æ$	aspect ratio $= b^2/S$	b	wingspan, semichord $= c/2$, width or depth of body, amplitude rise parameter defined by Eq. (5.2-4), width of plate
$Æ_e$	effective aspect ratio $= Æ \cdot e$		
$Æ_{e,0}$	original or reference aspect ratio defined by Eq. (3.2-25)		
$Æ_{e,H}$	equivalent aspect ratio in ground effect	$\bar{b}$	nondimensional wing span $= b/b_0$
(A, A_0)	parameters defined by Eqs. (4.3-30 and -31)	b'	projected wing span or wing span in deflected state
(A, B, C, D, E)	parameters defined in Table 3.2-3	b_b	maximum width of trunk or body
(A_0, A_1, A_2)	parameters defined in Table 4.2-2	b_e	effective wingspan $= \sqrt{Æ_e S}$
$(\bar{A}, \bar{M}, \bar{I})$	parameters defined by Eq. (3.6-13)	b_{max}	maximum height of body
$(Æ_h, Æ_v)$	aspect ratios of horizontal and vertical tails	b_0	reference wing span, height at the origin of coordinate
$(Æ_{e,w}, Æ_{e,h})$	equivalent aspect ratios of main wing and horizontal tail	b_v	distance between a pair of rolled-up tip vortices
$(Æ^f, Æ^h)$	aspect ratio of fore and hind wings	(b^f, b^h)	wing span of fore and hind wings
a	amplitude of undulation, lift slope, nondimensional distance along x axis, amplitude of transversal deflection of head, half-span of the major asis of ellipse, lapse rate defined by Eq. (4.3-27)	(b_h, b_v)	wing spans of horizontal and vertical tails
		(b_m, b_t)	heights of peduncle and of caudal fin
$\bar{a}$	nondimensional position of a point of interest about which the moment is considered, time ratio of lowering to raising, lift slope $= \partial C_l/\partial \alpha$ or mean sectional lift slope $= C_l/\alpha$	C	sound speed, wave speed, integration constant, parameter defined in Table 3.2-3, cost of transport defined by Eq. (4.1-23) $= P_n/WU$
a_0	radius of a spherical head of a microorganism, theoretical lift slope of thin wing $= 2\pi$	$\hat{C}$	Theodorsen function $= F + iG$
		C_D	drag coefficient $= D/\frac{1}{2}\rho U^2 S$
ac	aerodynamic center	C_{D_0}	parasite drag coefficient or drag coefficient at zero lift or minimum drag coefficient $= D_0/\frac{1}{2}\rho U^2 S$
(a, b)	semiaxes of axisymmetric ellipsoid, parameters defined by Eq. (6.3-61 and -62), longitudinal and lateral distances	$C_{D,A}$	drag coefficient caused by air
		C_{D_i}	induced drag coefficient $= D_i/\frac{1}{2}\rho U^2 S$
(a_c, a_s)	parameters defined in Table 4.2-2	$C_{D_{i,solo}}$	induced drag coefficient of solo wing
B	bending moment, buoyant force, tip loss factor defined in Table 7.3-1	$C_{D,m}$	drag increment of surface roughness
		$C_{D,w}$	drag coefficient based on wetted surface area S_w
B_e	bending moment of elliptic wing	$C_{D,|\gamma|_{min}}$	drag coefficient at $|\gamma|_{min}$
		$(C_{D,b}, C_{D,w})$	parasite drag coefficients caused by trunk and wing based on wing area S

$(C_{D,f}, C_{D,v})$	drag coefficients based on frontal area S_f and two-thirds of volume $V_B^{2/3}$
(C_{D_N}, C_{D_T})	drag coefficients of normal and tangential components
$(C_{D_{0,w}}, C_{D_{0,h}})$	profile drag coefficients of main wing and horizontal tail
C_d	drag coefficient of airfoil $= d/\frac{1}{2}\rho U^2 c$
C_{d_0}	drag coefficient of airfoil at zero lift or minimum drag coefficient of airfoil
C_f	friction drag coefficient
C_L	lift coefficient $= L/\frac{1}{2}\rho U^2 S$
C_{L_0}	lift coefficient defined in Table 4.2-6 $= L_0/\frac{1}{2}\rho U^2 S$
$C_{L,A}$	lift coefficient caused by air
$C_{L,(L/D)_{max}}$	lift coefficient at $(L/D)_{max}$
$C_{L,X_{max}}$	lift coefficient for the maximum range
$C_{L,|\gamma|_{min}}$	lift coefficient at $|\gamma|_{min} = C_{L,(L/D)_{max}}$
$C_{L,|w|_{min}}$	lift coefficient at $|w|_{min}$
C_{L_a}	lift slope $= \partial C_L/\partial \alpha$
$(C_{L,h}, C_{L,w})$	lift coefficients of horizontal tail and main wing
$(C_{L_{max}}, C_{L_{min}})$	maximum and minimum lift coefficients
C_l	lift coefficient of airfoil $= l/\frac{1}{2}\rho U^2 c$
$C_{l_{max}}$	maximum lift coefficient of airfoil
(C_{l_θ}, C_{l_h})	lift components given in Table 4.2-3, with subscript o, c, and s
C_m	moment coefficient of airfoil $= m/\frac{1}{2}\rho U^2 c^2$
$C_{m,ac}$	moment coefficient about aerodynamic center $= m_{ac}/\frac{1}{2}\rho U^2 c^2$
$C_{m,c/4}$	moment coefficient about a quarter chord $= m_{c/4}/\frac{1}{2}\rho U^2 c^2$
C_{m_a}	moment slope $= \partial C_m/\partial \alpha$
$C_{m_{max}}$	maximum moment coefficient
(C_{m_θ}, C_{m_h})	moment components given in Table 4.2-3, with subscript o, c, and s
(C_N, C_T)	normal and tangential components of drag coefficients defined by Eq. (5.1-6)
$(C_{N,A}, C_{T,A})$	normal and tangential components of aerodynamic force
C_n	normal force coefficient $= n/\frac{1}{2}\rho U^2 S$
C_P	power coefficient $= P/\frac{1}{2}\rho U^3 l^2$ in Table 6.2-2
C_p	pressure coefficient $= -\Delta p/\frac{1}{2}\rho U^2$, two-dimensional power
$C_{p_{min}}$	minimum power coefficient (two-dimensional)
$\bar{C}_p$	mean power coefficient (two-dimensional)
$C_p{}^*$	power coefficient including absolute value of negative work defined by Eq. (4.2-20)
$(C_{p_{\theta\theta}}, C_{p_{\theta h}}, C_{p_{hh}})$	power components given in Table 4.2-4, with subscripts o, c, and s
C_Q	torque coefficient $= Q/\rho SR(R\Omega)^2$
C_R	aerodynamic force coefficient $= R/\frac{1}{2}\rho U^2 S = \sqrt{C_L^2 + C_D^2}$
$C_{R_{max}}$	maximum aerodynamic force coefficient
C_T	thrust coefficient $= T/\rho S(R\Omega)^2$, or $T/\frac{1}{2}\rho U^2 l^2$ in Table 6.2-2, or $T/\frac{1}{2}\rho U^2 S$ defined by Eq. (6.3-34)
C_{T_0}	thrust coefficient defined in Table 4.2-6 $= T_0/\frac{1}{2}\rho U^2 S$
$(C_{T,t}, C_{P,t})$	coefficients defined in Table 6.2-2
$(C_{T,A}, C_{T,B})$	thrust coefficients for sail and hull
C_t	two-dimensional thrust coefficient $= t/\frac{1}{2}\rho U^2 c$
$C_{t_{max}}$	maximum thrust coefficient (two-dimensional)
$\bar{C}_t$	mean thrust coefficient
$\bar{C}_{t,given}$	a given thrust coefficient
$(C_{t_{\theta\theta}}, C_{t_{\theta h}}, C_{t_{hh}})$	thrust components given in Table 4.2-3, with subscripts o, c, and s
c	wing chord, fuel consumption rate or mass per unit energy, body height
$\bar{c}$	mean chord $= S_w/pl$ or $b/\!R$, mean aerodynamic chord $= \int_{-b/2}^{b/2} c^2 \, dy/S$
c_0	wing chord at center section
c_{max}	maximum height of elongated body
cg	center of gravity
cp	center of pressure
$(c_r, c_{3b/8})$	chord length at wing root and $y = 3b/8$
(c_1, c_4)	coefficient of added mass defined in Table 7.1-1
$(\bar{c}_h, \bar{c}_v)$	mean aerodynamic chords of horizontal and vertical tails
D	drag (three-dimensional) $= |\mathbf{D}|$, parameter defined in Table 3.2-3
$\mathbf{D}$	drag force vector $= -iD$
D_b	drag increment caused by inflated part of body
D_f	friction drag
D_{fp}	drag of flat plate
D_{head}	drag of head of microorganisms
D_i	induced drag
$D_{i,e}$	induced drag of elliptic wing
D_w	drag of wing, drag increment caused by extended wings
$\bar{D}_w$	mean wing drag
(D_N, D_T)	normal and tangential components of drag
(D, E)	parameters defined by Eq. (3.2-3)

d	diameter of circular column, diameter of sphere, two-dimensional drag, depth of curved wing or camber depth, depth of paddle, horizontal distance
E	Young's modulus or modulus of longitudinal elasticity, stored energy defined by Eqs. (3.3-12), total kinetic energy, kinetic energy loss of the fluid in unit time, parameter defined in Table 3.2-3, endurance, consumed energy
$\bar{E}$	mean value of rate of kinetic energy
E_c	energy cost, energy required for normal cruising
(E_c, E_l)	energy required for steady continuous swimming and porpoising defined by Eqs. (6.4-10 and -11)
E_D	energy dissipated by drag
E_r	energy fraction in intermittent flight
E_s	energy dissipated in the splash
E_w	energy obtained from wind
$(E_{w,L}, E_{w,D})$	energies generated by wind with lift and drag components
E_1	energy for full powered flight E_r $(r = 1)$
E_γ	energy consumed by two-stage locomotion
e	deficiency parameter $(e < 1)$ or efficiency $(e \geq 1)$ factor of wing $= A\!R_e/A\!R$
F	force, real part of Theodorsen function, functions defined by Eq. (3.5-4), Eqs. (4.3-20 and -25), and Eq. (5.2-28)
$\mathbf{F}$	force vector $= (F_X, F_Y, F_Z)^T$ or $= (F_x, F_y, F_z)^T$
$(\mathbf{F_A}, \mathbf{F_B}, \mathbf{F_G})$	aerodynamic force vector, buoyant force vector, and gravity force vector $= -\mathbf{K}W$
F_B	buoyant force $= |\mathbf{F_B}|$
$(F_{B,x}, F_{B,y}, F_{B,z})$	(x, y, z) components of buoyant force
Fr	Froude number $= U/\sqrt{gl}$
(F_j, F_s)	driving or jetting force and suction force
(F_L, F_D)	lift and drag components of aerodynamic force
(F_H, F_V)	horizontal and vertical components of aerodynamic force
$(\bar{F}_H, \bar{F}_V)$	mean horizontal and vertical forces
(F_H^f, F_H^h)	fore and hind wing components of horizontal force
(F_V^f, F_V^h)	fore and hind wing components of vertical force
(F_X, F_Y, F_Z)	(X, Y, Z) components of force
(F_x, F_y, F_z)	(x, y, z) components of force
(F_x, F_y, M_z)	tension force along x-axis, shear force along y-axis, and bending moment about z-axis
(F, G)	parameters defined by Eqs. (3.5-4 and -7) and Eqs. (3.5-23 and -24), real and imaginary parts of Theodorsen function
(F, G, H)	performance indices defined by Eq. (4.3-24, -38, and -39)
(F_1, F_2)	functions defined in Table 6.3-5
f	frequency of vibration or of locomotion, drag area $= SC_D$ or $= SC_{D,b} = S_f C_{D,f}$
$\mathbf{f}$	specific body force $= -\nabla\Omega$
$\bar{\mathbf{f}}$	nondimensional specific body force $= \mathbf{f}/g$
$f(x)$	spanwise position of midchord of wing
f_{tot}	total drag area defined by Eq. (3.2-8b)
(f_I, f_D)	inertial force distribution and hydrodynamic drag force distributions defined in Table 5.2-2
(f_I^B, f_I^W)	inertial forces of body and fluid elements defined in Table 5.2-2
(f_x, f_y)	distributed external forces along x- and y-axes
(f_x, f_y, f_z)	(x, y, z) components of distributed force of a body element
G	nondimensional acceleration based on the gravity acceleration, constraint given by Eq. (3.5-7), imaginary part of Theodorsen function, specific range defined by Eq. (4.3-35) or Eq. (4.3-42)
g	gravitational acceleration $= 9.807 \text{ m/s}^2$
H	total head, specific total head $= P_{\mathrm{tot}}/\rho g$, jumping height, altitude or height, maximum height $= Z_{\max}$, specific height defined by Eq. (4.3-36), side force
$\bar{H}$	nondimensional specific head $= H/\rho U^2$, mean side force
$H_{\min}$	minimum height
H_{limit}	limit height defined by Eq. (2.4-4)
(H_I, H_D)	side forces produced by inertia and drag
$(\bar{H}_{I,1}, \bar{H}_{I,2})$	inertial components of side force
$(\bar{H}_{D,1}, \bar{H}_{D,2})$	drag components of side force
$(\bar{H}_I, \bar{H}_D)$	mean side forces produced by inertia and drag
$(\bar{H}_{I,1}, \bar{H}_{I,2})$	inertial components of mean side force
$(\bar{H}_{D,1}, \bar{H}_{D,2})$	drag components of mean side force
h	nondimensional height $= H/l$, wave height, heaving height, maximum height of slender body, lateral deflection height of paddle, depth of mean wave surface
h_1	amplitude or the first sinusoidal component of harmonic heaving
h_b	maximum height of body
h^+	nondimensional height defined by Eq. (6.1-5a) $= \mathrm{Re}^*$

h_{max}	maximum heaving	l_w	wing length
$h^{\#}_1$	optimum heaving amplitude ($a \neq -\frac{1}{2}$) given by Eq. (4.2-24)	$(l_{ac,h}, l_{ac,v})$	distances of aerodynamic center of horizontal and vertical tails from the head
I	moment of inertia, second moment of inertia or moment of inertia of area	(l_h, l_v)	distances of horizontal and vertical tails from cg $= (l_{ac,h} - l_{cg}, l_{ac,v} - l_{cg})$
$\bar{I}$	nondimensional moment of inertia $= I/m_w R^2$	(l_{ac}, l_{cg})	distances of aerodynamic center of main wing and center of gravity from the head
I_w	moment of inertia of one wing	(l_m, l_n, l_t)	lengths of tail-base neck, nose, and tail measured from the point at the maximum height
(I_j, I_s)	impulses of jetting and sucking forces		
(I_x, I_y)	moments of inertia about x- and y-axis		
$(\bar{I}_x, \bar{I}_y)$	nondimensional moments of inertia about x- and y-axis $= (I_x/m_w R^2, I_y/m_w R^2)$	(l_1, l_2)	arm lengths of double hinge, distances defined in Fig. 4.3-1
$(\mathbf{I}, \mathbf{J}, \mathbf{K})$	unit vectors along (X, Y, Z) axes of inertial frame	M	Mach number $= U/C$
		M	moment about flapping hinge at cg
i	imaginary $= \sqrt{-1}$	$\mathbf{M}$	moment vector $= (M_x, M_y, M_z)^T$
$(\mathbf{i}, \mathbf{j}, \mathbf{k})$	unit vectors along (x, y, z) axes of moving frame	$\bar{M}$	parameter defined by Eq. (3.6-13)
		M_B	bending moment
J	performance function defined by Eqs. (3.5-6 and -28)	$\bar{M}_B$	mean bending moment
		M_{cg}	moment about cg
K	specific available power $= P_a/m_m$	(M_B^R, M_B^L)	bending moments of right and left halteres
K_i	modified Bessel functions of ith order ($i = 0, 1, 2$)	(M_x, M_y, M_z)	(x, y, z) components of moment
K_j	coefficients of added mass defined in Table 6.1-2	m	mass of whole body, mass of fluid flow related to a wing, two-dimensional moment
K_F	specific available energy given by Eq. (4.3-30)	m_0	initial mass of whole body
k	reduced frequency $= \omega l/2U$ or $= \pi f \bar{c}/U$, equivalent angular velocity defined in Eq. (4.4-6), ratio of the wetted surface area and two-thirds of the volume $= S_w/V_B^{2/3}$	m_{ac}	two-dimensional moment about aerodynamic center
		m_c	mass of fluid inside a cup of paddle
		m_f	mass of stored fuel
		m_j	mass of jetting fluid
k_β	spring stiffness about flapping hinge	m_m	mass of muscle
(k_s, k_1)	attenuate factors in Eqs. (6.4-7–13)	m_p	mass of paddle
(k_1, k_2, k_3)	parameters defined by Eqs. (3.5-15 and -20)	m_s	mass of sucked fluid, mass of skeleton
		m_w	mass of one wing, added mass
$(k_1, k_2, k_3, k_4, k_5)$	coefficients of added mass in Table 7.1-2	$m_{c/4}$	moment about a quarter-chord
(k_h, k_v)	parameters defined in Fig. 3.5-8	N	normal force, lifting force
L	lift (three-dimensional) $= |\mathbf{L}|$	$\bar{N}$	mean lifting force
$\mathbf{L}$	lifting force vector $= \mathbf{k}L$	N_0	steady normal force defined in Table 5.1-3
L_s	lifting force generated by surface tension		
$L_{s,max}$	maximum lifting force generated by surface tension	n	load factor, two-dimensional lifting force, $= l\cos\phi + d\sin\phi$, normal force component defined by Eq. (4.2-1a) and Eq. (6.3-8b), power index for density defined by Eq. (4.3-27), number of waves, number of birds in formation flight, thrust factor
(L_h, L_w)	lifts of horizontal tail and main wing		
(L_0, L_1)	amplitudes of lift defined in Table 4.2-6		
l	length of body, length of blade of rotary seed, length of muscle, reference length $= V_B/S$, air loading or two-dimensional lift	$(0, o)$	origins of coordinate frames (X, Y, Z) and (x, y, z)
l_0	arm length of single hinge	P	critical load, power released from stored energy $= \dot{E}$, necessary power
l_h	length of hand		

$\bar{P}$	mean power	p_{tot}	total pressure defined by Eq. (7.2-11), nondimensional power $= p/\rho_0 U^2$
P_A	aerodynamically consumed power	(p, q, r)	parameters of reflexed airfoils defined in Eq. (3.1-2)
P_a	available power		
P_{ab}	aerobic metabolic power $= P_t - P_B$	Q	torque
P_{aero}	aerodynamically consumed power	Q_{head}	torque required for rotating the head
P_B	basal metabolic rate	Q_g	gyroscopic torque
P_c	climbing power	Q_0	profile torque
P_{head}	power required for rotating the head defined in Table 5.1-5	(Q_A, Q_E, Q_I)	aerodynamic, elastic, and inertial torques
P_i	induced power	$(Q_{\dot{\beta} \neq 0}, Q_{\dot{\beta} = 0})$	torques caused by flapping and non-flapping wing
$P_{i,H}$	induced power for hovering flight		
P_{in}	inertial power	q	a fraction of wavelength $= 0.09\,\lambda$, parameter defined in Fig. 5.2-5a
P_m	miscellaneous power		
P_{min}	minimum power	$\mathbf{q}$	local flow velocity
P_n	necessary power	$\bar{\mathbf{q}}$	nondimensional velocity $= \mathbf{q}/U$
$\bar{P}_n$	mean necessary power	R	range of flight, turning radius, rotor radius, radius of helix, wing length from cg or half span or from beating hinge
$P_{n,0}$	necessary power required for trimmed buoyant condition		
$P_{n,H}$	necessary power for hovering flight	R_0	radius of blade root at dipped point
$P_{n,min}$	minimum necessary power	R_s	specific range defined by Eq. (4.1-45)
P_0	profile power, steady power defined in Table 5.1-3	Re	Reynolds number $= Ul/\nu$
$\bar{P}_0$	mean profile power	Re*	Reynolds number based on surface roughness, defined by Eq. (6.1-4)
$\bar{P}_{0,H}$	mean profile power for hovering flight	R_β	radius of another side of blade defined in Fig. 3.6-3a
P_p	parasite power		
P_t	total power, metabolic rate	$(R_{max,s}, R_{max,u})$	maximum ranges for steady and unsteady gliding flights
(P_D, P_I)	power required for drag and inertial forces		
$(P_{D,1}, P_{D,2})$	drag components of power given in Table 5.2-3	r	spanwise position, time fraction of intermittent locomotion, distance $= \sqrt{x^2 + y^2}$ defined in Fig. 4.3-1
$(\bar{P}_{D,1}, \bar{P}_{D,2})$	mean powers of drag components	r_{cg}	spanwise distance of cg from flapping hinge
$(P_{I,1}, P_{I,2})$	inertial components of power given in Table 5.2-3		
$(\bar{P}_{I,1}, \bar{P}_{I,2})$	mean inertial components of power	S	reference area, wing area, disc area $= \pi R^2$, extended fin area, Strouhal number $= fd/U$, cross sectional area of the exhaust nozzle
$(P_{i,L}, P_{i,T})$	induced powers caused by lift and thrust		
$(P_{i,in}, P_{i,out})$	induced powers in- and out-of-ground effect		
(P_0, P_1, P_2)	amplitudes of power defined in Table 4.2-6	S_0	reference area
		$\bar{S}$	nondimensional wing area $= S/S_0$
$(P_{p,b}, P_{p,w})$	parasite power caused by body and wing	(S_A, S_B)	sail area and centerboard area
		S_b	body surface area
$(\bar{P}_t, \bar{P}_m, \bar{P}_i)$	mean power caused by tail end section, vortex shedding and vortex interaction defined by Eq. (6.2-13b)	S_c	cross-sectional area
		S_f	frontal area
		S_e	sweeping area or effective operational area
p	pressure, number of wings (blades), two-dimensional power, nondimensional power $= P_0/W_0 V_0$	S_j	minimum cross-sectional area of jet flow
		S_s	cross-sectional area of sucking nozzle
$\bar{p}$	nondimensional pressure $= p/\rho_0 U^2$	S_w	blade area of rotary wings (total blades)
p_v	vapour pressure	S_w	wetted surface area
p_0	static pressure	(S^f, S^h)	wing area of fore and hind wings

(S_h, S_v)	surface area of horizontal and vertical tails
s	surface tension, distance along flight path, substantial length of body, distance between two wing tips of neighbouring birds, time fraction of deceleration
s^+	nondimensional space defined by Eq. (6.1-5)
T	thrust or propulsive force, period of a complete stroke, period of half stroke
T_0	steady thrust, absolute temperature at sea level $= 273.15°$K, steady thrust defined in Table 5.1-3
$\bar{T}$	mean thrust
T_s	suction force
T_t	total thrust $= |\mathbf{F}|$
(T_D, T_I)	driving forces produced by drag and inertia
$(T_{D,1}, T_{D,2})$	drag components of thrust given in Table 5.2-3
$(\bar{T}_{D,1}, \bar{T}_{D,2})$	drag components of mean thrust
$(T_{I,1}, T_{I,2})$	inertial components of thrusts given in Table 5.2-3
$(\bar{T}_{I,1}, \bar{T}_{I,2})$	inertial components of mean thrust
(T_1, T_2)	thrusts of two-stage swimming
$(\bar{T}_t, \bar{T}_m, \bar{T}_i)$	mean thrust caused by tail-end section, vortex shedding, and vortex interaction defined by Eq. (6.2-13a)
(T_0, T_1, T_2)	amplitude of thrust defined in Table 4.2-6
t	time, thickness, horizontal or driving force, tangential force component $= l \sin \phi - d \cos \phi$, tangential force defined by Eq. (4.2-1b) and (6.3-8a), nondimensional thrust $= T/W_0$, two-dimensional thrust
$\bar{t}$	nondimensional time $= Ut/l$, mean thickness
$(\bar{t}, \hat{t}, \tilde{t})$	nondimensional time; $\bar{t} = wt$, $\hat{t} = t/\sqrt{g/l}$, $\tilde{t} = t/(m/\rho S U)$
t'	intermediate time (or running variable of time)
t^*	retarded time or time at which a vortex is ejected from the trailing edge of wing defined by Eq. (6.2-5)
(t_1, t_2)	time spent two-stage swimming defined by Eq. (6.3-39)
(t_s, t_j)	sucking and ejecting time
(t_0, t_f)	initial and final time
$(\bar{t}, \check{t}, \hat{t}, \tilde{t})$	nondimensional time $= t/(l/U)$, $t\omega$, $t/\sqrt{g/l}$, and $t/(m/\rho S U)$
t_{max}	maximum thickness
U	speed or linear velocity $= |\mathbf{U}|$, relative speed with respect to surrounding fluid, longitudinal speed
$\mathbf{U}$	velocity vector, relative velocity with respect to surrounding fluid $= (u, v, w)^T$
U^*	critical speed
U_c	normal cruising speed
U_f	terminal speed
U_h	flow speed at horizontal tail
U_Q	optimal steady gliding speed at a tangential point Q
U_r	speed of intermittent flight
U_s	the lowest speed, stalling speed
$\mathbf{U}_w$	wind velocity $= (u_w, 0, w_w)^T$
U_w	wind speed $|\mathbf{U}_w|$
U_{limit}	limit speed defined by Eq. (2.4-4)
$U_{(L/D)_{max}}$	speed at maximum lift to drag ratio $= U_{|\gamma|_{min}}$
$U_{\phi=0}$	flight speed for zero bank angle
U_0	reference speed $= |\mathbf{U}|_o$, initial speed $= \sqrt{U_{x,0}{}^2 + U_{z,0}{}^2}$, speed for cruising flight
U_1	speed of full powered flight $= U_y (r = 1)$
(U, V)	(X, Y) components of a uniform velocity
$(U_{r,P_{min}}, U_{r,(P/U)_{min}})$	speed of intermittent flight at P_{min} and minimum (P/U)
$(U_{P_{min}}, U_{(P/U)_{min}}, U_{R_{max}})$	speeds at P_{min}, minimum (P/U), and maximum range
$(U_{1,P_{min}}, U_{1,(P/U)_{min}})$	speeds of continuous powered flight for minimum power and the minimum power-speed ratio
(U_N, U_T)	normal and tangential components of inflow velocity
$(U_{(P_n/W)_{min}}, U_{(P_n/WU)_{min}})$	optimal velocities for minimum power speed $(P_n/W)_{min}$ and maximum range speed $(P_n/WU)_{min}$
$(U_{|\gamma|_{min}}, U_{|w|_{min}})$	speeds at minimum gliding angle and minimum rate of descent
(U_0, V_0)	speeds defined by Eq. (4.3-22)
(U_1, U_2)	speeds of two-stage swimming
(U_1, U_2, U_3)	speeds of jet propulsing device
(U_A, U_w)	speeds related to air and water
(U_X, U_Y, U_Z)	(X, Y, Z) components of relative velocity
(U_x, U_y, U_z)	(x, y, z) components of relative velocity
u	backward component of induced velocity $= |u|$, nondimensional speed $= U/U_0$
$\mathbf{u}$	parallel flow component of flight velocity, induced velocity caused by thrust
(u, v, w)	(X, Y, Z) or (x, y, z) components of velocity

(u_w, v_w, w_w)	(X, Y, Z) components of wind velocity	w^*	crossflow velocity at the trailing edge and on the retarded time defined by Eq. (6.2-4)
$(\bar{u}_w, \bar{v}_w, \bar{w}_w)$	(X, Y, Z) components of nondimensional velocity defined by Eq. (3.3-7)	w_{max}	maximum width
u_{w_0}	horizontal wind velocity at $Z = 0$	$|w|_{min}$	minimum rate of descent
V	sinking rate, forward speed with respect to ground or ground speed, volume, lateral speed	$|w|_{|r|_{min}}$	rate of descent at $|r|_{min}$
		$w_{\phi=0}$	rate of descent at zero rolling angle
$\mathbf{V}$	velocity vector with respect to ground or ground velocity $= (\dot{X}, \dot{Y}, \dot{Z}) = \mathbf{I}\dot{X} + \mathbf{J}\dot{Y} + \mathbf{K}\dot{Z}$	X	distance along X axis
		X_1	horizontal shift of center of gravity
V_B	volume of whole immersed body	X_s	translational distance of sucking fluid
$V_{P_{min}}$	minimum power speed	(X_w, Y_w)	(X, Y) positions of an apparent element of wave
V_b	volume of body (except wing), volume of additional bladder	(X, Y, Z)	inertial coordinate frame
(V_{max}, V_{min})	maximum and minimum forward speeds	(X_B, Y_B, Z_B)	body fixed coordinate frame shown in Fig. 4.2-3, center of buoyant force
V_0	steady cruising speed		
V_t	maximum tip speed caused by beating motion defined in Eq. (4.2-35) $= R\dot{\psi}_1\omega$	(X_{max}, Z_{max})	maximum distance and maximum height defined by Eq. (2.4-8)
V^*_0	critical steady cruising speed	x	abscissa of airfoil, nondimensional distance $= r/R$ or $X/(c/2)$, spanwise position
(V_A, V_W, V_B)	apparent wind velocity, true wind velocity, and boat velocity defined in Fig. 7.5-1		
		x_0	nondimensional radius of black root $= R_0/R$
(V_N, V_T)	normal and tangential components of velocity	x^*	distance of the trailing edge
(V_h, V_v)	tail volumes of horizontal and vertical surface	x_j	distance between the first joints of fore and hind wings
$\hat{V}$	nondimensional forward speed $= V/R\omega$	x_β	nondimensional radius of other side of blade $= R_\beta/R$
v	downward component of induced velocity or induced velocity caused by lift $= |\mathbf{v}|$, nondimensional wind speed $= U_w/V_0$, jet speed	(x, y, z)	body coordinate frame
		(x_A, x_B)	abscissas of rachis
		(x_{ac}, x_{cp})	nondimensional distances of aerodynamic center and center of pressure
$\mathbf{v}$	normal flow component of flight velocity, induced velocity	(x_{cg}, z_{cg})	longitudinal and vertical positions of center of gravity
v_0	steady induced velocity		
v_1	the first harmonics of induced velocity	(x_{le}, x_t, x_r)	abscissas of leading edge, wing tip and wing root
(v_s, v_j)	mean sucking and ejecting velocities		
(v_t, v_s)	induced velocities generated by trailing and shed vortices	Y	translational deflection of elongated body
W	weight $= mg$, useful power	y	ordinate of mean camber line of airfoil, spanwise station
$\overline{W}$	nondimensional weight $= W/\frac{1}{2}\rho_0 U_0^2 S_0$	Z	height, altitude or distance along Z axis
We	Weber number $= \rho U^2 l/s$	$\dot{Z}_0$	initial vertical speed
W_f	weight of fuel	z	nondimensional height $= Z/l$, lateral deflection
w	climbing rate $= |\mathbf{w}|$, mass ratio defined by Eq. (4.3-21) $= m/m_0 = W/W_0$, width, crossflow velocity due to displacement of body $= dh/dt$, induced velocity $= |\mathbf{w}|$, maximum width of the slender body		
		z_w	surface ordinate of wing
		α	angle of attack, power index for wind profile in Eq. (3.3-21), inclined angle of water surface, aperture angle of wedge, angle of pronation about trailing edge
$\mathbf{w}$	induced velocity vector		
w_f	nondimensional fuel weight $= W_f/W_0$	α_A	angle of attack of sail
$\bar{w}$	nondimensional rate of descent $= w/\sqrt{W/\frac{1}{2}\rho S}$	(α, β)	parameters for defining wing configuration specified by Eq. (3.2-23)

(α_A, α_B)	angles of attack at which the moment about rachis is zero
$(\alpha_{C_L=0}$ or $\alpha_0)$	zero lift angle of attack, $\alpha_0 = \alpha_{C_L=0}$
(α_0, α_1)	steady and the first sinusoidal amplitudes of angle of attack
(α, γ)	angles defined in Fig. 4.4-3
β	camber factor $= d/(b'/2)$ defined in Fig. 3.4-1, flapping angle, coning angle, cant angle of blade, side slip of body element defined in Table 5.2-2
β_1	flapping amplitude of the first harmonics
β_{max}	maximum flapping angle
(β_A, β_W)	apparent wind and wind directions with respect to the boat axis
(β_1, β_2)	flapping angles about shoulder and wrist joints
$(\beta_{i0}, \beta_{i1}, \beta_{i2})$	steady and first harmonic flapping of ith $(i = 1, 2)$ wing
Γ	circulation, flight path angle defined in Fig. 4.5-5
γ	ratio of specific heat or adiabatic index, flight path angle, opening angle, beating angle $= \sqrt{\beta^2 + \zeta^2}$, correction factor of added mass given in Fig. 5.2-5, ratio of tangential drag to normal drag $= C_T/C_N$, inclination angle of buoyant force
γ_{max}	maximum beating angle defined in Table 7.6-1
$|\gamma|_{min}$	minimum gliding angle
γ'	flight path angle in wind
(γ_0, γ_f)	initial and terminal flight path angles
(γ^f, γ^h)	opening angles of front and hind wings in butterfly
(γ_x, γ_y)	vorticity of trailing and shed vortices
(γ_1, γ_2)	flight path angles of two-stage swimming
(γ, α)	angles defined in Fig. 4.4-3
Δ	small increment, Laplacian operator
$\Delta C_{D,f}$	drag increment defined by Eq. (7.1-10)
ΔP	power reduction
Δp	pressure difference $= p - p_0$
ΔV	small increment of speed
$(\Delta T_0, \Delta N_0, \Delta P_0)$	additional steady terms of thrust, normal force and power given in Table 5.1-3
$(\Delta T_1, \Delta N_1, \Delta P_1)$	additional sinusoidal terms of thrust, normal force and power given in Table 5.1-3
δ	height of surface roughness, reference height, mean drag coefficient of wing or blade $= \bar{C}_d$
δ_A	setting angles of sail
δ_ψ	phase lag of flapping angle defined by Fig. (4.5-1)
$(\delta_1, \delta_2, \delta_3)$	coefficients of drag coefficient defined by Eq. (3.6-11) and in Table 7.3-1
$(\delta_\psi^f, \delta_\psi^h)$	phase lag in flapping of fore and hind wings
ε	downwash angle behind a wing, strain, excess length ratio $= (l - c)/c$, slenderness ratio
ζ	lead-lag angle, vertical position of a particle
ζ_1	lead-lag amplitude of the first harmonics
ζ_{max}	maximum lead-lag angle
(ζ_{nc}, ζ_{ns})	lead-lag amplitudes of nth cosine and sine harmonics defined in Table 7.3-1
η	Froude efficiency, nondimensional spanwise distance $= y/(b/2)$
η_E	mechanical efficiency defined by Eq. (4.1-25)
η_F	efficiency of power producing mechanism defined by Eq. (4.1-27)
η_H	hovering efficiency
η_I	overall efficiency defined by Eq. (4.1-42)
η_c	center of air loading
$\eta_{c,e}$	lateral centroid of elliptic wing $= 4/3\pi$
η_e	efficiency defined in Table 6.2-2
η_h	ratio of dynamic pressure defined by Eq. (3.5-2)
η_{max}	maximum efficiency
η_0	mechanical efficiency excluding the contribution of induced power
(η_h, η_θ)	efficiencies for heaving and feathering motion alone
(η_1, η_2)	mechanical efficiency in two-stage swimming
Θ	elevation angle, body pitch angle, spiral angle defined by Eq. (5.1-14)
Θ_s	inclination angle of stroke plane
(Θ_s^f, Θ_s^h)	inclination angles of stroke plane of fore and hind wings
θ	pitch or feathering angle, local pitch angle of a plane wave defined by Eq. (5.1-15), angle of pronation
θ^*	parameter defined in Table 6.2-2
θ_n	amplitude of nth harmonic
θ_s	sine component of pitch angle in fanning motion defined by Eq. (6.1-6)
θ_0	steady feathering angle, amplitude of sinusoidal feathering
θ_1	the first sinusoidal amplitude of feathering motion
θ_{max}	maximum feathering

θ_n	amplitude of nth Fourier expansion series $= \sqrt{\theta_{nc}{}^2 + \theta_{nc}{}^2}$
$(\theta_{nc}, \theta_{ns})$	amplitudes of nth cosine and sine harmonics defined in Table 7.3-1
(θ_1, θ_2)	feathering angles of double hinge
$(\theta_{1s}, \theta_{2s})$	sine components of pitch angle at the first and second hinges in fanning
$(\theta_{0.75}, \theta_\pi)$	pitch angles of the wing at spanwise station $x = 0.75$ and $x = 0$
κ	wave number $= 2\pi/\lambda$, angular rate, circumference length ratio or circumference length of ellipse to that of circle
Λ	additional spinning rate about x-axis, wavelength in spiral wave defined in Table 5.1-5, sweep angle, stride or wavelength, local sweep angle, nondimensional wave length $= (\lambda/C)U\lambda$, sweep angle of halter, distance of travel in one complete tail beat cycle $= (\lambda/C)V$
$\Lambda_{c/2}$	swept angle at mid chord
λ	wavelength, inflow ratio defined by Eq. (3.6-3), wing stroke to chord length ratio $= R\beta_{max}/c$ or $= R\zeta_{max}/c$ defined by Eq. (4.4-1), Lagrange multiplyer
μ	coefficient of fluid viscosity
$\bar\mu$	nondimensional viscosity $= \mu/\mu_0$
μ_0	reference fluid viscosity
(u_1, μ_b, μ_c)	density parameters based on reference length, wingspan, and wing chord $= (m/\rho Sl, m/\rho Sb, m/\rho S\bar c)$
ν	kinematic viscosity $= \mu/\rho$
$\bar\nu$	nondimensional kinematic viscosity $= \nu/\nu_0$
ν_0	reference kinematic viscosity
(ξ, ζ)	positions of a particle on a wave defined by Eq. (7.6-3)
$\bar{\bar\xi}$	mean position of a fluid particle
Π	elastically stored energy
π	the ratio of circumference of a circle to its diameter, $= 3.14159$
ρ	fluid density, air density $= \rho_0\sigma$
$\bar\rho$	nondimensional density $= \rho/\rho_0$
ρ_B	density of whole body
ρ_b	density of trunk (body without wings), density of bladder
ρ_m	density of muscle
ρ_w	density of wing, mass distribution along the wingspan
ρ_0	reference fluid density, air density at sea level $= 1.2250$ kg/m³
ρ_{ij}	added mass of unit volume of fluid $(i, j \leq 3)$
(ρ_A, ρ_W)	density of air and water
σ	density ratio for standard atmosphere $= \rho/\rho_0$, density ratio $= (\rho_B/\rho) - 1$, specific weight $= \rho_B/\rho_W$, solidity $= S_w/S$ or $= b\bar c/\pi R$ defined by Eq. (4.3-8), stress, reduced frequency $= \omega l/U = 2k(l/\bar c)$ or $= 2k(l/d)$, mass of membrane per unit area defined in Fig. 6.1-5
$\sigma^\#$	density ratio defined by Eq. (6.3-63)
σ_B	tensile strength
$\sigma_{X_{max}}$	air density ratio for the maximum range
σ_c	cavitation number $= (p - p_v)/\frac{1}{2}\rho U^2$
(σ_2, σ_3)	geometrical parameters defined by Eqs. (4.2-37) and (4.2-35)
τ	running parameter of volume
ϕ	bank angle, inflow angle defined by Eqs. (3.6-3), (4.2-3), and (6.3-4), phase difference between the first and second hinges, potential function defined by Eq. (7.6-1)
$\phi_1^\#$	optimum phase shift of heaving $(a \neq -\frac{1}{2})$ given by Eq. (4.2-24)
ϕ_n	phase shift of nth harmonics
$\phi_{h\theta}$	phase difference defined in Table 4.2-3
$\phi_{|w|_{min}}$	optimal bank angle for a given turning radius
ϕ_{12}	phase difference between inner wing and outer wing flappings
$(\phi_h, \phi_\alpha, \phi_\beta, \phi_\zeta, \varphi_\theta, \phi_v)$	phase shifts of heaving, angle of attack, flapping, lead-lag, feathering motions, and induced velocity
$(\phi_{1,0.75R}^f, \phi_{1,0.75R}^h)$	phase shifts of the first harmonic feathering at $r = 0.75R$
(ϕ_s, ϕ_c)	phase shifts defined in Table 6.3-1
(ϕ_1, ϕ_2, ϕ_3)	parameters defined in Table 4.2-4
Ψ	azimuth angle of circular cylindrical coordinate
ψ	parameter defined in Table 3.3-1, azimuth angle, yaw angle of body element defined in Table 5.2-2 $= \tan^{-1}(dY/dX)$, phase difference, skewed angle beating amplitude of zero and first harmonics in stroke plane
$\psi_{h\theta}$	phase difference defined in Table 4.2-4
$\psi_{\gamma_{max}}$	azimuth angle of maximum sweep angle
(ψ_0, ψ_1)	beating amplitude of zero and first harmonics in stroke plane
(ψ^f, ψ^h)	azimuth angles of fore- and hindwings
Ω	gravity potential, spin rate $|\mathbf{\Omega}|$, rotation speed of head, reduced frequency based on the half span defined in Fig. 6.3-12
$\mathbf{\Omega}$	angular velocity $= (\Omega_{X_B}, \Omega_{Y_B}, \Omega_{Z_B})^T$ or $= (\Omega_X, \Omega_Y, \Omega_Z)^T$

$\bar{\Omega}$	nondimensional gravity potential = Ω/U^2
$(\Omega_{X_B}, \Omega_{Y_B}, \Omega_{Z_B})$	(X_B, Y_B, Z_B) components of the angular velocity
ω	angular velocity = $2\pi f$, undamped natural frequency, spin or spiral rate
ω_b	relative spin rate of flagellum with respect to head
ω_i	rotational component of induced velocity defined by Eq. (4.2-50)
$\omega_{i,1}$	first harmonic component of rotational speed of induced velocity

OPERATOR

∇	gradient operator
$\bar{\nabla}$	nondimensional gradient operator = $l\nabla$
$[(\ )]$	dimension of ( )
$|(\ )|$	absolute value of ( )
$\Delta(\ )$	difference of ( ) between both sides of wing or plate, small increment of ( )
d	differential operator
∂	partial differential operator
0( )	order of ( )
$\Re(\ )$	real part of ( )
$\Im(\ )$	imaginary part of ( )

SUPERSCRIPT

$(\bar{\ })$	nondimensional value of ( ), mean value of ( ) $= \dfrac{\omega}{2\pi}\displaystyle\int_0^{2\pi/\omega}(\ )\,\mathrm{d}t$
$(\dot{\ })$	derivative by time = d( )/dt
$(\ddot{\ })$	double derivative by time = d^2( )/dt^2
$(\ )'$	quantity at intermediate time, derivative by length = $\partial(\ )/\partial s$, derivative by nondimensional time d( )/d$\bar{t}$, ( ) in left hand side coordinate frame
$(\ )^*$	value of ( ) at retarded time = ( )(t^*)
$(\ )^\#$	modified optimum value of ( ), nondimensional value
$(\hat{\ })$	complex value
$(\ )^f, (\ )^h$	fore and hind wing components of ( )
$(\ )^T$	transposed matrix of ( )
$(\ )^R, (\ )^L$	( ) of right and left hand side

SUBSCRIPT

$(\ )_{limit}$	limit value of ( )
$(\ )_{max}$	maximum value of ( )
$(\ )_o$	initial or steady value of ( ), amplitude of ( )
$(\ )_f$	terminal value of ( )
$(\ )_H$	( ) related to hovering flight
$(\ )_I$	inertial component of ( )
$(\ )_i$	ith phase of two-stage locomotion (i = 1, 2, etc.)
$(\ )_{(X,Y,Z)}$	(X, Y, Z) components of ( )
$(\ )_{(x,y,z)}$	(x, y, z) components of ( )
$(\ )_i$	i component of ( ) (i = X, Y, Z etc.)
$(\ )_w$	related to apparent element of wave
$(\ )_A, (\ )_a$	aerodynamic components of ( )
$(\ )_B, (\ )_b$	whole body and body components of ( )
$(\ )_T, (\ )_H$	( ) related to thrust and horizontal force components
$(\ )_H, (\ )_V$	( ) related to horizontal and vertical components
$(\ )_T, (\ )_N$	( ) related to tangential and normal components
$(\ )_A, (\ )_w$	( ) related to air and water
$(\ )_f, (\ )_w$	( ) related to frontal and wetted surface areas
$(\ )_t, (\ )_h$	( ) related to thrust and horizontal force components
$(\ )_w, (\ )_b$	( ) related to wing or wind and trunk or body without wings
$(\ )_h, (\ )_v, (\ )_t$	( ) related to horizontal tail, vertical tail and tail
$(\ )_L, (\ )_D$	lift and drag component of ( )
$(\ )_l, (\ )_d$	lift and drag components of two-dimensional ( )
$(\ )_s, (\ )_u$	steady and unsteady components of ( )
$(\ )_c, (\ )_s$	cosine and sine components of ( )
$(\ )_{nc}, (\ )_{ns}$	cosine and sine components of nth order higher harmonics of ( ), n = 1, 2, 3, etc.
$(\ )_e, (\ )_e$	( ) related to elliptic wing or equivalent elliptic wing and equivalent or effective value of ( )
$(\ )_{le}, (\ )_t$	( ) at leading edge and trailing edge
$(\ )_{max}, (\ )_{min}, (\ )_{limit}$	maximum, minimum and limit values of ( )

CHAPTER 1

Introduction

This chapter introduces some fundamental concepts regarding the environmental conditions of living creatures and how their forms and modes of locomotion have been shaped by the evolutionary process. In order to understand the difference in size among organisms of various species, their dynamic similarity is also discussed. By applying a "similarity analysis," it is possible to surmise the ecology of extinct creatures and even of unknown creatures living on planets other than Earth, provided that information can be obtained on the prevailing environmental conditions.

1.1 Form and Locomotion

As shown in Fig. 1.1-1, in living creatures the body size and shape (morphology), the manner of movement (locomotion), the way of life (ecology), and the environmental conditions (habitat) are strongly related to one another.

Body shape and size is a natural result of the adaptation of the manner of movement to environmental conditions; conversely the manner of movement is severely restricted by the body shape, and by the way of life or the method of feeding and mating. Throughout nature, the adaptation of body shape, way of life, and locomotion to environmental conditions seems to be ideal, even though evolution is still going on in response to changes in the habitat. The truth of this statement will become clearer in subsequent chapters and sections.

1.1.1 EVOLUTION OF PLANTS AND ANIMALS

Earth's developing atmosphere 3.8 billion years ago probably consisted mainly of carbon dioxide, water vapor, nitrogen, carbon monoxide, hydrogen sulfide, and hydrogen. The temperature of the planet's solid surface had by then fallen from about the melting point of iron (1,500°C) to a mean temperature between the boiling and freezing points of water.

In the seas of Earth at the very beginning of its history, complex molecules, believed to be the building blocks of life, were generated. The molecules were concentrated, complex interactions took place, and eventually living organisms developed, containing deoxyribonucleic acid (DNA), which (a) acts as a blueprint for the manufacture of amino acids and (b) replicates and recombines with some probability of imperfect reproduction. These characteristics of DNA have allowed the evolution of organisms through the process of natural selection, one that often involves violent interactions of organisms with each other and with their surroundings.

Throughout evolution, the food chain has played an important role. Green marine plants, for example, supplied with nutrient minerals from the reserves of the sea in which they drift, convert light energy from the sun into chemical energy in the form of organic substances (Campbell 1983). These plants are a source of food for small fishes or for drifting animals (filter-feeding zooplankton). The zooplankton becomes the prey of many mobile species, some of which become in turn the prey of larger fishes and mammals, including many land species. The "rain" of organic debris and vertical migration facilitate the feeding of the varied inhabitants of the mesopelagic, bathypelagic, and benthic zones (Isaacs 1969). Similar food chains also operate in the air and on land. Tiny creatures such as bacteria and insects feed on plants and in turn are the prey of larger insects and birds. Both prey and predator have had to evolve to preserve their species in such food chains.

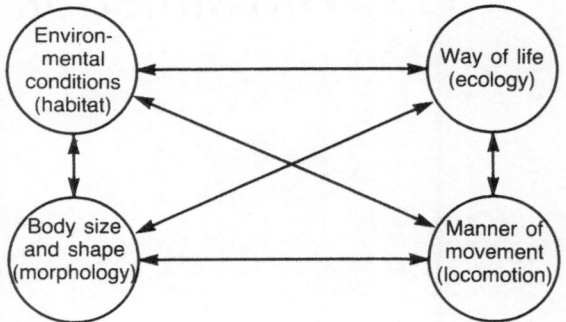

FIG. 1.1-1. Strong ties among the habitat, ecology, morphology, and locomotion.

1.1.2 FORMS OF LOCOMOTION

The development of locomotion stems from four categories of instincts: (1) for perpetuating the species, e.g., mating, breeding, nursing, and dispersing; (2) for obtaining food, e.g., searching, preying, and homing; (3) for escaping from enemies; and (4) for migrating, which affects all the other categories, but also economizes on energy consumption.

Animals moving in water are usually either longitudinally slender (thin) or vertically (rarely laterally) flat, which gives them a higher flexibility for propulsion, a lower drag to move against, and hence, less need for exertion. Animals moving in air, on the other hand, are laterally flat or have widely extended wings, can support their weight in this extremely low density medium, and are capable of driving their body against the drag at much higher speeds than their counterparts in water. Animals running on land are supported by limbs and can move with a speed intermediate between

The generation time of living creatures is also an important factor in their evolution. Smaller creatures probably have more chances to adapt themselves to a change in environmental conditions than larger ones. McMahon and Bonner (1983) pointed out the very interesting relationship between generation time t and body size l, as shown in Fig. 1.1-2.

FIG. 1.1-2. Length of organism versus generation time. (Redrawn from McMahon and Bonner 1983).

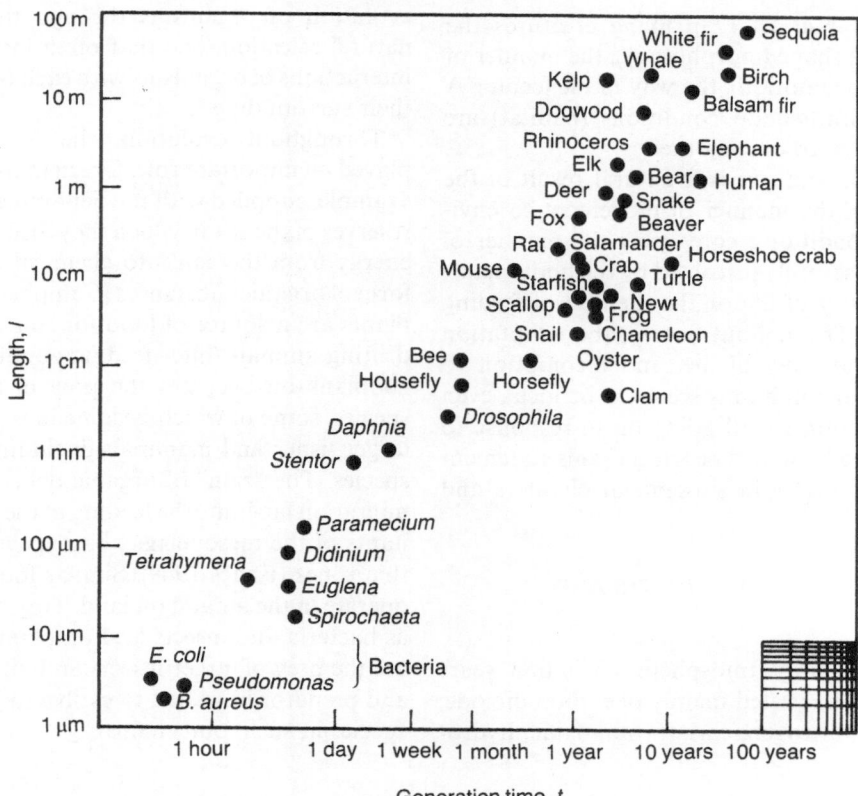

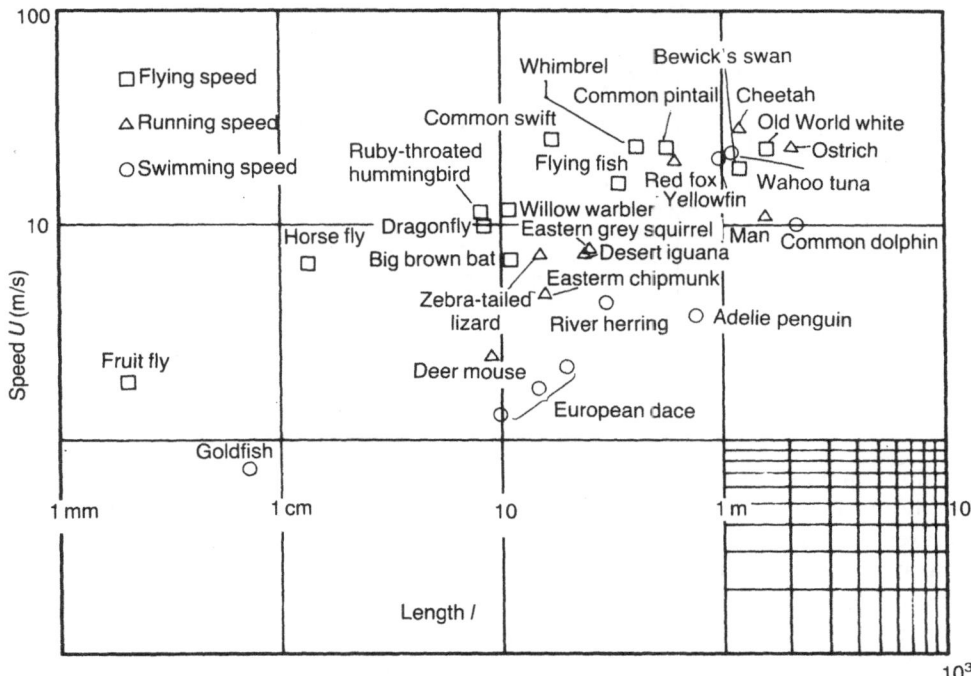

FIG. 1.1-3. Speed of animals. (Data from Alexander 1977; Goldspink 1977c; McMahon and Bonner 1983).

those of swimming and flying. However, the "energy cost" or "cost of transport" for a unit mass traveling for a unit distance is, as discussed in the following section, higher for land animals, and various kinds of obstacles confine their movements within a range narrower than that for swimming and flying animals.

1.1.3 SPEED AND ENERGY COST IN ANIMALS

Speed. Speed of movement U is the most important factor and is essentially dependent on environmental conditions, manner of locomotion, and body shape. Measurements of the speeds of animals, obtained either by calculating time for a specified distance moved or by measuring distance traveled in a specified time, do not always produce correct data because of miscellaneous errors introduced by, for example, wind or currents in the test area, or conditions affecting muscles, such as the temperature of the body and surroundings. Shown in Fig. 1.1-3 are examples of the maximum speeds of animals. The data are from several sources and are shown as a function of organism length l. These data are, actually, dependent on the distances

traveled and thus are not necessarily precise. Generally, it can be seen that larger (or heavier) animals can move more rapidly.

Energy cost. The energy required to move one kilogram a distance of one kilometer is called the "energy cost" or "cost of transport"; it is measured by the rate of oxygen consumption. Figure 1.1-4 plots the cost of transport E_c against the body mass m for birds, fish, and terrestrial animals. It can be seen that: (1) smaller animals have a higher energy cost; (2) for a given body size running is the most expensive in terms of energy cost, and of the other two ways swimming is a far more economical way to move to a distant point than flying; and (3) birds cease to fly when they reach a body mass in the neighborhood of 15 kg. Since an animal loses heat through the surface of its body, while its capacity for heat production is related to its volume, the larger the terrestrial animal, the lower its energy cost should be.

Since many animals are almost neutrally buoyant in water, they need to exert little effort to keep from sinking, but require energy to propel themselves against the drag of the fluid. Flying animals must, on the other hand, expend energy not only to keep from falling to the ground, but also to propel their bodies against both the parasite and induced drags in the air. Running animals have firm support and need perform no external work

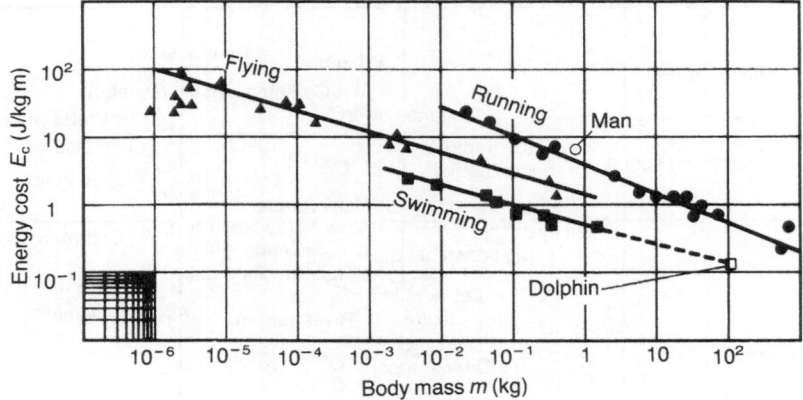

in their motion on horizontal ground. Energy
in their motion on horizontal ground. Energy losses due to both frictional resistance against the ground and exertion against the air resistance are small except at very high running speeds or in windy conditions. It is known that most energy is dissipated internally by (1) the frictional and viscous resistance in the joints and muscles, and (2) the work required for the continual acceleration and deceleration of the mass of the body and of the limbs (Schmidt-Nielsen 1972).

However, terrestrial animals are, like flying animals, subjected to the effects of gravity. When an animal runs uphill, it must inevitably spend additional energy for positive work. But it is still in question whether an animal running downhill is able to recover the negative work as perfectly as a flying animal on a descending course.

Because of the inefficiency of walking and running, some animals like the cetaceans have returned to the ocean rather than dwell on land. They have thus been able to develop large bodies, because their enormous bulk is supported by distributed pressure in the water. It is known that a whale's daily food intake is about 2% of its body weight, which is smaller than the corresponding figure of 3.6% for a man[1]. However, regardless of their size, cetaceans maintain a normal mammalian internal temperature of about 37°C in the Arctic, where the water temperature may be −2°C. As will be seen later in Fig. 7.4-2, the cetaceans, specifically those dwelling in cold water, have an intrinsically high rate of basal metabolism (Kanwisher and Ridgway 1983).

Flying birds migrate great distances. The journey of insectivorous birds from England to Ugan-

FIG. 1.1-4. Energy cost of movement for living creatures. (From Goldspink 1977a with permission, data from Schmidt-Nielsen 1972).

da spans about 50° of latitude or roughly 5,500 km and requires 3–6 weeks, calling for an average rate of progress of 120–240 km/day (Pennycuick, 1972). This slow rate of progress suggests that they spend many hours for refueling and break the distance into at least three stages. This agrees with the maximum stage distance for insectivorous birds in calm conditions, which is roughly 2,000 km, as will be explained later in sect. 4.3.

1.2 Dynamic Similarity

Because of the great diversity in their body shape and size, living creatures have developed a wide variety of forms of locomotion. However, in order to understand the fundamental characteristics of their locomotion it is necessary to organize body shapes and of locomotion in a convenient frame work for analysis, through a concise and systematic description of the dynamics related to locomotion. The nondimensional parameters that appear in the process of formulating nondimensionalized equations for describing the motion of a living creature and of the surrounding fluid are very useful for an understanding of the "dynamic similarity" among creatures with similar shapes but different sizes.

Since an object moving through air or water in a continuous fluid in opposition to universal gravitation, the force and moment acting on the moving object are related to: (1) physical properties of the fluid such as density, viscosity, pressure, and temperature; (2) the configuration of the

---

[1] Human being at rest, 60–70 kcal/h or 1.4–1.7 × 10³ kcal/day; in locomotion, 1,200 kcal/h or 3 × 10⁴ kcal/day

object; (3) its velocity and acceleration relative to the fluid; (4) its elasticity; (5) its mass and moment of inertia; and (6) the gravity acceleration determined by its position.

In the study of bio-fluid and bio-flight dynamics, many kinds of living creatures with a great variety of sizes, shapes, temperatures, and body motion behaviors must be considered. They can be classified, in some cases by size and in others by shape, into several groups, from which representative examples will be discussed.

1.2.1 GEOMETRIC SIMILARITY

Any expression or equation describing the behavior of the motion must be dimensionally homogeneous so that the equation remains true whatever units are used. By dividing any such equation of motion by a common dimensional parameter, a dimensionless equation of motion can be derived, in which dimensionless parameters determine a physical similarity of motion. Thus, it is useful first to estimate the order of magnitude of the inertial, gravitational, fluid viscous, and dynamic forces. The fundamental quantities which characterize the motions of flying or swimming objects and the surrounding fluid are: physical length l, linear velocity of the object U, angular velocity ω, fluid density ρ, gravitational acceleration g, sound speed in the fluid C, and coefficient of fluid viscosity μ or kinematic viscosity $v = \mu/\rho$. In addition, the following nondimensional parameters are important for understanding the behavior or motion of living creatures moving in a fluid:

Reynolds number

$$= \frac{\text{Inertial force}}{\text{Viscous force}} = \frac{\rho U^2 l^2}{\mu U l} = \frac{Ul}{v} \equiv \text{Re}$$

Froude number

$$= \sqrt{\frac{\text{Inertial force}}{\text{Gravity force}}} = \sqrt{\frac{\rho U^2 l^2}{\rho g l^3}} = \frac{U}{\sqrt{gl}} \equiv \text{Fr}$$

Reduced frequency

$$= \frac{\text{Angular velocity}}{\text{Linear velocity}} = \frac{l\omega}{U} \equiv k \tag{1.2-1}$$

The Reynolds number is the ratio of inertial force to viscous force. Hence, a low value of this parameter indicates a predominance of viscous effects. It is a well-known fact that many fluid-dynamic coefficients, such as the drag coefficient

and the maximum lift coefficient, are strongly related to the Reynolds number. This point is discussed later, with specific reference to stream-lined bodies.

The Froude number is the ratio of inertial force to gravity force and is important in simulating the flight dynamics of the object in question. In hydrodynamic simulation, the Froude number can be considered as the ratio of the hydrodynamic or inertial pressure to the gravitational or hydrostatic pressure of the free surface of a fluid. This parameter is essential for the reduplication of the surface wave profile of long waves.

The reduced frequency is the dimensionless angular velocity parameter. When the angular velocity is replaced with the frequency, this parameter is called the Strouhal number. In scale model testing for unsteady fluid dynamics, the reduced frequency or Strouhal number must be the same for the model as at full scale. The reciprocal form of the reduced frequency or $1/k = U/l\omega$ is called the Rossby number.

The Mach number parameter, which is the ratio of the inertial force to the elastic force of the fluid or $\text{M} = \sqrt{\rho U^2 l^2/\rho C^2 l^2} = U/C$, has been discarded here for the following reason: Usually compressibility effects of more than 10% will arise in the local Mach number beyond $\text{M} \simeq 0.4$. The most predominant effect is an increase in the drag due to the accompanying shock waves. However, no living creature has been found to move at speeds approaching the sound speed of the surrounding fluid, 340 m/s and 1,500 m/s in standard air and water, respectively.

Now, as an example, let us consider the flow of a uniform stream past a body of given shape and given orientation at velocity U parallel to the longitudinal axis of the body, and exerted by the "specific body force" $\mathbf{f} = -\nabla \Omega$, where Ω is the gravity potential. By introducing the following nondimensional variables based on the steady state values or reference values with subscript 0,

$$\begin{aligned}
\bar{\mathbf{q}} &= \mathbf{q}/U \\
\bar{\nabla} &= l\nabla \\
\bar{t} &= Ut/l \\
\bar{p} &= p/\rho_0 U^2 \\
\bar{\rho} &= \rho/\rho_0 \\
\bar{\mu} &= \mu/\mu_0, \bar{v} = v/v_0 = v(\rho_0/\mu_0) \\
\bar{\mathbf{f}} &= \mathbf{f}/g
\end{aligned} \tag{1.2-2}$$

where l is a reference length, the fluid motion of local velocity q can be described by the Navier-Stokes equation (Lamb 1930, Pankhurst 1964) as follows:

$$\left.\begin{array}{c}(\partial\bar{\mathbf{q}}/\partial\bar{t}) + \bar{\mathbf{q}}\cdot\bar{V}\bar{\mathbf{q}} - (1/\mathrm{Fr}^2)\bar{\mathbf{f}} + \bar{V}\bar{p}/\bar{\rho} \\ - (1/\mathrm{Re})\{(1/3)\bar{V}(\bar{V}\cdot\bar{\mathbf{q}}) + \bar{V}^2\bar{\mathbf{q}}\} = 0\end{array}\right\} \quad (1.2\text{-}3)$$

where by using the adiabatic index γ, the non-dimensional density $\bar{\rho}$ in the above equation may be replaced by

$$\bar{\rho} = [1 + \{(\gamma - 1)/2\}\mathrm{M}^2]^{-1/(\gamma-1)} \quad (1.2\text{-}4)$$

If, however, the nondimensional time is replaced by

$$\bar{\bar{t}} = \omega t \quad (1.2\text{-}5)$$

then the first term may be replaced by

$$\partial\bar{\mathbf{q}}/\partial\bar{t} = k(\partial\bar{\mathbf{q}}/\partial\bar{\bar{t}}) \quad (1.2\text{-}6)$$

It is thus clear that the fluid motion surrounding the body can be characterized by the above four nondimensional physical parameters. Similarly, the motion of a body flying or swimming in fluid can be described by nondimensional equations of motion with these nondimensional parameters. For a high reduced frequency ($k > 1$), the inertia given by the first term in Eq. (1.2-3) or Eq. (1.2-6) governs the fluid-dynamic force acting on the body. For a small Froude number (Fr < 1), the gravitational force given by the second term in Eq. (1.2-3) makes an important contribution, whereas for a small Reynolds number (Re < 1), the viscous force specified by the last two terms of Eq. (1.2-3) makes a large contribution. For a high moving speed or large Froude and Reynolds numbers, the gravitational and viscous forces may be neglected, in which case Eq. (1.2-3) becomes the Euler equation for ideal fluid.

The equation of continuity and Bernoulli's equation in nondimensional form can be expressed respectively as:

$$(1/\bar{\rho})(\mathrm{d}\bar{\rho}/\mathrm{d}\bar{t}) + \bar{V}\cdot\bar{\mathbf{q}} = 0 \quad (1.2\text{-}7)$$

$$\int(\mathrm{d}\bar{p}/\bar{\rho}) + \bar{\Omega} + \tfrac{1}{2}(\bar{\mathbf{q}})^2 = \bar{H} \quad (1.2\text{-}8)$$

where

$$\bar{H} = H/\rho U^2 \quad \text{and} \quad \bar{\Omega} = \Omega/U^2 \quad (1.2\text{-}9)$$

and H is called total head.

Although any effect of compressibility related to the longitudinal or elastic wave of fluids and represented by the Mach number has been discarded as insignificant, the effects of lateral waves generated by the undulatory body motion of living creatures, and of the surface wave at the boundary between two fluids, are considered in this book. Nondimensional parameters related to wave motion are as follows:

$$\left.\begin{array}{l}\text{Nondimensional wave length} \\ = \dfrac{\text{wavelength}}{\text{body length}} = \dfrac{\lambda}{l} \\[2mm] \text{Nondimensional speed} \\ = \dfrac{\text{Speed of body}}{\text{Wavespeed}} = \dfrac{U}{C} \\[2mm] \text{Nondimensional amplitude} \\ = \dfrac{\text{Amplitude}}{\text{Wavelength}} = \dfrac{a}{\lambda} \\[2mm] \quad\quad \text{or} \quad = \dfrac{\pi a}{\lambda}\end{array}\right\} \quad (1.2\text{-}10)$$

In undulatory motions, the wavelength λ, frequency f, angular velocity ω, wave number κ, wavespeed C, and reduced frequency k are mutually related as follows:

$$\left.\begin{array}{l}\text{Angular velocity: } \omega = 2\pi f \\[2mm] \text{Wave number: } \kappa = 2\pi/\lambda \\[2mm] \text{Wave speed: } C = f\lambda \\[2mm] \text{Reduced frequency:} \\ \quad k = \pi l f/U = \omega l/2U = \pi/(U/C)(\lambda/l)\end{array}\right\} \quad (1.2\text{-}11)$$

Other related parameters arising from an examination of the mechanical aspects of the motion of living creatures in a fluid are as follows:

$$\left.\begin{array}{l}\text{Cavitation number} \\ = \dfrac{\text{Pressure difference from vapor pressure}}{\text{Dynamic pressure}} \\[3mm] = \dfrac{p - p_v}{\frac{1}{2}\rho U^2} \equiv \sigma_c \\[3mm] \text{Weber number} = \dfrac{\text{Inertial force}}{\text{Surface-tension force}} \\[3mm] = \dfrac{\rho U^2 l^2}{sl} = \dfrac{U^2 l}{s/\rho} \equiv \mathrm{We}\end{array}\right\}$$

$$(1.2\text{-}12)$$

The cavitation number or cavitation inception coefficient is a measure of the likelihood of cavitation or the pressure difference between a characteristic pressure p in the fluid, such as the hydrostatic pressure at a given depth, and the "vapor pressure" p_v relative to the properties of the fluid and its temperature, each of which is nondimensionalized by the dynamic pressure $\frac{1}{2}\rho U^2$. Cavitation will occur only when the cavitation number is made sufficiently small beyond some critical value of the fluid, for example, by sufficiently increasing the speed of an immersed body. The Weber number is the ratio of the inertial force to the surface tension force sl where s is the surface tension in unit length.

Whenever the form of equations of motion and the value of related parameters of a model are equal to those of the actual object with respect to the surrounding fluid, the motion of the model and surrounding fluid are said to be "dynamically similar." Such dynamic similitude is necessary for evaluating the actual problem, using a geometrically scaled model. However, the simultaneous scaling of all nondimensional parameters is not always possible. In such cases, the extent of deviation of the partly simulated model must be evaluated by estimating the possible behavior of the model in the state described by the actual parameters.

The most distinct feature in the locomotive behavior of living creatures is the fact that the form of locomotion is strongly dependent on size. Since the size of living creatures ranges from the order of microns to several tens of meters, and their flying or swimming speed is roughly proportional to their size, the values of the Reynolds number is directly related to the size of the body. For instance, a human swims in water, at a kinematic viscosity of $v = O(10^{-6} \text{ m}^2/\text{s})$, and at a Reynolds number of 10^6, whereas bacteria and cetacea swim at Reynolds numbers of 10^{-3} or less, and 10^8 respectively. These creatures have adapted to swimming in fluid in ways reflecting these wide differences in Reynolds number. If a man wants to know how it feels to be a swimming bacterium, he should swim at normal speed in a vat of warm pitch having 10^9 times the kinematic viscosity of water (Azuma 1980, 1986). Similarly, the cetacea could mimic human swimming by flying in the air, at $v = O(10^{-5} \text{ m}^2/\text{s})$, with a tenfold reduced speed (1 m/s rather than 10 m/s).

Given in Table 1.2-1 are two important parameters, Reynolds number Re and reduced frequency k, based on the body length l, swimming speed U, and angular frequency of locomotion $\omega = 2\pi f$ for typical creatures in two realms, the "Stokesian realm" with Re $\ll 1$, and the "Eulerian realm" with Re $\gg 1$ (Childress 1981).

1.2.2 ELASTIC SIMILARITY AND EFFECT OF ADDED MASS

The previous section discussed geometric similarity, or "isometry." Now, let us consider another scaling rule, "elastic similarity," which uses two length scales instead of one as in isometry (McMahon and Bonner, 1983).

Since leg and arm bones are compressed by muscles during locomotion, a critical load P, beyond which a bone of diameter d and length l will buckle, can be given by the equation

$$P = \pi^2 EI/l^2 \propto d^4/l^2 \qquad (1.2\text{-}13)$$

where E and I are the Young's modulus and the second moment of inertia respectively. The load P is further considered to be proportional to the mass of the body m which is, in turn, proportional to $d^2 l$, or

$$\left.\begin{array}{r} P \propto m \\ m \propto d^2 l \end{array}\right\} \qquad (1.2\text{-}14)$$

Then the above Eqs (1.2-13) and (1.2-14) yield for diameter d, length l, and cross-sectional area S,

$$d \propto l^{3/2} \propto m^{3/8} \qquad (1.2\text{-}15a)$$

$$l \propto d^{2/3} \propto m^{1/4} \qquad (1.2\text{-}15b)$$

$$S_c \propto ld \propto m^{5/8} \simeq m^{0.63} \qquad (1.2\text{-}15c)$$

In the dynamics of living creatures in air or water, the following nondimensional parameters are important for understanding the effects of body size and mass:

Nondimensional time: $\hat{t} = t/\sqrt{g/l}$ (1.2-16a)

or $\tilde{t} = t/(m/\rho SU)$ (1.2-16b)

Density parameter: $\mu_l = m/\rho Sl$ (1.2-17)

where l and S are respectively a "reference length" such as chord c or wingspan b and "reference area" such as wing area. The nondimensional times $\hat{t}$ and $\tilde{t}$ are used to learn the effect of size or mass in the dynamics of buoyant bodies and cruising bodies, respectively.

The "density parameter" μ_l is the ratio of the densities of the body and the surrounding fluid, or the ratio of the mass of the body to the mass of a

Table 1.2-1. Reynolds numbers and reduced frequencies for typical creatures. (Childress 1981)

Species	Items	Reference length l (m)	Speed U (m/s)	Angular frequency $\omega = 2\pi f$ (s^{-1})	Reynolds number[a] $Ul/\nu = Re$	Reduced frequency $\omega l/U = k$	Remarks
Stokesian realm	Bacterium	10^{-7}	10^{-4}–10^{-5}	10^4	10^{-5}	10–10^2	Limit of Navier–Stokes theory; Brownian motion affects smaller organisms
	Spermatozoan	10^{-4}–10^{-5}	10^{-4}	10^2	10^{-2}–10^{-3}	10–10^2	Flagellar diameter $\simeq 10^{-7}$ m
	Ciliated protozoan	10^{-4}	10^{-3}	10	10^{-1}	1	Cilium length $\simeq 10^{-5}$ m
	Small wasp	6×10^{-4}	1	400	40	0.25	U is wingtip speed while hovering
Eulerian realm	Locust	4×10^{-2}	4	20	10^4	0.2	Re appropriate to wing $\simeq 2000$
	Pigeon	2.5×10^{-1}	1–10	5	10^5	0.25	Re appropriate to wing $\simeq 10^4$
	Medium-sized fish	5×10^{-1}	1	2	5×10^5	1	

[a] by taking $\nu = 1.5 \times 10^{-5}$ m²/s for air and $= 1.1 \times 10^{-6}$ m²/s for water.

volume of fluid equal to the volume of the body. This ratio shows the effect of density on the form of locomotion; for example, the difference between the movements of an ant lion (low density) and that of a dragonfly (high density) with similar size. The density parameter reflects creature size, because generally the larger the creature, the larger the value of μ_1, and vice versa.

On the other hand, the "added mass" of the surrounding fluid (Yih 1961; Landweber 1961) may have some effect on the flight dynamics of living creatures. The most important component of the added mass will result from the vertical acceleration of extended wings. By assuming that the wing is approximately a flat plate of rectangular planform with span b, chord c, and thickness t, the real mass m and the added mass A_{33} in the normal direction are given respectively by[1]

$$m \simeq \rho_w bct + \rho_b V_b \qquad (1.2\text{-}18a)$$

$$A_{33} \simeq \tfrac{1}{4}\rho\pi c^2 b \qquad (1.2\text{-}18b)$$

where ρ_w, ρ_b and ρ are the densities of the wing, the body, and the surrounding fluid respectively, and V_b is the volume of the remaining part of the creature. Then, the mass ratio can be expressed by

$$A_{33}/m = 1/[(4/\pi)(\rho_w/\rho)(t/c) + (4/\pi)(\rho_b/\rho)(V_b/bc^2)] \qquad (1.2\text{-}19)$$

The above equation states that the added mass cannot be neglected for cases in which (1) the density of the wing and body becomes small by approaching that of the surrounding fluid, and (2) both the thickness of the wing and the volume of the body become small in comparison with the wing dimension.

This mass ratio is very small for almost all living creatures flying in the air. One exception may be the dayfly. Hang gliders and man-powered airplanes are also exceptional because their structural weight is made so light as to be of a comparable order with the body weight.

In water, the mass ratio for swimming creatures is close to one. In the longitudinal direction of the body, however, the added mass is again very small in comparison with the mass of the body.

---

[1] The detailed expression of the added masses will be presented in Table 7.1-1 and 7.1-2 of Sect. 7.1.

CHAPTER 2

Dragging, Floating, and Jumping

Primitive ways of flying and swimming are introduced in this chapter. Extremely small creatures can move through very viscous fluids without utilizing any special motive device (Azuma 1979). The force of gravity acting on their bodies does not prevent them from floating easily in turbulent air or water, using the "drag force," or simply "drag." The magnitude of drag depends on three things: effective surface area of the body, relative speed of the flow, and density of the fluid. Jumping is also easier for smaller creatures than for larger ones. Large creatures are unable to fly in air, and a fall from a high altitude to the ground or the surface of water can injure them severely. Without wings, they are confined to water or land.

2.1 Terminal Speed of Free-falling Bodies

Here, drag is considered to be a fluid-dynamic force acting on any body moving in a continuous medium of fluid, in a direction opposite and parallel to the motion of the body. Drag force, such as felt when running against the wind is strongly related to the relative speed of the body with respect to the surrounding fluid and also to the external shape and dimensions of the frontal area of the body.

Drag Acting on a Body. Drag acting on a body is composed of:

1. "Friction drag" due to the shearing stress or tangential force acting on the "wetted surface" of the body, caused by fluid viscosity;
2. "Pressure drag" due to a normal force or pressure on the body surface, caused by

a) the viscous flow at very low speed for a small body,
b) flow separation, or "form drag,"
c) The flow deviation resulting from the normal force to the flow, called "induced drag,"
d) The existence of a longitudinal wave or shock wave generated by the compressibility of the fluid and of a lateral wave generated at the free surface of the fluid (surface wave), both of which cause "wave drag".

The drag D of a fully immersed body can be expressed as a product of: "dynamic pressure" $\frac{1}{2}\rho U^2$; the reference area S; the nondimensional drag coefficient C_D, which is a function of the "angle of attack" α; Reynolds number $Re = Ul/v$; and the "surface roughness ratio" δ/l. Thus,

$$D = \tfrac{1}{2}\rho U^2 S C_D(\alpha, Re, \delta/l) \qquad (2.1\text{-}1)$$

for a given shape of the body. Examples of drag coefficients are found in many related books and well summarized by Hoerner (1965).

Falling Rate of Sphere and Drag of Cylinder. The vertical sinking or falling rate of a body immersed in any fluid is determined by a balance among the vertical components of the fluid-dynamic force or drag, fluid-static force or "buoyant force," and gravity force acting on the body, such that

$$\tfrac{1}{2}\rho U^2 S C_D = (\rho_B - \rho)g V_B \qquad (2.1\text{-}2)$$

where ρ and ρ_B are the densities of the fluid and whole body respectively. There are many kinds of particles suspended in the atmosphere or in the ocean. These include tiny particles educed by chemical reactions among suspended substances; smoke from volcanoes, forest fires, or industrial activities; dust from soil; pollen and seeds blown by the wind; plankton drifting in water.

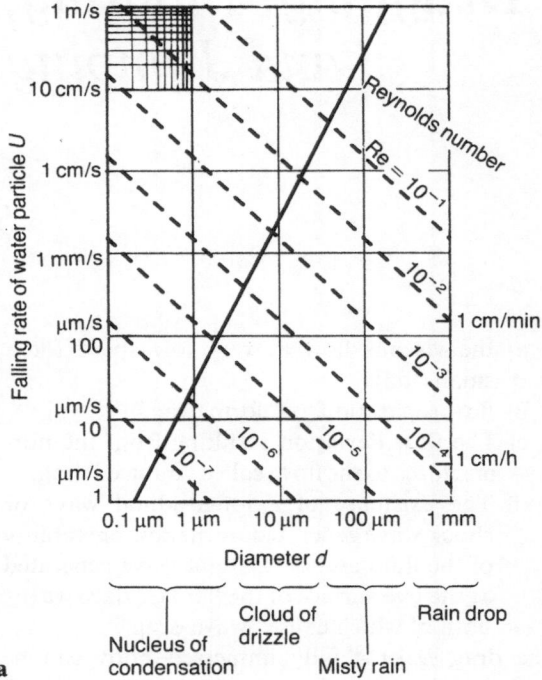

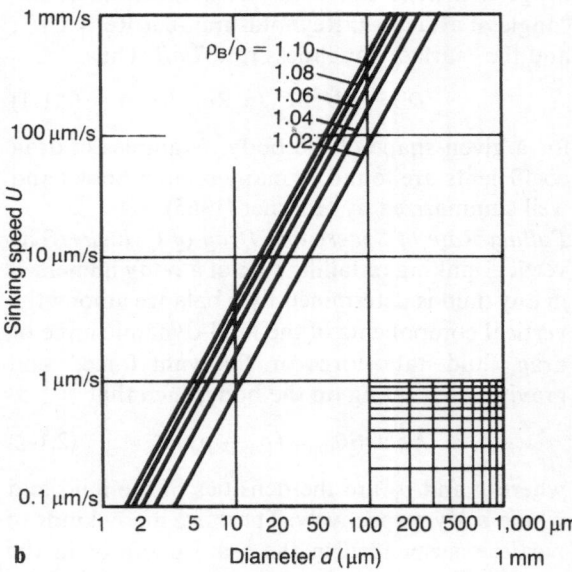

FIG. 2.1-1. Sinking speed of a spherical body. **a** Sphere of water in air (standard atmosphere at sea level). **b** Sphere of material heavier than surrounding sea-water (temperature 15°, salinity 35 ‰).

In this section, let us focus our attention on the falling rate of small bodies with an idealized form, such as spheres and circular cylinders.

Small Sphere. For a very small rigid spherical body by taking frontal area $S_f = \pi d^2/4$ as the reference area and volume $V_B = \pi d^3/6$, the drag coefficient C_D based on S_f is

$$C_{D,f} = 24/\text{Re} \qquad (2.1\text{-}3)$$

in "Stokes' approximation" (Oseen 1913; Goldstein 1929). Substituting this into Eq. (2.1-2) gives the trimmed rate of fall as a function of the diameter:

$$U = \tfrac{1}{18}\{(\rho_B/\rho) - 1\}gd^2/\nu \qquad (2.1\text{-}4)$$

It is interesting to note that the falling rate increases in proportion to the square of the diameter.

Figure 2.1-1 shows this falling rate for (a) a sphere of water in air and (b) a sphere of denser material sinking in sea water. For higher altitudes the falling rate is not so different. The above velocity should, however, be limited to a small value because of the validity of the above equation within a small range of Reynolds number, such as Re < 1.0, which yields $U \le \nu \text{Re}/d = 1.46 \times 10^4(\text{Re}/d)$ mm/s (d in microns) for standard atmosphere at sea level. In the case of a large rain drop, as the effect of surface tension is small, the drop is not a solid sphere but a deformed particle. Furthermore, it has been determined experimentally that owing to internal circulatory flow of the water in the drop, the drag coefficient is approximately twice that of a solid sphere of the same size. It is also suspected that like airplanes rain and snow falls constitute a hazard for many small creatures living on the ground (Haines and Luers 1983; Luers and Haines 1987; Luers 1983).

Fine cylinder. The normal and tangential components of the drag of a fine circular cylinder in an oblique flow are respectively given by

$$D_N = \tfrac{1}{2}\rho\nu U_N l\{10.9/(0.87 - \log \text{Re})\} \quad (2.1\text{-}5\text{a})$$

$$D_T = \tfrac{1}{2}\rho\nu U_T l\{4/(2.5\text{-}2.3\log \text{Re})\}, \qquad (2.1\text{-}5\text{b})$$

where U_N and U_T are normal and tangential components of the inflow velocity respectively, and Re is the Reynolds number based on the diameter of the cylinder (Hoerner 1965).

Figure 2.1-2 shows the normal and tangential drags per unit length versus speed in the air of standard atmosphere for cylinders of various diameters.

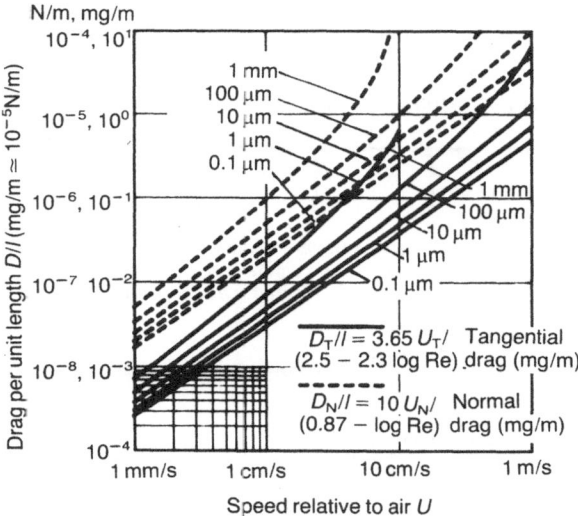

FIG. 2.1-2. Drag per unit length for circular cylinder.

2.2 Drifting in Air

Flight utilizing drag is performed by small particulate bodies and by larger bodies with thin drag lines. The particulate materials treated here are less than 1 mm in diameter. For such bodies, the surrounding fluid can feel very sticky; they can thus fly in wind by using the drag acting on them without need for any flying device.

In daytime, wind exhibits turbulence close to the ground and the updrafts experienced in many places on rough and open fields can be stronger than the downdrafts in forest areas or over water. Even though air is prevented from moving vertically close to the ground by the ground surface, its horizontal motion assists in the generation of vortices or eddies, as can be observed from the motion of small seeds or dust, or the fluttering of leaves in the wind. It is well known that in dense forests only the taller trees disperse seeds by wind scatter, whereas in grass fields many plants use this method of seed dispersion.

Airborne materials. Fall rates of typical airborne materials are shown in Fig. 2.2-1 together with the range of their diameters. Most biological particles have diameters between 0.5 and 100 µm. Significant proportions of plants and crops are lost annually due to airborne diseases, mostly caused by fungi, bacteria, and viruses. The transmission of human diseases by aerosols containing pathogens is also significant. On the other hand, pollen

dispersion is absolutely necessary for plant reproduction.

Airborne materials are injected into the atmosphere through wind and rain action on soil and plants, and by the bursting of bubbles on water. When bubbles burst at the surface, the virus- and bacteria-rich water of the bubble surface layer is ejected into the air as small droplets. The diameter of these droplets ranges from 9 to 400 µm. The mean drop diameter ranges from 20 to 40 µm, approximately 10% of the diameter of the bubble itself (Baylor et al. 1977).

After becoming airborne, these materials float in the air or fall at such a low sinking rate that they are strongly affected by local air currents or turbulence. Biological particles are subjected to deleterious physical and chemical conditions while in the air.

Fungus, moss, and fern spores. Spores of these plants are always in the air around us and along with pollen grains and other suspended particles, often cause allergies. Fungal spores are typically spherical or ovoid, but vary greatly. In mushrooms and other fleshy fungi, spores are produced and released by their spore-bearing structures (Azuma 1986). In a wind, the mushroom or horsetail generates a pair of trailing vortices or shed vortices, and releases spores into the vortex cores. The shapes of the mushroom and the horsetail allow dispersal of spores by wind from any direction. In some members of the lower fungi, cellular turgor pressure forcibly expels the spores or spore sacs (Buller 1934). The puffball (*Lycoperdon*) is one example. A pore opening at the top is regulated to close in humid air and open in dry air through a hygroscopic mechanism consisting of teeth surrounding the pore. The teeth are sensitive to the moisture level and curve outward in dry air to release the seeds, or inward in humid air to prevent their escape.

Pollen. The pollen grains of most seed plants are transported by either animals (zoophily) or the atmosphere (anemophily) from the stamen to the pistils of the same species. For effective cross-fertilization, the timing and method of transportation of pollen and the quantity of pollen produced must be skillfully controlled.

The pollen of many flowers is almost spherical, but the form varies widely among different species. The mean diameter of airborne pollen ranges from 10 to 100 µm, whereas most protozoa cysts sampled from the atmosphere range from 2 to

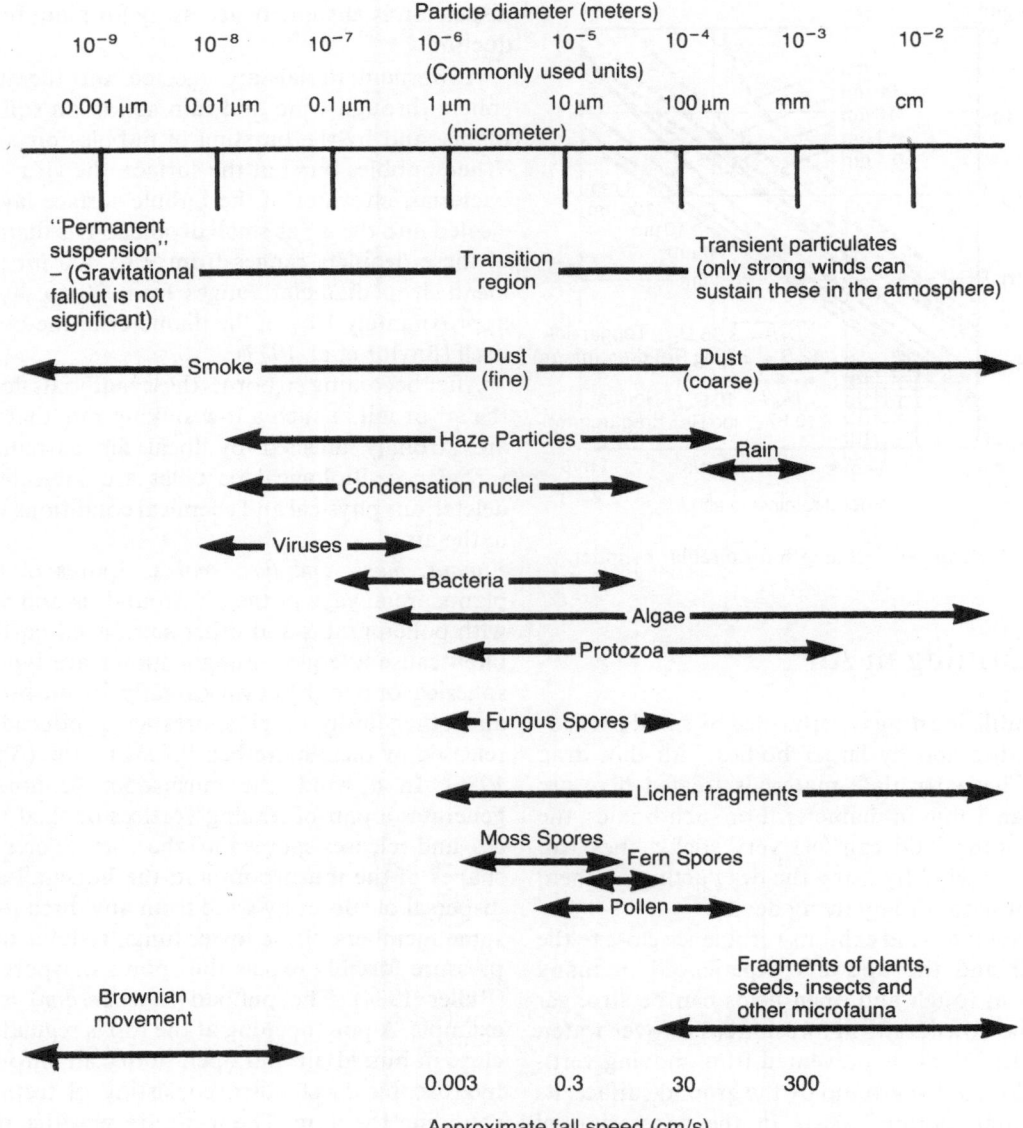

Fig. 2.2-1. Terminal fall speeds of spherical bodies in standard air. (From Bryson and Kutzbach 1968 with permission).

50 μm. Many flowering plants are aerodynamically designed to generate turbulence to facilitate the capture of pollen from the wind.

Seeds. Splitting mechanisms for the active discharge of seeds are found in the fruits of many legumes such as sweet pea, tare, and soybean, and of various other plants such as oxalis, geranium, and the balsam known as "touch-me-not." The squirting cucumber (*Ecballium elaterium*) of Mediterranean regions cannot depend on the wind to carry its seeds, as they are bigger and heavier than

the entire gun of the moss plant. It shoots a jet of juice containing the seeds at an inclination angle of 50°–55° by the sudden release of an internal pressure six times greater than atmospheric pressure. A range of over 12 m and a muzzle velocity of some 10 m/s are attained by the cucumber (Paturi 1974).

Insects and other microfauna. Other than spores and pollen, some fauna also take advantage of air drag for transportation. The smallest airborne creatures are the tiny arthropods that float about in the wind like grains of pollen. Their masses are usually less than 1 mg. They include small mites and scale insects (length ≤ 1 mm) whose presence in the aerial plankton is well documented (Washburn and Washburn 1984).

Spiders. On clear and calm days in late autumn, one can often see many thin, clear threads flowing in glistening streams against the sky. These threads, known as "gossamer," are the work of spiders traveling through the air. This dispersive process plays an important role in the population dynamics of spiders. When a spider, living on the ground, is stimulated by one or more meteorological parameters on a warm, sunny, and calm day after a freezing night, it climbs to the top of the grass or some other elevated starting point and, stretching its legs, produces streams of silk from its spinnerets, as shown in Fig. 2.2-2. A pull on these threads by an upward air current is sufficient to lift the spider into the air in what is called ballooning (Vugts and Wingerden 1976).

The length of the threads required for ballooning depends on the weight of the spider and the strength of the upcurrent, as can be seen from Eq. (2.1-5) or Fig. 2.1-2. Usually the length averages a few meters for one or two threads produced by small spiders whose mass is about 10 mg.

Pappous seeds. In species that propagate by wind scatter, the plant must be taller than the surrounding vegetation, and the seeds must be either small enough to fly without any special organ or have some flying device or mechanism enabling them to ride on air currents. The flight distance will be increased if the falling height is great and if the seeds are light for their size, so that the rate of fall is smaller than the speed of the upcurrent.

Flying mechanisms can be observed in (a) the capsules of *Staphyleaceae*, (b) the long silklike fibers of *Anemone* or windflower, (c) the featherlike fibers of Traveler's Joy, and (d) the parachutelike fibers of dandelions and milkweed.

In a species of dandelion (*Taraxacum officinale*) one large solitary flower head includes 88 oblong-ovate to fusiform fruits, or achenes, which have 120 fuzz hairs 5 mm in length and weigh $W = 0.5$ mg. Their mean sinking velocity in calm air is about 0.3 m/s, which is similar to that of misty rain (Azuma 1979).

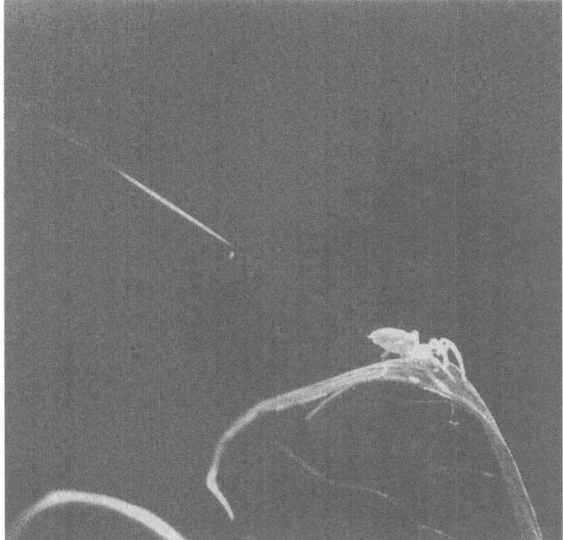

FIG. 2.2-2. A spider ready for takeoff. (Courtesy of Nishiki 1983).

2.3 Drifting in Water

Water is also an important medium for the locomotion of living creatures. Many floating seaweeds are carried by water currents, while young fish and some insects use such currents to migrate. Some species of clams can also migrate in water, using the flow with the help of mucus strings.

2.3.1 VIABILITY OF MICROORGANISMS IN WATER

Algae. Algae are chlorophyll-bearing plants that live in salt- and freshwater, on moist stones, wood and other surfaces, and in the soil. Marine algae encompass almost all seaweeds. The occurrence of extensive floating colonies of the brown alga *Sargassum* in the North Atlantic Ocean has resulted in that region called the "Sargasso Sea" (Encyclopedia Americana 1963). Algae vary in size from single cells, whose dimensions range between 5 µm and 3 cm, to the large plant bodies of certain Phaeophyceae measuring 30 m.

The horizontal distribution of plankton and floating species like seaweeds is not random. There are, as shown in Fig. 2.3-1, convection cells in which "Langmuir convection" or "Langmuir circulation" (Langmuir 1938) is generated in the water when wind blows across the surface. The cells sweep organic matter into narrow filmlike

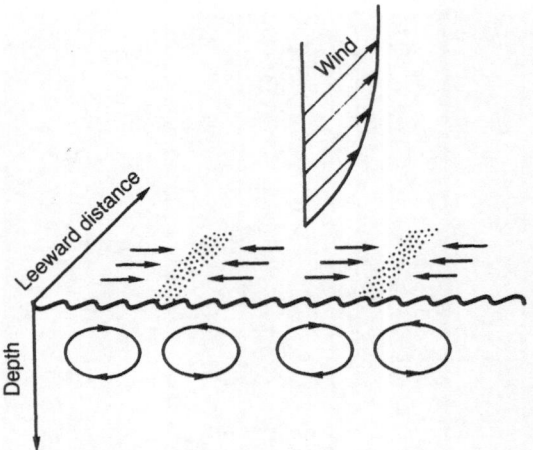

FIG. 2.3-1. Langmuir convection induced by the wind.

areas at the sea surface along lines of convergence, making "streets" of floating matter aligned parallel to the direction of the wind. Some of the small material particles are carried downward by the sinking water. The organic matter concentrated by Langmuir convection is a food source for many zooplankton and fish that congregate here. Seaweeds are also good hiding places for young fish such as yellowtail (*Seriola quinqueradiata*), and insects such as *Micralymma marinum*, which are dispersed widely by this method: (Nishimura 1976).

Phytoplankton. It is advantageous for phytoplankton to have a large surface area because it obtains nutrition by the passage of nutrients through its surface. A large surface area is attained by extending many small appendages or spines. The relative length of the appendages is often related to water viscosity, which varies with temperature. In warmer water where the kinematic viscosity is low, the appendages are longer and more elaborate than in colder water (Anikouchine and Sternberg 1973).

Zooplankton. Since many zooplankton species depend upon marine plants for food, they must remain close to the phytoplankton, and they exhibit similar adaptations to those of the phytoplankton. The surface area-to-weight ratio is high, because of their small bodies with elaborate appendages. Zooplankton is known to make daily journeys between the surface and deeper water, probably traveling to depths of more than several hundred meters. The upward migration usually begins in late afternoon and the downward migration begins after midnight (Anikouchine and Sternberg, 1973).

2.3.2 DRAGGING AND FLOATING MODES

Asiatic clam. Prezant and Chalermwat (1984) reported that small freshwater Asiatic clams such as the small *Corbicula fluminea* are capable of floating, both in gentle water currents produced by an aquarium filtration system (current speeds of 10–20 cm/s), and in a field test. Before lifting off, the clam produces long, relatively viscous mucous threads that extend upward into the water flow from within the exhalant siphon. This produces sufficient drag to pull the clams and sometimes to lift them from the substratum when the current has an upward velocity component. Another species of clam, *Meretrix lusoria*, can also travel by means of a string of mucus which is colorless, transparent, homogeneous, and lighter than the seawater.

Coconut palm. Coconuts falling from palms inclined over water may drift on the current until they are cast ashore. When the fruit is washed up on an island by the breaking waves as explained later in Sect. 7.6-2, it is protected against damage by collision with rocks and abrasive sand. The seed has enough flesh to remain capable of germination until it reaches a place suitable for growth.

Similar floating voyages over several thousand kilometers are observed for other fruits and seeds. The fruit of the false buckwheat is known to be extremely buoyant.

Shellfish. In the ocean, there are many floating shellfish that use their shell to help them float. One example is *Ianthina globosa*, a small snail that inhabits the oceans of the temperate zones. It drifts with the current by hanging from floating bubbles of mucus and feeds on jellyfish such as the Portuguese man-of-war. The gastropod *Ianthina ianthina* also clings to a bubble and floats at the sea surface. Other floating animals such as waterfowls and siphonophores, will be treated in Sect. 6.1.3 and Sect. 7.3.2, respectively.

2.4 Jumping and Free Falling

In most cases, jumping or hopping is for an animal a form of escape reaction. It is performed by storing energy in the flexed legs and by releasing it impulsively at the initial stage of a jump rather than by continuous locomotion. After the legs have lifted off from the ground surface, the animal undergoes "free fall" due to the force of gravity, although its velocity is still directed upward and it continues to climb. The jumping animal also encounters the effect of the aerodynamic drag. The interaction of these forces causes the animal to fall to the ground after passing the top of its trajectory. Jumping animals usually have poor control over their posture during jumping flight.

2.4.1 MECHANICS OF JUMPING

If the elastically-stored energy in an animal's legs Π is completely converted into kinetic energy, the takeoff velocity U of the animal just after the impulsive takeoff

$$U = \sqrt{2\Pi/m} \qquad (2.4\text{-}1)$$

where m is the mass of the body.

If the animal's kinetic energy is then completely converted into potential energy, in the form of an increase in height without dissipation by any fluid-dynamic force as if in a vacuum, then the maximum height H obtained is given by

$$H = \Pi/mg = U^2/2g \qquad (2.4\text{-}2)$$

where gravity acceleration g is 9.81 m/s^2.

Usually, since the stored energy per unit of body mass (Bennet-Clark 1977, 1980) is roughly given by

$$\Pi/m = 20\,\text{J/kg} = 20\,\text{m}^2/\text{s}^2 \qquad (2.4\text{-}3)$$

and the jumping velocity and the obtained height in an impulsive or standing takeoff are, respectively

$$\left. \begin{array}{l} H \simeq 2\,\text{m} = H_{\text{limit}} \\ U = 6.3\,\text{m/s} = U_{\text{limit}} \end{array} \right\} \qquad (2.4\text{-}4)$$

which are probably upper limits for animals performing a standing jump from the ground. It is interesting to find that these limits are not affected by the mass or length of the animal, but are only dependent on the "energy mass ratio" Π/m. It is therefore not surprising that the ratio of jumping height H to body length l takes a large value for small animals. However, for very small insects, air drag becomes significant in comparison with gravitational force. Because of their low Reynolds number, insects jump through "stickier" air than larger creatures. As a result, the upper limit of height is reduced, as shown later in Fig. 2.4-2.

TABLE 2.4-1. Jumping performance of insects and mammals. (R. H. J. Brown 1963; Alexander 1971)

Species	Height of jump H [m]	Distance of acceleration s [m]	Time of acceleration t [s]	Mean acceleration $\dot{U}$ [m/s^2] ( ) in [G][a]	Calculated jumping speed U [m/s]	Calculated specific peak power P/m [W/kg]	Calculated specific energy Π/m [J/kg]	Source of data
Insects								
Locust adult	4.5×10^{-1}	4×10^{-2}	2.6×10^{-2}	1.10×10^2 (11)	3.07	3.38×10^2	4.40	Bennet-Clark (1975)
Locust 1st instar	1.7×10^{-1}	7×10^{-3}	8×10^{-3}	2.40×10^2 (24)	1.85	4.44×10^2	1.67	Bennet-Clark (1977)
(*Xenopsylla*) Rat flea	1×10^{-1}	5×10^{-4}	7×10^{-4}	2.00×10^3 (200)	1.43	2.86×10^3	0.98	Bennet-Clark and Lucey (1967)
Ctenocephalus Pulex	2.5×10^{-1}							
Mammals								
Antelope	2.5	1.5	0.43	1.6×10^1 (1.6)	6.9	1.15×10^2	2.47×10^1	Hill (1950)
Lesser galago	2.25	0.16	0.047	1.4×10^2 (14.3)	6.6	9.15×10^2	2.15×10^1	Hall-Craggs (1965)

[a] G is the acceleration based on the gravity acceleration, G = acceleration/gravity acceleration.

In order to minimize the mass of the skeleton, it is advantageous for the jumping force F to be constant throughout the jump impulse (Bennet-Clark 1977). Then the velocity U and the mean acceleration $\dot{U}$ can be given by

$$\left.\begin{array}{l} U = Ft/m \\ \dot{U} = U/t = F/m = U^2/2s \end{array}\right\} \quad (2.4\text{-}5)$$

where t and s are the time and distance over which the acceleration occurs. The power P is also given by

$$P = FU = m\dot{U}U \quad \text{or} \quad P/m = \dot{U}U = U^2/t$$
$$(2.4\text{-}6)$$

and

$$\left.\begin{array}{l} t = 2s/U = \dfrac{U^2}{P/m} = \dfrac{2(\Pi/m)}{P/m} \\[2mm] s = tU/2 = \dfrac{U^3}{2(P/m)} = \dfrac{(2\Pi/m)^{3/2}}{2(P/m)} \\[2mm] H = \dfrac{(2sP/m)^{2/3}}{2g} = \dfrac{\Pi/m}{g} \end{array}\right\} \quad (2.4\text{-}7)$$

Since the power-mass ratio or the "specific power" P/m is considered almost constant, and the distance of acceleration s is proportional to the animal's length l or to the cubic root of the mass $m^{1/3}$, the height H is proportional to $l^{2/3}$ or $m^{2/9}$. This means that the height ratio H/l is proportional to $l^{-1/3}$ or $m^{-1/9}$; therefore, smaller insects have a higher jumping height-to-body-length ratio.

Table 2.4-1 gives the data on the jumping performance of typical insects and mammals and the results calculated by using Eqs. (2.4-6) and (2.4-7). As will be described in Sect. 4.1.2, the specific power based on the muscular mass m_m is of the order of $P/m_m = 200$ W/kg, which is equivalent to the specific power based on the total body mass P/m ranging 10–40 W/kg for 5%–20% (respectively) muscle mass as a percentage of body mass. A comparison of these expected (or conventional) values with those given in Table 2.4-1 shows that the calculated specific power (P/m) is about 10–100 times higher than the expected values, whereas the calculated specific energy is considerably lower than that given by Eq. (2.4-3). This suggests that the jumping mechanism of their legs enables small jumping insects to release stored muscle energy rapidly. For this purpose the hindlegs of jumping insects are articulated outgrowths from between the pleura and the sternum with typically six segments: coxa, trochanter, femur, tibia, tarsus, and pretarsus or claws. They are furthermore powered by strong muscles.

In a running jump, factors contributing to high jumps exceeding 2 m include changes in the height of the center of gravity, rotation of the body to a horizontal position, use of energy stored as elastic energy in the legs, and conversion of horizontal kinetic energy to vertical kinetic energy by means of elastic deformation of the legs.

If jumping is performed in a vacuum, or if the aerodynamic force is negligible in comparison with the gravitational force during free fall, and if

Fig. 2.4-1. Jumping heights for various jumping animals. (Redrawn from McMahon and Bonner 1983).

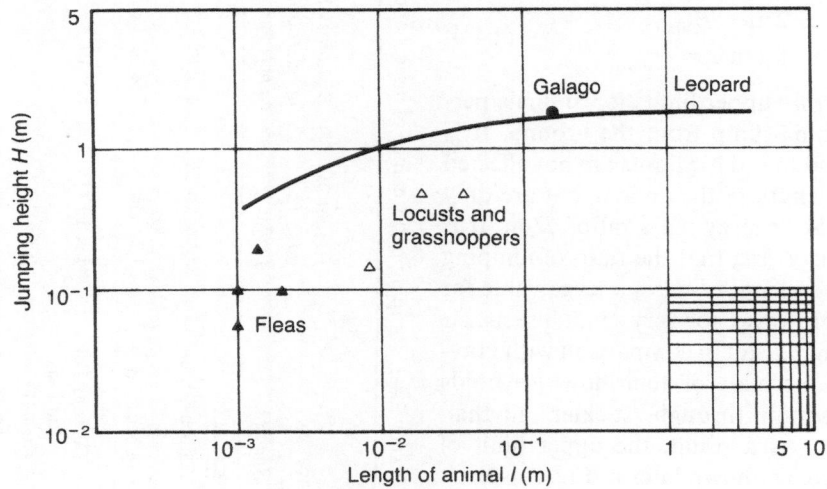

TABLE 2.4-2. Velocity and height in free fall [based on Eq. (2.4-9)]

Items	Equations	Notice
Upward projection Velocity	$u = \sqrt{1 + \{2(\rho_B/\rho)/C_D \mathrm{Fr}^2\}\{1 - e^{(\rho/\rho_B)C_D z}\}}$ $\quad \cdot e^{[-\frac{1}{2}(\rho/\rho_B)C_D z]}$	U_0 and z are positive for upward motion
Height	$h = \{(\rho_B/\rho)/C_D\}\ln\{1 + \frac{1}{2}(\rho/\rho_B)C_D \mathrm{Fr}^2\}$ $\quad = \{(\rho_B/\rho)/C_D\}\ln\{1 + (\rho/\rho_B)C_D(\Pi/mgl)\}$ $\quad \simeq \frac{1}{2}\mathrm{Fr}^2\{1 - \frac{1}{2}(\rho/\rho_B)C_D \mathrm{Fr}^2\}$	
Downward projection Velocity	$u = [\sqrt{1 + \{2(\rho_B/\rho)/C_D \mathrm{Fr}^2\}\{e^{(\rho/\rho_B)C_D z} - 1\}}]$ $\quad \cdot e^{[-\frac{1}{2}(\rho/\rho_B)C_D z]}$	U_0 and z are positive for downward motion
Terminal velocity	$\dot{U}_\infty = \sqrt{2(m/\rho f)g} = \sqrt{2(\rho_B/\rho)/C_D}\sqrt{lg}$	$\rho/\rho_B \ll 1, 0$
Notation	$u = U/U_0, \qquad h = H/l, \qquad z = Z/l$ $l = V_B/S, \qquad \rho_B = m/V_B, \qquad C_D = f/S$ $\mathrm{Fr} = \sqrt{U_0{}^2/lg} = \sqrt{2\Pi/mgl}$	C_D may be replaced with $C_{D,f}$ $C_{D,f} = f/S_f$

the gravitational force is taken as unidirectional by assuming the earth surface to be flat instead of spherical, then the flight path is a parabola determined only by the initial velocity $\mathbf{U}_0 = (U_{X,0}, U_{Z,0})$. The maximum height $Z_{max} = H$ and distance X_{max} are attained with an initial projection angle of 45° or $U_{X,0} = U_{Z,0} = U_0/\sqrt{2}$ as follows:

$$\left.\begin{array}{l} Z_{max} = H = \frac{1}{4}(U_0^2 g) \\ X_{max} = 4Z_{max} = U_0^2 g \end{array}\right\} \qquad (2.4\text{-}8)$$

However, if the aerodynamic drag and lift (upward) cannot be discounted, the initial path angle should be less than 45° to attain the maximum distance.

Shown in Fig. 2.4-1 are observed jump heights for various jumping animals (McMahon and Bonner 1983). The broken line indicates the theoretical height attainable assuming a specific energy of 20 J/kg, taking air resistance into account.

2.4.2 FREE FALL IN AIR

Mechanics of free fall. Let us consider the vertical motion of a free-falling object thrown up with an initial speed of U_0. When the drag force is taken into account, the equations of motion governing this object are

$$m(dU/dt) + \tfrac{1}{2}\rho f|U|U + mg = 0 \quad (2.4\text{-}9)$$

$$U = dZ/dt \qquad (2.4\text{-}10)$$

where the "drag area" is $f = SC_D$, and Z is the distance above the ground. For flight either in a

vacuum (fluid density $\rho = 0$) or with zero drag area ($f = 0$), these equations give Eq. (2.4-2).

If the drag area can be considered constant, Eq. (2.4-9) gives the nondimensional velocity $u = U/U_0$ and the maximum nondimensional height $h = H/l = Z_{max}/l$ (where u is zero) as given in Table 2.4-2.

It can be said from Table 2.4-2 that: (1) The effect of the drag force is strongly dependent on the density ratio between the falling body and the fluid, i.e., the maximum height increases in proportion to the density ratio ρ_B/ρ and in inverse proportion to the drag coefficient C_D; (2) as the initial speed or the Froude number increases, the maximum height also increases; and (3) the falling velocity is proportional to the square root of the density ratio ρ_B/ρ times the representative body length $l = V_B/S$, inversely proportional to the square root of the drag coefficient.

In the case of insects, for example, the drag coefficient based on the body frontal area S_f without wing opening is roughly approximated by $C_{D,f} \simeq 1.0$. If, further, the specific weight of the insect can be considered to be one and the stored energy per unit mass is limited by Eq. 2.4-3, then the jumping height and the terminal falling velocity are respectively approximated by

$$h = 816 \ln\{1 + 2.5/l\} \quad (l \text{ is measured in mm}) \tag{2.4-11}$$

$$U_f = 4\sqrt{l}. \quad \left(\begin{array}{l} l \text{ is measured in mm and} \\ U_f \text{ is measured in m/s} \end{array}\right) \tag{2.4-12}$$

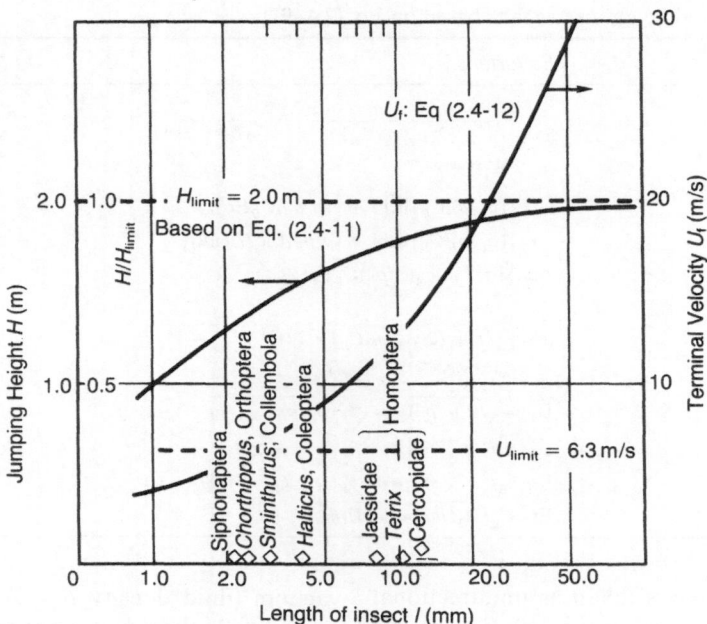

Figure 2.4-2 shows the above results for insects of various lengths and, in addition, the actual lengths for insects that are good jumpers. It can be seen that as the length or volume of an insect increases, the jumping height approaches the limit value given by $H_{limit} = 2.0$ m, and the terminal falling rate increases appreciably. Since the maximum starting speed is limited by $U_{limit} = 6.3$ m/s, a falling rate beyond this limit could be attained only if the insect should be dropped from a higher altitude than the limit altitude.

If the drag coefficient and/or the density ratio are different from the assumed values, i.e., $C_{D,f} \neq 1.0$ and $\rho_B/\rho \neq 816$ (or the specific weight is other than one or $\rho_B/\rho_W \neq 1$ where ρ_W is the density of water), then the height and the falling rate should be multiplied approximately by $(\rho_B/\rho_W)/C_{D,f}$ and $\sqrt{(\rho_B/\rho_W)/C_{D,f}}$ respectively. For instance, if the drag coefficient is twice that of the given drag or $C_{D,f} = 2$, then the height and the terminal falling rate should be reduced to approximately 50% and

FIG. 2.4-2. Jumping height and terminal falling rate of insects ($C_{D,f} = 1.0$ and $\rho_B/\rho = 816$).

70% of those given in Fig. 2.4-2. The solid line shown in Fig. 2.4-1 was probably obtained by taking some higher values of the drag coefficient than that of Eq. 2.4-11.

It is thus assumed that jumping insects will tend to fold their wings in order to attain higher altitude and that, as shown in Fig. 2.4-2, a smaller insect is safer from violent impact on touchdown, should it accidentally fall from a dangerous height.

In very small insects, the energy stored for jumping is consumed more for overcoming the aerodynamic drag than for attaining height or potential energy. It is known that larvae of *Ricania japonica* Melichar have a bundle of hairs that is closed during jumping and opens like a parachute at touch down.

CHAPTER 3

Flight By Gliding

In this chapter we will look mainly at creatures that fly by gliding. Typical of such creatures are large land and sea birds, flying squirrels, flying fish, and flying squid. Flying seeds also disperse in this manner. In gliding flight, the gravity force and the aerodynamic force are major factors. Extreme structural economy is needed to minimize the effects of gravity, especially in a large flying animal. Flight modes and performances in relation to body weight and configuration, environmental conditions and way of life are discussed. Optimal flight will also be described, as well as the auto-rotational flight of the samara, or winged fruit of some hardwood species.

3.1 Wing Characteristics

A wing is a flying device or a kind of thin plate which generates "lifting force" or "lift" perpendicular to its moving direction. While all animals that fly by gliding have one or more wings, the wing size, configuration, and material vary greatly from species to species. Since the gliding performance is strongly dependent not only on the wing area (or wing loading) but also on the wing configuration, each has a characteristic flying mode matched to its living environment and way of life.

3.1.1 STRUCTURE OF A BIRD'S WING

The wings of any flying animal must generate lift and thrust to support the animal's weight and drive the body forward against drag. The wing structure is, therefore, designed to bear the aerodynamic force and moment without diminishing its performance during either gliding or beating flight. Although there are a few exceptions, such as the penguin's wing, wings are usually flexible and foldable.

In comparing the lifting or thrusting surface between birds and other flying creatures such as mammalia (including bats and flying squirrels), insects, flying reptiles, fish, and seeds, the most important morphological differences are in the construction of the wings. Specifically, in contrast with a bird's wing, which consists of many feathers that can slide over each other, the wings of insects, mammals, fish, and seeds are made of a membrane reinforced with skeleton and/or fibers.

General configuration. The external configuration of a bird's wing can be seen in Fig. 3.1-1. The wing is shaped by bony structures, muscles, and plumage consisting of (1) primary feathers or "primaries," (2) secondary feathers or "secondaries," (3) "tertiaries," (4) humeral and auxiliary feathers such as "scapulars," (5) "wing coverts," and (6) "bastard wings" or "alulae." A bird's wing would thus seem to be more resistant to damage than a membranous wing like the bat's.

For good performance, a bird's body is shaped into a streamlined form, and the airfoil or wing section is also streamlined in the manner of an airplane wing, because the Reynolds number of the wing tip is usually of the order of 10^5. The thickness and camber of the wing section increases from tip to root. The bird can alter the wing camber to some degree, either actively by adjusting the muscle, the tendon, or passively by aeroelastic action of the feathers.

At the shoulder joint and two other joints (elbow and wrist), the wing can make "feathering" (pitch change), "flapping" (flatwise), and "lagging" (chordwise) motions in some restricted conditions. The shoulder is involved in all motions, whereas the elbow is mostly used to shorten the wing by

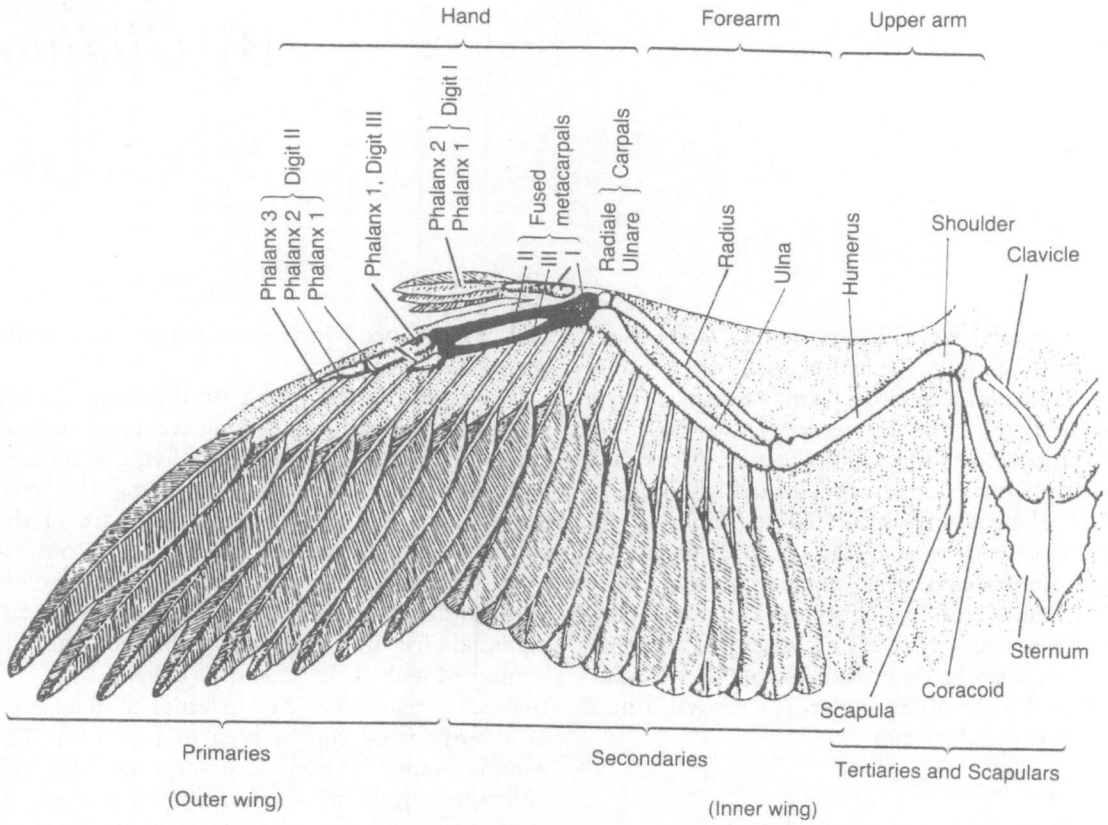

FIG. 3.1-1. Birds' wing structures. (From Berger 1961).

folding it compactly in the lagging direction or in the shape of the letter Z (Whitfield and Orr 1978). On the contrary, the wrist joint appears to be responsible for all additional motions of the hand or outer wing, by which the outer wing attains a widened angle of attack as shown in Fig. 3.1-2. A more detailed description on the beating motions will be presented in Sect. 4.2.1.

The mean mass of the skeleton m_s accounts for only 6% of the total mass, compared with about 10% or $m_s = 0.1 \, m^{1.13} \simeq 0.1 \, m$ in the average mammal (Schmidt-Nielsen 1977). The frigate bird, for instance, has a wingspan of 2 m, but its 0.11 kg skeleton weight is less than the weight of its feathers (Welty 1955).

According to Kirkpatrick (1990), the mass and moment of inertia of one wing of birds, m_w and I_w, are statistically given by

$$m_w = 9.74 \times 10^{-2} m^{1.10} \qquad \text{(3.1-1a)}$$

$$I_w = 9.23 \times 10^{-4} b^{5.08} \cong 3.76 \times 10^{-3} m^{2.05}$$
$$\text{(3.1-1b)}$$

The last expression is analytically deduced from the fact that $I_w \propto m_w b^2 \propto m_w^{5/3} \propto m^{1.8} \cong m^2$.

Musculature. The most important muscles for flight are two pairs of muscles that run between the upper arm and the keel. The larger pair is called pectoralis major which, in some species, is divided into slow and fast fibers. It provides the powerful downstroke of the wing when it contracts. The upstroke requires far less energy and is achieved by contraction of the second pair of flight muscles, the smaller pectoralis minor. This second pair of muscles runs between the humerus and the keel, and lies between the pectoralis major and sternum. They are not attached directly to the humerus, but terminate in a tendon which runs through a hole between the bones of the pectoral girdle to the upper side of the humerus (Perrins and Cameron 1976).

Since ample space for the pectoralis major is necessary for obtaining the powerful downstroke

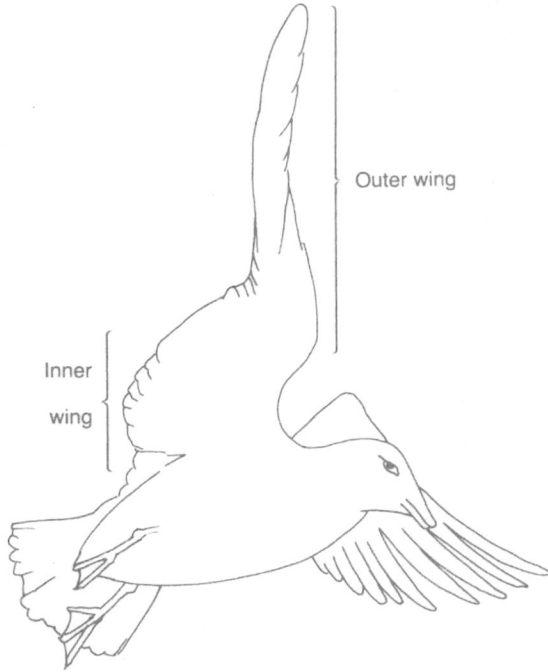

FIG. 3.1-2. Additional feathering of outer wing in a flying gull.

of the wing, the high-wing configuration cannot be avoided. From the aerodynamic-performance point of view, the wing should be attached to the body where the body width is at its maximum, in middle-wing configuration. This would enable the wing surface to extend at right angles to the body surface so that adverse interference such as flow separation between the wing and the body could be avoided. It is interesting to find that many birds have "gull-type wings," i.e., wings attached normally at the upper side of the body and with a dihedral at the wing root and capable of reducing the angle by making a downward deflection outside the elbow. However, excess dihedral at the wing root causes unfavorable characteristics, such as the "Dutch roll" for birds lacking a vertical tail fin (Perkins and Hage 1965).

Plumage. A bird's plumage provides insulation, protection from water, streamlining, and camouflage.

PRIMARY FEATHERS. In flying birds, 9 to 12 primaries are attached to the bones of the hand. The primaries are capable of various independent movements and form an outer wing with the triangular surface of the manus, accounting for 30%–40% of the whole area of the wing.

SECONDARY FEATHERS. The secondaries, numbering from 6 in some hummingbirds up to 20 in land-soaring birds, and more in sea-soaring birds, are attached to the ulna of the forearm, parallel to one another. They are controlled not individually but in small groups by the motion of joints and by an elastic membrane running from the first primary back to the elbow (Storer 1948).

TERTIARIES. The feathers rising from the upper-arm bone or humerus are known as tertiaries or tertials. They are considered extensions of the secondaries and close the gap between the active wing and the body. In most birds these feathers are few in number, but in those species in which the upper-arm bone is long, such as gulls, herons, or albatrosses, they are fully developed.

SCAPULAR FEATHERS. These feather are found on a bird's shoulders and constitute a separate group for "tailoring" at the connection of the body and wing. In soaring flight they fulfill the role of an aircraft's wing fairing.

COVERTS. The humeral groups of feathers are covered on the outside by a triple row of small coverts and on the inside by one or two rows of finer feathers that are easily lifted. These coverts play an important role in fairing the profile of the wing.

At a high angle of attack, flow separation which results in a loss of lift or stalling of the wing can probably be sensed by the upward deflection of the coverts, as will be explained later in Sect.3.2.1.

BASTARD WING OR ALULA. The feathers of the bastard wing or alula are thumb quills attached to the first digit of the manus and are capable of independent movement effected by a special system of muscular connections. It has been reported by Storer (1948) that some birds cannot take off or land without them.

Structure of feathers. The vane of a feather is, as shown in Fig. 3.1-3, made up of parallel rows of barbs projecting obliquely from either side of a shaft. The bare end of the shaft, the quill, and the distal portion, the rachis, are corneous tubes, the material of which, keratin, has a specific gravity of only 1.15 g/cm^3. It has been reported that the modulus of elasticity and the tensile strength of keratin are $E = 9.0 \times 10^3$ MPa or 920 kgf/mm^2, and $\sigma_B = 3.5 \times 10$ MPa or 36 kgf/mm^2, respectively (Hertel 1966).

The quill is a hollow elliptical tube with an approximately constant wall thickness. The rachis has a rectangular section filled with foam material

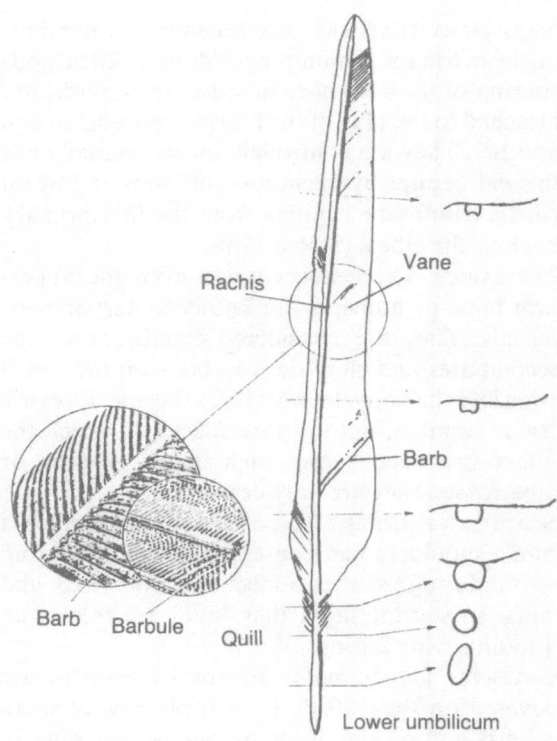

FIG. 3.1-3. Structure of a feather.

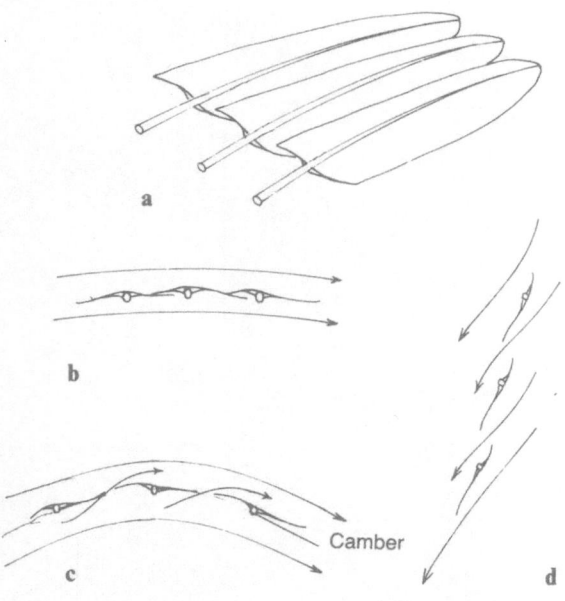

FIG. 3.1-4. **a–d. a** Arrangement. **b** Tight position.
c Spread position. **d** Reversed position.

and can more easily accommodate bending distortion than the quill, specifically in directions normal to the vane. In order to maintain aerodynamic smoothness on the upper surface of the vane, most of the rachis projects downward from the vane surface. Its shape is almost square at the tip of the vane, oblong in the middle part, and elliptical with a little dimple at the bottom near the root.

The barbs are flattened (about 0.08 mm wide) and are very flexible with respect to bending toward either end of the feather, but are fairly stiff and rigid against bending up or down. A row of very fine fibers or "barbules" (length $\simeq$ 120 μm, diameter = 3–5 μm) runs along either side of these barbs. Barbs and barbules are interlocked with microscopic hooked "barbicels" (length $\simeq$ 20 μm and diameter $\simeq$ 1.5 μm) and make the surface of the vane. Thus, the wing, with these feathers and coverts, does not have a completely smooth surface but a somewhat rough or grooved surface having what may be called "riblets." The riblets are, as we will see later in Sect. 6.1.2, devices to reduce drag.

The flight feathers of the wing are so arranged that each feather is overlapped from the wing tip

to the root by the one next to it. Together, they form an adaptable lifting surface as shown schematically in Fig. 3.1-4.

Tail wing. A bird's tail wing is made up of "rectrices" and can be regarded as a stabilizer with variable area and angle of attack, and also as an organ which creates moment for the control of rotation about the horizontal and vertical axes. Whenever the tail wing shares the lift, the lateral tilt of the lifting surface generates a lateral force in the direction of tilt and thus induces yawing moment. There are many types of tail wing as seen from the many birds shown in the figures of this and the following chapter.

3.1.2 Aerodynamic Characteristics of a Feather

The "center of pressure" $x_{cp}c$ is the distance from the leading edge to the point on the chord at which the resultant of all pressure forces on the airfoil section is assumed to act. If, for a change of lift coefficient or angle of attack, the corresponding pitching moment coefficient around a point is a constant, then the point is called the "aerodynamic center" $x_{ac}c$. The relation between the nondimensional distances x_{cp} and x_{ac} for a thin airfoil is shown in Fig. 3.1-5. Usually the aerodynamic center lies very close to a quarter-chord, specifi-

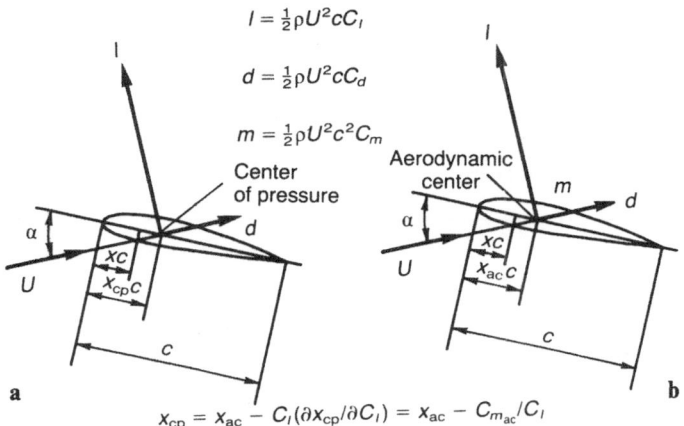

$$l = \tfrac{1}{2}\rho U^2 c C_l$$

$$d = \tfrac{1}{2}\rho U^2 c C_d$$

$$m = \tfrac{1}{2}\rho U^2 c^2 C_m$$

$$x_{cp} = x_{ac} - C_l(\partial x_{cp}/\partial C_l) = x_{ac} - C_{m_{ac}}/C_l$$

FIG. 3.1-5. **a, b.** Center of pressure and aerodynamic center of a thin airfoil. **a** Center of pressure. **b** Aerodynamic center.

cally between 22% and 26% of the chord length from the leading edge ($x_{ac} = 0.22$–0.26).

Since a feather is very thin, it can be analyzed by the thin wing theory (Theodorsen and Garrick 1933; Moriya 1959). Let us assume that, as shown in Fig. 3.1-6, the camber configuration can be approximated by the following formulas with three parameters, p, q, and r:

$$y = px(x - 1)(x - q) \quad \text{(for reflexed airfoil)} \tag{3.1-2a}$$

$$\pm y = -r\sqrt{1 - (1/2r)^2} + \sqrt{r^2 - (x - 1/2)^2}$$

$$\text{(for circular airfoil)} \tag{3.1-2b}$$

FIG. 3.1-6. Configuration of exempified thin airfoils.

Then the "zero-lift angle of attack" $\alpha_0 = \alpha_{C_L=0}$, the moment coefficient about the aerodynamic center, $C_{m,ac}$ and the center of pressure x_{cp} are as given in Table 3.1-1.

As seen from Fig. 3.1-6, in the reflexed airfoils the camber increases with the parameter p, whereas the point of reflection moves forward as the reflection parameter q increases. In the circular airfoils, the upward and downward convexes are respectively represented by the $\pm$ signs in Eq. (3.1-2b).

As Table 3.1-1 indicates, the zero-lift angle of attack α_0 decreases as the reflection parameter increases. When $q = 0.75$, the lift coefficient is zero at $\alpha = 0$ for every airfoil with reflection. As would be assumed from the geometrical resemblance, the zero-lift angle of attack takes a similar value for both the reflexed airfoil of $p = 0.5$ and $q = 0$ and the circular airfoil or $r = 1.30$. A positive value of α_0 indicates negative lift at zero angle of attack, whereas a negative value of α_0 or $q > 0.75$ indi-

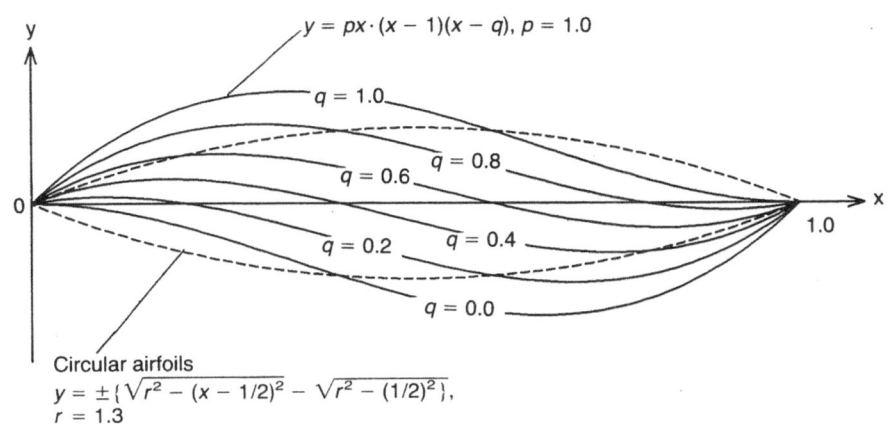

TABLE 3.1-1. Aerodynamic characteristics of a reflected airfoil

Item	Symbol	Formula
zero-lift angle of attack	$\alpha_{C_L=0} = \alpha_0$	$\frac{1}{8}p(3-4q)$
moment coefficient about ac	$C_{m,\text{ac}}$	$\frac{\pi}{32}p(7-8q)$
nondimensional distance of CP	x_{cp}	$\frac{1}{4}\{1 - (\frac{p}{16} + \alpha_0)/(\alpha - \alpha_0)\}$

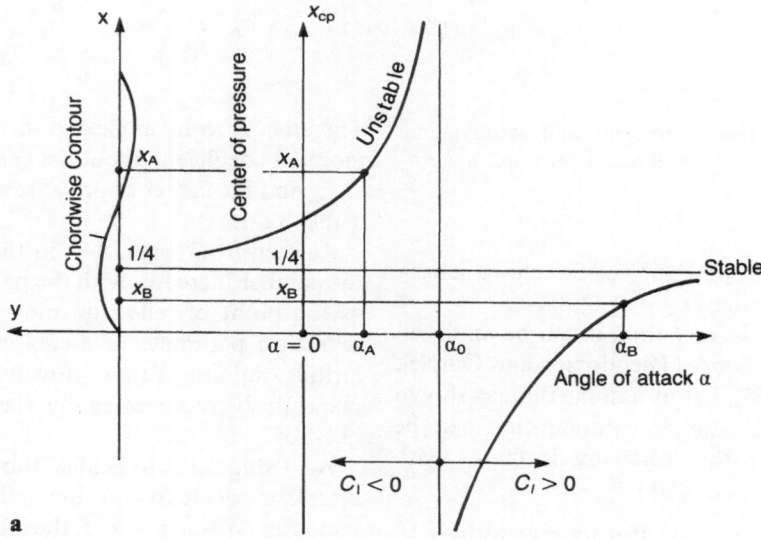

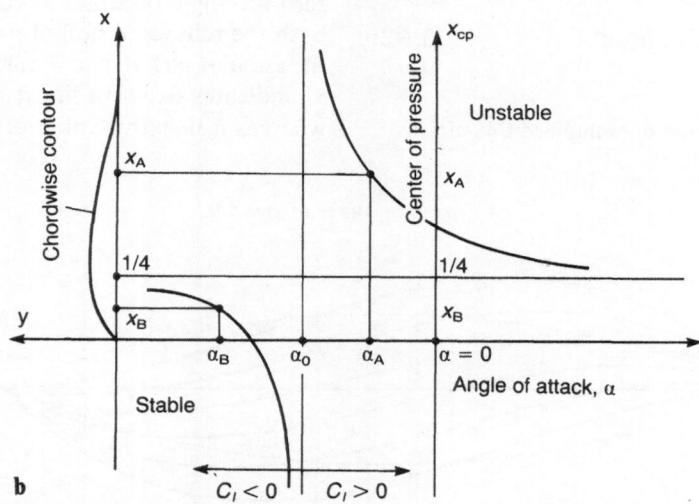

FIG. 3.1-7. **a, b.** Change of center of pressure. Although the range of angle of attack α is given as $\pm 20°$ for better understanding, α should be restricted within the unseparated flow range. **a** Reflexed airfoil. **b** Unreflexed airfoil.

cates positive lift at zero angle of attack. The moment coefficient about the aerodynamic center of the airfoil $C_{m,ac}$ is almost or fully positive (head-up) for a reasonable range of angles of attack, excluding very large ones ($q > 7/8$) and the case of a circular airfoil with upward convex.

Figure 3.1-7a shows the locus of the center of pressure for a reflexed airfoil within a reasonable range of angles of attack. When the rachis is located at $x = x_A > 1/4$, the moment about the rachis is zero at the angle of attack α_A. However, this is not a trimmed point since increasing (or decreasing) the angle of attack within the range $\alpha_0 > \alpha > \alpha_A$ (or $\alpha < \alpha_A$) causes a head-up (or head-down) moment that will destroy the trim balance if the elastic twisting moment is weaker than the aerodynamic moment. When the rachis is located at $x = x_B < 1/4$, the moment about the rachis is zero at the angle of attack α_B. This trimmed point is stable because if the angle of attack increases beyond this point, a head-down moment will result as the center of pressure is shifted backward. At this trimmed point, α_B, the lift is positive and adequate for the feather of a wing. As the quill position moves forward or x_B moves toward the leading edge of the feather, the trimmed angle of attack decreases. This aerodynamic characteristic helps the primary feathers make the wing tip upwardly convex. That is to say, as shown in Figs. 3.1-1 and 3.1-3, the fact that the rachis position gradually shifts from the most forward position of the primaries to the backward position of primaries makes the trimmed angle of attack of the respective feathers increase from a smaller value at the outermost primary to a larger value at the innermost primary and, with proper elasticity distribution through the feathers, makes the camber of the wing tip convex (Fig. 3.1-4c). For a short period during the upstroke part of the beating motion, almost all primaries operate at a negative angle of attack. Since in this state a feather with a reflexed airfoil is aerodynamically unstable in the torsional direction, the torsional rigidity of the quill will prevent further deformation of the feather. This stress on the quill will, as already shown in Fig. 3.1-4d, be experienced only for a short time during the upstroke in hovering or very low speed flight, such as at takeoff and landing as shown later in Fig. 4.2-8.

Figure 3.1-7b shows the locus of the center of pressure for an unreflexed airfoil. A stable trimmed point can be obtained at a forward position greater than a quarter-chord and a negative lift coefficient.

It is also interesting that if the primary feathers are a kind of swept wing, i.e., the rachis is swept back (or forward) to the tip, then, as the lift of the feather increases, for example in the downstroke, the aerodynamic force acting at an arbitrary point generates a negative (or positive) pitching moment about the quill. This moment acts to reduce (or increase) the angle of attack or the lift of the feather at any spanwise station and, in combination with the torsional rigidity of the rachis and quill, adjusts the angle of attack of the respective feathers within an appropriate range of the lift coefficient.

On the other hand, the flapwise bending moment increases as the point of interest on the spanwise station approaches the root of the rachis and quill. Thus, the rachis and quill must be constituted to withstand both twisting and bending moments. This mechanism is also effective for storing a part of the beating (kinetic) energy in the rachis and quill as elastic (potential) energy.

3.2 Gliding Flight in Birds

The flight of birds can be generally classified into non-flapping flight (including gliding, soaring or sailing) and flight by means of a flapping or beating of the wings. The latter type of flight will be discussed in depth in the following chapter.

The flight performance of birds is strongly related to their body configuration, specifically wing configuration and arrangement, in which there are clear differences between land birds and sea birds, depending on their ecology or way of life and the environmental conditions. Each bird is also able to alter its wing configuration in response to flight conditions. With this adaptive ability the bird can optimize performance in any flight condition and select the degree of stability and control or maneuverability, which are inversely related to each other.

3.2.1 FLIGHT MODES

A bird can adopt different flight modes depending on its purpose in flying. It can adjust the configuration of its wings as well as their profile by either extending or folding them against the body. This allows for the optimal wing for each flight mode

a

b

and each phase of a stroke movement. The following are typical types of flight observed in the sky (Storer 1948; Vinogradov 1951; Terres 1968):

Cruising flight. In steady level flight the wing acts to give a lifting force mainly at the inner part of the wing, and a thrusting force at the outer part of the wing or oscillating manus. In this mode, the wing is almost fully extended to give the best performance for minimum power.

Ducks, geese, swans, flamingos, storks, and cormorants always fly with their head and neck

Fig. 3.2-1. **a, b.** Gliding flight. **a** Buzzard, *Buteo buteo*. (Courtesy of M. Tanaka 1976). **b** Albatross (*Diomedea albatrus*). (Courtesy of Asahi News Paper).

stretched out to the fullest extent. On the other hand, herons, egrets, and pelicans, though also long-necked birds, draw their head back till it rests almost on their shoulders.

Gliding or soaring flight. Gliding is performed without supplying any beating energy to the wings other than for maneuvering action; therefore, either the flight altitude or the flight speed gradually decreases during flight in calm air. The lost potential or kinetic energy is equal to that consumed by the drag of the body in forward motion. Gliding for a long duration or great distance is called soaring.

Soaring flight is shown in Fig. 3.2-1 for the buzzard (*Buteo buteo*), a land bird, and the albatross (*Diomedea albatrus* Pallas), a seabird. The wings are usually fully extended to obtain maximum area and span, and to get either a minimum gliding angle or a minimum rate of descent. During this flight, maneuvering is performed gently without wasting energy.

Diving flight. A steep descent is executed with immobile and partly drawn wings to control speed and direction with the help of the tail. This flight mode is used for preying and sometimes for pinpoint landing by small birds, such as the lark.

Bounding flight. Small birds employ an undulating form of flight called "bounding flight" in which, with wings periodically or intermittently folded, they fly like an arrow, first losing height and then swooping up again.

Hovering flight. Before touching down on a tree or on land, many birds can temporarily remain in one place even in calm air by raising their body and beating their wings in what is called the "avian stroke." This is illustrated in Fig. 3.2-2a. On the other hand, the hummingbird is able to stay at one point in the air for a prolonged period by beating its wings (mostly the manus) almost horizontally in what is called the "insect stroke," shown in Fig. 3.2-2b, at an exceptionally high rate of more than 20 strokes per second.

Takeoff. For fight, birds must acquire enough speed with respect to the air to utilize the aerodynamic force. Since a bird does not have forward velocity at takeoff, it needs some lift assistance before gaining speed for normal flight.

Long-legged birds like the heron get a big boost with their first leap into the air. Many small birds also can use their legs for projecting themselves upward at the initial stage of takeoff. Large and thus heavy birds like the flamingo, coot, and swan take long runs against the wind on water or land before they gain enough speed to raise themselves. In this sprint, the pelican repeatedly kicks the water surface with both legs simultaneously, whereas the swan uses its legs alternately. The albatross runs downhill for takeoff. The gannet launches itself into the air from its perch and glides downward until flying speed is attained. Most birds of prey, such as buzzards, hawks, and kestrels, jump out from their aeries with wings folded and fall for a while.

Landing. In landing, birds must drive their wings to get a large drag force to kill their speed and to maintain the lift against their weight at low speed. As shown in Fig. 3.2-3, in a highly lifted wing, stall is prevented by extension of the alulae, wide spreading of the primaries, and use of the tendon to form a concave wing section that is tailored by the coverts on the underside of the wing in the manner of a "Krueger flap" (see Sect 3.4.2). This change in section configuration is important. In addition, the wings elevated in a V shape, increase the maximum aerodynamic coefficients by reducing the aspect ratio, as we will see later in Sect. 3.2.3.

Either of the high-lift devices mentioned above will prevent flow separation of the wing at high angle of attack and increase the maximum lift at slow flight speed. The effect on the tail surface cannot be ignored either. The spread tail forms an auxiliary surface behind and slightly below the main wings, like the flap of an airplane wing, and prevents flow separation of the main wing.

The legs of the landing bird are fully extended to increase drag and reduce speed, and to obtain an adequate stroke to cushion the impact at touchdown.

In water birds, touchdown on water is accomplished either on the breast or on the feet. Breast landing is sometimes observed in diving birds. Most water birds land on their feet. They stretch their webbed feet forward and slide on them as though on water skis.

Many birds can reduce their flying speed by assuming an ascending flight path or can convert their kinetic energy to potential energy during the landing approach by changing their flight attitude, wingspan, and wing area.

Great tit

Recovery stroke → time

Power stroke → time

Red start

a

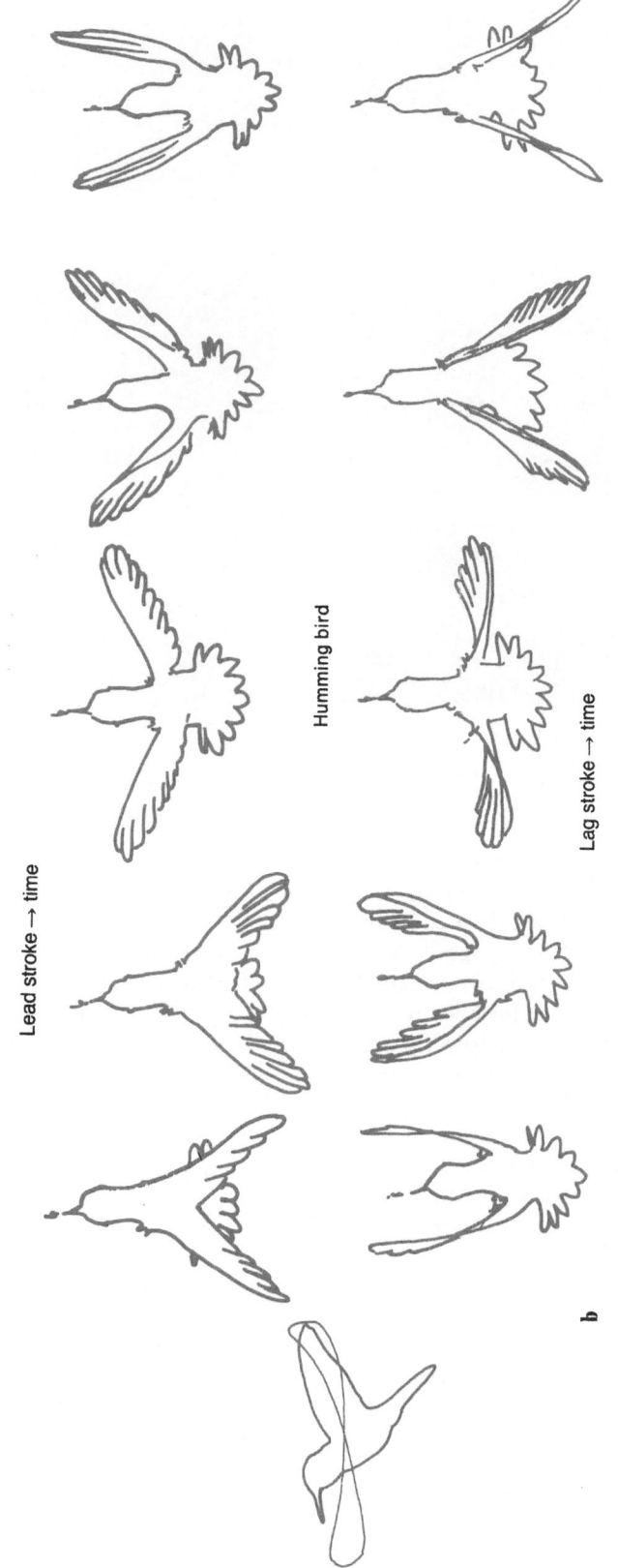

Lead stroke → time

Humming bird

Lag stroke → time

b

FIG. 3.2-2. **a, b.** Hovering flight. (Sketched from Rüppell 1977). **a** Avian stroke. **b** Insect stroke.

3.2.2 Statistical Data on Wing Configuration

Valuable statistical data on wing configuration in birds and insects were given by Greenewalt (1962). Some of the data presented in this section have been taken from his work. The wing span and area are defined to include the part of the wing extended into the trunk and are measured for fully extended planform in soaring, from wing tip to wing tip.

Shown in Fig. 3.2-4 are statistical data on the wingspan (spread) b versus body mass m, and the length of the hand l_h for various birds. When the wing length is given by the length of the hand, the distance from the wing tip to the first articulated joint, the wingspan can be expressed as 1.61 times the length of the two hands.

The data on wing area S versus body mass and wingspan are shown in Fig. 3.2-5. It can be seen that as the mass of the body increases the wingspan and the wing area also increase. This is due to the "square-and-cubic rule" which states that if one doubles the linear dimensions of a body while maintaining a similar configuration, the surface area increases by a factor of four (2^2) while the volume (or mass) increases by a factor of eight

Fig. 3.2-3. Landing flight (about to strike) of red-shouldered hawk (*Buteo lineatus*). (From Gilliard 1967).

(2^3) (dimensional characteristics will be discussed again in chap. 4). Thus, as shown in Figs. 3.2-4 and 3.2-5, in which data on flying fish and flying squid are also included, the weight-to-wing-area ratio or wing loading will increase roughly as a linear function of the length or as a cubic root function of the mass.

The straight lines given by Greenewalt (1962) and Tucker (1973) in Figs. 3.2-4 and 3.2-5,

$$\left. \begin{array}{l} b = 1.1m^{1/3} = 3.23l_h \\ S = 0.2m^{2/3} \end{array} \right\} \begin{array}{l} b \text{ and } l_h \text{ in m, } S \text{ in m}^2, \\ \text{and } m \text{ in kg} \end{array}$$

$$(3.2\text{-}1)$$

illustrate the relationships among wingspan, wing area, hand length, and body mass of the flying animal considered. As shown in Fig. 3.2-5a, these can also be extended to the range of insects and other flying animals. Since the different animals are all made of the same materials and are similar in structural configuration, they roughly follow the straight line prescribed by the similarity law. Divergence from the above general trend reflects

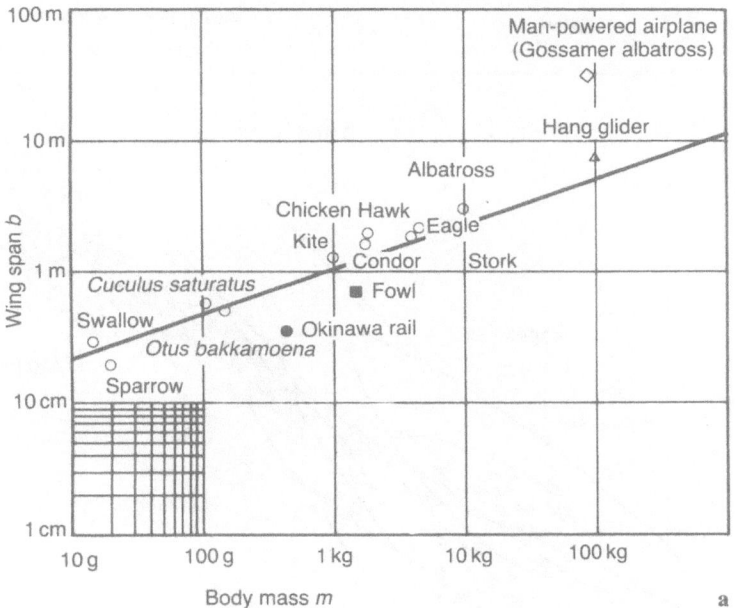

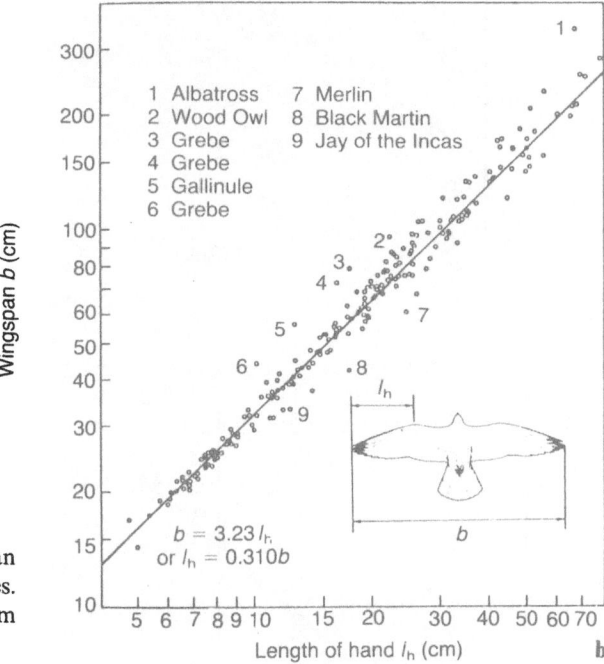

FIG. 3.2-4. **a, b.** Statistical data on wing span. **a** Winspan versus body mass of birds and man-made planes. **b** Wingspan versus length of hand. (Redrawn from Greenewalt 1962).

the morphological and functional differences among representative types of flying animals (Kokshaysky 1977). For instance, birds of prey have a comparatively larger wing area, and poor cruisers such as the gallinaceous birds have a smaller wing area. It is interesting to note in Fig. 3.2-4a that the Yanbaru rail or Okinawa rail (Rallus okinawae), which was discovered recently on Okinawa Island, Japan, is believed to be incapable of flight, and domestic fowls, which are poor flyers, are an exception, while hummingbirds also fall into a very special group.

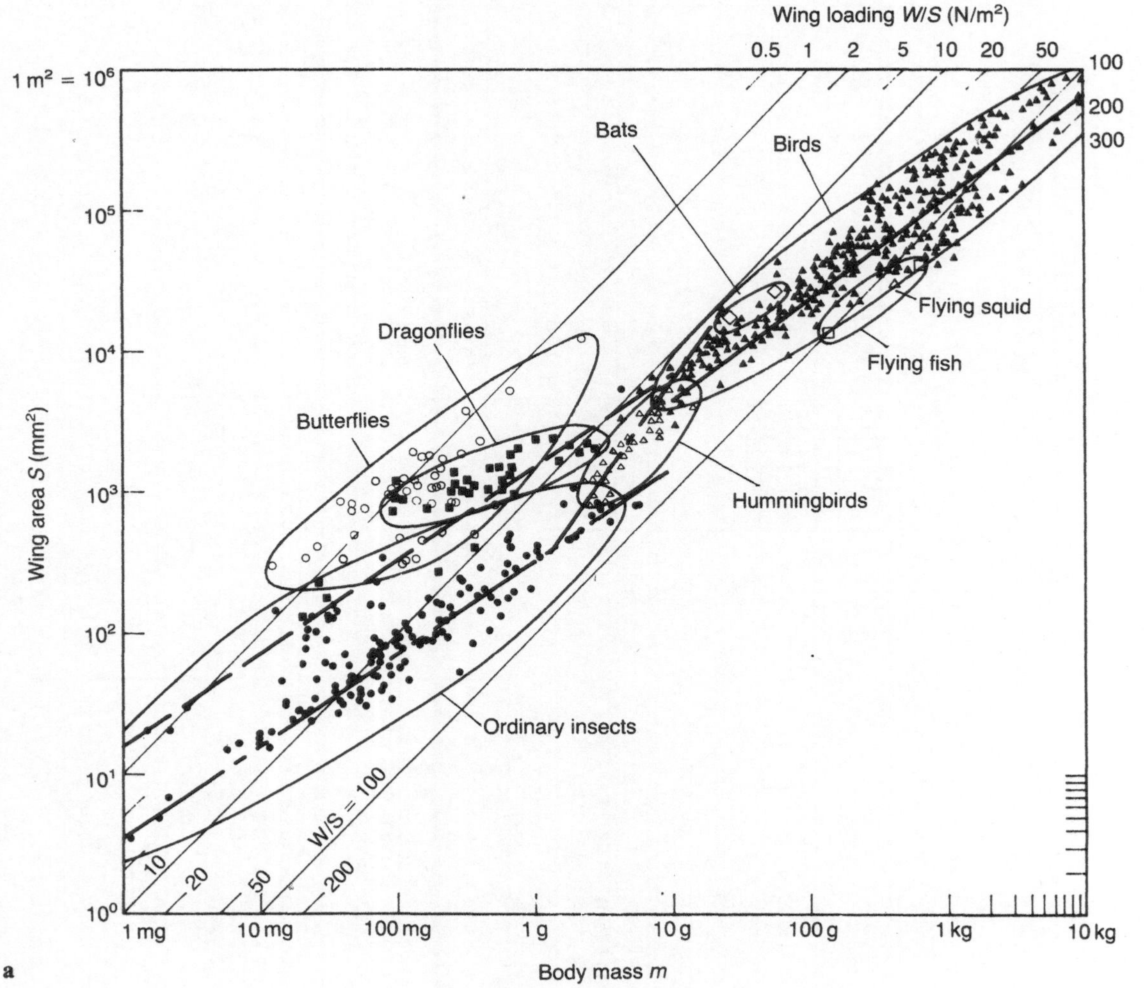

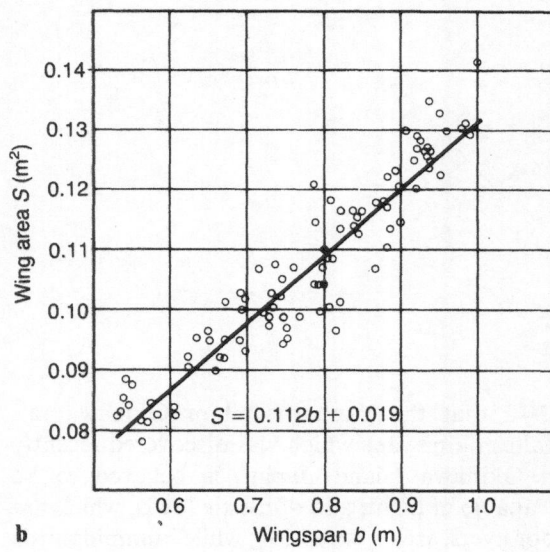

FIG. 3.2-5. **a, b.** Wing area. **a** Wing area versus body mass. (Redrawn from Greenewalt 1962). **b** Wing area versus wingspan. (From Tucker and Parrott 1970 with permission of the Company of Biologists Ltd.).

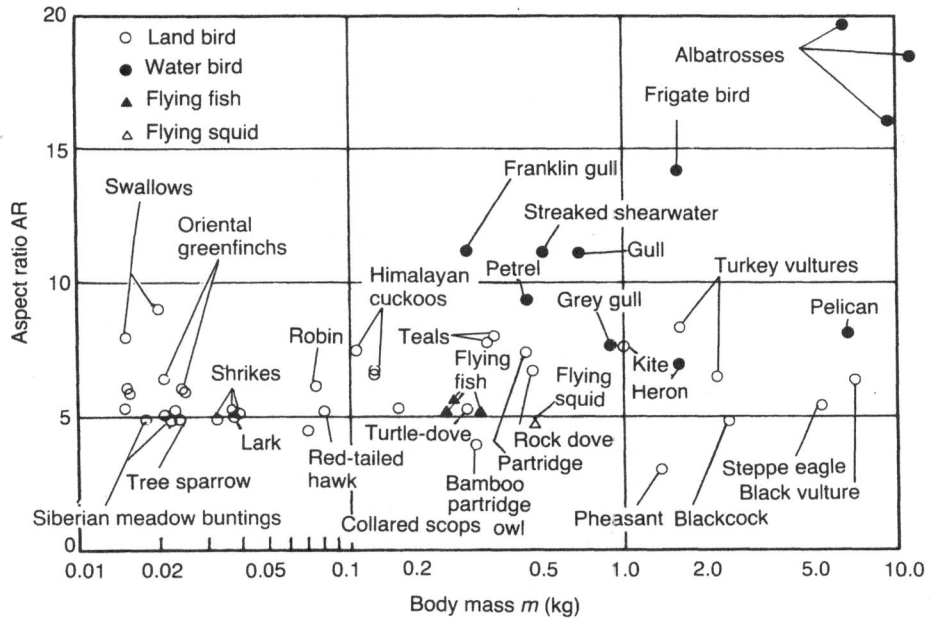

FIG. 3.2-6. Aspect ratio versus body mass.

Figure 3.2-6 shows statistical data on "aspect ratio," which is a ratio of square wingspan and wing area ($AR = b^2/S$), versus mass for various species of bird. The aspect ratio is, as explained later in Sect. 3.2-3, a parameter indicating the flight performance of any flyer. It can be seen that most land birds have a wing with an aspect ratio of between 5 and 7, and birds making soaring flight, such as sea birds like the albatross and the frigate bird, invariably have a wing with a large aspect ratio of more than 10.

3.2.3 Performance in Steady Gliding Flight

Let us consider a trimmed or steady gliding flight, as shown in Fig. 3.2-7, in which the parallel and normal components of the weight W to the glide path $-\gamma$ are respectively equal to the drag D and lift L acting on a flying creature as follows:

$$W \sin(-\gamma) = D = \tfrac{1}{2}\rho U^2 S C_D \quad (3.2\text{-}2a)$$

$$W \cos(-\gamma) = L = \tfrac{1}{2}\rho U^2 S C_L \quad (3.2\text{-}2b)$$

where the air density ρ may be replaced with

$$\rho = \rho_0 \sigma \quad (3.2\text{-}3)$$

σ being the "density ratio" based on the standard atmosphere at sea level. The coefficients C_L and C_D are respectively the lift coefficient and the drag coefficient both of which include the contribution of not only the main wings themselves but also of the tail wing, the trunk, and the extended spatulate toes.

As clearly seen from Eq. (3.2-2b), the required lift coefficient decreases with an increase of speed and/or decrease of wing loading W/S. Thus, it should be noted that the larger and heavier birds must fly at higher velocity and need more power in level flight.

Equation (3.2-2a,b) yields:

$$D/L = C_D/C_L = \tan(-\gamma) \cong |\gamma| \quad (3.2\text{-}4a)$$

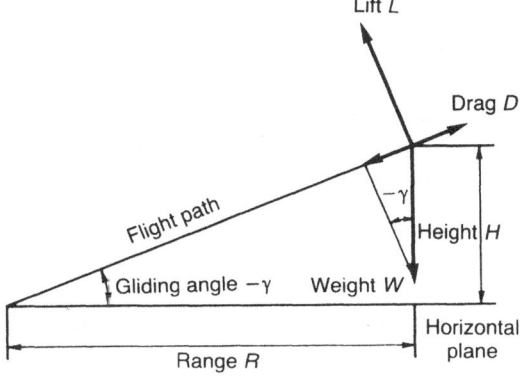

FIG. 3.2-7. Steady gliding flight.

or

$$L/D = C_L/C_D = \cot(-\gamma) \cong 1/|\gamma| \quad (3.2\text{-}4b)$$

Then, by referring to Fig. 3.2-7, the range of steady gliding flight R from a given altitude H can be expressed by

$$R = H/\tan(-\gamma) = H(L/D) = H(C_L/C_D) \quad (3.2\text{-}5)$$

Equations (3.2-4) and (3.2-5) state that the gliding angle $-\gamma$ is inversely proportional to the arctangent of the "lift-to-drag ratio" L/D, and that under the assumption of a constant flight speed U, the minimum gliding angle which gives the maximum range of flight from a given altitude can be obtained by taking an optimal angle of attack at which the lift-to-drag ratio is maximum.

Drag consists of the "parasite drag" $C_{D,b}$ from body portions other than the wing, and the "profile drag" $C_{D,w}$ and "induced drag" C_{D_i} from the wing, as follows:

$$C_D = C_{D_0} + C_{D_i} \quad (3.2\text{-}6)$$

where

$$C_{D_0} = C_{D,b} + C_{D,w} = f/S + C_{D,w} \quad (3.2\text{-}7a)$$

$$C_{D_i} = C_L^2/\pi A\!R_e \quad (3.2\text{-}7b)$$

and where $f = C_{D,b}S$ is called the drag area of the body exclusive of the wing profile drag. The method for estimating the parasite drag will be given later. In Eq. (3.2-7b), the "effective aspect ratio" ($A\!R_e = A\!R \cdot e$, where e is the "deficiency factor") has been used instead of the geometrical aspect ratio ($A\!R = b^2/S \geq A\!R_e$), which is a measure of the wingspan to the wing chord. In a wing with an elliptic planform, called an "elliptic wing," the effective aspect ratio is equal to the geometrical aspect ratio or $A\!R = A\!R_e$, and the induced drag is minimum (Perkins and Hage 1965). A more detailed explanation is presented later in Sect. 4.1.1. *Flight of minimum path angle.* Let us further define the effective wing span b_e and the total drag area of the drag, including body and wing, as follows:

$$b_e = \sqrt{A\!R_e S} \quad (3.2\text{-}8a)$$

$$f_{tot} = C_{D_0}S = f + SC_{D,w} \quad (3.2\text{-}8b)$$

Then, the lift coefficient giving the minimum flight path angle or minimum gliding angle $|\gamma|_{min}$ is determined by making

$$\partial|\gamma|/\partial C_L \simeq (\partial/\partial C_L)(C_D/C_L) = 0 \quad (3.2\text{-}9)$$

which provides some interesting relations among aerodynamic parameters as given in Table 3.2-1.

The ratio of the effective span to the geometrical span b_e/b is considered a measure of the aerodynamic efficiency of the wing. In gliding birds, the ratio can be expressed by $b_e/b = 0.96$ (Spedding 1987a), whereas the tip loss factor of a rotary wing in helicopter engineering is $b_e/b = 0.97$ (Gessow and Myers 1952).

The minimum gliding angle or "optimal range of flight" can be obtained by adjusting the lift or angle of attack of the wing in such a way that the induced drag is just equal to the parasite drag of the bird. It can be improved by increasing the aspect ratio of the wing, or wingspan, by reducing the zero-lift drag coefficient C_{D_0} as much as possible.

The optimal flight speed at the minimum gliding angle is dependent on the aspect ratio through $C_{L,|\gamma|_{min}}$ and is proportional to the wing loading W/S over the lift coefficient at $|\gamma|_{min}$ or the square root of span loading $\sqrt{W/b_e}$. The requirement for high speed dictates a low product of drag area f_{tot} and square of equivalent span b_e^2, whereas the high lift-to-drag ratio requires a low ratio of f_{tot} to b_e^2. Usually, soaring birds fly with a wing of large span at relatively low speed, whereas birds in beating flight fly with a wing of small span at high speed.

When the drag is expressed by

$$D = \tfrac{1}{2}\rho U^2 S\{C_{D_0} + C_L^2/\pi A\!R_e\}$$
$$\simeq \tfrac{1}{2}\rho U^2 SC_{D_0} + [(W/S)^2/\{(\pi A\!R/S)(\tfrac{1}{2}\rho U^2)\}] \quad (3.2\text{-}10)$$

the minimum drag for a given configuration in gliding flight can be obtained by $\partial D/\partial U = 0$. Thus, it can be said that the speed for the minimum gliding angle is the speed for the minimum drag. Without adequate maneuvering birds cannot glide stably at speeds much below their minimum-drag speed, because a further decrease in speed would produce an increase in drag and lead to more deceleration. The requirement for stable flight is a speed between $\sqrt{W\sin|\gamma|/\rho SC_{D_0}}$ and $\sqrt{2W/\rho SC_{D_0}}$, the former of which being the speed of minimum drag or minimum gliding angle and the latter the diving speed in flight without lift. In this speed range the parasite drag is always larger than the induced drag, and the lift coefficient is smaller than that of the maximum lift-to-drag ratio or $C_L < C_{L,(L/D)_{min}}$.

TABLE 3.2-1. Optimal flight conditions

Items	Symbols	Minimum gliding angle $\lvert\gamma\rvert_{min}$ or maximum lift-to-drag ratio $(L/D)_{max}$	Minimum sinking speed $\lvert w\rvert_{min}$
Lift coefficient	C_L	$C_{L,\lvert\gamma\rvert_{min}} = C_{L,(L/D)_{max}} = \sqrt{\pi\mathcal{R}_e C_{D_0}}$	$C_{L,\lvert w\rvert_{min}} = \sqrt{3C_{D_0}\pi\mathcal{R}_e}$
Drag coefficient	C_D	$C_{D,\lvert\gamma\rvert_{min}} = C_{D,(L/D)_{max}} = 2C_{D_0} = 2C_{D_i}$ $= 2C_{L,\lvert\gamma\rvert_{min}}^2/\pi\mathcal{R}_e$	$C_{D,\lvert w\rvert_{min}} = 4C_{D_0} = \frac{4}{3}C_{D_i} = \frac{4}{3}C_L^2/\pi\mathcal{R}_e$ $C_{D_i} = 3C_{D_0}$
Lift-to-drag ratio	C_L/C_D or $C_L^{3/2}/C_D$	$(C_L/C_D)_{max} = \frac{1}{2}\sqrt{\pi\mathcal{R}_e}/C_{D_0} = \frac{1}{2}\sqrt{\pi b_e^2/f_{tot}}$ $= \frac{1}{2}\pi\mathcal{R}_e/C_{L,\lvert\gamma\rvert_{min}} \simeq 1/\lvert\gamma\rvert_{min}$	$(C_L^{3/2}/C_D)_{max} = \frac{1}{4}(3\pi\mathcal{R}_e)^{3/4}/C_{D_0}^{1/4}$ or $(C_L/C_D) = (\sqrt{3/4})\sqrt{\pi\mathcal{R}_e}/\sqrt{C_{D_0}}$
Gliding angle	$\lvert\gamma\rvert$	$\lvert\gamma\rvert_{min} \simeq 2\sqrt{C_{D_0}/\pi\mathcal{R}_e} = 2\sqrt{(f_{tot}/b_e^2)/\pi}$	$\lvert\gamma\rvert_{\lvert w\rvert_{min}} \simeq 4\sqrt{C_{D_0}/3\pi\mathcal{R}_e}$
Flight speed	U	$U_{\lvert\gamma\rvert_{min}} = U_{(L/D)_{max}}$ $= \{(2/\rho)(W/S)/C_{L,\lvert\gamma\rvert_{min}}\}^{1/2}\sqrt{1 + (2C_{L,\lvert\gamma\rvert_{min}}/\pi\mathcal{R}_e)^2}\,^{1/2}$ $= \{(2/\rho)(W/S)\}^{1/2}/\{\pi\mathcal{R}_e C_{D_0}\}^{1/4}$ (for wing loading) $= \{(2/\rho)(W/b_e)\}^{1/2}/\{\pi C_{D_0} S\}^{1/4}$ (for span loading) $= \{2W/\rho\}^{1/2}/\{\pi b_e^2 f_{tot}\}^{1/4}$ (for drag area)	$U_{\lvert w\rvert_{min}} \simeq \{(2/\rho)(W/S)\}^{1/2}/\{3C_{D_0}\pi\mathcal{R}_e\}^{1/4}$ $= \{(2/\rho)(W/b_e)\}^{1/2}/\{3\pi C_{D_0}S\}^{1/4}$ $= 3^{-1/4}U_{\lvert\gamma\rvert_{min}} \simeq 0.760 U_{\lvert\gamma\rvert_{min}}$
Sinking rate	$-w$	$\lvert w\rvert_{\lvert\gamma\rvert_{min}} \simeq 2\{(2/\rho)(W/S)\}^{1/2}C_{D_0}^{1/4}/\{\pi\mathcal{R}_e\}^{3/4}$	$\lvert w\rvert_{min} \simeq 4\{(2/\rho)(W/S)\}^{1/2}C_{D_0}^{1/4}/(3\pi\mathcal{R}_e)^{3/4}$ $= \{4\sqrt{2/\rho}(3\pi)^{3/4}\}\{W/b_e^2\}^{1/2}\{f_{tot}/b_e^2\}^{1/4}$ (for drag area) $= \frac{4}{3}\{(2/\rho)(W/S)C_{L,w_{min}}\}^{1/2}/\pi\mathcal{R}_e$ (for lift coefficient)

Flight of minimum rate of descent. The rate of descent or sinking rate, $-w$, is given by

$$-w = U \sin(-\gamma) = U(C_D/C_L)/\sqrt{1 + (C_D/C_L)^2}$$

$$\simeq U(C_D/C_L) = U\{C_{D_0} + C_L{}^2/\pi R_e\}/C_L \tag{3.2-11}$$

and, by substituting Eq. (3.2-2a,b) into the above equation, can be rewritten as follows:

$$-w = \sqrt{2/\rho}\sqrt{(W/S)}(C_D/C_L{}^{3/2})/\{1 + (C_D/C_L)^2\}^{3/4}$$

$$\cong \sqrt{(2/\rho)(W/S)}(C_D/C_L{}^{3/2}) \tag{3.2-12}$$

It can be said from Eq. (3.2-12) that the sinking rate $-w$ is proportional to the square root of wing loading $\sqrt{W/S}$ and is optimized approximately by taking the minimum value of $C_D/C_L{}^{3/2}$ for a given wing loading of the creature flying at a given altitude. This is, obtained by making

$$(\partial/\partial C_L)(C_L{}^{3/2}/C_D) = 0 \tag{3.2-13}$$

which gives some other interesting relations as shown in Table 3.2-1: The sinking rate is proportional to the square root of the wing loading times the lift coefficient, and is inversely proportional to the aspect ratio.

Estimation of performance. The drag coefficient of body portions other than the wings is estimated by assuming a slender body with an elliptic cross section as described later in Sect. 6.1.2. Pennycuick (1969) and Tucker (1973) gave the following formula to estimate the drag area:

$$f = (2.84 \sim 3.34) \times 10^{-3} m^{2/3}$$

$$\text{for } f \text{ in m}^2 \text{ and } m \text{ in kg;} \tag{3.2-14}$$

in which $f = 3.34 \times 10^{-3} \ m^{0.66}$ was obtained from wind tunnel tests using wingless bird bodies (Tucker 1973).

As examples, wind tunnel measurements on frozen bodies of waterfowl and raptors gave the drag coefficient of $C_{D,f} = f/S = (f/S_f)(S_f/S) = (0.25-0.39)(S_f/S)$, where S_f is the frontal area of the body at the Reynolds number of $(1.45-4.62) \times 10^5$ and is approximately computed (Pennycuick et al. 1988) at

$$S_f = (8.13 \times 10^{-3})m^{0.666}$$

$$\text{for } m \text{ in kg and } S_f \text{ in m}^2 \tag{3.2-15}$$

The parasite drag coefficient of the wing $C_{D,w}$ results mostly from the friction drag of the wing wetted surface S_w which is roughly twice the wing

area S or estimated by Tucker and Parrott (1970) to be;

$$S_w \simeq 2.04S \tag{3.2-16}$$

Then, the drag coefficient based on the wing area can be expressed by

$$C_{D,w} = C_f S_w/S \simeq 2.04 C_f \tag{3.2-17}$$

The friction drag coefficient of a flat surface C_f is given in many aerodynamic and hydrodynamic books for various Reynolds numbers.

For small birds and insects, the following values of the drag coefficient are used:

$$\begin{aligned}
C_{D,w} &= 0.35 \text{ for drosophila (Vogel, 1967b)} \\
&= 0.07 \text{ for dragonfly} \\
&\quad \text{(Azuma and Watanabe 1988)} \\
&= 0.04 \text{ for hummingbird} \\
&\quad \text{(Horton-Smith 1938)} \\
&= 0.02 \text{ for small birds in general} \\
&\quad \text{(Terres 1968)}
\end{aligned} \tag{3.2-18}$$

Figure 3.2-8 shows optimal flight speed versus wing loading for two values of the parameter σC_L. A very rough set of optimal flight speeds for a soaring bird can be approximated by assuming $\sigma C_L = 0.500$ for $|\gamma|_{\min}$ and $\sigma C_L = 0.866$ for $|w|_{\min}$.

Performance of various gliding objects. Modern man-made sailplanes have a wing with an extremely high aspect ratio of over 20 and a beautifully smooth surface. However, the fairing is not always good and the tail volume, which is a product of tail surface area and tail length from the center of gravity (see Sect. 3.5-2), is too large for good stability and control. As a result, the optimal lift coefficient is about $C_L = 0.6-0.8$, and the lift-to-drag ratio barely achieves $(L/D)_{\max} \simeq 50$. This is a little better than that attained by an albatross with a wing of $R \leq 20$.

Figure 3.2-9 shows the steady gliding performance or "glide polar" for several typical examples of flying creatures, such as a seabird with excellent soaring performance (the wandering albatross), birds that glide well (the black vulture, soaring vulture, and juvenile golden eagle), and creatures which beat powerfully (the pigeon, long-eared bat, falcon, and migratory locust). Some of their aerodynamic characteristics are assumed to be those

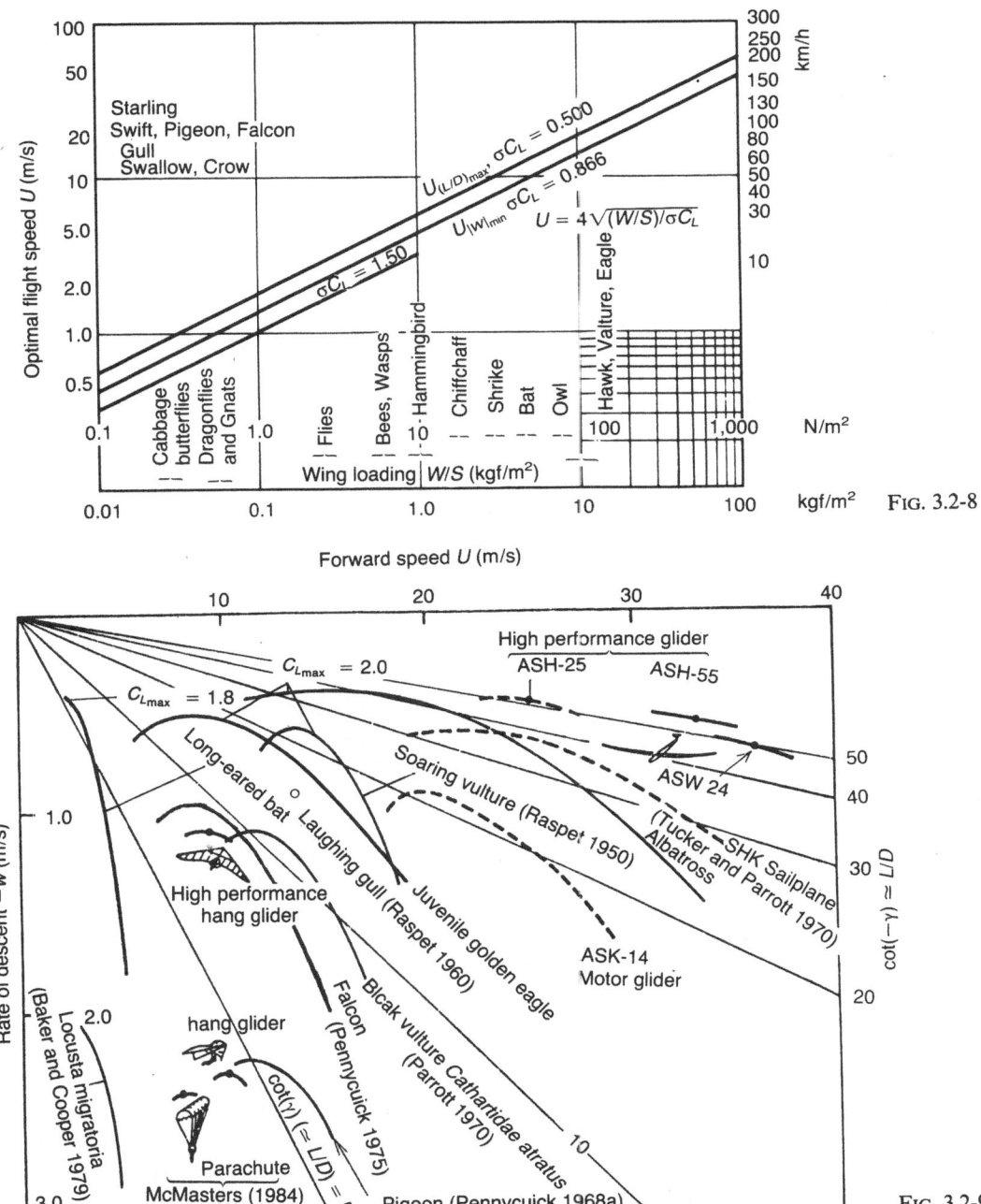

FIG. 3.2-8

FIG. 3.2-9

FIG. 3.2-8. Optimal flight speed versus wing loading.

FIG. 3.2-9. Rate of descent against forward speed in steady gliding performance.

given in Table 3.2-2. It is interesting to find that the sea birds make use of a large aspect ratio and high wing loading to obtain a small gliding angle and low rate of descent, whereas the land birds obtain a low rate of descent by having small wing loading.

These data, shown in Fig. 3.2-9 are comparable against several additional examples of the gliding

TABLE 3.2-2. Geometrical characteristics of exemplified birds

Items	Symbol	Unit	Albatross	Juvenile golden eagle	Long-eared bat	Humming-bird
Wingspan	b	m	3.00	1.63	0.270	0.0956
Wing area	S	m²	0.667	0.431	0.0123	0.00113
Aspect ratio	$\mathcal{R}$	—	18	6.17	5.93	8.10
Mass	m	kg	10	1.8	0.00901	0.00370
Weight	$W = mg$	N	98	17.6	0.0882	0.0362
Wing loading	W/S	kgf/m²	15.0	4.18	0.733	3.28
		N/m²	147	40.8	7.17	32.0
Density parameter	$\mu_c = m/\rho S\bar{c} = m\mathcal{R}/\rho Sb$	—	73.5	12.9	13.2	228
Lift slope	a	—	5.20	4.40	4.37	4.67
Minimum drag coefficient	C_{D_o}	—	0.009[a]	0.013	—	—

[a] Here, the minimum drag coefficient has been estimated to be extremely low in consideration of the fact that the flow around the wing and body is laminar.

performance of man-made devices: a sailplane, a motor glider, a high-performance glider, a high-performance hang glider, a primary hang glider, and a parachute.

Flight time. The flight speed at minimum sinking rate is $3^{-1/4} \simeq 0.76$ times the speed at the maximum lift-to-drag ratio. By flying at this speed the bird can make long-duration flight over a specified area, because the time spent in steady gliding flight can be given as

$$t = R/U\cos(-\gamma) = H/|w| \quad (3.2\text{-}19)$$

In steep descending flight such as diving flight, the approximate expression in Table 3.2-1 cannot be applied. The least time for loss of height can, then, be given by making zero lift, $C_L = 0$, or vertical flight, $\gamma = -90°$, as follows:

$$t = \{H/\sqrt{(2/\rho_0)(W/S)/\sigma}\}\sqrt{C_{D_0}} \quad (\gamma = -90°) \tag{3.2-20}$$

Thus, it can be seen that quick diving flight should be performed by reducing the drag coefficient and making the wing area as small as possible.

Lowest speed. Now, let us consider the lowest speed U_s in steady gliding flight. Any flying creature has to reduce this speed, rather than the speed of the minimum rate of descent, to reduce the impact at landing on a spot such as a branch or a field of limited length.

Equation (3.2-2a,b) gives the following relation

$$U = \sqrt{(2/\rho)(W/S)/C_R}, \quad (3.2\text{-}21)$$

where C_R is the aerodynamic coefficient defined by

$$C_R = \sqrt{C_L{}^2 + C_D{}^2}. \quad (3.2\text{-}22)$$

Usually, except at very small speed such as on landing approach, the drag coefficient is neglected as being a small quantity, $C_L \gg C_D$, as in the preceding discussions.

The above equation reveals that the lowest speed U_s can be attained by (1) making the wing area maximum and reducing the wing loading, (2) reducing altitude, and (3) making the total aerodynamic coefficient maximum, $C_{R_{max}}$.

In an airplane, which has a large lift-to-drag ratio, the maximum aerodynamic coefficient can be obtained approximately at the maximum lift coefficient at which the flow separation is spread over the wing surface. However, if a sudden decrease in the lift coefficient does not appear after the flow separation, and if the drag coefficient is increased by extending legs, alulae and tail wing, then the maximum aerodynamic coefficient, $C_{R_{max}}$ can be obtained at a higher angle of attack than the stalling angle of attack where the lift coefficient is $C_{L_{max}}$. Figure 3.2-10 shows that this maximum value can be obtained at the tangential point between the polar curve A (high $\mathcal{R}$ wing) or B (low $\mathcal{R}$ wing) and a circle (shown by the broken curve) enclosing the polar curve with the minimum radius from the origin of the coordinate axes. Specifically, in a low aspect ratio wing, such as that of a flying squirrel, the importance of the drag component C_D is great, comparable to that of the lift component C_L. The glide angle is, of course, steep at such low speed because of the low aspect ratio wing.

For a high-performance bird with a large aspect ratio wing, the $C_{R_{max}}$ may be approximated as $C_{R_{max}} \simeq C_{L_{max}}$. Then, the lowest speed is defined as the stalling speed beyond which trimmed flight obviously cannot be maintained.

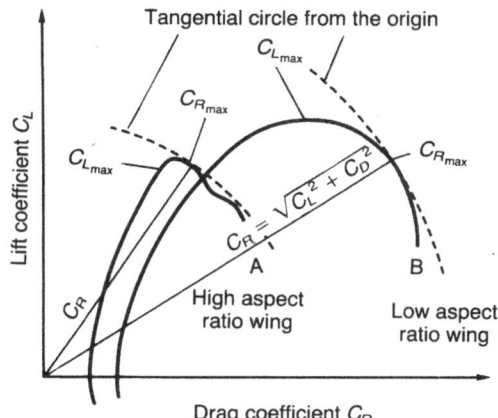

Fig. 3.2-10. Maximum aerodynamic coefficient.

As stated before and shown in Fig. 3.2-3, many birds make a large dihedral in approaching flight by raising the wing tips. This not only increases the lateral stability but, as a result of the reduced span, shifts the aerodynamic characteristics of the wing and body from curve A to curve B in Fig. 3.2-10. The result is an improvement in the landing characteristics, especially in the case of land birds, in spite of a reduction in projected wing area.

3.2.4 Effects of Change in Geometrical Configuration

Birds can change their geometrical configuration appreciably. During gliding flight, as experimentally checked in a wind tunnel test (Pennycuick 1968a) they can vary the wingspan and area from a state of full extension to one of nearly complete retraction. They can also change the vertical and longitudinal position of the aerodynamic center of the main wing. The case where the reduction of wingspan is performed progressively and the gliding speed becomes faster and faster is called flex gliding (Hankin 1913). They can also change the span or area of the tail wing by a fanwise spreading of the rectrices. Because the tail wing is close to the main wing, it can act not only to produce a control moment for change of attitude but also as stated before, as a kind of flap or slotted flap of the main wing, to assist in increasing the lift.

For a bird in trimmed gliding flight with an optimal flight speed U_0, the force balance is as given by Eq. (3.2-2a,b). If the bird changes the geometrical configuration of the main wing by either extending or retracting the wingspan, then, according to Tucker (1987) the wing area S varies with the wingspan b as follows:

$$\bar{S} = S/S_0 = \alpha\bar{b} + \beta \qquad (3.2\text{-}23a)$$

$$\bar{b} = b/b_0 \qquad (3.2\text{-}23b)$$

and the profile drag area and the induced drag area are respectively

$$\left.\begin{array}{l} SC_{D,\text{w}} = 2.04\,SC_\text{f} = 2.04\,S_0 C_\text{f}(\alpha\bar{b} + \beta) \\ SC_{D_\text{i}} = SC_L{}^2/\pi A\!R_e = S_0(\alpha\bar{b} + \beta)^2 C_L{}^2/\pi A\!R_{e,0}\bar{b}^2 \end{array}\right\} \\ (3.2\text{-}24)$$

where

$$A\!R_{e,0} = (b_0{}^2/S_0)e \qquad (3.2\text{-}25)$$

and where the profile drag is assumed to be

Table 3.2-3. Effect of change in wing configuration

Items	Equations
Total drag area	$f_{\text{tot}} = SC_d = SC_{d,\text{w}} + SC_{D_\text{i}} + f$
	$= 2.04C_\text{f}S_0(\alpha\bar{b} + \beta) + \{S_0(\alpha\bar{b} + \beta)^2 C_L{}^2/\pi A\!R_{e,0}\bar{b}^2\} + f$
Trimmed gliding angle	$\gamma = -\sin^{-1}[\frac{1}{2}(\bar{b}^2/B)u^2 + \sqrt{\frac{1}{4}(\bar{b}^2/B)^2 u^4 + (A/B)(\alpha\bar{b} + \beta)\bar{b}^2 u^4 + (C/B)\bar{b}^2 u^4 + 1}]$
Optimal flight speeds	
Minimum gliding angle	$\bar{U}_{|\gamma|_{\min}} = [4A^2(\alpha\bar{b} + \beta)^2 + 4C^2 + (A/B)(\alpha\bar{b} + \beta)\bar{b}^2 + (C/B)\bar{b}^2 + 8(AC)(\alpha\bar{b} + \beta)]^{-1/4}$
Minimum rate of descent	$\bar{U}_{|w|_{\min}} = \{(-E + \sqrt{E^2 - 16D})/2D\}^{1/4}$
Notations	$\sigma = \rho/\rho_0, \qquad u = U/U_0, \qquad \bar{W} = W/\frac{1}{2}\rho_0 U_0{}^2 S_0, \qquad \bar{f} = f/S_0$
	$A = 2.04\sigma C_\text{f}/\bar{W}, \qquad B = \bar{W}/\sigma\pi A\!R_{e,0}, \qquad C = \sigma\bar{f}/\bar{W}$
	$D = 9\{4(A/B)^2(\alpha\bar{b} + \beta)^2\bar{b}^4 + 4(C/B)^2\bar{b}^4 + (1/B^2)(A/B)(\alpha\bar{b} + \beta)\bar{b}^6$
	$\qquad + (1/B^2)(C/B)\bar{b}^6 + 8(A/B)(C/B)(\alpha\bar{b} + \beta)\bar{b}^4\}$
	$E = 3\{8(A/B)(\alpha\bar{b} + \beta)\bar{b}^2 + 8(C/B)\bar{b}^2 - (1/B^2)\bar{b}^4\}$

invariant throughout the change in lift coefficient and the deficiency factor e for the nonelliptic wing is constant. The former assumption is realized when the bird can, as stated earlier in Sect. 3.1.1, adjust the wing camber to keep the minimum profile drag for any given lift coefficient.

By assuming further that the parasite drag area of the body exclusive of the wing, $SC_{D,b} = f$, is invariant through this change in wing configuration, the total drag area f_{tot} can be given as in Table 3.2-3.

Optimal gliding. (a) Under a given $\bar{b}$, the speed for the minimum gliding angle can be obtained from the following equation:

$$\partial \sin(-\gamma)/\partial u = 0 \qquad (3.2\text{-}26)$$

which yields a solution given by $U_{|\gamma|_{min}}$ in Table 3.2-3. (b) Since the nondimensional rate of descent is given by $u\sin(-\gamma)$, the speed for the minimum rate of descent under a given $\bar{b}$ is obtained from the following equations:

$$\partial \sin(-\gamma)/\partial u = \sin(-\gamma) + u\{\partial \sin(-\gamma)/\partial u\} = 0 \qquad (3.2\text{-}27)$$

Then the solution is given by $U_{|w|_{min}}$ and also shown in Table 3.2-3.

Shown in Table 3.2-4 are geometrical characteristics of gliding birds and estimated data on performance. Figure 3.2-11 shows the optimal gliding performance for a retracted wingspan. The deviation from the optimal condition caused by changing the wingspan is clear, and the differences among the different species of birds nondimensional speed are small because of their nondimensional expression.

However, if the bending moment at the wing root is taken into consideration and is presumed to be invariant during extension of the wing, then, as proposed by Jones (1950, 1980) and further discussed in Sect. 4.2.7, the induced drag can be reduced more than in the case of an elliptic wing having shorter span. Adopting the elliptic load distribution for the wing of longer span further decreases the induced drag, but in that case the bending moment at the wing root increases beyond that of the original wing. By making the wing in a tapered plan form with pointed tip, the spanwise position of the mean aerodynamic chord, the forces and moments on which represent all the aerodynamic forces and moments, can be shortened and thus the bending moment at the wing

TABLE 3.2-4. Given characteristics of gliding birds. (Geometrical data obtained from Tucker 1987)

Item Species	Weight W (N)	Maximum wingspan b_0 (m)	Maximum wing area S_0 (m²)	Aspect ratio $\mathcal{R}$	Wing configuration α	β	Drag coefficient Body $C_{D,b}$	Wing $C_{D,w}$	Skin friction C_f	Velocity U_0 (m/s) at $|\gamma|_{min}$	at w_{min}
Falcon	5.60	1.01	0.132	7.73	0.857	0.144	0.0162	0.0906	0.444×10^{-2}	9.40	7.15
Black vulture	17.5	1.37	0.336	5.59	1.060	−0.060	0.0145	0.0602	0.295×10^{-2}	11.9	9.04
African white-backed vulture	52.8	2.18	0.689	6.90	1.060	−0.060	0.0147	0.0492	0.241×10^{-2}	13.9	10.5
Fulmar	7.20	1.09	0.144	10.42	0.700	0.299	0.237	0.0922	0.452×10^{-2}	9.87	6.74

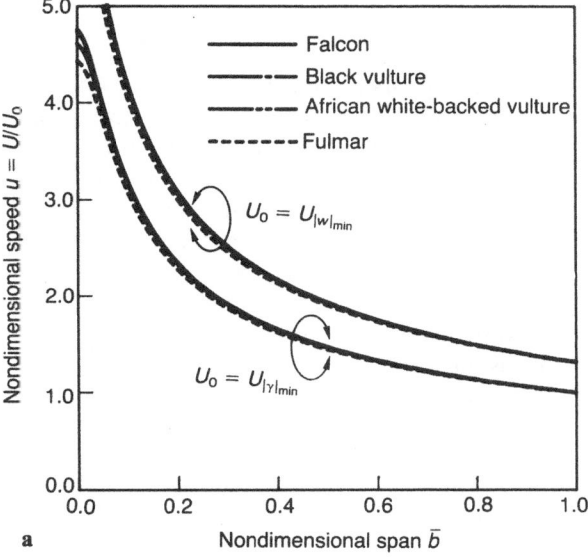

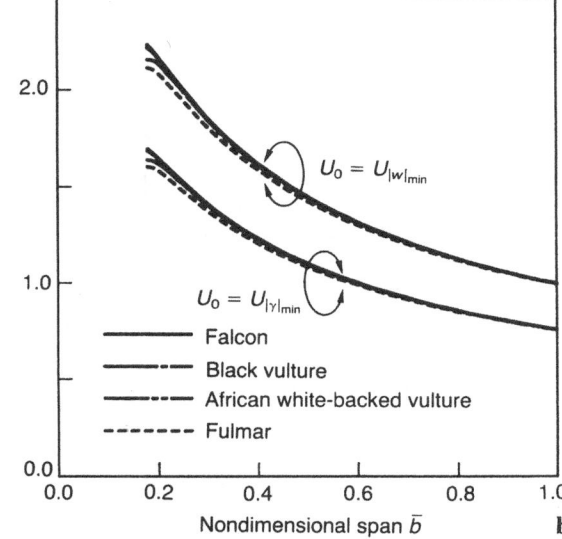

Fig. 3.2-11. **a, b.** Optimal gliding performance for retracted wingspan. (The reference speed U_0 can be selected in two ways, either $U_0 = U_{|\gamma|_{min}}$ or $U_0 = U_{|w|_{min}}$ at $\bar{b} = 1.0$). **a** The speed for the minimum gliding angle under given $\bar{b}$. **b** The speed for the minimum rate of descent under given $\bar{b}$.

TABLE 3.2-5. Effects of wingspan for a restricted bending moment

	Original elliptic wing		*Extended optimal wing*		*Extended elliptic wing*
Load distribution (See Fig. 4.2-15a)	Elliptic b_e		$b > b_e$		Elliptic $b > b_e$
Lift	$L(b_e)$	=	$L(b > b_e)$	=	$L(b)$
Bending moment at wing root	$B_e(b_e)$	=	$B(b > b_e)$	<	$B_e(b)$
Induced drag	$D_{i,e}(b_e)$	>	$D_i(b > b_e)$	>	$D_{i,e}(b)$

root can be reduced. These situations are illustrated in Table 3.2-5.

3.3 Wind Effects on Gliding Flight

All flying creatures are inevitably surrounded by a body of moving fluid. In some cases the winds or currents of the moving fluid bring inestimable benefit to the life of the flying creatures, and in others, they wreak havoc (Alexander and Camp 1983). Most flying creatures skillfully utilize the winds or currents of their fluid medium to make

their lives easier and more enjoyable. For instance, in order to compensate for the loss of height in gliding flight, a bird can glide in upward currents of air or in gentle wind shear to save itself the work of beating its wings.

3.3.1 STATIC WIND EFFECTS

If any wind is blowing, the actual velocity of a bird with respect to the ground, **V**, is the vector sum of the wind velocity $\mathbf{U}_w$ and the relative velocity of the bird with respect to the air, **U**. Referring to Fig. 3.3-1, where the respective components of the above velocity are given by

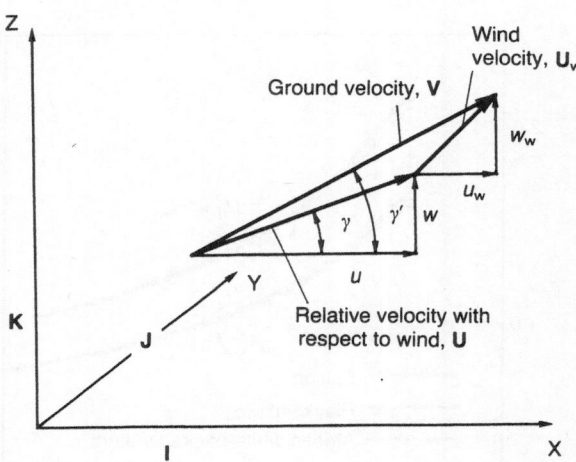

FIG. 3.3-1. Effect of wind.

$$\mathbf{V} = (\dot{X}, \dot{Y}, \dot{Z})^{\mathrm{T}} = \mathbf{I}\dot{X} + \mathbf{J}\dot{Y} + \mathbf{K}\dot{Z}$$

$$\mathbf{U}_{\mathrm{w}} = (u_{\mathrm{w}}, 0, w_{\mathrm{w}})^{\mathrm{T}}$$

$$\mathbf{U} = (u, v, w)^{\mathrm{T}} \qquad (3.3\text{-}1)$$

the forward, sideward, and upward components of the absolute velocity can be expressed by

$$\mathbf{V} = \mathbf{U} + \mathbf{U}_{\mathrm{w}} \quad \text{or}$$

$$(\dot{X}, \dot{Y}, \dot{Z})^{\mathrm{T}} = (u, v, w)^{\mathrm{T}} + (u_{\mathrm{w}}, 0, w_{\mathrm{w}})^{\mathrm{T}} \quad (3.3\text{-}2)$$

Shown in Fig. 3.3-2 are changes in steady gliding slope due to steady horizontal and vertical winds. It can be seen that either or both tail wind and updraft will extend the horizontal distance, whereas either or both head wind and downdraft will diminish the horizontal distance. In horizontal wind, the flight time is not affected by the wind and is thus equal to the normal or no-wind flight time. In a vertical current, on the other hand, the flight time is increased in an updraft, and when the velocity of the updraft is equal to, or greater than, the rate of descent, the bird can glide without any loss of altitude.

The range of a steady gliding flight ($\gamma' < 0$) in a steady wind can be given by

$$R = H/\tan(-\gamma') \qquad (3.3\text{-}3)$$

FIG. 3.3-2. **a, b.** Effect of wind on the gliding slope. **a** Horizontal wind. **b** Vertical wind.

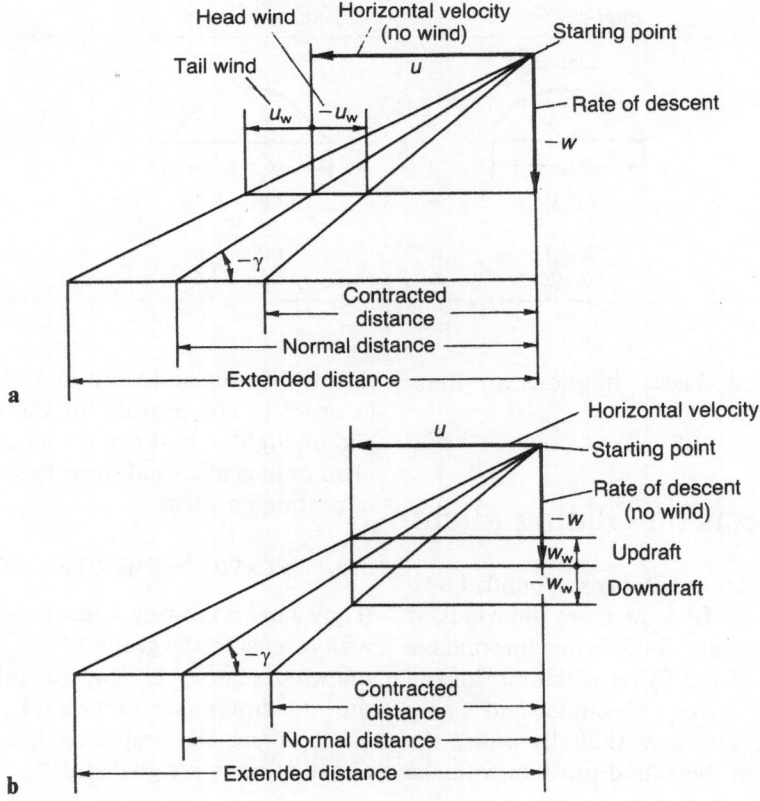

FIG. 3.3-3. Optimal gliding in wind.

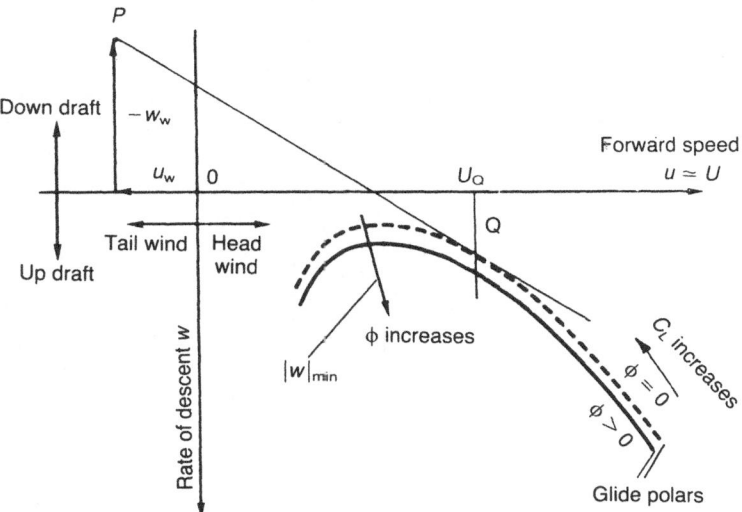

where, from the trimmed condition, $-\gamma'$ is

$$\tan(-\gamma') = -(w + w_\mathrm{w})/(u + u_\mathrm{w})$$
$$= -(-C_D/C_L + w_\mathrm{w}/u)/(1 + u_\mathrm{w}/u)$$
$$\text{(3.3-4)}$$

and where the lateral velocity v is assumed zero. Then the range-to-height ratio or "gliding ratio" is given by

$$R/H = \cot(-\gamma') = (C_L/C_D)\left\{\frac{1 + (u_\mathrm{w}/u)}{1 + (C_L/C_D)(-w_\mathrm{w}/u)}\right\}$$
$$\text{(3.3-5)}$$

Thus, contrary to Eq. (3.2-5), it can be said that the maximum range of gliding flight from a given height $(R/H)_{\max}$ is not the same as the flight of maximum lift-to-drag ratio $(C_L/C_D)_{\max}$ in steady wind.

Since, under the assumption of $u_\mathrm{w} \ll u$, the gliding ratio can be rewritten as

$$R/H = \{C_L + \bar{u}_\mathrm{w} C_L^{3/2}\}/\{C_{D_0} + \frac{C_L^2}{\pi A\!R_e} - \bar{w}_\mathrm{w} C_L^{3/2}\}$$
$$\text{(3.3-6)}$$

where

$$\bar{u}_\mathrm{w} = u_\mathrm{w}/\sqrt{W/\tfrac{1}{2}\rho S}, \quad \bar{w}_\mathrm{w} = w_\mathrm{w}/\sqrt{W/\tfrac{1}{2}\rho S} \quad \text{(3.3-7)}$$

the maximum range-to- height ratio $(R/H)_{\max}$, can be determined from the following equation:

$$\partial(R/H)/\partial C_L = 0 \qquad \text{(3.3-8a)}$$

or

$$C_{D_0} + \frac{3}{2} C_{D_0} \bar{u}_\mathrm{w} C_L^{1/2} + \frac{1}{2} \bar{w}_\mathrm{w} C_L^{3/2}$$

$$-\frac{1}{2}\bar{u}_\mathrm{w} \frac{C_L^{5/2}}{\pi A\!R_e} - \frac{C_L^2}{\pi A\!R_e} = 0 \qquad \text{(3.3-8b)}$$

Then, it can be said that (1) the lift coefficient should be increased for tail wind $(u_\mathrm{w} > 0)$ and updraft $(\bar{w}_\mathrm{w} > 0)$, and (2) the lift coefficient should be decreased for head wind $(u_\mathrm{w} < 0)$ and downdraft $(w_\mathrm{w} < 0)$. The dotted line in Fig. 3.3-3 shows the "glide polar." An optimal steady-gliding flight at the speed U_Q can be determined by finding a tangential point Q on the glide polar from the point P, shifted from the origin by the amount of u_w and $-w_\mathrm{w}$.

For steady turning flight in an updraft of thermal, with a bank angle ϕ and a turning radius R, the equations of motion can, by referring to Fig. 3.3-4, be modified from Eq. (3.2-2) to the following equations:

$$\left.\begin{array}{l} L\cos\phi = W\cos(-\gamma) \\ L\sin\phi = mU^2/R \\ D = W\sin(-\gamma) \end{array}\right\} \qquad \text{(3.3-9)}$$

where $W = mg$. For high-performance birds having a large lift-to-drag ratio, $L/D \gg 1$, the above equations yield approximate expressions of several related quantities and their optimal values at the minimum sinking rate as given in Table 3.3-1.

It suggests that the rate of descent $-w$ increases in inverse proportion to the cosine of the bank angle. This relation has been shown in Fig. 3.3-3 by a solid line of $\phi > 0$ and a dotted line of $\phi = 0$. The turning radius R is inversely proportional to the sine of the bank angle and proportional to the square power of the flight speed U^2 or $U_{\phi=0}^2$.

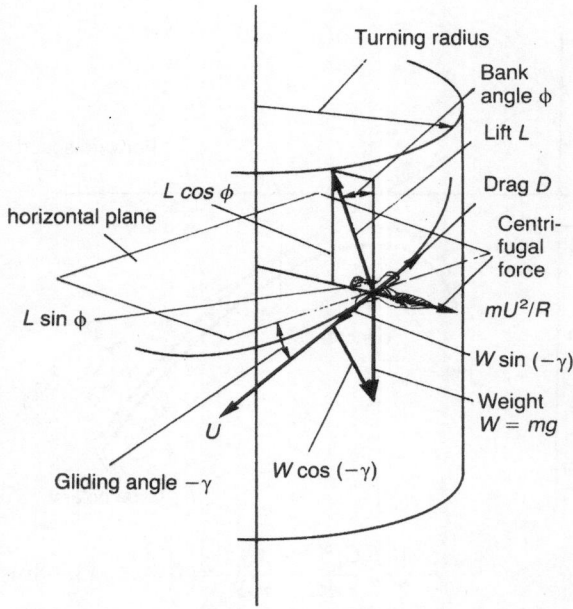

FIG. 3.3-4. Force balance in a steady turning flight.

$$\mathbf{F} = m\,\mathrm{d}\mathbf{V}/\mathrm{d}t \qquad (3.3\text{-}10)$$

where the external force $\mathbf{F}$ is made up of the aerodynamic force $\mathbf{F}_A$, which is a function of the relative velocity $\mathbf{U}$, and the gravity force $\mathbf{F}_G$, which is a function of the altitude Z. Multiplied by $\mathbf{V}$ and integrated with time, the above equation becomes

$$\int \mathbf{F} \cdot \mathbf{V}\mathrm{d}t = \int \mathbf{F}_A \cdot \mathbf{V}\mathrm{d}t + \int \mathbf{F}_G \cdot \mathbf{V}\mathrm{d}t = \frac{1}{2}mV^2 + C$$
$$(3.3\text{-}11)$$

where C is an integration constant dependent on the initial condition. Substituting $\mathbf{F}_G = -\mathbf{K}W$ (where $\mathbf{K}$ is a unit vector in the vertical direction) yields

$$\int \mathbf{F}_A \cdot \mathbf{V}\mathrm{d}t = \frac{1}{2}mV^2 + WZ + C = E \quad (3.3\text{-}12)$$

Here E is an energy consisting of the kinetic energy $\frac{1}{2}mV^2$ and the potential energy WZ in this dynamic system.

By referring to Fig. 3 3-1 and by using Eq. (3.3-2), the left-hand side of Eq. (3.3-12) can be rewritten as

$$\int \mathbf{F}_A \cdot \mathbf{V}\mathrm{d}t = \int \mathbf{F}_A \cdot \mathbf{U}\mathrm{d}t + \int \mathbf{F}_A \cdot \mathbf{U}_w \mathrm{d}t = E_D + E_w$$
$$(3.3\text{-}13)$$

The first term on the right-hand side of the above equation gives the work accompanying flight in the air. The lift contribution has been discarded because the aerodynamic force $\mathbf{F}_A$ is composed of the lift $\mathbf{L}$ and the drag $\mathbf{D}$, which are respectively normal and parallel to the relative flight velocity $\mathbf{U}$, or

$$\mathbf{L} = \tfrac{1}{2}\rho S C_L U\mathbf{U} \times \mathbf{j}, \qquad \mathbf{D} = -\tfrac{1}{2}\rho S C_D U\mathbf{U}$$
$$(3.3\text{-}14\mathrm{a,b})$$

At the minimum sinking rate a small turning radius increases the bank angle and the lift coefficient and thus the minimum sinking speed. A more detailed discussion on soaring flight in a "bubble of thermal" of warmer air than the surroundings is presented by C.D. Cone (1964).

3.3.2 DYNAMIC WIND EFFECTS

The motion of a flying bird, which is assumed to have a mass m without rotation and to fly with a ground velocity $\mathbf{V}$, can be expressed by the equation

TABLE 3.3-1. Approximate expressions of related quantities in turning flight

Item	Symbol	Expression
Bank angle	ϕ	$\tan^{-1}\{U^2/Rg\cos(-\gamma)\} \simeq \tan^{-1}\{U^2/Rg\}$
Glide path angle	$-\gamma$	$\tan^{-1}\{D/L\cos\phi\} \simeq D/L\cos\phi = (C_D/C_L)/\cos\phi$
Rate of descent	$-w$	$U(D/L\cos\phi)\cos(-\gamma) \simeq U(C_D/C_L)/\cos\phi = |w_{\phi=0}|/\cos^{3/2}\phi$
Radius of turn	R	$U^2/g\tan\phi\cos(-\gamma) \simeq U^2/g\tan\phi = U_{\phi=0}^2/g\sin\phi$
Flight speed	U	$\sqrt{mg\cos(-\gamma)/\tfrac{1}{2}\rho S C_L\cos\phi} = U_{\phi=0}/\sqrt{\cos\phi}$
At minimum sinking rate	$|w|_{\min}$	$\{C_{D_0} + C_{L,|w|_{\min}}^2/\pi\!A\!R_e\}\sqrt{gR\psi}/\{C_{L,|w|_{\min}}^2 - \psi^2\}^{3/4}$
	$C_{L,|w|_{\min}}$	$\sqrt{3C_{D_0}\pi\!A\!R_e + 4\psi^2}$
	$\phi_{|w|_{\min}}$	$\sin^{-1}\{\psi C_{L,|w|_{\min}}\}$
Notation	ψ	$(W/S)(2/\rho g)/R = 2\mu_b(b/R)$
	μ_b	$m/\rho Sb$

where $\mathbf{j}$ is a unit vector perpendicular to the plane in which the flight path is included. If, in Eq. (3.3-13), the second term given by $E_w = \int \mathbf{F_A} \cdot \mathbf{U_w} \, dt$ is equal to or larger than the first term given by $E_D = \int \mathbf{F_A} \cdot \mathbf{U} \, dt$, or $E = \int \mathbf{F_A} \cdot \mathbf{U} \, dt > 0$, then gliding flight can be continued without losing flight speed or height.

Thus, the first term can be rewritten as

$$E_D = \int \mathbf{F_A} \cdot \mathbf{U} dt = -\frac{1}{2} \int \rho S U^3 C_D dt \quad (3.3-15)$$

That is to say, the work performed by the aerodynamic force is a negative energy or energy loss caused by drag and is therefore dissipated into the air as heat energy.

The second term on the right hand side of Eq. (3.3-13) can also be rewritten as

$$
\begin{aligned}
E_w &= \int \mathbf{F_A} \cdot \mathbf{U_w} dt \\
&= \frac{1}{2} \int \rho S U \{C_L (\mathbf{U} \times \mathbf{j}) \cdot \mathbf{U_w} - C_D \mathbf{U} \cdot \mathbf{U_w}\} dt \\
&\simeq \frac{1}{2} \int \rho S U \{C_L (u w_w - w u_w)\} dt \ (\text{for } C_L \gg C_D)
\end{aligned}
$$

$$(3.3-16)$$

where U and u should be considered always positive. The approximation has been obtained by assuming that (1) the rolling angle of the bird is small or $\mathbf{j} = \mathbf{J}$, where $\mathbf{J}$ is the unit vector along the Y-axis, and (2) both the vertical components of the velocities and the drag coefficients are small, i.e., $(u u_w + w w_w) C_D \cong 0$. The above equation suggests that (1) level flight in updraft ($u w_w > 0$) brings a positive energy gain, and (2) either climbing flight against the wind or descending flight with the wind velocity ($-w u_w > 0$) provides another energy gain.

3.3.3 DYNAMIC SOARING

It is known that continuous vertical wind motions over a flat surface like an ocean are virtually nonexistent below an altitude of about 100 m. Sea birds with high-performance gliding ability can use the horizontal wind to acquire energy to maintain continuous flight.

The utilization of wind shear in the gliding flight of sea birds was initially pointed out by Idrac (1932). Such sea birds, usually called "shearwaters" (Procellariidae), are known to have high wing loading, a clean body with small drag coefficient and high aspect ratio wings (Rayleigh 1983). They are also capable of taking off from and touching down on the sea surface.

Shown in Fig. 3.3-5 is a typical example of a flight path in "dynamic soaring," which is characterized by a series of flight courses: (1) descending flight in a tail wind; (2) high-speed, almost horizontal turning close to the sea surface; (3) ascending flight against a head wind; and (4) low-speed, almost horizontal turning at the highest altitude attained.

If there is no vertical gust ($w_w = 0$) but a horizontal gust ($u_w \neq 0$), and the vertical velocity w is small (w $\ll U$), then the energy given by the wind, E_w, for a high-performance bird can, from Eq. (3.3-16), be approximated by

$$
\begin{aligned}
E_w &\simeq -\frac{1}{2} \int \rho S U^3 \{C_L (w/U) + C_D (u/U)\} (u_w/U) dt \\
&= E_{w,L} + E_{w,D}
\end{aligned}
$$

$$(3.3-17)$$

By replacing the independent variable of time t with the altitude Z or $dZ = w \, dt$, the first term on the right hand side of the above equation, which is related to the lift, becomes

$$E_{w,L} = -\frac{1}{2} \int \rho S U C_L u_w dZ. \quad (3.3-18)$$

By assuming a flight path starting at zero altitude and returning to zero altitude, and a wind speed, as shown in Fig. 3.3-5, of nearly zero (or $u_w \simeq 0$) at $Z = 0$, and by applying partial integration, the above energy caused by the wind velocity can be expressed as

$$E_{w,L} = \oint \left[\left\{ \frac{1}{2} \int^z \rho S U C_L dZ \right\} (du_w/dZ) \right] dZ$$

$$(3.3-19)$$

In addition to supporting the statement made in the preceding section that energy is gained in climbing flight against the wind and descending flight with the wind, this equation further shows that in flight along a closed path in a vertical plane energy can be drawn from the "wind shear" du_w/dZ rather than the wind itself u_w, and that in level flight no energy gain is obtained from the lift.

Under the same assumption of $u_w(0) = 0$ and $\mathbf{V}(0) = \mathbf{U}(0)$, the second term on the right-hand side of Eq. (3.3-17), which is related to the drag, can be rewritten as

$$E_{w,D} = -\frac{1}{2} \int \rho S U^3 (u/U)(u_w/U) C_D dt$$

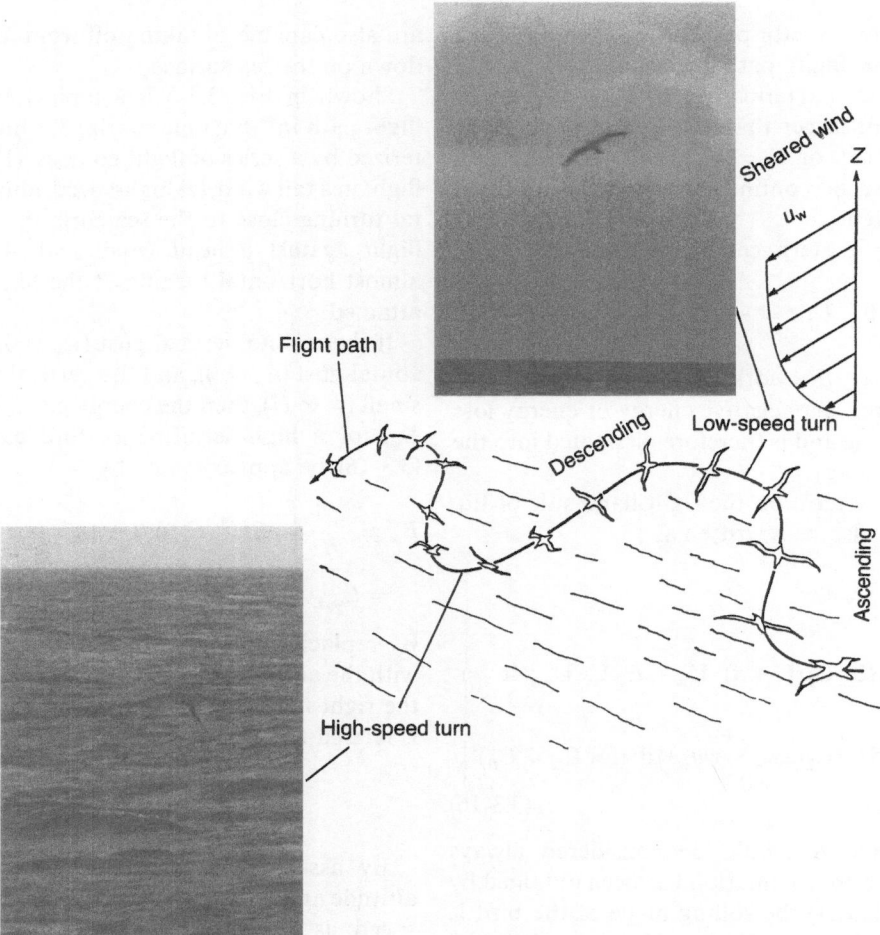

FIG. 3.3-5. A flight path in dynamic soaring. (Redrawn from Nelson 1980).

$$= \oint s(Z)\frac{d}{dZ}\left\{\frac{1}{2}\rho U^2 S C_D(u/U)(u_w/U)\right\}dZ$$

$$= \oint \frac{1}{2}\rho S C_D s(Z)\left\{\frac{du}{dZ}u_w + \frac{du_w}{dZ}u\right\}dZ$$

$$(3.3\text{-}20)$$

where $s(Z)$ is the length of the flight path. The above energy can be positive for the tail wind rather than the head wind.

Thus, it can be said that if the energy gain caused by the wind $E_w = E_{w,L} + E_{w,D}$ is larger than, or equal to, the energy loss caused by the drag E_D, or $E = E_w + E_D \geq 0$ in Eq. (3.3-13), then the bird can continue gliding flight indefinitely. All in all, tail wind flight is rather more advantageous than head wind flight. It is known that the greatest of the albatrosses are only found in the southern oceans where they soar around Antarctica several times

a year in the persistently prevailing westerlies (Scorer 1958).

Information on the strength of wind shear is not fully available yet. From the power law, it can be expressed thus

$$\partial u_w/\partial Z = u_{w,0}(\alpha/\delta)(Z/\delta)^{\alpha-1} \quad (3.3\text{-}21)$$

where α is given by $\alpha = 1/6\text{--}1/7$ on the sea surface and δ is the reference height ($\delta \cong 10$ m). Data for wind shear on the ground has been provided by Alexander and Camp (1985), but only for stormy wind.

3.4 Auxiliary Devices

Birds in soaring flight over land sometimes require very high maneuverability for short range flight, to change course in response to complicated ground configurations and to cope with wind gusts or atmospheric turbulence, which are stronger over rough land than over the ocean. In addition, birds flying over land have to take off without running long distances against the wind and to make pinpoint landings on the ground or in a tree. They also need to catch small prey that move about swiftly in air or on land, and to evade attack from other animals.

Hence, most soaring land birds have developed wings with high-lift devices, such as highly effective bastard wings and deeply slotted tips, as we saw in Sect. 3.1.1. On the other hand, since most water birds live in an environment that is vast and uncomplicated, they have high-performance wings with large aspect ratio and pointed tips. Some water birds that soar over land, such as the white pelican, have slotted tips too. Birds with slotted wing tips can adjust the size of the slots (Storer 1948).

The wings and body thus have many auxiliary devices for performance improvement, specifically either to increase the lift-to-drag ratio for a given bending moment of the wing or to reduce the speed of loitering flight and landing approach by increasing the maximum aerodynamic coefficient.

3.4.1 Winglet

The aerodynamic effects of the nonplanar wing were analyzed by C.D. Cone (1962b). He found that (1) there is an appreciable performance improvement in the efficiency factor, which is the ratio of effective aspect ratio to the geometrical aspect ratio of the elliptic wing, $\mathcal{R}_e/\mathcal{R} = e$; and (2) if the longitudinal component of the induced velocity generated by the bound vortex is negligible, the values of e depended only on the spanwise curvature of the wing regardless of either upward or downward concave. For instance, Fig. 3.4-1 shows the efficiency factor of two exemplified wings of upward concave semielliptic arc and circular arc. Some gains in the effective aspect ratio can be obtained by relatively minor alterations to the tip region of flat wings for equal spans, rather than by radical modification of the entire span. This principle can be applied similarly to the

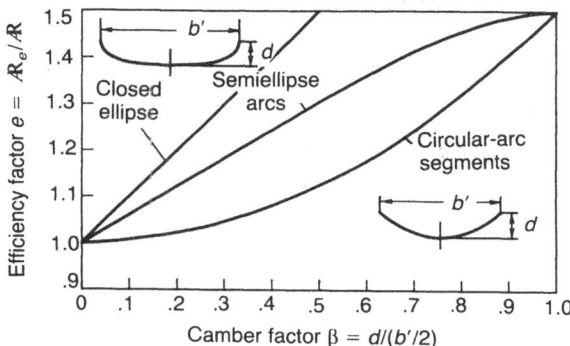

FIG. 3.4-1 Variations of the efficiency factor. b', projected wing span; d, camber depth. (Redrawn from C.D. Cone 1962b).

nonplanar wing of downward concave arc. The important thing is to obtain a nonplanar vortex wake, specifically near the wing tips.

Many sea birds that have wings with pointed tips adopt the upward concave arc wing configuration instead of the downward concave arc wing configuration during gliding flight. This probably results from both the performance gain of the nonplanar wing and the lateral stability of birds with high-wing configuration, because the dihedral effect, which plays a role in stopping the side slip without vertical tail (as explained in Sect. 3.1.1), must be reduced by taking a negative dihedral.

Now let us examine the local flow behavior near the wing tip more closely. With a positive angle of attack at the wing tip, there is, as shown in Fig. 3.4-2, an outflow component of velocity on the underwing surface near the tip, an upflow component around the tip from the lower surface to the upper surface, and an inflow component on the top surface of the tip. These flow components make a spiral vortex flow around the tip, called tip vortex, which trails from the respective wing tips and concentrates a "vortex core." As a result, the local flow direction at the tip has a greater upwash component than that of the free stream (Spillman 1978).

Let us concentrate here on the effects of a small auxiliary wing or "winglet" that has smaller chord than the tip chord. The winglet is mounted in such a way that (1) its span projects radially in any direction from the wing tip, not necessary in parallel with the main wing, and (2) its chord forms an adequate angle of attack with respect to the local flow, as shown in Fig. 3.4-3. With this

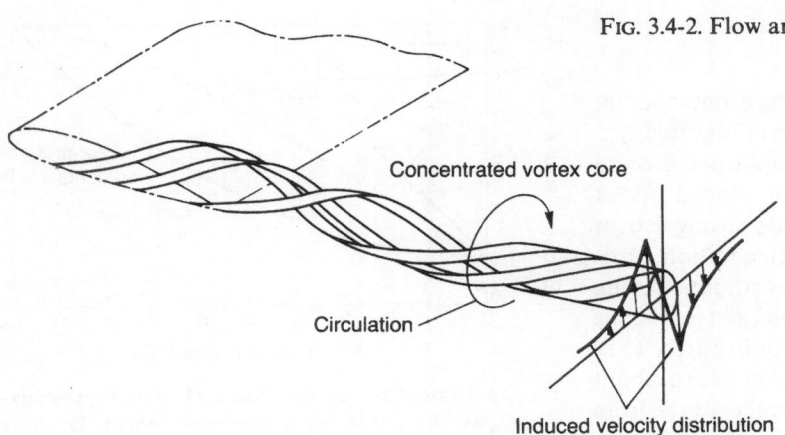

FIG. 3.4-2. Flow around a wing tip.

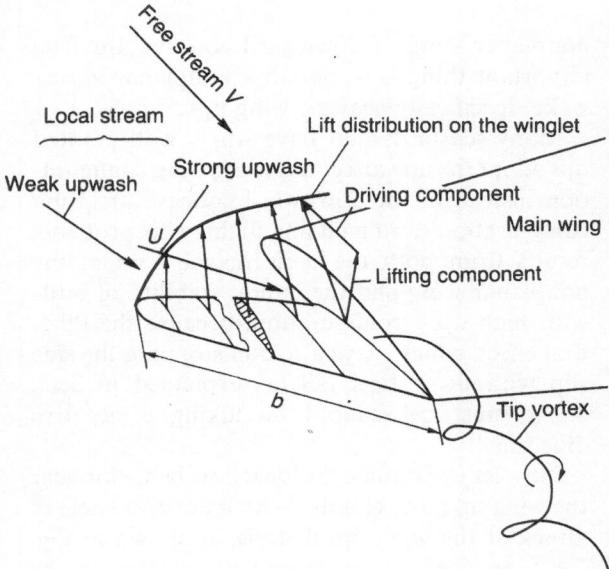

FIG. 3.4-3. Lift distribution on a winglet.

arrangement, the resultant aerodynamic force acting on the winglet can have not only a lifting component but also a driving component or thrust in the free stream direction. Although because of the small winglet area the lifting component causes little increase in the lift of the whole wing system, the driving component contributes to reducing the induced drag of the wing. This concept was pioneered by Whitcomb (1976).

As pointed out earlier, the outer primaries in stretched wings are separated in almost all land birds and even in many water birds. In such primaries, the distal part of the vane is more or less narrowed in comparison with the proximal part, which remains covered by the overlapping nearby primaries. As shown in Fig. 3.4-4, such primary feathers at the tips of a bird's wing particularly in the case of split tips, are bent upwardly during flight (Oehme 1977). This configuration produces an aerodynamic improvement in the flight performance and lateral stability of birds. The airloading near the tip increases, but the induced drag decreases appreciably for the same root bending moment, eliminating the need for any additional structural reinforcement that might detract from the aerodynamic gain.

Winglets frequently cause an increase in maximum lift coefficient (Eliraz and Ilan, 1977; Johnson et al. 1979). The winglets act to prevent the wing tip from stalling early, and thus reduce the tendency of the wing to roll off at the initial stage of a stall (Dam et al. 1981). These performance improvements were also confirmed experimentally by Montoya et al. (1977), and theoretically by C.D. Cone (1962b), Heyson et al. (1977), and De Young (1977).

Directional stability is virtually unaffected on unswept wings, but the lateral-directional dynamic response can be critically affected (Flechner 1979; Cornelis 1981; Dam et al. 1981).

3.4.2 HIGH-LIFT DEVICES

There are several ways to get high lift at slow speed. The conventional way of obtaining high lift without increasing the lift coefficient itself is to spread the wing in both the spanwise and chordwise directions, so as to obtain the largest possible wing area. In cruising, it is also necessary to achieve uniform lift coefficient distribution over the span, irrespective of whether the wing geometry is straight or swept. In addition, it is preferable

FIG. 3.4-4. Winglet of condor. (From Wildlife 1983 with permission).

to adopt a cambered wing to increase the maximum lift coefficient.

Aside from these fundamental ways of improving low-speed performance, it is also necessary to employ additional or auxiliary devices to raise the lift coefficient. The following are examples of such devices observed in flying creatures:

1. Leading edge comb. Laminar flow separation can be prevented by making the flow turbulent through the use of turbulence generators or small protuberances at the leading edge, such as feathers with filaments, comb teeth, or hairs standing perpendicular to the wing surface.

By observing the flow around the comb of an owl wing in a low-speed wind tunnel, Kroeger et al. (1972) found that at a high angle of attack, the comb generated a stationary spanwise vortex sheet which extended from the bastard wing at midspan out to the tip of the first primary feather and prevented flow separation on the outer half of the wing. This vortex sheet also has the effect of quieting flight by destroying laminar separation bubbles and reducing the peak sound pressure resulting from violent fluctuation of the bubbles (Graham 1934; Schwind and Allen 1973; Hersh et al. 1974). The spanwise vortex sheet generated by the hooked comb was also found to work in conjunction with a sharp leading edge to produce nonlinear lift on the outer half of the wing (G.W. Anderson, 1973).

2. Bastard wing. As mentioned earlier (see also Fig. 3.2-3), projecting forward from the top of the wrist area are two or three short feathers called bastard wings or alulae. Their spanwise location is considered to be the point at which the maximum lift coefficient is attained, and thus the point at which most of the flow separation would be expected. The bastard wing is shaped and arranged to function as a conventional leading edge slat that prevents flow separation by passing air through the slit between the leading edge of the wing and the trailing edge of the extended bastard wing on the upper surface of the wing. In this respect, it is like the aircraft wing shown in Fig. 3.4-5a. In addition to the above slat effect, longitudinal vortex lines, which are twisted with respect to the main flow, are produced by the leading edge of the bastard wing. They supply flow energy from the general flow to the weakened flow of the boundary layer and, acting as a vortex generator, prevent flow separation.

3. Flaps or split wings. The cascade arrangement of the split wings of land birds is also effective for delaying stall and obtaining a large lift coefficient for the whole wing system. The camber of the cascade wings is so large, and the slots or gaps among the flaps are so arranged, that the flow on the upper surface of the wing is accelerated by effectively introducing the active air from the underside of the wing through the gaps, and thus obtaining the highest lift coefficient. Holst and Küchemann (1942) pointed out the effect of the deeply forked tail of the frigate bird and the fork-tailed falcon: stall is delayed by spreading out the tail toward the wings as a slotted trailing edge flap (Fig. 3.4-5b).

Another type of leading-edge flap that enhances airflow diversion along the upper surface without separation is the Krueger flap shown in Fig. 3.4-5c. This flap permits higher angles of attack and lift coefficients. A similar device is often seen in birds of prey, like the hawk shown in Fig. 3.2-3 and the snipe shown in Fig. 3.4-5c, where a forwardly extended tendon acts like the Krueger flap.

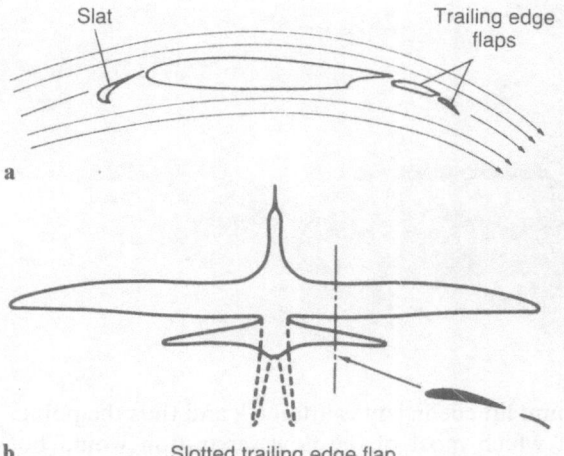

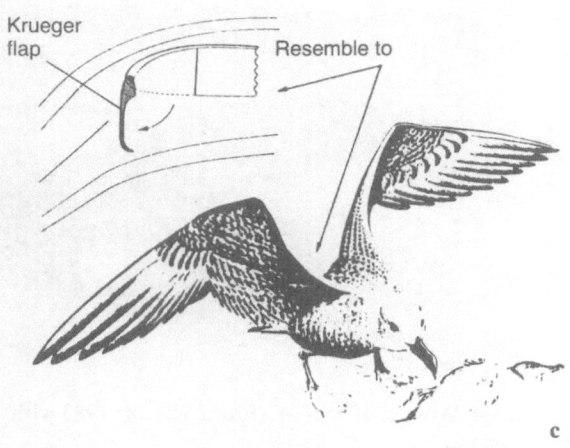

FIG. 3.4-5. **a–c.** Airfoil section of flap systems. **a** multi-slotted flaps. **b** deeply forked tail. (From Holst and Küchemann 1942). **c** forwardly extended tendon. (Redrawn from Perrins and Middleton 1985).

3.4.3 DYNAMIC STALL

In very low speed flight, such as loitering flight or landing approach, many large birds, specifically sea birds having wings with pointed tips and of large aspect ratio like the gannet and the albatross, perform mainly sinusoidal wing-pitching motions in addition to either slight flapping or lead-lag oscillations of small amplitude. High-speed-movie observation reveals that as the angle of attack of the wing increases, some of the coverts rise from the upper surface of the wing, clearly indicating a trace of flow separation. Then, the same coverts are immediately returned to their normal position by a reduction of the angle of attack of the wing. This kind of wing motion is continued sinusoidally during low-speed flight.

When a quick oscillation is introduced in either pitching or heaving motion, and thus when an angle of attack oscillation is imposed on an airfoil,

the instantaneous values as well as the mean values of all of the lift, drag, and moment coefficients increase well beyond their equivalent steady values (Favier et al. 1982). This favorable effect of unsteadiness on the mean lift coefficient is called "dynamic stall" and is obtained as a result of a cyclic boundary-layer separation and reattachment phenomenon on the upper side surface (Victory 1943; Halfman et al. 1951; A.G. Rainey 1956; Carta 1967).

A detailed analysis of the process of dynamic stall (see Fig. 3.4-6) was made by Carr et al. (1977), based on experimental tests with the NACA 0012 airfoil for a wide range of frequencies, Reynolds numbers, and amplitudes of oscillation, and using

FIG. 3.4-6. **a, b.** Dynamic stall events on NACA 0012 airfoil. (From Carr et al. 1977). **a** Normal force coefficient. **b** Pitching moment coefficient.

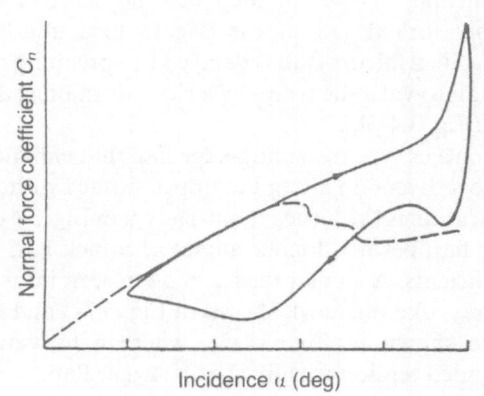

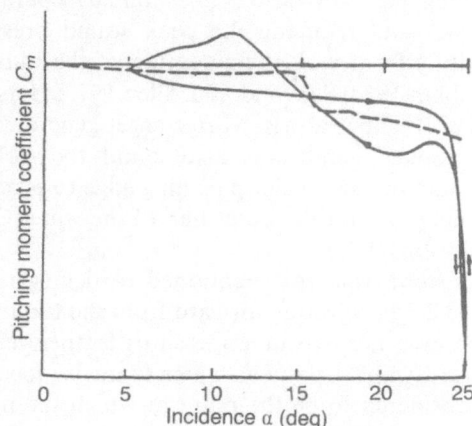

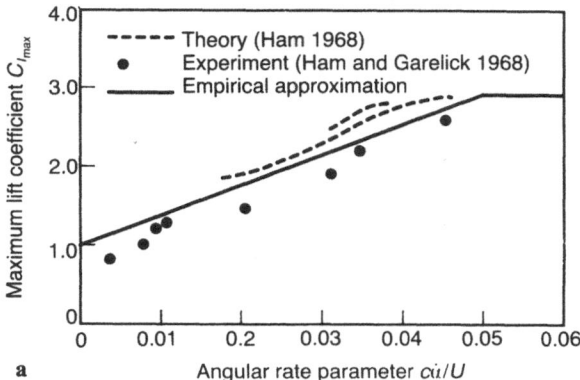

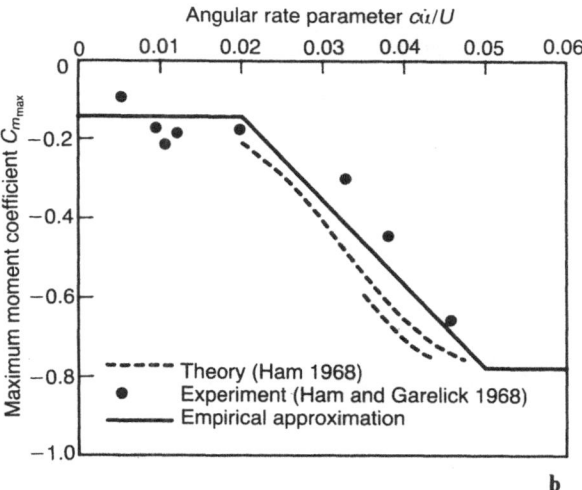

Fɪɢ. 3.4-7. Maximum lift and moment coefficient based on chord c versus rate of change at angle of attack. (From Johnson 1969 with permission; Noll and Ham 1982). **a** Lift coefficient. **b** Moment coefficient.

a combination of smoke-flow visualization, hot-wire anemometry, and normal-force and pitching-moment data obtained from surface pressure transducers. It can be seen from Fig. 3.4-6 that the normal force increases beyond its stationary value over the stalling angle of attack. The area inside the hysteresis indicates work per cycle or cycle damping. This area is positive for a counter-clockwise circuit, while the area of a clockwise loop represents energy added to the airfoil motion from the surroundings and indicates a reduction in damping.

Shown in Fig. 3.4-7 are the maximum lift and moment coefficients as a function of the non-dimensional rate of angle of attack of the airfoil at the instant of dynamic stall. The empirical representation used is also shown. It can be seen clearly that dynamic stall is very helpful in increasing the maximum lift coefficient and decreasing the lowest flight speed without consuming power for the feathering motion of the wing.

3.5 Other Forms of Gliding Flight

There are many flying animals other than insects and birds. In these animals, flight usually serves as a means of escape from predators. Apart from bats, which rely on flight exclusively, for such animal flight is not a principal mode of locomotion. Nor is it for winged plant seeds, which fly only once in the life cycle of the plant. Even so, for these animals and seeds, as for birds, insects, and bats, everything seems to be designed to ensure optimal flight.

The wings treated in this section are mostly membranous. They are characteristically simple, flexible, and sometimes foldable. Thus the camber of their airfoil configuration is principally determined by the pressure distribution and the tension imposed on the membrane, while the pressure distribution is also related to the airfoil configuration.

3.5.1 GLIDING SEEDS

A seed or fruit sometimes has a filmlike wing that is extremely light and has a large surface. If the center of gravity of such a seed is located adequately near but in front of the aerodynamic center of the wing, then the seed may be able to fly by using the lift generated by the wing in a gliderlike translation during fall, and, if it is windy, to make a long journey through the air. Such seeds are called samaras and it is rare to find one that will fly without autorotation. Many winged seeds, for example, those of maple, black pine, Santalaceae, linden, hornbeam, phoenix tree, and ash, autorotate while falling in the wind (Norberg 1973; McCutchen 1977; Azuma and Yasuda 1985, 1989). This autorotational flight will be discussed, in Sect. 3.6.1−3, as we concentrate now on gliding seeds.

A few examples of gliding seeds found in Japan are the yam (*Dioscorea japonica*), white birch (*Betula platyphylla*), and great trumpet flower (*Campsis chinensis*). The gliderlike flight of samaras was described by Hertel (1966), and the flight mechanics were analyzed by Kimura (1943) and Azuma and Okuno (1987). In this section the flight mechanics of *Alsomitra macrocarpa* are analyzed.

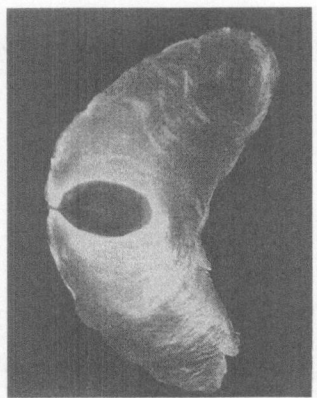

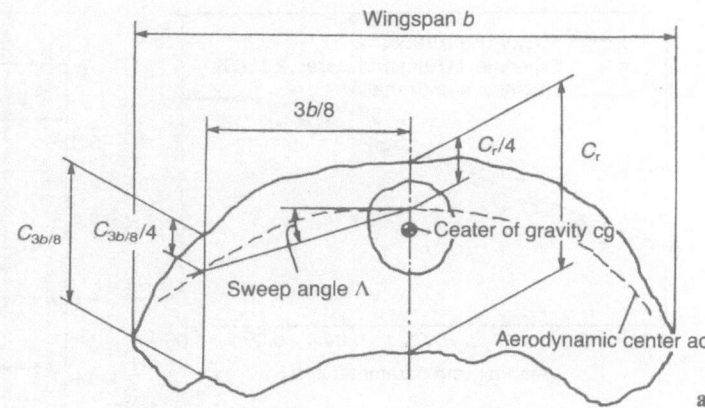

FIG. 3.5-1. Seed of *Alsomitra macrocarpa*. (Courtesy of Satoshi Kuribayashi).

Configuration of seed and wing. The seed of the *Alsomitra macrocarpa*, a vine species from Java, has a sail wing (Fig. 3.5-1) and performs stable gliding flight without any tail surface. The geometrical characteristics of the mean and standard deviation of ten seeds are given in Table 3.5-1. The dimensions are defined in Fig. 3.5-2.

The seed itself is very thin, about 1 mm in thickness, and is located nearly at the center of gravity, which is slightly forward of the wing center. The wing is also very thin (from a few microns to some tens of microns) and has a swept and tapered planform, a twisted ("wash out") angle, and a reflected trailing edge. Moreover, the center of gravity (cg) is at an appropriate position. An example of the mean camber line is shown in Fig. 3.5-2c.

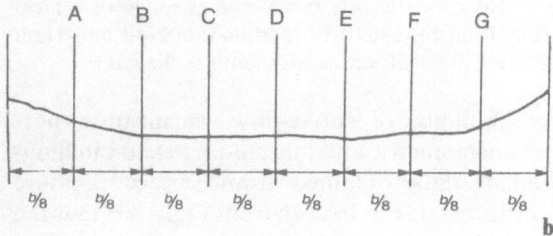

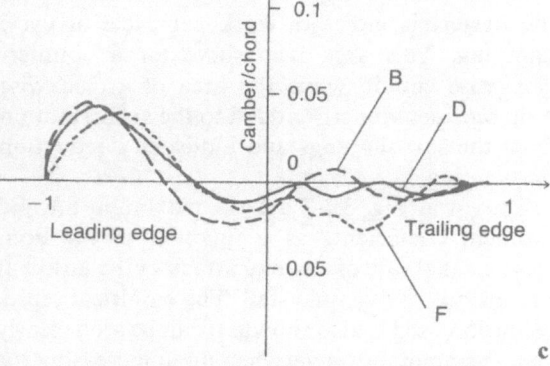

FIG. 3.5-2. **a–c.** Geometrical configuration of the seed. (From Azuma and Okuno 1987 with permission). **a** Plan view. **b** Frontal view. **c** Profile.

TABLE 3.5-1. Geometrical configuration of seeds (*Alsomitra macrocarpa*). (Azuma and Okuno 1987)

Item	Number of samples Mean value of 10 samples	Standard deviation
Mass m (mg)	212.1	70.9
Wingspan b (cm)	14.3	0.9
Wing area S (cm^2)	59.7	8.7
Aspect ratio $\mathcal{R}$	3.5	0.4
Geometrical sweep angle Λ (deg)[a]	13.7	3.1
Wing loading mg/S (N/m^2)	0.357	0.115
Center of gravity cg (cm) (distance from leading edge)	1.6	0.1
Aerodynamic center cx_{ac} (cm)	2.2	0.1

[a] Swept angle of a line passing through one-quarter chords at wing root and $3b/8$ spanwise position.

Consideration of the seed in light of the thin wing theory (Abbott and Doenhoff 1959 or see Sect. 3.1.2) reveals that the reflexed airfoil has a positive moment at a positive angle of attack, and shifts the center of pressure (cp) backward as the angle of attack increases. This moment thus tends to stabilize the pitching motion of the seed. The pitching stability is further strengthened by the sweep angle of the wing if the center of gravity is located in front of the mean aerodynamic center. The washout of the wing and the dihedral, which

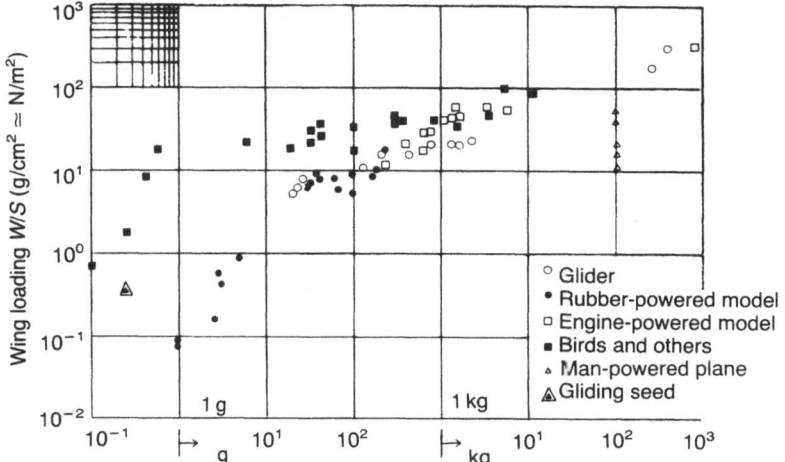

FIG. 3.5-3. Wing loading of *Alsomitra macrocarpa* in comparison with other subjects.

in both cases results from the elastic deformation of the wing during flight, prevents "spiral instability" (Perkins and Hage 1965).

Many seeds are piled up one on top of another in a husk that has a human headlike form and hangs from a vine. In windy conditions, the husk is swung by a "Karman vortex" and scatters its seeds in regular sequence. A hole underneath the husk has a lip that lets the wind get into the husk and assists in the separation of the seeds. After separation, each seed sets out in a gliding flight with a performance determined by its geometrical configuration and the wind conditions at the time. *Performance.* It is interesting to compare the mean wing loading of the seed of this species with other flying subjects shown in Fig. 3.5-3. The low wing loading of this seed clearly guarantees a small rate of descent. Flight tests conducted on a number of samaras (Azuma and Okuno 1987) gave very scattered flight data, even for the same speed. This probably resulted from the nonlinear aerodynamic characteristics of the wing attributable to the low Reynolds number [(4–5) × 10³ based on the mean wing chord], the occurrence of nonlinear deformation owing to the thinness of the wing, the surrounding atmospheric conditions (temperature, humidity, etc.), unnoticeable air currents in the room, and other factors.

The seed has a mean lift-to-drag ratio of $L/D = 3.7$ at a lift coefficient of $C_L = 0.34$, and a mean rate of descent of $|w| = 0.41$ m/s. This guarantees

higher gliding performance for the seed than that of rotary seeds, whose rate of descent is about 1 m/s (Azuma and Yasuda 1985, 1989).

Typically, when a seed is released at a height of 40 m on a calm day, it flies at a speed of 1.5 m/s, remains in flight for 100 s, and lands at a horizontal distance of 150 m from the point of release. (Actually, the *Alsomitra macrocarpa* vine climbs tall trees, such as *Pterocarpus indicus*, *Ficus albifila*, *Shorea leprosula*, *Pometia pinnata* and *Brownea grandiceps*, and bears seeds at a height of 30–50 m.) In the presence of a horizontal 10 m/s wind, the horizontal flight distance increases to 10 m/s × 100 s = 1 km. From the viewpoint of aeronautical engineering, it is very interesting that the flight is performed at a lift coefficient as small as $C_L = 0.34$. By conducting flight tests on natural samaras and samaras modified to have their center of gravity shifted from the natural position, Azuma

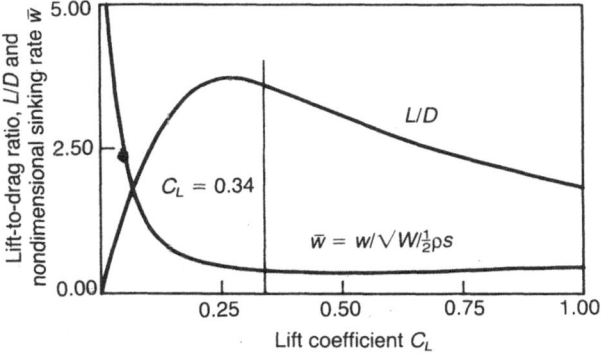

FIG. 3.5-4. Performance of the estimated wing. (From Azuma and Okuno 1987 with permission).

and Okuno (1987) obtained a mean polar curve of the seeds and determined their performance. They found that (1) the natural CG position of every tested samara was the best for the maximum lift-to-drag ratio, (2) the minimum drag coefficient C_{D_0} was about twice the skin friction drag of a flat surface at the same Reynolds number. The wrinkles on both surfaces of the wing seem to be unrelated to the drag rise at this small Reynolds number, and (3) the three-dimensional profile drag coefficient C_{D_0} was not a constant but a function of the lift coefficient C_L because of the flow separation even in low C_L range. It must be mentioned that the drag coefficient of the two-dimensional wing $C_{d_0} (= C_{D_0})$ is very large with a high angle of attack. This explains why gliding flight is performed at such a low lift coefficient. If C_{d_0} is assumed to be constant, $C_{d_0} = 0.037$, then the equivalent aspect ratio of this wing is approximated by a very low value such as $Æ_e = 0.4$.

Shown in Fig. 3.5-4 are the lift-to-drag ratio and the rate of descent of the three-dimensional wing as a function of lift coefficient, calculated by the "Local Circulation Method" LCM (Azuma et al. 1983, Azuma and Okuno 1987) from the estimated two-dimensional aerodynamic characteristics. The maximum lift-to-drag ratio and the minimum rate of descent are respectively obtained at $C_L = 0.27$ and $C_L = 0.53$. However, the rate of descent is almost constant in a speed range of less than 1.7 m/s.

If no wind is present, the seed can attain the greatest distance by flying with a lift coefficient of $C_L = 0.27$, which produces the maximum lift-to-drag ratio and thus the maximum gliding ratio (the ratio of horizontal distance to height loss). However, if there is a wind, then the flight is strongly dependent on the time before the seed reaches the ground after starting to fall. This suggests that flight should occur at a lift coefficient close to the minimum rate of descent. Thus, it is interesting that the actual lift coefficient of $C_L = 0.34$ lies between two optimal lift coefficients, the maximum lift-to-drag ratio (or the maximum gliding ratio) and the minimum rate of descent, but is closer to the former because the minimum rate of descent is almost constant beyond the selected value, if $C_L > 0.34$. Here also, as seen in other cases of flight and swimming, the seed functions in such a way as to get optimal performance.

3.5.2 Fish and Mollusks

Some fish and mollusks can become airborne by using pectoral and pelvic fins, or their fins and arms. Two examples, the flying fish and flying squid, are shown in Fig. 3.5-5.

Flying fish. In the case of marine flying fish of the families of Thoracocharax, Exocoetus, Halocypselurus, Cypselurus, and so on, which are

a

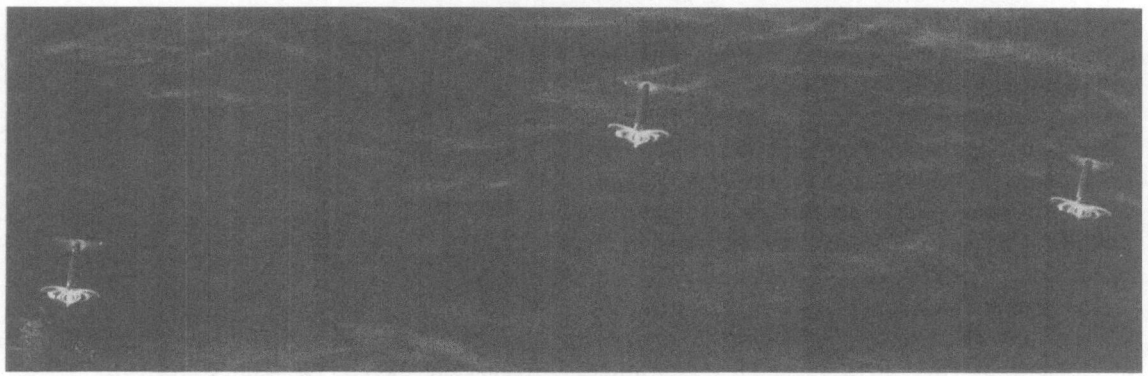

b

Fig. 3.5-5. **a, b.** Flying fish and flying squid. (Courtesy of Mitsuaki Iwago). **a** Flying fish. **b** Flying squid.

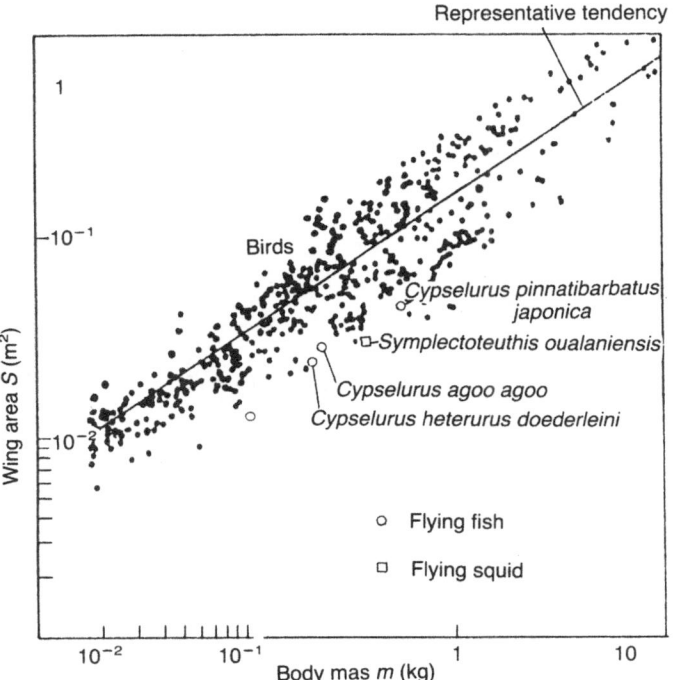

FIG. 3.5-6. Wing area of flying fish and flying squid in comparison with birds. (Modified from Greenewalt 1962).

common throughout tropical and subtropical seas, the mode of flight is as follows: after approaching the sea surface and pushing the anterior part of the body above the surface with an angle of attack of about 20°, the flying fish skims to accelerate its speed by spreading and raising its pectoral fins as a pair of main wings with aerodynamic lift, while the lower part of the caudal fin, called the "hypobatic tail" (Breder 1926), continues to beat or fan the water. Sometimes, a series of slappings of the caudal fin on the water surface can be observed from the ripples left. In the case of Cypselurus, this taxiing may average 9 m in length with a tail beat frequency of 70 Hz (Hubbs 1933). After attaining enough speed for flight, the fish jumps from the water with a speed of more than 15 m/s (Franzisket 1965). Then it makes almost horizontal gliding flight with both its pectoral and pelvic fins extended. When flight power becomes weak, the fish slaps the wave crest with the caudal fin and continues flying: having touched the water surface, the fish can repeat the takeoff. It can also maneuver in the air. For example, it can make turning flight

and then apply rapid deceleration, probably by utilizing the stall of the wing.

The reason for flight in fish is probably airborne evasion from predators that are high-speed swimmers, such as dolphin, tuna, skipjack, or marlin. Flying fish can apparently see as well in air as in water. It should be noted that the wing loading of flying fish (probably flying squid too) is slightly larger than that of birds of similar size, as shown earlier in Fig. 3.2-5 as well as Fig. 3.5-6. Birds of this size are not gliding birds but beating birds.

Since the main and tail fins have a good planform as a fixed wing with high aspect ratio, and the body is slender and lighter than that of other fish of similar size, the fish can fly more than 100 m, or on rare occasions 200 m, with an estimated lift-to-drag ratio of $L/D \simeq 15$, and can stay in the air for many seconds. Actually, this kind of flight is not steady gliding flight of constant speed but almost horizontal flight with decelerating speed. A mathematical explanation for such unsteady gliding flight will be presented in Sect. 3.5.4.

The geometrical configurations of three exemplified species of flying fish are shown in Table 3.5-2, and their estimated performances in Table 3.5-3. The geometrical configuration of a typical flying fish is shown in Fig. 3.5-7a. It is interesting to find that both the pectoral and pelvic fins are

TABLE 3.5-2. Dimensions of flying fish and squid

Items	Symbol	Unit	Flying Fish			Flying squid
			Cypselurus heterurus doederleini	*Cypselurus agoo agoo*	*Cypselurus pinnatibarbatus japonicus*	*Symplectoteuthis oualaniensis*
Wingspan						
Main wing	b	m	0.347	0.383	0.418	0.370
Horizontal tail or canard	b_h	m	0.162	0.166	0.161	0.260
Vertical tail	b_v	m	0.072	0.068	0.066	—
Wing area						
Main wing	S	m²	2.25×10^{-2}	2.62×10^{-2}	3.41×10^{-2}	3.48×10^{-2}
Horizontal tail or canard	S_h	m²	6.62×10^{-3}	7.36×10^{-3}	7.82×10^{-3}	1.83×10^{-3}
Vertical tail	S_v	m²	1.82×1^{-3}	1.85×10^{-3}	2.05×10^{-3}	—
Aspect ratio						
Main wing	$A\!R$	—	5.35	5.59	5.14	5.1
Horizontal tail or canard	$A\!R_h$	—	3.96	3.74	3.32	3.7
Vertical tail	$A\!R_v$	—	2.81	2.49	2.10	—
Length						
Total length	l	m	0.320	0.328	0.346	0.470
Center of gravity from the head	l_{cg}	m	0.133	0.116	0.128	0.253
Aerodynamic center from the head						
Main wing	l_{ac}	m	6.9×10^{-2}	8.3×10^{-2}	7.8×10^{-2}	34.1×10^{-2}
Horizontal tail or canard	$l_{ac,h}$	m	17.4×10^{-2}	17.8×10^{-2}	16.7×10^{-2}	2.4×10^{-2}
Vertical tail	$l_{ac,v}$	m	26.6×10^{-2}	27.4×10^{-2}	29.2×10^{-2}	—
Maximum width of trunk	b_b	m	4.5×10^{-2}	3.42×10^{-2}	3.72×10^{-2}	8×10^{-2}
Maximum height	h_b	m	3.14×10^{-2}	4.38×10^{-2}	4.90×10^{-2}	—
Volume	V_b	m³	2.34×10^{-4}	2.26×10^{-4}	2.90×10^{-4}	—
Mean aerodynamic chord						
Main wing	$\bar{c}$	m	7.28×10^{-2}	7.96×10^{-2}	9.25×10^{-2}	9.40×10^{-2}
Horizontal tail or canard	$\bar{c}_h$	m	4.58×10^{-2}	4.93×10^{-2}	5.78×10^{-2}	6.70×10^{-2}
Vertical tail	$\bar{c}_v$	m	2.79×10^{-2}	2.96×10^{-2}	3.34×10^{-2}	—
Tail volume						
Horizontal tail or canard	$l_h S_h / \bar{c}S$	—	0.165	0.218	0.0966	1.22
Vertical tail	$l_v S_v / bS$	—	0.037	0.029	0.0236	—
Mass	m	kg	0.226	0.243	0.303	0.450
Wing loading	W/S	N/m²	98.44	90.99	87.08	114.05
Specific weight	σ	—	0.97	1.08	1.04	—

TABLE 3.5-3. Performance of flying fish and squid

Item	Species	Flying fish			Flying squid
		Cypselurus heterurus doederleini	Cypselurus agoo agoo	Cypselurus pinnatibarbatus japonicus	Symplectoteuthis oualaniensis
Mass ratio[a]	$\mu_b = m/(\rho Sb)$	23.6	19.7	17.5	28.6
Minimum drag coefficient	C_{D_0}	0.020	0.018	0.020	0.025
Effective aspect ratio[b]	$Æ_e$	6.28	6.43	5.75	6.35
Maximum lift-to-drag ratio	$(L/D)_{max}$	15.0	15.6	15.0	12.2
Steady gliding speed at $(L/D)_{max}$	$U_{(L/D)_{max}}$ (m/s)	16.9	16.6	16.3	18.6
Lift coefficient at $(L/D)_{max}$	$C_{L,(L/D)_{max}}$	0.56	0.54	0.54	0.66
Minimum rate of descent	$\|w\|_{min}$ (m/s)	1.06	0.97	1.06	1.34
Steady gliding speed at $\|w\|_{min}$	$U_{\|w\|_{min}}$ (m/s)	12.85	12.60	12.36	14.13
Lift coefficient at $\|w\|_{min}$	$C_{L\|w\|_{min}}$	0.97	0.93	0.93	1.14

[a] See Sect. 3.5.4.   [b] See Sect. 3.5.3.

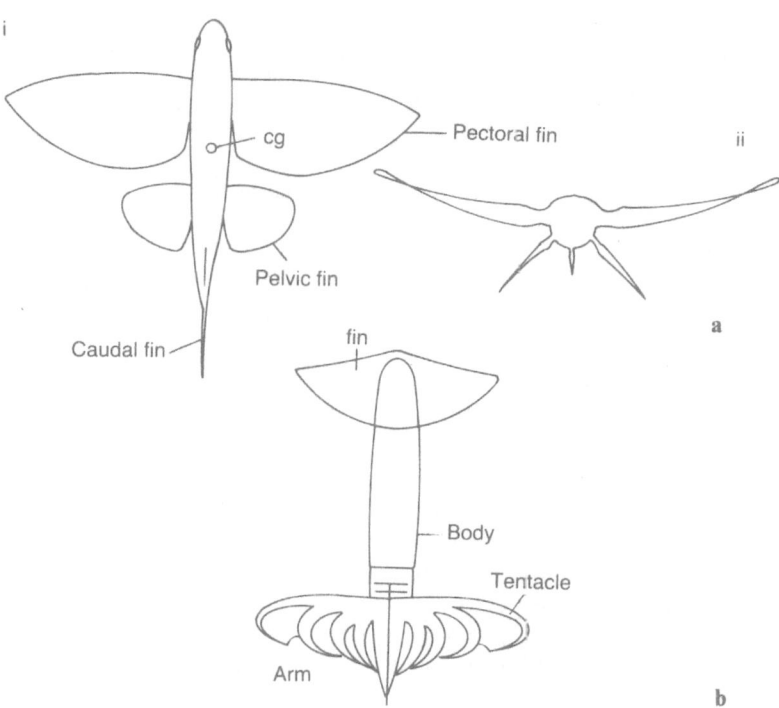

FIG. 3.5-7. **a, b.** Plan views of a flying fish and a flying squid. **a** Flying fish. *i* Plan view *ii* Front view. **b** Flying squid.

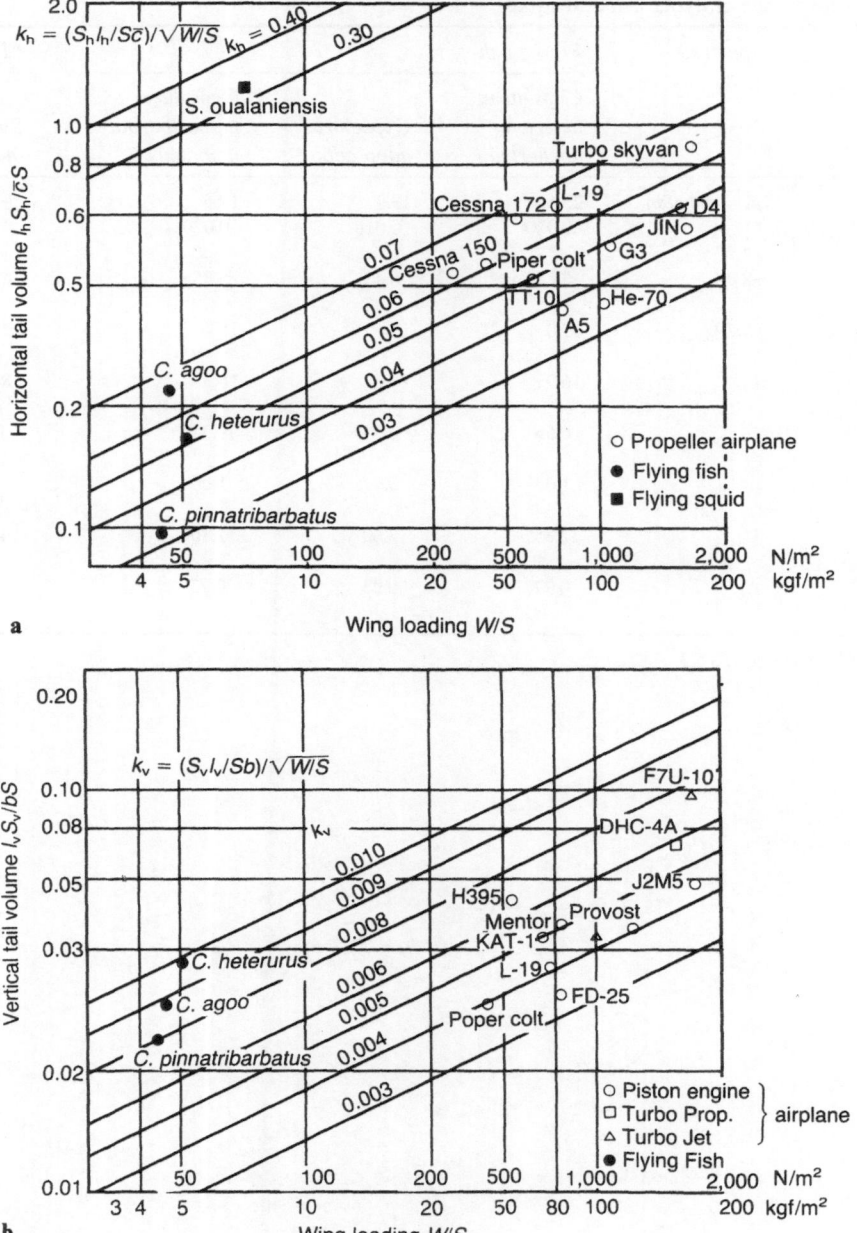

FIG. 3.5-8. **a, b.** Tail volumes of airplanes and flying creatures. (Redrawn from Yamana and Nakaguchi 1968, for airplane data with permission). **a** Horizontal tail volume. **b** Vertical tail volume.

attached normally to the body surface at their roots as middle wings (see sect. 3.1.1) in order to reduce interference drag. The strong downward extension or negative dihedral of the pelvic fins can be freed from the downwash of the pectoral fin or the main wing. As can be seen in Fig. 3.5-5a, the main wing has a positive dihedral for the live fish in flight.

While some flying fish are known to flap their wings (see Sect. 8.2), the pectoral fins of flying fish that carry out high performance gliding flight cannot execute beating flight because they do not have musculature powerful enough for this purpose. The center of gravity of the fish is usually located between the aerodynamic centers of the pectoral and pelvic fins and is more rearward than in other fish. Flight is inherently stable.

The "tail volume," which is, as defined in Table 3.5-2, the nondimensional product of tail area and the distance from the tail to the center of gravity, is a good index for estimating the stability of the flying subject. In the case of flying fish, the vertical and horizontal tail volumes of caudal fins and pelvic fins fall within an appropriate range for stable flight, comparing favorably with those of airplanes, as shown in Fig. 3.5-8.

Flying squid. The group of flying squids shown in the cover and in Fig.3.5-5b shoots out of the water under acceleration produced by the jetting of water. As seen in this impressive picture, taken by Mitsuaki Iwago (1981) in the Indian Ocean, flight is performed by a pair of fins and laterally extended arms like those of a canard, which is a kind of airplane having control surfaces in front of the main wings. The fins in front are bent slightly upward by lift, causing a positive pitch-up moment around the center of gravity, no doubt located at some point in the rear part of the body. The assumed dimensions and configuration of the flying squid are given in Table 3.5-2 and shown in Fig.3.5-7b, and its performance is given in Table 3.5-3.

Higher lift than that produced by fins is generated by a main wing formed by arms, and is balanced with the pitch-up moment. Therefore, in order to sustain the necessary lift, the main wing has to have a membranous lifting surface and be reinforced by four pairs of arms and a pair of tentacles. The tentacles probably form the leading edge of the membranous wing and maintain the elliptic planform required for high-performance flight. Without a membrane, it would be impossible to support the pressure because the vortex wakesheet trailing from the tentacles is a discontinuous surface of the velocity but not of the pressure.

What is the membrane made of in the species shown here? A species of flying squid observed in the Japan current (or Black current), *Symplectoteuthis oualaniensis*, has thin skin membranes each one of which is hemmed and usually retracted at one side of the respective arms, except for one pair of tentacles. During flight, the skin membranes may be fully extended among all arms and supported at their open ends by adjacent arms, probably by suction forces generated by sucking discs on the arms. However, the wing area covered with these skin membranes is so small that this type of flight would not be as good as is actually observed. Another possibility is that either the form and area of the skin membranes are well adapted to the planform observed in the photograph or the wing is made of both skin membranes and membranes made of sticky mucus. This is reinforced by the arrangement of the tentacles, specifically at their tips, which can be clearly seen in the cover photo and Fig. 3.5-7b to be rounded and to make narrower gaps than those found in the middle parts of the arms and tentacles. This formation may prevent the open trailing edge of the mucus membrane from shrinking and thus help to maintain the extended wing form (Azuma 1981a, 1983). Since the pressure peak can be observed only near the leading edge of the wing where the tentacles and the skin membranes can support the high pressure difference near the leading edge, the tension of the mucus must bear the pressure difference acting on the remaining parts of the wing.

It is important to note the difference between the planform of the main wing and that of the front wing. The elliptical main wing guarantees the best performance for a given span and is bending-moment free. On the other hand, the strong upward deflection of the pointed tips of the front wing of fins shows the hard bending moment generated by the aerodynamic force. As mentioned in Sect. 3.2.4, pointed tips of this type relax the wing bending moment. Pointed tips are also observed in flying fish and ray as well as in many sea birds with high-performance gliding ability.

Interestingly, the planform of the fin's leading edge of this flying squid is similar to that of the wing of Concorde, the supersonic transport airplane (SST). Specifically, the wing is ogive-shaped and characterized by a large apex angle and a concave backward form along the span. This form of wing generates longitudinal vortices which, as described later, increase the nonlinear lift for a low-aspect-ratio wing of this type at a high angle of attack.

3.5.3 OPTIMAL LIFT ALLOCATION

The wing area of the tail fin of flying fish and that of the canard fin of the flying squid are both large relative to that of the main wing. This is probably because (1) the larger main wing is not adequate for the creature's activities in the sea; and (2) for longitudinal stability except canard configuration, the tail volume of the horizontal wing must be maintained by adopting a large tail wing because of its small distance from the center of gravity. What then would be the optimal allocation of lift between the main and tail wings?

In steady gliding flight with a conventional tail configuration, the force balance in the vertical direction can be given by

$$\left. \begin{aligned} W = L = \tfrac{1}{2}\rho U^2 S C_L = L_w + L_h \\ = \tfrac{1}{2}\rho U^2 S\{C_{L,w} + \eta_h(S_h/S)C_{L,h}\} \end{aligned} \right\} \quad (3.5\text{-}1)$$

where η_h is the ratio of dynamic pressure at the tail wing and that of the free stream, or

$$\eta_h = (U_h/U)^2 \qquad (3.5\text{-}2)$$

In the case of the canard wing, it is the main wing lift, not the tail wing lift, that should be multiplied by η_h.

The drag force is given by

$$\left. \begin{aligned} D = \tfrac{1}{2}\rho U^2 S C_D \\ = \tfrac{1}{2}\rho U^2 S\{C_{D_{0,w}} + C_{L,w}{}^2/\pi A\!R_{e,w} + f/S \\ + \eta_h S_h(C_{D_{0,h}} + C_{L,h}{}^2/\pi A\!R_{e,h})\} \end{aligned} \right] \quad (3.5\text{-}3)$$

where f is the drag area of the fish or squid other than that of the main and horizontal tail wing.

When the lift-to-drag ratio given by $L/D = W/\tfrac{1}{2}\rho U^2 S C_D$ can be optimized by taking the minimum value of $F = U^2 C_D$ or

$$\begin{aligned} F = U^2 C_D = U^2 C_{D_0} + U^2\{C_{L,w}{}^2/\pi A\!R_{e,w} \\ + \eta_h(S_h/S)C_{L,h}{}^2/\pi A\!R_{e,h}\} \end{aligned}$$

$$(3.5\text{-}4)$$

where

$$C_{D_0} = C_{D_{0,w}} + \eta_h(S_h/S)C_{D_{0,h}} + f/S \quad (3.5\text{-}5)$$

for a given weight. As a mathematical model of a variational problem, the "performance function" to be optimized is thus given by

$$J = F + \lambda G \qquad (3.5\text{-}6)$$

where λ is a Lagrange multiplier and G is a "constraint" derived from Eq. (3.5-1), or

$$G = W - \tfrac{1}{2}\rho U^2 S\{C_{L,w} + \eta_h(S_h/S)C_{L,h}\} = 0$$

$$(3.5\text{-}7)$$

By assuming that the effect of the downwash generated by the front wing on the rear wing is negligible, solutions can be obtained by solving

$$\partial J/\partial C_{L,w} = \partial J/\partial C_{L,h} = \partial J/\partial U = \partial J/\partial \lambda = 0$$

$$(3.5\text{-}8)$$

as follows:

$$\begin{aligned} (C_{L,w})_{(L/D)_{max}} \\ = \sqrt{\pi A\!R_{e,w} C_{D_0}/\{1 + \eta_h(S_h/S)(A\!R_{e,h}/A\!R_{e,w})\}} \end{aligned}$$

$$(3.5\text{-}9a)$$

$$(C_{L,w}/C_{L,h})_{(L/D)_{max}} = A\!R_{e,w}/A\!R_{e,h} \quad (3.5\text{-}9b)$$

$$\begin{aligned} U_{(L/D)_{max}} = (2W/\rho S)^{1/2}/[\pi A\!R_{e,w}C_{D_0}\{1 + \eta_h(S_h/S) \\ \cdot(A\!R_{e,h}/A\!R_{e,w})\}]^{1/4} \end{aligned}$$

$$(3.5\text{-}9c)$$

$$\begin{aligned} (L/D)_{max} = \tfrac{1}{2}[(\pi A\!R_{e,w}/C_{D_0})\{1 + \eta_h(S_h/S) \\ \cdot(A\!R_{e,h}/A\!R_{e,w})\}]^{1/2} \end{aligned}$$

$$(3.5\text{-}9d)$$

It is interesting to find from Eq. (3.5-9b) that the ratio of lift allocation $C_{L,w}/C_{L,h}$ should be proportional to the ratio of the aspect ratio $A\!R_{e,w}/A\!R_{e,h}$. By using the data given in Table 3.5-2, the optimal lift distributions are given by

$$\left. \begin{aligned} (C_{L,w})_{(L/D)_{max}} = 0.559 \qquad (C_{L,h})_{(L/D)_{max}} = 0.414 \\ (L/D)_{max} = 15 \end{aligned} \right\}$$

$$(3.5\text{-}10)$$

These values agree well with the previously estimated performance data in Table 3.5-3. In the following analyses on the performance of flying birds and insects, the lift contributed by the body or trunk is neglected because every animal will take an optimal attitude of the body, usually a slender or small aspect ratio.

In the case of the above optimal lift allocation, the effective aspect ratio, which is the equivalent aspect ratio for the main wing and is thus specified by

$$C_D = C_{D_0} + C_{L,w}{}^2/\pi A\!R_e \qquad (3.5\text{-}11)$$

can be given as

$$A\!R_e = A\!R_{e,w}/\{1 + (S_h/S)(A\!R_{e,h}/A\!R_{e,w})\} \quad (3.5\text{-}12)$$

As the tail or canard wing area is reduced, this effective aspect ratio approaches the equivalent aspect ratio of the main wing.

3.5.4 OPTIMAL FLIGHT TRAJECTORY

In Sect. 3.2.3, the optimal flight for maximizing the horizontal distance from a given height was presented for steady gliding at constant speed. However, this flight may not give the actual longest distance for a given initial altitude and speed if the flight is unsteady gliding with variable speed in close to the water surface.

Simple analysis. It was found in Sect. 3.2.3 that for steady gliding the maximum range $R_{max,s}$ and the flight velocity $U_{(L/D)max}$ from a given altitude H can be obtained from the formulas:

$$R_{max,s} = \tfrac{1}{2}\sqrt{\pi A\!\!R_e/C_{D_0}} \cdot H = \tfrac{1}{2}k_1 H \quad (3.5\text{-}13)$$

$$U_{(L/D)max} \simeq \sqrt{2(W/S)/\rho\sqrt{\pi A\!\!R_e C_{D_0}}} \quad (3.5\text{-}14)$$

where parameter k_1 is defined by

$$k_1 = \sqrt{\pi A\!\!R_e/C_{D_0}} \quad (3.5\text{-}15)$$

and where the approximation is valid for a small gliding angle, $|\gamma| \simeq D/L \ll 1$.

Let us next consider the optimal flight trajectory in the case where gliding is not steady, i.e., flight is performed with variable speed, but the height is kept the same to enable utilization of the same initial energy.

From Fig. 3.5-9, it seems reasonable to derive the initial kinetic energy of unsteady gliding, $\tfrac{1}{2}mU_0^2$, by equating the total energy, composed of the kinetic energy in steady gliding, $\tfrac{1}{2}m(U_{(L/D)max})^2$, and the potential energy of the altitude difference, mgH, and to assume that the final speed is determined by the same gliding speed, $U_{(L/D)max}$. Then, the initial speed of unsteady gliding can be given by

$$U_0 = \sqrt{2gH + (U_{(L/D)max})^2} \quad (3.5\text{-}16)$$

In unsteady gliding, the equations of motion are

$$m\dot{U} = -D = -\tfrac{1}{2}\rho U^2 S(C_{D_0} + C_L^2/\pi A\!\!R_e) \quad (3.5\text{-}17)$$

$$0 = \tfrac{1}{2}\rho U^2 SC_L - W \quad (3.5\text{-}18)$$

Combining the above two equations yields

$$dU/dt = -(k_2 U^2 + k_3/U^2) \quad (3.5\text{-}19)$$

where parameters k_2 and k_3 are defined by

$$k_2 = \tfrac{1}{2}(\rho SC_{D_0}/m) \qquad k_3 = 2mg^2/\rho S\pi A\!\!R_e \quad (3.5\text{-}20a, b)$$

By changing the independent variable from time dt to the distance dR or the speed dU, the maximum distance $R_{max,u}$ is

$$\begin{aligned}
R_{max,u} &= \int_0^{R_{max,u}} dR \\
&= -\int_{U_0}^{U_{(L/D)max}} \{U/(k_2 U^2 + k_3/U^2)\}dU \\
&= (1/4k_2)\ln\{(k_2 U_0^4 + k_3)/ \\
&\quad (k_2 U^4_{(L/D)max} + k_3)\}
\end{aligned} \quad (3.5\text{-}21)$$

The ratio, $R_{max,u}/R_{max,s}$, is then given by

$$\begin{aligned}
R_{max,u}/R_{max,s} &= (1/4k_2)\ln\{(k_2 U_0^4 + k_3)/ \\
&\quad (k_2 U^4_{(L/D)max} + k_3)\}/\tfrac{1}{2}k_1 H \\
&= \ln\{(k_2 U_0^4 + k_3)/ \\
&\quad (k_2 U^4_{(L/D)max} + k_3)\}/\ln e^{2k_1 k_2 H} \\
&= \ln F/\ln G
\end{aligned} \quad (3.5\text{-}22)$$

where, by combining Eqs. (3.5-13–17) and (3.5-20–22), F and G become respectively

$$F = 1 + \sqrt{\pi A\!\!R_e C_{D_0}}(H/b\mu_b) + \tfrac{1}{2}\pi A\!\!R_e C_{D_0}(H/b\mu_b)^2 \quad (3.5\text{-}23)$$

$$G = \exp\{\sqrt{\pi A\!\!R_e C_{D_0}}(H/b\mu_b)\} \quad (3.5\text{-}24)$$

where μ_b is the density parameter based on the wingspan

$$\mu_b = m/\rho Sb. \quad (3.5\text{-}25)$$

It can be seen from Eqs. (3.5-23 and -24) and Fig. 3.5-10 that the functions F and G are represented by a polar curve (solid line) and an exponential curve (broken line) as a function of nondimensional height $H/b\mu_b = \rho g H/(W/S)$, and the function F is always smaller than the function G except at zero height difference or $H/b\mu_b = 0$ where the two curves have the same tangent and curvature,

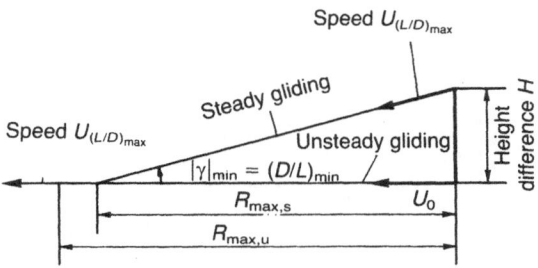

FIG. 3-5-9. Comparison of flight paths between steady and unsteady gliding.

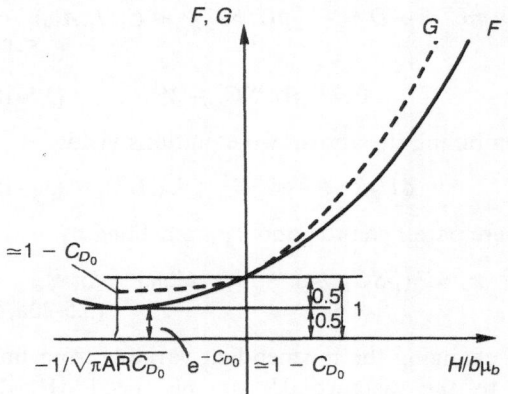

FIG. 3.5-10. Functions F and G.

$$\left.\begin{array}{l} \partial F/\partial (H/b\mu_b)_{H=0} \\[4pt] \quad = \partial G/\partial (H/b\mu_b)_{H=0} = \sqrt{\pi \mathcal{R} C_{D_0}} \\[8pt] \partial^2 F/\partial (H/b\mu_b)_{H=0}{}^2 \\[4pt] \quad = \partial^2 G/\partial (H/b\mu_b)_{H=0}{}^2 = \pi \mathcal{R}_e C_{D_0} \end{array}\right\} \quad (3.5\text{-}26)$$

Since, if the ground effect is out of consideration, the difference is negligible in the reasonable range of $H/b\mu_b$, the flying fish or squid can actually take either one of the above two courses: steady gliding or unsteady horizontal flight. However, if the value of $H/b\mu_b$ is large, such as in the case of an artificial (or high-performance) glider, the ratio $R_{\max,u}/R_{\max,s}$ becomes small (<1). This fact shows that any artificial glider should use steady gliding flight to get the maximum flight range.

Ground Effect. As will be explained in Sect. 4.3.2, the effective aspect ratio increases as the wing approaches the ground as follows:

$$\mathcal{R}_{e,H}/\mathcal{R}_e = \{1 + 33(H/b)^{3/2}\}/33(H/b)^{3/2} \quad (3.5\text{-}27)$$

Near the ground, therefore, unsteady horizontal flight gives higher performance than steady gliding because the difference is very small without the ground effect.

However, a still better mode of flight than this unsteady horizontal flight may be possible. This possibility will now be analyzed.

Variational analysis. Let us define a performance function of the maximum flight distance between time t_0 and time t_f by

$$J = \int_{t_0}^{t_f} U \cos \gamma \, dt \quad (3.5\text{-}28)$$

which should be maximized for any initial condi-

tion at takeoff. Equations of motion are again obtained from Eq. (3.3-16) as follows:

$$\left.\begin{array}{l} \dot{U} = -g \sin \gamma - \tfrac{1}{2}(\rho S/m)U^2(C_{D_0} + C_L{}^2/\pi \mathcal{R}_{e,H}) \\[4pt] \dot{\gamma} = -(g/U)\cos \gamma + \tfrac{1}{2}(\rho S/m)UC_L \\[4pt] \dot{Z} = U \sin \gamma \end{array}\right\}$$
$$(3.5\text{-}29)$$

where Z is the height from the ground as a variable. The constraints for variables can be given by

$$C_{L,\min} \le C_L \le C_{L,\max}: \begin{Bmatrix} C_{L,\min} = -0.5 \\ C_{L,\max} = 1.4 \end{Bmatrix} \quad (3.5\text{-}30)$$

$$Z \ge H_{\min}: \begin{Bmatrix} H_{\min} = 0, \text{ out of ground effect (OGE)} \\ H_{\min} = b/10, \text{ in ground effect (IGE)} \end{Bmatrix}$$
$$(3.5\text{-}31)$$

Instead of using Eq. (3.5-27), the ground effect is represented by

$$\left.\begin{array}{ll} \mathcal{R}_{e,H} = \mathcal{R}_e: & \text{OGE} \\[4pt] = \mathcal{R}_e/\{1 - 1/0.35(Z/b - 0.35)^2\}: & 0 < Z/b < 0.35 \\[4pt] = \mathcal{R}_e & : 0.35 < Z/b \end{array}\right\}$$
$$\text{IGE}$$
$$(3.5\text{-}32)$$

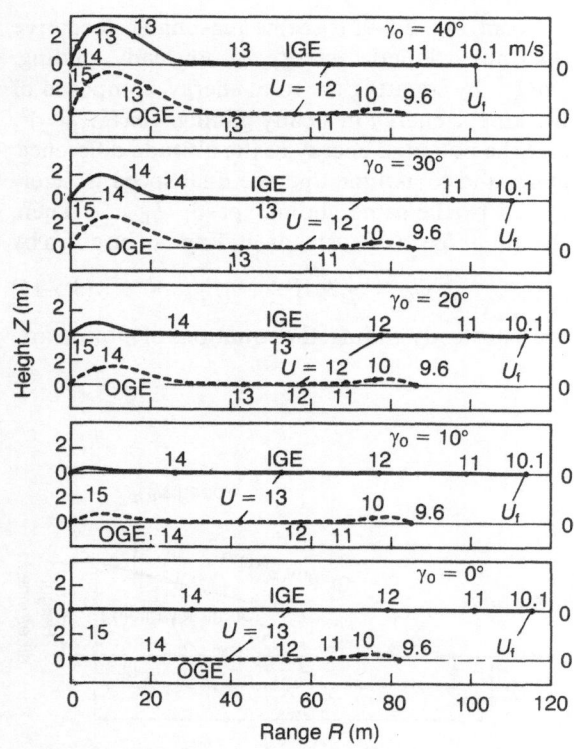

FIG. 3.5-11. Flight path and speed variation (Numbers show the speed U in m/s).

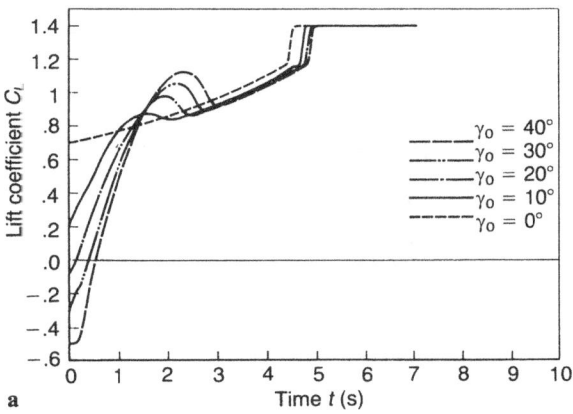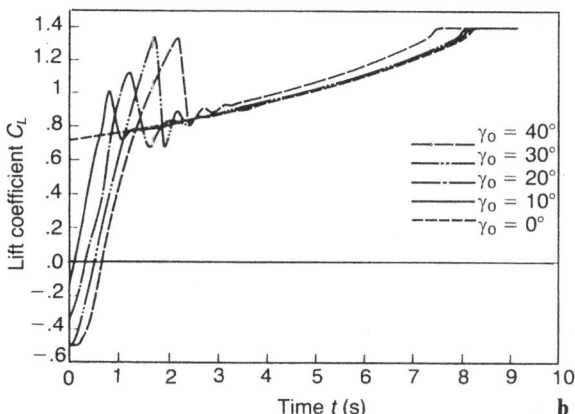

FIG. 3.5-12. **a, b.** Change of lift coefficient, $U_0 = 15$ m/s. (From Kawachi et al. 1989 with permission). **a** OGE. **b** IGE.

This modification is a good approximation of Eq. (3.5-27) and experimental results, and is also effective for avoiding difficulty related to numerical computation.

Initial and final conditions are given respectively by (U_0, γ_0, H_{min}) and (U_f, γ_f, H_{min}). The above variational problem can be solved by applying the "SCGR method" (Wu and Miele 1980) for three unknown dependent variables (U, γ, Z) and one unknown input C_L. The results for a flying fish, *Cypselurus heterurus doederlieini* (Kawachi et al. 1989), are shown in Fig. 3.5-11 for the speed and the flight path, and in Fig. 3.5-12 for the input C_L. These figures show that (1) as the takeoff angle γ_0 increases, the maximum range R_{max} increases for

out of ground effect (OGE) but decreases for in ground effect (IGE); (2) near the final approach to the ditching, the lift coefficient C_L is adjusted to the maximum value to allow jumping flight for OGE but to maintain level flight for IGE; and (3) the ground effect helps to increase the flight distance by about 20%–30% over that in OGE flight.

However, as shown in Fig. 3.5-13, for either OGE or IGE, the link between the initial takeoff angle and the maximum range is so weak that the maximum range can be attained for any takeoff angle by making the flight height as small as possible after takeoff.

3.5.5 MAMMALS, AMPHIBIANS, AND REPTILES

The membranous wing discussed in this section is usually of low aspect ratio and is, therefore, relatively rigid. Thus, the aerodynamic characteristics are quite different from those of the pterosaur and bats, which have high aspect ratio wings.

Fig. 3.5-14 shows (a) the lift coefficient of three wings with different planforms, two rectangular and one triangular (a type of "Rogallo wing"), as a function of angle of attack, and (b) the polar curves of these wings. It can clearly be observed that (1) the low aspect ratio rectangular wing has the lowest lift slope $a = \partial C_L/\partial \alpha$ but the highest maximum lift coefficient, $C_{L_{max}}$; (2) the flow separation (shaded area) starts early at the middle part of the low aspect ratio rectangular wing, whereas it begins later at the wing tips of the triangular wing; (3) the triangular wing has the highest lift slope but the lowest maximum lift coefficient; (4) as the aspect ratio increases, the drag rise in high angle-of-attack region is reduced; (5) the full stall

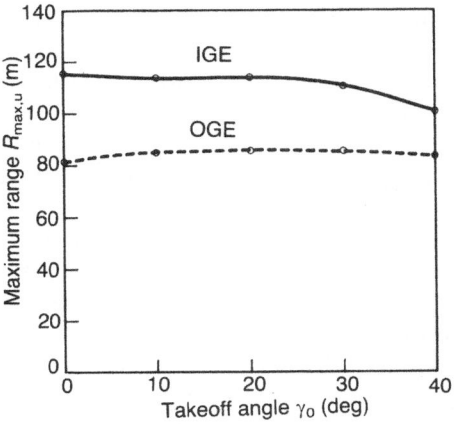

FIG. 3.5-13. Maximum range versus initial takeoff angle.

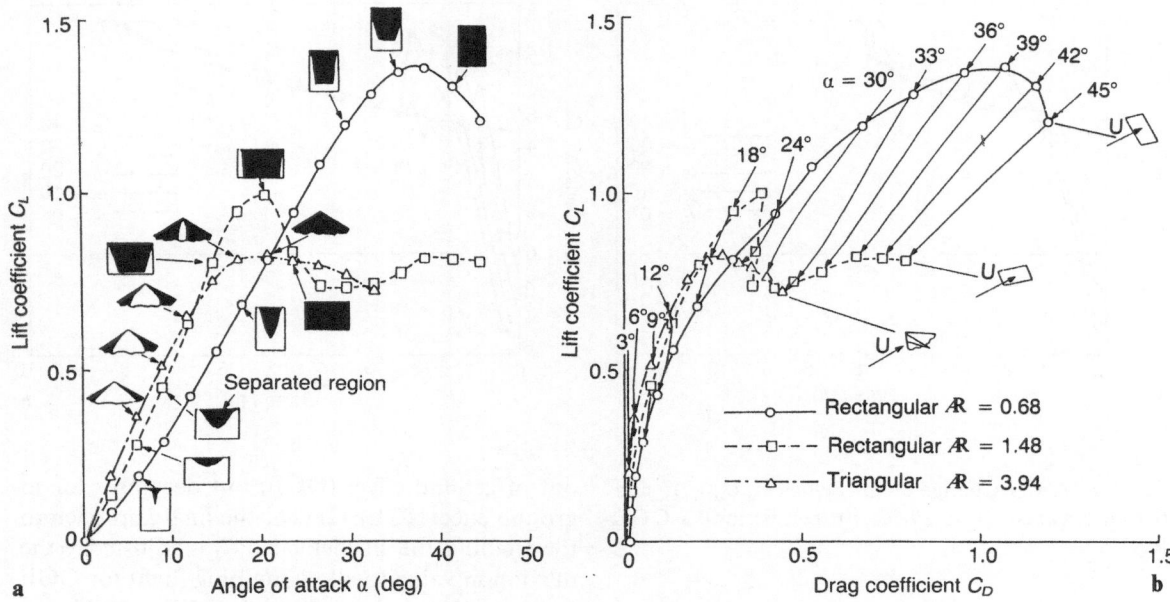

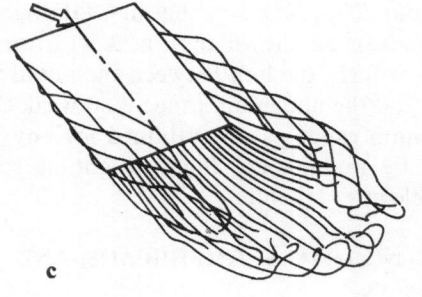

FIG. 3.5-14. **a–c.** Aerodynamic characteristics of three wings. **a** Lift coefficient. **b** Polar curve. **c** Rolled up vortices.

of the low aspect ratio rectangular wing is milder than that of the medium aspect ratio rectangular wing; and (6) the maximum value of the aerodynamic coefficient defined by Eq. (3.2-22) increases as the aspect ratio becomes small. The nonlinear lift, called vortex lift (Polhamus 1966, 1968), of the low aspect ratio wings results from strong rolled-up trailing vortices, as shown in Fig. 3.5-14c.

Flying squirrel. Petaurista leucogenys shown in Fig. 3.5-15 a squirrel belonging to Sciuridae. It lives in trees and executes gliding flight by extending a pair of patagia arranged between fore and hind legs on opposite sides of the body. The animal uses flight as a way of quick, economical and safe movement. It maneuvers in flight by changing the positions of its four legs and tail. The initial acceleration is obtained by kicking the tree with the hind legs. On landing on a tree, it flares its body

to assume a large angle of attack and reduce its speed, so as to land on the vertical surface of the tree. The low aspect ratio of the wing enables the squirrel to maintain a high aerodynamic force at slow speed with high angle of attack. The maximum lift-to-drag ratio can probably be given by $(L/D)_{max} \simeq 3$ at a flight speed of $U \simeq 10$ m/s.

Other flying animals. Several other kinds of flying mammals, such as the phalanger and the colugo are known to be good flyers. All forms have a membranous wing of furry skin extending between front and rear limbs. Specifically, during gliding flight, a flying lemur (*Cynocephalus*) beats its tail up and down in a fanning motion described in Sect 7.1.3. This locomotion mode is effective in reducing drag or improving gliding performance. Special gliding wings, in the form of membranes, have also evolved in certain frogs and geckos, and in the flying dragon, a kind of lizard (Mertens 1960; Losos et al. 1989). It is not clear whether the wings of these animals are still in the developmental stage for further airborne life, or whether they are already well-adapted for their present ecology. A flying lizard, *Draco fimbriatus*, for example, has a slender body with a dorsoventral flattening that is extended laterally as a slender wing. As discussed

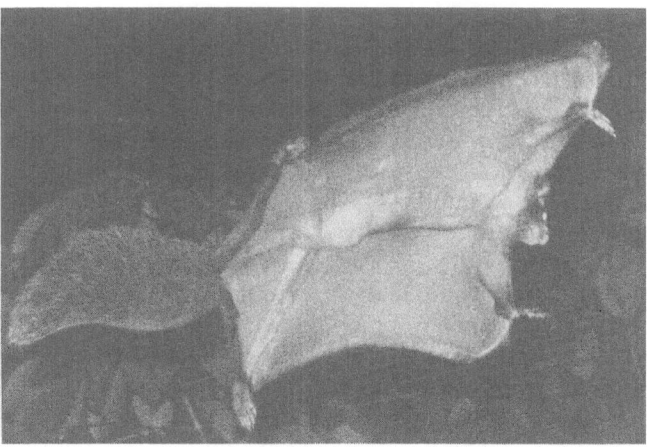

FIG. 3.5-15. Gliding flight of the flying squirrel *Petaurista leucogenys*. (Courtesy of Yukiyo Kira).

further in Sects. 5.2-2 and 7.2-1, this configuration generates a large aerodynamic force based on the linear potential lift and nonlinear vortex lift; thus, it guarantees a low sinking speed for safe landing in spite of a steep descent path.

3.6 Rotary Seeds

Many samaras or winged seeds execute autorotational flight as they fall from their trees. Generally, the seeds of simple configuration, like those of maple and black pine, perform well: they achieve a low rate of descent in spite of high wing loading (or small wing area but large seed weight) because of their high rotational speed and thus low "coning angle."

Observation of the flight of samaras rotating on the vertical or spinning axis during fall and detailed analyses of their aerodynamics and flight dynamics were conducted by Norberg (1973) and followed up by McCutchen (1977) and Augspurger (1986). In his analysis, Norberg made two major assumptions: that the samara has a flat wing, and that the mass of the wing lies on the long axis from the wing root to the wing tip. Then, by combining the momentum theory based on the homogeneous (or constant) induced velocity with the blade element theory, he analyzed the mechanism of autorotation and its stability. On the other hand, by applying the local circulation method (LCM), Azuma and Yasuda (1989) further clarified the

aerodynamic characteristics of rotary seeds with nonhomogeneous induced velocity distribution.

3.6.1 FLIGHT MODE AND GEOMETRICAL CONFIGURATION

Flight mode. Many plants and animals are spread from their place of origin by the wind. The flight time and range are increased in fruits or seeds that are very light in weight and are provided with flight organs such as wings. As stated in Sect. 3.5.1, if the center of gravity of a winged seed or samara is located at the center of wingspan and just in front of the aerodynamic center of the wing, then its flight will be a gliderlike translation without any rotation. If, however, the center of gravity (or the seed itself) is located near a terminal end of the wing, then its flight will be a screwlike rotation. Many winged seeds take the latter mode of flight, which enables them to reduce their falling rate and increase their dispersion distance by the wind. The spinning motion is initiated as soon as the seed begins to fall from its tree. Then the downward acceleration caused by gravity is stopped, and a steady (or self-stabilized) autorotational flight with a constant speed of rotation, constant coning angle, and constant rate of descent is attained by balancing the lifting force acting on the wing with the weight of the seed, the driving torque with the resistant torque, and the aerodynamic moment about the flapping axis with the centrifugal and gravitational moments about the same axis.

It is interesting to find that the seeds of the tulip tree and the ash tree not only execute a spinning motion but also rotate on the feathering (or long) axis of their wing at a slightly different speed from the speed about the spinning (autorotating) axis.

Geometrical configuration. The configuration of winged seeds is usually simple in a structural sense. Typical examples are shown in Table 3.6-1 with their fundamental geometrical parameters (the definition of the parameters is shown later in Fig. 3.6-3). These data are mean values of twenty samples for each of the ten species.

Many seeds have a single flat wing or blade, whereas a few have multiple and/or curved wings that are geometrically bent (convex upward) to reduce the outside "preconing angle" or to make the wing tip horizontal. This is reasonable in the aerodynamic sense because the airloading is concentrated near the tip: the heavy tip reduces the coning angle or "flapping angle" in flight and thus increases the effective disc area ($S = \pi R^2$ where R is the radius of blade), whereas the blade root helps to enlarge the drag area so that the horizontal force is extended widely and the seed is carried easily by horizontal wind.

Although in all rotary seeds the "wing loading" W/S_w is larger than in gliding seeds such as *Alsomitra macrocarpa*, the wing loading of which is about 0.5 N/m^2 (Sect. 3.5.1), the "disc loading" W/S is smaller than the wing loading of all gliding seeds except Santalaceae.

The wing is usually narrow in chord and its aspect ratio, $Æ = l^2/(S_w/p)$, (where p is the number of wings,) is nearly equal to or a little larger than, the $Æ = 4.0$ of the gliding seed. The small "solidity" $\sigma = S_w/S$, guarantees good performance, as will be explained later. There is great variety in wing planform, as illustrated in Table 3.6-1. Some seeds, like those of Santalaceae, linden, hornbeam, and phoenix tree, are tapered at the wing tip, but others, such as those of maple and pine, are inversely tapered as the wing chord widens to the tip.

Examples of wing cross section at the three-quarter-radius $\frac{3}{4}R$ are also shown in Table 3.6-1. Interestingly, each section is very thin except at the leading edge, which is reinforced by strands and thus does not necessarily have a smooth surface. In the case of hornbeam, Santalaceae, and phoenix tree, a vein spans from wing root to tip at either the quarter- or the half-chord. Furthermore, the mean upper and lower surfaces of such seeds consist of two convex lenses. This is clearly the result of the fact that since the Reynolds number based on chord length and circumferential speed

at the three-quarter-radius is of the order of 10^3, a sharp leading edge with a somewhat rough surface is better than a streamlined shape with a smooth surface for aerodynamic performance (Pope and Harper 1966). It is also interesting to find that for stability of flight, the mean camber line of airfoil within the three-quarter radius of the wings is configured either to be a downward convex or to have an upwardly reflexed trailing edge, with the exception that their center of gravity is located downward. Furthermore, the chordwise location of the center of gravity is very delicate for stable flight. That is to say, if the leading edge is made incomplete artificially, for example, by cutting vertically and making square, then the center of gravity for stable autorotation shifts backward and the rate of descent increases because of a spin rate lower than the original rotation. However, if the leading edge is left as it is, then the aerodynamic force acting on the leading edge is greate and the center of gravity for optimal autorotation is located near the leading edge (Yasuda and Azuma 1992).

Performance. Measured data on flight performance are also given in Table 3.6-1. These are mean values for twenty samples of each species. In the subsequent figures, the standard deviation of the data is shown by the length of the cross lines ($+$), the origin of which is the mean value for the species concerned.

The rate of descent ($|w| = V$) increases with the disc loading W/S, but in most seeds it is scattered around 1 m/s (Fig. 3.6-1). Augspurger (1986) reported nearly the same results. This is important because updrafts exceeding this magnitude can be observed in the turbulent atmosphere near the ground only when it is windy. Thus, the rotary seeds are attached firmly to their tree so as not to take off when it is calm. The solid line shown in this figure is, as described later, obtained from the momentum theory and is considered an optimal or lower limit.

As shown in Fig. 3.6-2, the spin rate Ω or the tip speed $R\Omega$ generally increases with disc loading W/S. The number affixed to each species is the equivalent drag coefficient $\sigma\delta$ obtained later by solving Eq. (3.6-2c). The solid lines here are obtained later from Eq. (3.6-9) for various values of equivalent drag coefficient $\sigma\delta$ in the optimal rotor. It can be seen that the tested species are not far from the optimal rotor.

More detailed information on the performance

TABLE 3.6-1. Geometrical configuration and performance of rotary seeds. (Azuma and Yasuda 1989)

Item	Maple: *Acer diabolicum* Blume	Maple: *Acer palmatum* Thunb. var. *matsumurae* Makino	Maple: *Acer palmatum* Thunb.	Black pine *Pinus thunbergii* Parlatore	Santalaceae *Buckleya joan* Makino	Linden *Tilia miqueliana* Maxim.	Hornbeam *Carpinus tschonoskii* Maxim.	Phoenix tree *Firmiana platanifolia* Schott et Endl	Ash *Fraxinus japonica* Blume	Tulip tree *Liriodendron tulipifera* L.
Profile at r/R = 0.75										
Mass, $m \times 10^4$ (kg)	0.58	0.38	0.13	0.23	2.29	1.39	0.23	6.98	0.58	0.50
Wing area, S_w (cm^2)	3.04	1.67	0.56	1.09	3.91	5.57	1.40	21.97	1.84	3.39
Span, l (cm)	3.62	2.52	1.48	2.19	2.55	5.24	2.30	9.01	3.64	4.51
Radius of rotation, R (cm)	2.84	2.09	1.20	1.79	2.65	3.69	1.76	6.73	2.66	3.44
Disc area, S (cm^2)	25.5	13.8	4.49	10.1	22.3	43.6	9.73	142.9	22.4	37.4
Number of wings, p	1	1	1	1	4	2	1	1	1	1
Solidity, σ	0.12	0.12	0.12	0.11	0.18	0.13	0.15	0.15	0.08	0.09
Aspect ratio, $\mathcal{R}$	4.33	3.83	3.96	4.43	6.70	5.07	3.79	3.73	7.22	6.04
Mean chord, $\bar{c} = S_w/lp$ (cm)	0.84	0.66	0.38	0.50	0.38	1.08	0.61	2.43	0.51	0.75
Maximum thickness ratio, $t_{max}/\bar{c}$	0.05	0.03	0.04	0.02	0.05	0.03	0.03	0.01	0.05	0.05
Wing loading, W/S_w (N/m^2)	1.87	2.24	2.32	2.05	5.76	2.51	1.61	3.13	3.10	1.46
Disc loading, W/S (N/m^2)	0.22	0.27	0.29	0.22	1.02	0.32	0.23	0.48	0.26	0.13
Rotational speed, Spinning axis Ω (rpm)	977.0	1101.5	1805.8	1472.8	1517.8	832.6	965.0	717.7	888.7	498.3
Feathering axis Λ (rpm)	—	—	—	—	—	—	—	—	877.4	493.9
Rate of descent, V (m/sec)	0.82	1.04	1.09	0.98	1.58	1.34	1.02	1.14	1.72	1.19
Coning angle, β (deg)	23.7	27.6	15.0	20.8	15.6	16.5	12.9	17.2	31.7	34.2
Pitch angle at 0.75 R, θ (deg)	-1.17	-1.39	-0.90	-1.43	-1.34	—	-2.16	-2.67	—	—
Thrust coefficient, $\bar{C}_T \times 10^2$	2.11	3.84	4.76	2.47	5.18	3.00	6.59	1.70	3.72	3.51
Reynolds number at 0.75 R, $Re \times 10^{-3}$	1.37	0.97	0.50	0.75	0.85	1.81	0.63	5.91	0.79	0.88
Tip speed, $R\Omega$ (m/sec)	2.91	2.41	2.26	2.75	4.21	3.19	1.77	5.05	2.48	1.79
Tip speed ratio, $V/R\Omega$	0.28	0.44	0.49	0.36	0.38	0.43	0.58	0.23	0.69	0.66
Thrust coefficient over solidity C_T/σ	0.18	0.32	0.38	0.23	0.29	0.23	0.45	0.11	0.44	0.38
Nondimensional parameter $\rho R^2 \bar{c}/m_w$	0.15	0.12	0.05	0.10	—	—	0.11	—	—	0.23
Nondimensional mass $\bar{M} \times 10^2$	6.69	6.88	5.99	4.52	—	—	7.19	—	—	15.3
Nondimensional moment of inertia about y-axis $\bar{I}_y \times 10^2$	4.04	3.98	3.64	2.73	—	—	4.05	—	—	9.48
Nondimensional moment of inertia about x-axis $\bar{I}_x \times 10^2$	—	—	—	—	—	—	—	—	—	0.37

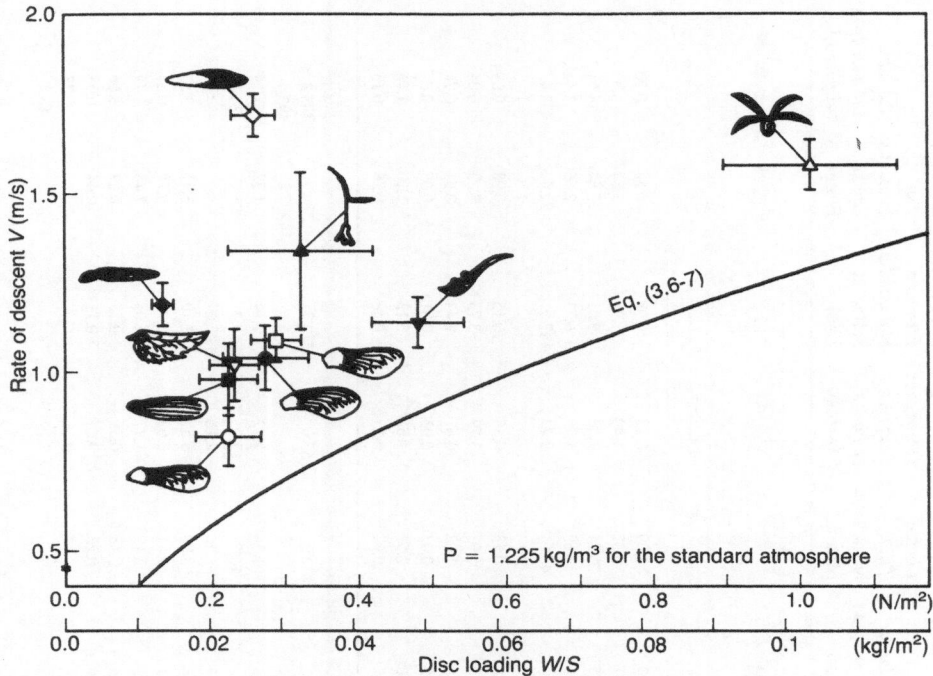

FIG. 3.6-1. Rate of descent and disc loading. (From Azuma and Yasuda 1989 with permission).

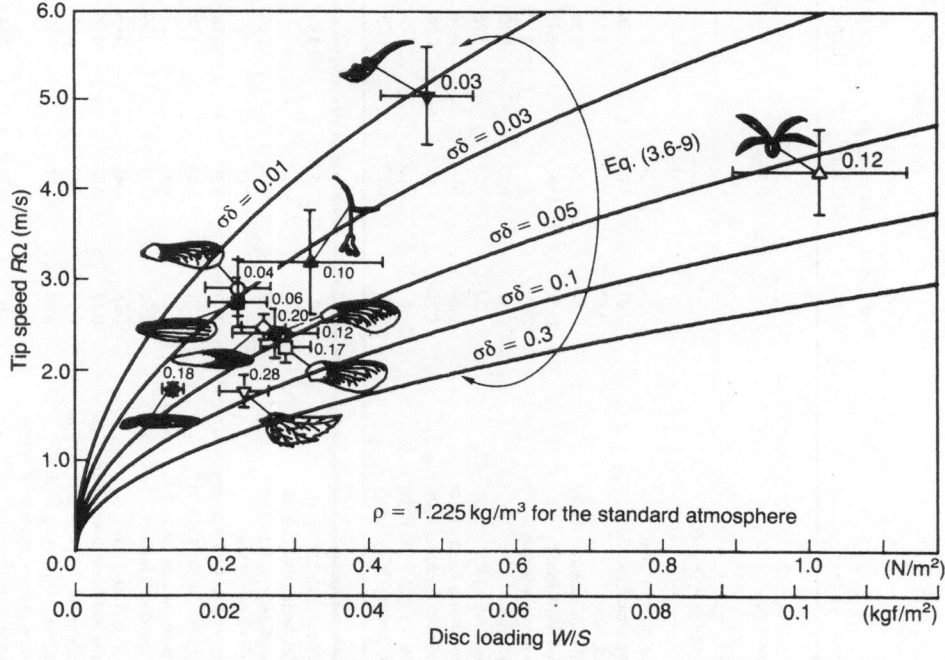

FIG. 3.6-2. Tip speed and disc loading. (From Azuma and Yasuda 1989 with permission).

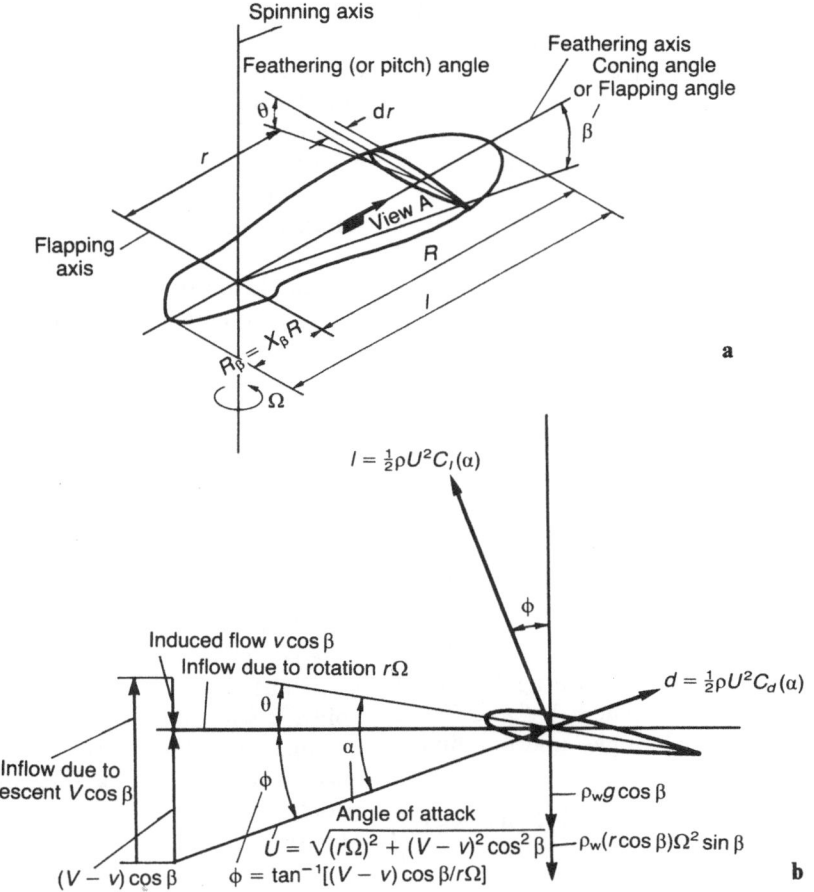

FIG. 3.6-3. **a, b.** Dimension and velocity profile of a rotary wing. **a** Perspective. **b** View A. (From Azuma and Yasuda 1989 with permission).

data, such as tip speed, thrust coefficient, drag coefficient, coning angle, and pitch angle, have been presented by Azuma and Yasuda (1989).

3.6.2 SIMPLE ANALYSIS BASED ON THE MOMENTUM THEORY

In the steady vertical flight of an autorotational seed or a rotor shown in Fig. 3.6-3, the thrust T must be balanced with the weight of the seed W, and the profile or resistant torque Q resulting from the drag (d) of the rotary wing must be equal to the driving torque generated by the lift (l) in vertical descent (or the rate of height loss) of the thrusting rotor. By assuming a small coning angle β and a constant and homogeneous (or uniform)

induced velocity distribution v over the whole disc area ($S = \pi R^2$) swept by the rotary seed in vertical flight (velocity V and rotational speed Ω), the momentum theory and the blade element theory (Gessow and Myers 1952) present the following relations by using two-dimensional lift and drag coefficients C_l and C_d and nondimensional distances of $x = r/R$ and $x_\beta = R_\beta/R$ (Fig. 3.6-3).

Vertical force trim

$$T = \int_{-R_\beta}^{R} (l\cos\phi + d\sin\phi)\cos\beta\,dr$$

$$\simeq \int_{-R_\beta}^{R} (l + d\phi)dr = W \qquad (3.6\text{-}1a)$$

or

$$C_T = T/\rho S(R\Omega)^2 = \frac{1}{2}\sigma \int_{-x_\beta}^{1} (U/R\Omega)^2 (c/\bar{c})$$

$$\cdot \{C_l(\alpha) + (\lambda/x)C_d(\alpha)\}dx$$

$$= W/\rho S(R\Omega)^2 \qquad (3.6\text{-}1b)$$

where $\bar{c}$ is the mean chord, $\bar{c} = S_w/pl$, and the integration along the spanwise distance r is performed from root $-R_\beta$ to tip R.

Torque trim

$$Q = \int_{-R_\beta}^{R} (l\sin\phi - d\cos\phi)(\cos\beta)r\,dr$$

$$\simeq \int_{-R_\beta}^{R} (l\phi - d)r\,dr = 0 \qquad (3.6\text{-}2a)$$

or

$$C_Q = Q/\rho SR(R\Omega)^2$$

$$= \frac{1}{2}\sigma \int_{-x_\beta}^{1} (U/R\Omega)^2 (c/\bar{c})\{(\lambda/x)C_l(\alpha) - C_d(\alpha)\}x\,dx$$

$$= 0$$

$$(3.6\text{-}2b)$$

or, for a constant chord

$$C_T\lambda = \tfrac{1}{8}\sigma\delta \qquad (3.6\text{-}2c)$$

where ρ, $\bar{c}$, σ, and δ are air density, mean chord, solidity or $\sigma = b\bar{c}/\pi R$, and mean drag coefficient or $\delta = \bar{C}_d$ respectively, and where λ, ϕ and $U/R\Omega$ are "inflow ratio", "inflow angle" and nondimensional speed respectively;

$$\left.\begin{array}{l} \lambda = (V/R\Omega) - (v/R\Omega) \simeq (V/R\Omega) - \tfrac{1}{2}\{C_T/(V/R\Omega)\} \\ \phi = \tan^{-1}\{(V-v)\cos\beta/r\Omega\} \end{array}\right\}$$

$$(3.6\text{-}3)$$

$$U/R\Omega = \sqrt{x^2 + \lambda^2} \qquad (3.6\text{-}4)$$

Momentum balance

$$T = 2\rho S(V-v)v \quad \text{or}$$

$$(V/R\Omega) = \{\tfrac{1}{2}C_T/(v/R\Omega)\} + (v/R\Omega) \quad (3.6\text{-}5)$$

The angle of attack α is given by the sum of the pitch angle θ and inflow angle ϕ:

$$\alpha = \theta + \phi \simeq \theta + \lambda/x \qquad (3.6\text{-}6)$$

The approximate expressions given in Eqs. (3.6-1 through -5) result from the assumption of a small and constant induced velocity distribution.

Optimal rotation: Then, the minimum rate of descent in an optimal state of operation can be given by

$$V/R\Omega = \sqrt{2C_T} \quad \text{or} \quad V = \sqrt{(2/\rho)(W/S)} \quad (3.6\text{-}7)$$

when the induced velocity is selected to be

$$v/R\Omega = \sqrt{C_T/2} \qquad (3.6\text{-}8)$$

Equation (3.6-7) states that the rate of descent,

which is a measure of the performance of autorotational flight, is proportional to the square root of the disc loading W/S if the rotor is operating in the optimal state.

However, as illustrated in Fig. 3.6-1, all the experimental data are scattered above the minimum rate of descent shown by the solid line. The difference between the theory and the experimental data probably results from the limitations of the simple momentum theory.

By combining Eqs. (3.6-1) and (3.6-5 through -7), the rotor tip speed $R\Omega$ in the optimal state can be derived approximately as:

$$R\Omega = 2^{5/6}\sqrt{W/S}/\sqrt{\rho}(\sigma\delta)^{1/3} \qquad (3.6\text{-}9)$$

The above equation states that the tip speed $R\Omega$ is also proportional to the square root of the disc loading, and the proportional constant is inversely proportional to the cubic root of the equivalent profile drag coefficient $\sigma\delta$ of the seed as a rotor. The above theoretical results for the optimal rotor were shown earlier in Fig. 3.6-2. Again, it can be seen that the experimental data do not always coincide with the theoretical calculations for the optimal rotor, but the relationship among the tip speed, the disc loading W/S, and the equivalent profile drag coefficient $\sigma\delta$ is well described by Eq. (3.6-9).

3.6.3 APPLICATION OF THE LOCAL CIRCULATION METHOD

As Fig. 3.6-3 and Eqs. (3.6-1 and 3.6-2) show, the spanwise loading determines not only the lifting force T and the spinning torque Q of the wing but also the coning angle β of a rotary seed in falling flight. Since, as indicated in Eq. (3.6-2), the rotational speed of the seed is determined in a delicate balance of the driving and resistant torques, the induced velocity distribution, which determines the inflow ratio λ (or the inflow angle ϕ) as well as the airloading, must be calculated precisely along the spanwise station of the rotary wing. This can be done by applying the Biot-Savart law (Stepniewski 1979, Stepniewski and Keys 1984) for the trailing vortices distributed on the wake of the rotary wings. The detailed formation of trailing vortices can be observed in Fig. 3.6-4 (Onda et al. 1986). However, the calculation is laborious, specifically for free wake analysis (Landgrebe 1972), because of the divergent tendency of the computation.

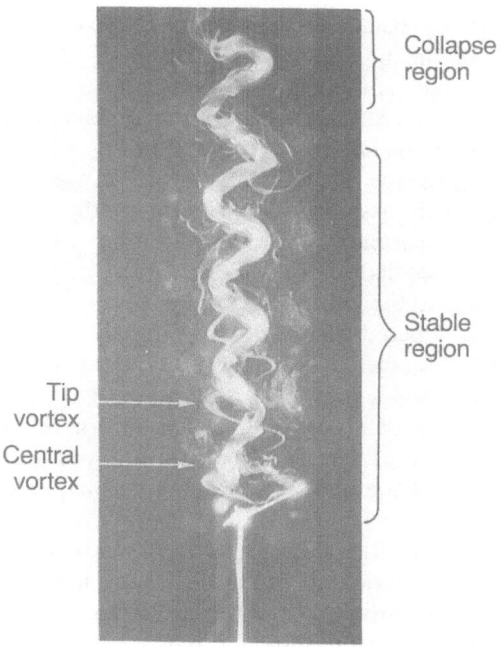

Collapse region

Stable region

Tip vortex

Central vortex

FIG. 3.6-4. Trailing vortices from a rotary seed, maple. (From Onda et al. 1986 with permission).

The local circulation method (LCM) (Azuma et al. 1983), which is an extension of the local momentum theory (Azuma and Kawachi 1979), is a simple, yet useful method of computation for analyzing the airloading and the driving and resistant force distribution along the span of the rotary wing, with the same degree of accuracy as the vortex theory.

Because of the small Reynolds number and the complex configuration of the airfoil section of the wing, the estimation of the aerodynamic lift and drag coefficients is not easy. We shall therefore postulate the following equations:

$$C_l = a\alpha \quad \text{in the range of } C_l < C_{l_{\max}} \brace = C_{l_{\max}} \quad \text{in the range of } C_l \geq C_{l_{\max}}} \quad (3.6\text{-}10)$$

$$C_d = C_{d_0} + \delta_1\alpha + \delta_2\alpha^2 + \delta_3\alpha^3 \brace = C_{d_0} - 0.008\alpha + 4.27\alpha^2 - 2.50\alpha^3} \quad (3.6\text{-}11)$$

where the coefficients δ_1, δ_2, and δ_3 are used commonly for all species, and the lift slope $a = \partial C_l/\partial\alpha$, the maximum lift coefficient $C_{l_{\max}}$, and the minimum drag coefficient $C_{d_0} = C_d \,(\alpha = 0)$ are unknown values or parameters for making the computation.

Referring to Fig. 3.6-3, and assuming that the blade is rigid, the center of mass is very close to one terminal end (or $x_\beta = R_\beta/R \ll 1$), and the coning angle β is small, the coning angle is determined in the balance of the flapping moment (about the center of mass) derived from the aerodynamic force, centrifugal force, and gravitational force distributions along the wingspan by the formula

$$M = -\int_{-R_\beta}^{R} (l\cos\phi + d\sin\phi)r\,dr \brace + \int_{-R_\beta}^{R} \rho_w r^2\,dr\Omega^2\cos\beta\sin\beta + g\int_{-R_\beta}^{R} \rho_w r\,dr\cos\beta \\ = 0} \quad (3.6\text{-}12a)$$

or

$$\beta \simeq \tfrac{1}{2}(\rho R^2\bar{c}/m_w)(\bar{A}/\bar{I}) - (g/R\Omega^2)(\overline{M}/\bar{I}) \quad (3.6\text{-}12b)$$

where

$$\bar{A} = \int_{-x_\beta}^{1} (c/\bar{c})(U/R\Omega)^2\{C_l(\alpha) + C_d(\alpha)\phi\}x\,dx \brace \overline{M} = \int_{-x_\beta}^{1} (R\rho_w/m_w)x\,dx \\ \bar{I} = \int_{-x_\beta}^{1} (R\rho_w/m_w)x^2\,dx} \quad (3.6\text{-}13)$$

and where m_w and ρ_w are the mass of one wing (which is equal to the mass of single-winged seeds) and the mass distribution along the wingspan (or line density) respectively.

Usually the rotational (or spin) rate Ω is so high (about 50–100 rad/s) that the gravitational contribution given by the second term in Eq. (3.6-12) may be neglected. A heavy wing, specifically a heavy tip, decreases the ratio of $\bar{A}/\bar{I}$ and thus reduces the flapping angle as stated earlier.

By combining the three Eqs. (3.6-1, 3.6-2 and 3.6-12) and by using the observed data Ω, θ, and V for several samples of the respective seeds in every species listed in Table 3.6-1, the LCM can determine three unknown coefficients, a, $C_{l_{\max}}$, and C_{d_0}, which are assumed to be invariant along the span.

The results are given in Table 3.6-2 for several samples of the various species listed in Table 3.6-1, and two exemplified results of maple (*Acer diabolicum* Blume) and phoenix tree (*Firmiana platanifolia* Schott et End) are shown in Fig. 3.6-5. In

TABLE 3-6-2. Estimated airfoil characteristics. (Azuma and Yasuda 1989)

Items	Maple — Acer diabolicum Blume	Maple — Acer palmatum Thunb. var. matsumurae Makino	Maple — Acer palmatum Thunb.	Black pine Pinus thunbergii Parlatore	Santalaceae Buckleya joan Makino	Hornbeam Carpinus tschonoskii Maxim.	Phoenix tree Firmiana platanifolia Schott et Endl	Ash Fraxinus japonica Blume	Tulip tree Liriodendron tulipifera L.
Plan view									
Profile at $r/R = 0.75$									
Lift slope a (rad^{-1})	4.80/4.40[a]	4.42	3.90	4.14	4.30	3.32	4.20	—	—
Maximum lift coefficient $C_{l_{max}}$	1.63	1.78	1.40	1.40	1.25	1.55	1.38	—	—
Minimum drag coefficient C_{d_0}	0.07/0.05[a]	0.21	0.06	0.08	0.06	0.22	0.03	—	—
Mean lift coefficient at 0.75 R, C_l	1.02	1.57	1.38	1.16	1.24	1.53	0.77	1.62	1.60
Mean drag coefficient at 0.75 R, C_d	0.25	0.65	0.51	0.38	0.42	0.95	0.16	1.48	1.41

[a] Data obtained from the gliding flight test.

almost all species, the wing is considered to operate with the maximum lift coefficient $C_{l_{max}}$ over the whole span of the wing. However, as can be seen from the operational range of the angle of attack shown by the horizontal segment in Fig. 3.6-5, the outer (tip) side of the wing operates in the linear range of the assumed lift coefficient. Although the data are scattered over a range of a, $C_{l_{max}}$, and C_{d_0} for a specified species, their mean values are peculiar to the individual species.

The tapered planform and nearly flat, but slightly cambered, configuration of the wing of the hornbeam may give a favorable $C_{l_{max}}$ under the penalty of a large drag coefficient, whereas the highly three-dimensional configuration of Santalaceae and the phoenix tree may be unfavorable to $C_{l_{max}}$.

Figure 3.6-6 presents a typical example of vertical force distribution, $n = l \cos\phi + d \sin\phi$, horizontal force distribution consisting of driving and resistant force distributions, $t = l \sin\phi - d \cos\phi$, and angle of attack distribution α of an exemplified seed, *Acer diabolicum* Blume, in steady autorotational flight. The vertical force n is mostly concentrated near the three-quarter-radius because the velocity against the wing chord is proportional to the radius, and the induced velocity increases (and thus the angle of attack α decreases) near the wing tip. Since the driving force $l \sin\phi$ is widely distributed along the span, whereas the resistant force $d \cos\phi$ is, like the vertical force, concentrated near the three-quarter-radius, the horizontal force t is negative near the tip and positive within about 60% radius. Thus, it can be seen that the driving torque of the rotor arises near the midpoint of the wingspan.

Effects of additional rotation. As stated earlier, the seeds of the ash tree and the tulip tree have an additional angular rate Λ along the spanwise axis (called the feathering axis) during fall, with a spinning rate of Ω. This rotation results from the symmetric configuration of these seeds in front of and behind the feathering axis, as seen in Fig. 3.6-7. Thus the aerodynamic force acting approximately on a quarter-chord generates a pitch-up moment and gives an additional rotation Λ about the feathering axis (or center of gravity located at the half-chord). Thus, the aerodynamic coefficients of the airfoil section are changed by the rotation as a function of nondimensional rotation $(c\Lambda/U)$ under the "Magnus effects" (any book of fluid-dynamics such as Massey 1979), and simultaneously the

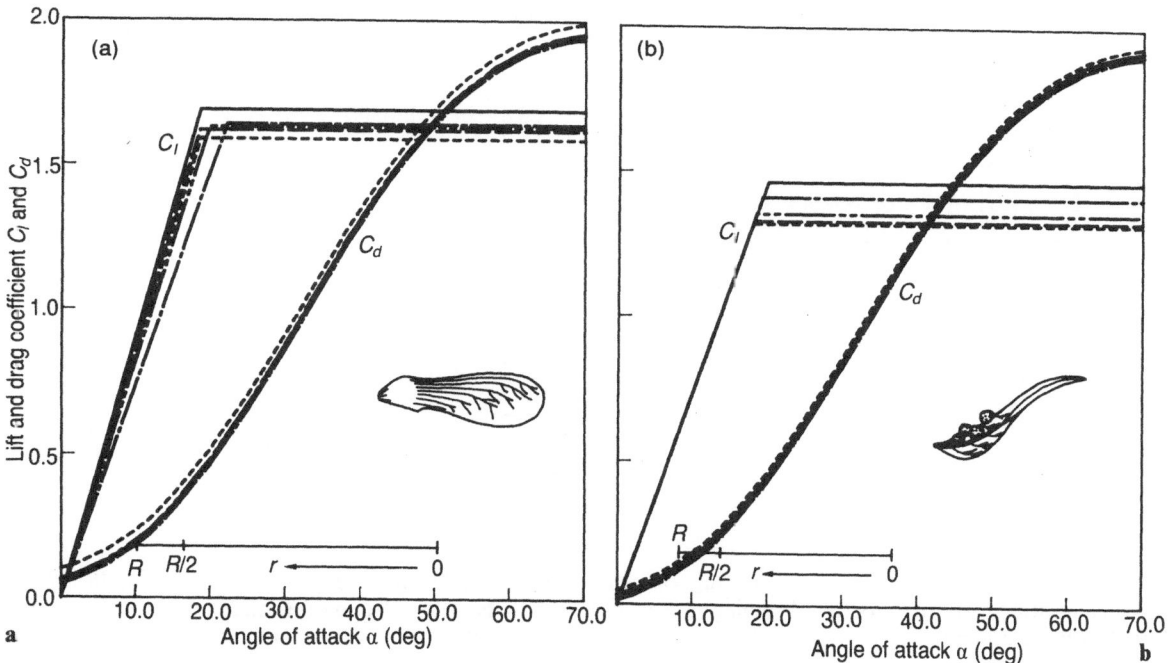

FIG. 3.6-5. **a, b.** Two-dimensional aerodynamic characteristics of airfoil section of rotary seed. (From Azuma and Yasuda 1989 with permission). **a** Maple (*Acer diabolicum* Blume). **b** Phoenix tree (*Firmiana platanifolia*).

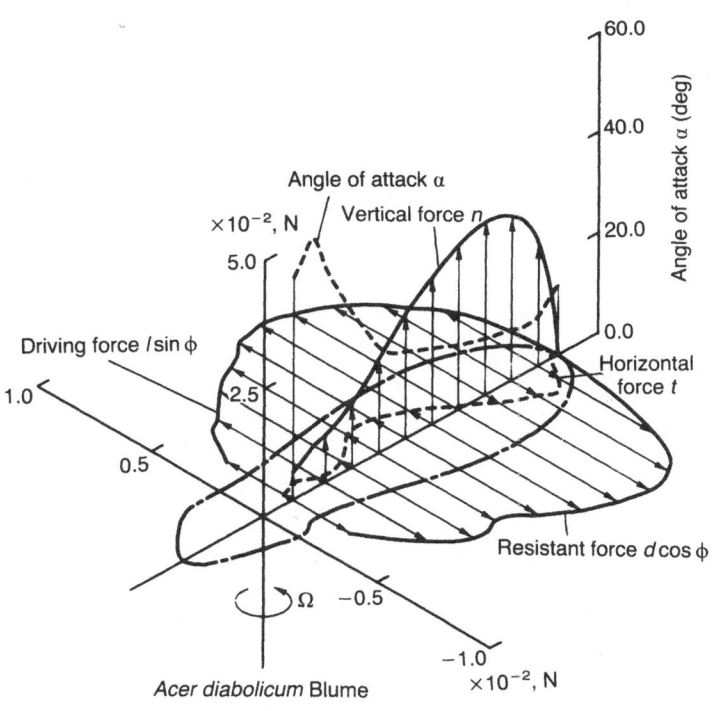

FIG. 3.6-6. Force distribution. (From Azuma and Yasuda 1989 with permission).

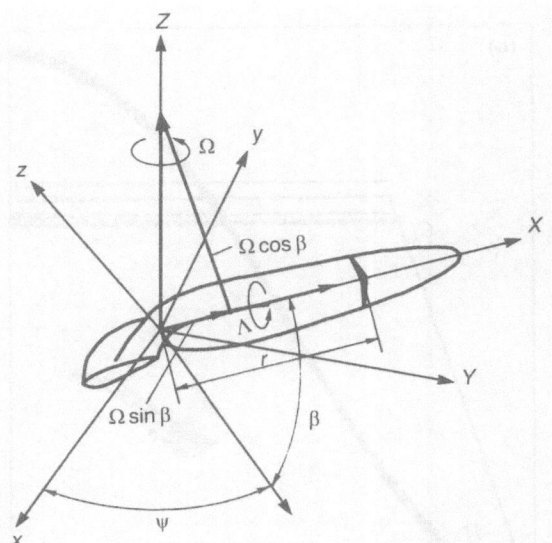

FIG. 3.6-7. Coriolis force acting on a wing element having additional rotation. (From Azuma and Yasuda 1989 with permission).

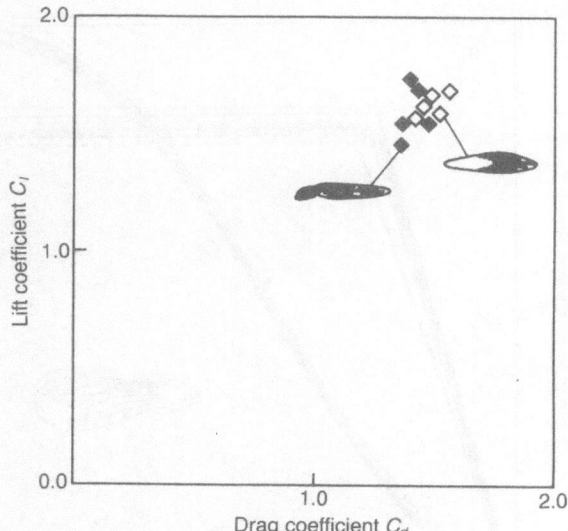

FIG. 3.6-8. The lift and drag coefficients obtained from the seeds of the ash and the tulip tree. (From Azuma and Yasuda 1989 with permission).

inertial moment is also generated by the "Coriolis force" acting on the wing element.

Referring to Fig. 3.6-7, let us introduce an inertial frame (X, Y, Z) and a body frame (x, y, z), the x-axis of which is fixed to the feathering axis of the seed and rotates with the angular velocity $\Omega = d\psi/dt$ about the Z-axis of the inertial frame, and with the angular velocity Λ about the x-axis itself. Then, an additional inertial torque about the flapping axis due to the Coriolis force is given by

$$Q_I = -I_x \Omega \Lambda \cos \beta \qquad (3.6\text{-}14)$$

where

$$I_x = \int_0^R (y^2 + x^2)\rho_w \, dr = m_w R^2 \bar{I}_x \quad (3.6\text{-}15)$$

Then, Eq. (3.6-12) should be modified to:

$$\beta \simeq \tfrac{1}{2}(\rho R^2 \bar{c}/m_w)(\bar{A}/\bar{I}) - (g/R\Omega^2)(\bar{M}/\bar{I})$$
$$+ (\bar{I}_x/\bar{I})(\Lambda/\Omega) \qquad (3.6\text{-}16)$$

This equation states that the Coriolis force increases the coning angle.

In the case of the tulip tree, the contribution of the Coriolis force is about 16%. As shown in Table 3.6-1 and Fig. 3.6-8, the airfoil characteristics of the ash tree and the tulip tree are obtained for only the mean value of the lift and drag coefficients. The lift coefficient is similar to that of other seeds, but the

drag coefficient is very large and is approximately equal to that of a circular cylinder of the same diameter as the chord. This is to be expected because the frontal area of the seed rotating about the feathering axis resembles that of a circular cylinder.

With a few exceptions, the autorotational flight performed by the present examples of winged seeds is very close to optimal, i.e., the rate of descent is near the minimum. The seeds rely on horizontal wind for their dispersion, rather than the upcurrent.

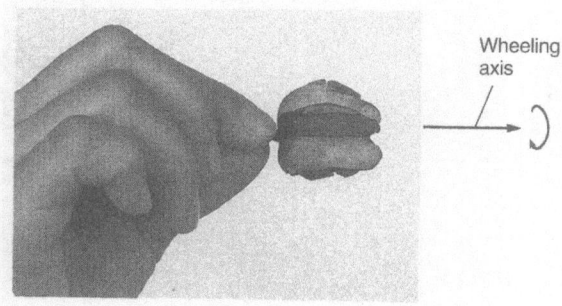

FIG. 3.6-9. Flight of a wheeling seed (Caconia trifoliatum Vent).

3.6.4 WHEELING

When a wing rotates about a spanwise axis through the midchord of the wing, like the seed of a tulip tree, all the while moving forward, the motion may be called "wheeling" or "tumbling". A "wheeling wing", generates lift through the Magnus effect due to the induced circulation caused by the wing rotation (Skews 1990).

An actual application of the wheeling wing can be observed in the dispersion of Cacouia trifoliatum Vent., as shown in Fig.3.6-9 and *Cavanillesia platanifolia* (Augspurger 1986). By executing steady gliding flight, the seed can attain a sinking speed of about 1m/s.

CHAPTER 4

Flight by Beating

This chapter discusses beating or "ornithopter." The beating motion of wings is exclusively used in the powered flight of birds and insects. In flying, this is the only way these creatures can counter the gravity force and propel themselves against aerodynamic drag. Other methods of locomotion, adopted widely in the swimming motion of fish and mammals, are inappropriate for flying because they are incapable of generating an aerodynamic force close to the center of gravity and maintaining trimmed flight without tumbling.

The propulsive force is produced by giving a "heaving" (normal) velocity in addition to a forward (parallel) velocity to the wings. This kind of motion can be generated principally by a flapping (up and down) motion of the wing, but not by a feathering (pitch-up and pitch-down) motion. The mode and frequency of the beating motion differ among different species and are strongly dependent on body size, shape, and flight mode. They seem, however, always to be selected for the optimal power consumption of the respective flight modes, except in a case of emergency.

A typical difference in beating motion between birds and insects is observed in the way they use the aerodynamic forces, lift and drag. Birds rely entirely on lift because the Reynolds number of their wings is high enough, whereas insects use drag as well as lift, which, because of the low Reynolds number and high frequency beating of low aspect ratio wings, includes the unsteady effects of these forces.

4.1 Powered Flight

The mechanics of natural powered flight is a topic addressed by many famous mechanical and aero-nautical investigators such as Leonardo da Vinci, Sir George Cayley, and Otto Lilienthal in their research into aeronautics. Following are very useful studies of the elemental mechanics of bird and insect flights: Greenewalt (1962, 1975), Tucker (1970, 1971, 1987), Pennycuick (1972), Weis-Fogh (1975a), Lighthill (1975a,b), Rüppell (1977), Nachtigall (1976, 1980b) and Ellington (1984a,b).

It is a well-known fact that larger flying creatures fly principally by gliding or slow beating, whereas smaller ones fly by strong beating at high frequency. Thus, the range of beating frequency and Reynolds number varies greatly.

The power required for beating flight can be analyzed simply as an extension of the analysis of gliding flight, without considering any detailed mechanism of wing motion. Here, the momentum theory can be applied to obtain the aerodynamic force and the related power, because the theory is simple but effective and does not require detailed knowledge of the pressure distribution on the wings. Then, by taking the power (or energy) balance between the necessary power and the available power derived from the muscles of locomotion, the performance of the flight can be roughly but quickly estimated.

4.1.1 NECESSARY POWER

The mass, the beating frequency, and the Reynolds numbers of flying creatures are distributed over a wide spectrum in creatures ranging from small insects to large birds, (see Table 4.1-1). The power required for steady flight, namely, the "necessary power," also varies widely and can be explained as follows:

Power for forward flight. In forward flight, the air must be accelerated rearward and downward to

77

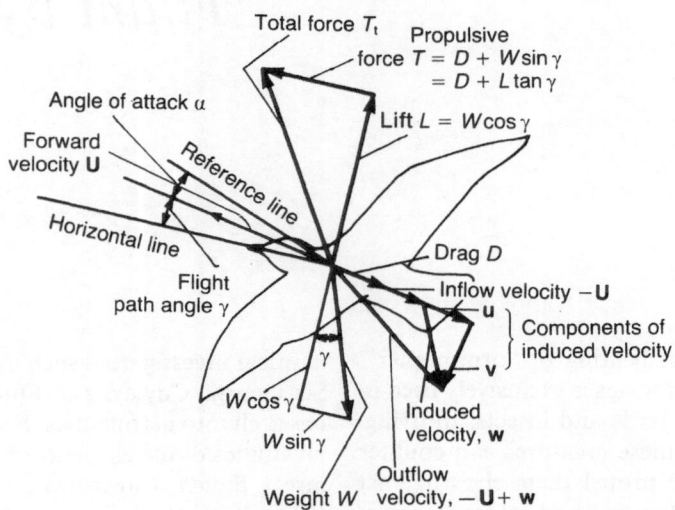

TABLE 4.1-1. Variety of beating frequency and Reynolds number of living creatures

FIG. 4.1-1. Balance in forces and flow velocities in steady climbing flight.

Species	Mass m (kg)	Beating frequency f (Hz)	Reynolds number Re
Chalcid wasp (*Encarsia formosa*)	2.5×10^{-8}	370	2×10^1
Fruit fly (*Drosophila virilis*)	2×10^{-6}	240	2×10^2
Bumblebee (*Bombus terrestris*)	8.8×10^{-4}	156	4×10^3
Hummingbird (*Patagona gigas*)	2×10^{-2}	15	1.5×10^4
Sparrow	3×10^{-2}	13	1×10^5
Pigeon	3.5×10^{-1}	6	2×10^5
Stork	3.5	2	4×10^5

sustain a thrust component balanced by the drag force and a lift component balanced by the creature's weight. As dictated by Newton's law and illustrated in Fig. 4.1-1, the momentum change related to such motion of the air generates a driving force acting on the creature as a reaction force. By assuming that the inflow velocity is constant across the span and throughout the flapping cycle, the driving force $\mathbf{F}$ is given by the product of the mass of air related to the wing motion m and the induced flow velocity in the fully developed state $2\mathbf{w}$ as:

$$\mathbf{F} = -2m\mathbf{w} \qquad (4.1\text{-}1)$$

where the "induced velocity" $\mathbf{w}$ is composed of a parallel flow component $\mathbf{u}$ and a normal flow component $\mathbf{v}$ with respect to the forward velocity $\mathbf{U}$, or

$$\mathbf{w} = \mathbf{u} + \mathbf{v} \qquad (4.1\text{-}2)$$

and the mass is, for an elliptic wing of span b to which the spanwise induced velocity distribution is constant, defined by

$$m = \rho(\pi b^2/4)U \qquad (4.1\text{-}3)$$

where U is the flight speed or $U = |\mathbf{U}|$.

As stated previously, the lifting force L trimmed with the weight component $W\cos\gamma$ in forward flight can be determined from

$$W\cos\gamma = L = 2mv \qquad (4.1\text{-}4)$$

where v is the downward component of the induced velocity, or $v = |\mathbf{v}|$. Then, the induced velocity v and the "induced power" $P_{i,L}$ required to support the weight in air can respectively be given by

$$v = 2L/\rho\pi b^2 \sqrt{(U\cos\alpha)^2 + (-U\sin\alpha + v)^2}$$

$$\simeq UC_L/\pi\!R \qquad (4.1\text{-}5)$$

$$P_{i,L} = Lv = 2L^2/\rho\pi b^2 U = 2(W\cos\gamma)^2/\rho\pi b^2 U$$

$$\simeq 2W^2/\rho\pi b^2 U \qquad (4.1\text{-}6)$$

If the wing is not elliptical, then, by introducing a deficiency parameter e, the induced power may be formulated as

$$P_{i,L} = 2L^2/\rho\pi b^2 Ue = (\tfrac{1}{2})\rho U^3 S(C_L{}^2/\pi \mathcal{R}_e)$$
$$= W\cos\gamma\, U(C_L/\pi\mathcal{R}_e) \qquad (4.1\text{-}7)$$

where the deficiency parameter e is less than one.

Similarly, the propulsive force T, which is trimmed with the aerodynamic drag D and the backward component of the weight $W\sin\gamma$, can be expressed as

$$T = D + W\sin\gamma = D + L\tan\gamma = 2mu \qquad (4.1\text{-}8)$$

where u is the backward component of the induced velocity or $u = |\mathbf{u}|$, and D is the total drag, including the induced drag for generating the lift against the weight, the profile drag of the wings, as well as the drag of other body parts. If, as seen in the swimming of some fish like the shark, propulsion is assumed to be performed by a separate propulsive system other than the lifting system, then the induced power related to the propulsive force T can be expressed by

$$\left.\begin{aligned}
P_{i,T} = Tu &= 2T^2/\rho\pi b^2 Ue \\
&= 2(D + L\tan\gamma)^2/\rho\pi b^2 Ue \\
&= \tfrac{1}{2}\rho U^3 S\{C_L{}^2\tan^2\gamma/\pi\mathcal{R}_e + C_D{}^2/\pi\mathcal{R}_e \\
&\quad + 2C_L C_D\tan\gamma/\pi\mathcal{R}_e\} \\
&= WU(C_L\tan\gamma\sin\gamma/\pi\mathcal{R}_e) \\
&\quad + DU(C_D/\pi\mathcal{R}_e) + 2WU(C_D\sin\gamma/\pi\mathcal{R}_e)
\end{aligned}\right\}$$
$$(4.1\text{-}9)$$

It must, however, be mentioned here that in flying birds and insects the propulsive force is generated by the wings, which also generate the lifting force by the beating motion. Then the induced power must be written as

$$P_i = T_t w \qquad (4.1\text{-}10)$$

where $T_t = |\mathbf{F}|$ is the total force given by vectorial summation of the lifting and propulsive forces, or

$$\begin{aligned}
T_t &= \sqrt{(W\cos\gamma)^2 + (W\sin\gamma + D)^2} \\
&= W\sqrt{1 + (C_D/C_L)^2\cos^2\gamma + 2(C_D/C_L)\cos\gamma\sin\gamma}
\end{aligned}$$
$$(4.1\text{-}11)$$

and $w = |\mathbf{w}|$ is the induced velocity thereof, or

$$w \simeq T_t/2\rho(\pi b^2/4)Ue = 2T_t/\rho\pi b^2 Ue \qquad (4.1\text{-}12)$$

Thus, the induced power can be given again by

$$\left.\begin{aligned}
P_i &= 2W^2\{1 + (C_D/C_L)^2\cos^2\gamma \\
&\quad + 2(C_D/C_L)\cos\gamma\sin\gamma\}/\rho\pi b^2 Ue \\
&= \tfrac{1}{2}\rho U^3 S(C_L/\cos\gamma)^2\{1 + C_D/C_L)^2\cos^2\gamma \\
&\quad + 2(C_D/C_L)\cos\gamma\sin\gamma\}/\pi\mathcal{R}_e \\
&= P_{i,L} + P_{i,T}
\end{aligned}\right\}$$
$$(4.1\text{-}13)$$

The above equation shows that (1) the induced powers are additive in the present approximation analysis, and (2) the induced power can be reduced by increasing the aspect ratio of the wing.

Power for hovering flight. When an animal is executing a hovering flight, the air is accelerated downward. The mass of air required to generate the lifting force is

$$m = \rho S_e v \qquad (4.1\text{-}14)$$

where S_e is the effective operational area or sweeping area of both wings. Usually the sweeping area can be considered to be about 70% of the circular disc swept by both wings. Since the lifting force balances with the weight,

$$W = 2mv = 2\rho S_e v^2 \qquad (4.1\text{-}15)$$

the induced velocity and the induced power in hovering flight can be given respectively by

$$v = \sqrt{W/2\rho S_e} = \sqrt{(W/S)/2\rho(S_e/S)} \qquad (4.1\text{-}16)$$
$$P_i = W\sqrt{W/2\rho S_e} = W\sqrt{(W/S)/2\rho(S_e/S)} \qquad (4.1\text{-}17)$$

In powered flight, the bird needs other powers for driving the wing against the profile drag and for the feathering motion. The former will be discussed later. The power required for the feathering motion of the wing, is usually small and thus negligible in comparison with that of the flapping motion.

Fig. 4.1-2 shows the ripples generated on a water surface by the strong induced velocity from a hovering wagtail (*Motacilla grandis*) and from a hovering SA-341 helicopter. It can be seen that in order to sustain the body in the air, the beating or rotary wings must push the air strongly downward to obtain lift as a reaction force.

Shown in Fig. 4.1-3 is the induced velocity versus the disc loading, $W/\tfrac{1}{4}\pi b^2 = W/S$. The induced velocity ranges from about 0.5 m/s for insects, to 8 m/s for birds, and it can be seen that large or heavy birds use strong induced velocity to perform hovering flight. The mass of airflow induced by the beating motion of a hovering bird

a

b

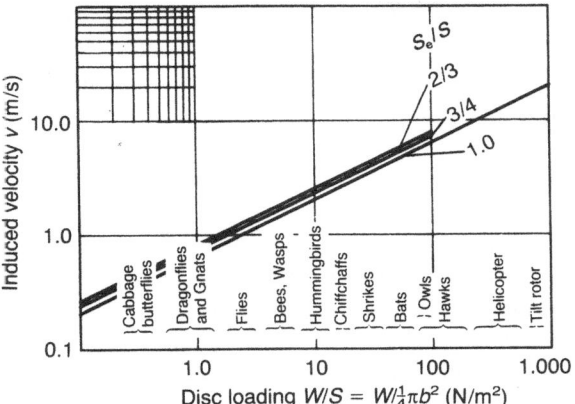

FIG. 4.1-3. Induced velocity versus disc loading ($\rho_0 \sigma =$ 1.225 kg/m³; at sea level, 15°C, 1 Pa = 1 N/m² = 0.102 kgf/m² = 0.0209 lb/ft²).

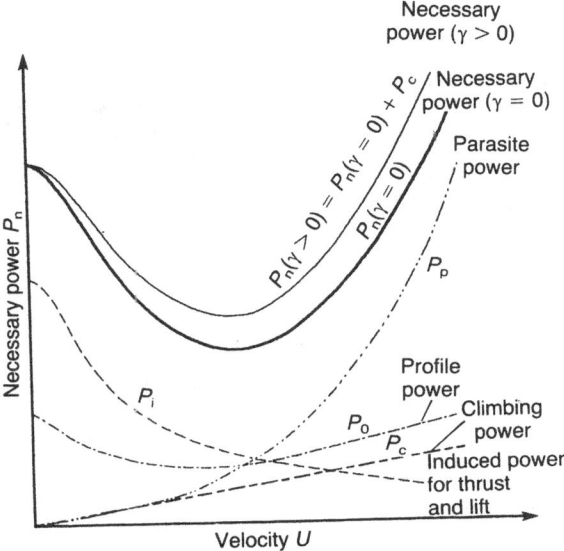

FIG. 4.1-4. Composition of necessary power.

is much smaller than in forward flight. Thus, the induced velocity has to be large.

Total aerodynamic power. Since the propulsive force T drives the bird against the drag D and a parallel component of the weight along the flight path $W \sin \gamma$, consumptions of parasite power P_p and climbing power P_c are required. These are given respectively by

$$P_\mathrm{p} = DU = \tfrac{1}{2}\rho U^3 S C_{D_0} = \tfrac{1}{2}\rho U^3 S (C_{D,\mathrm{b}} + C_{D,\mathrm{w}})$$
$$= P_{\mathrm{p,b}} + P_{\mathrm{p,w}}$$
$$= W \cos \gamma\, U (C_{D_0}/C_L)$$
$$= \sqrt{2/\rho S}\, C_{D_0} \{W \cos \gamma/C_L\}^{3/2}$$

$$\hspace{10cm}(4.1\text{-}18)$$

$$P_\mathrm{c} = WU \sin \gamma \qquad (4.1\text{-}19)$$

where C_{D_0} is the parasite drag coefficient including the profile drag of both wings, or $C_{D_0} = f/S + \delta$.

The total aerodynamic power or necessary power P_n is given by a summation of (1) the profile power P_0, which is required to beat the wing against the profile drag of the wing and is described later, in Sect 4.2.5, (2) the induced power ($P_{i,L} + P_{i,T} = P_i$) which accompanies the generation of lift and propulsive force (or thrust), (3) the parasite power P_p which consists of the parasite

power of parts of the body other than the wing $P_{\mathrm{p,b}}$ and that of the wing $P_{\mathrm{p,w}}$ and (4) the climbing power P_c, which is required to drive the body against the weight:

$$P_\mathrm{n} = P_0 + P_{i,L} + P_{i,T} + P_\mathrm{p} + P_\mathrm{c}$$
$$= P_0 + \tfrac{1}{2}\rho_0 \sigma U^3 [C_{D_0} + (C_L/\cos\gamma)^2$$
$$\cdot \{1 + (C_D/C_L)^2 \cos^2\gamma$$
$$+ 2(C_D/C_L)\cos\gamma\sin\gamma\}/\pi \mathcal{R}_e + C_L \tan\gamma]$$
$$\simeq P_0 + \tfrac{1}{2}\rho_0 \sigma U^3 S [C_{D_0} + (C_L{}^2 + C_D{}^2)/\pi \mathcal{R}_e$$
$$+ \{C_L + 2(C_D/C_L)/\pi \mathcal{R}_e\}\gamma] : \text{for small } \gamma$$

$$\hspace{10cm}(4.1\text{-}20)$$

Figure 4.1-4 shows the above individual power components and the total aerodynamic power for a given configuration and beating condition of a pair of wings. The necessary power shown by the thick curve has a concave or U-shaped valley at the point where the minimum power can be attained. Actually, in addition to the above aerodynamic powers, the inertial power P_I is required for beating against the inertial force of the wing if it is not eliminated by the elastic force. However, this power can be neglected. More specifically, while the inertial work has to be done to accelerate the wing at the beginning of the downstroke, at the end of the downstroke the downward motion can be stopped and reversed by the aerodynamic lift. For most birds in very slow flight, the angular

$\triangleleft$ ———————————————————————

FIG. 4.1-2. **a, b.** Ripples generated by a hovering bird and a hovering helicopter. **a** Wagtail (*Motacilla grandis*). (Courtesy of Hiroshi Uchida 1983). **b** SA341 (Gazelle). (Courtesy of Aerospatiale).

velocity remains almost constant until nearly the end of the downstroke (R.H.J. Brown 1948, 1953). Through the initial acceleration and the downstroke, the primary feathers are bent upward, and a part of the energy of the downstroke is stored as elastic energy in the primary feathers. At the end of the downstroke this stored energy is released to straighten the feather and to prevent reduction of the lift at the tip of the wing. This assists in stopping the downward motion of the wing. The resilience of other components in the beating system, such as the muscles (Maruyama 1984), helps in reducing considerably the inertial power. A detailed explanation will be given later in Sect. 4.4.4.

Figure 4.1-4 and Eq. (4.1-20) also show that (1) the profile power is almost constant and independent of speed in the usual range of flight speed, (2) the induced power attains a maximum value at hovering flight and then decreases as the speed increases, (3) the parasite power increases with the cube of the forward speed and is proportional to the drag area of the body, and (4) the total U-shaped power curve increases (or decreases) for ascending (or descending) flight. The final statement (4) has been actually observed from oxygen consumption in budgerigars as demonstrated by Tucker (1968).

Mechanical efficiency. In cruising flight ($\gamma = 0$), the ratio of useful work performed by the flight, $P_\mathrm{p} = DU = WU(D/W) = WU(C_D/C_L)$, to the necessary power P_n, or $P_\mathrm{p}/P_\mathrm{n}$ is termed mechanical efficiency η or Froude efficiency (Massey 1979) and can be derived from Eq. (4.1-20):

$$\eta = P_\mathrm{p}/P_\mathrm{n} = DU/P_\mathrm{n}$$
$$= (C_D/C_L)/\{(P_0/WU) + C_L/\pi\mathcal{R}_e$$
$$+ (C_D/\pi\mathcal{R}_e)(C_D/C_L) + (C_{D_0}/C_L)\} \quad (4.1\text{-}21)$$

In the case of a maximum lift-to-drag ratio in which $C_L = \sqrt{\pi\mathcal{R}_e C_{D_0}}$ and $C_D = 2C_{D_0}$ (see Sect 3.2.3), this expression can be rewritten as follows:

$$\eta \simeq 1 - 2(C_{D_0}/\pi\mathcal{R}_e) - (P_0/WU)/2\sqrt{C_{D_0}/\pi\mathcal{R}_e}$$
$$(4.1\text{-}22)$$

where $(P_0/WU)/2\sqrt{C_{D_0}/\pi\mathcal{R}_e}$ may usually be approximated at 0.1–0.2 to take into consideration miscellaneous loss of power. The above equation states that the mechanical efficiency in cruising flight is large for high-performance birds or for a small value of $C_{D_0}/\mathcal{R}_e$.

Cost of transport. The power (or energy) required to carry a unit weight at a unit velocity over a unit distance is called "cost of transport" and can be computed as follows:

$$C = P_\mathrm{n}/WU = \{(P_0/WU) + C_L/\pi\mathcal{R}_e$$
$$+ (C_D/\pi\mathcal{R}_e)(C_D/C_L) + (C_{D_0}/C_L)\} \quad (4.1\text{-}23)$$

In the case of a maximum lift-to-drag ratio, the above equation becomes:

$$C = 2\sqrt{C_{D_0}/\pi\mathcal{R}_e}\{1 + 2(C_{D_0}/\pi\mathcal{R}_e)$$
$$+ (P_0/WU)/2\sqrt{C_{D_0}/\pi\mathcal{R}_e}\}$$
$$\simeq 1/\eta(C_L/C_D) \quad (4.1\text{-}24)$$

For a given weight of the bird, the speed at which the minimum cost of transport is realized is the same as the speed at which the power consumption is minimum.

Shown in Fig. 4.1-5 are the values of the minimum cost of transport versus the body mass for various species of birds, as estimated by Rayner (1979a) It can be seen that the cost of transport at the speed of the maximum lift-to-drag ratio

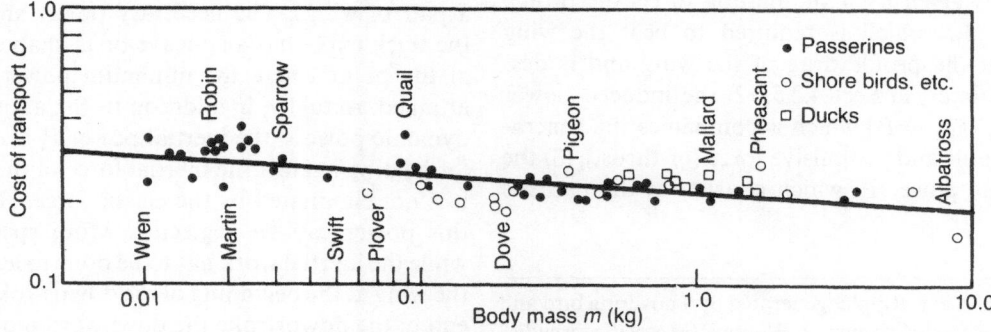

Fig. 4.1-5. The estimated cost of transport at $U_{(L/D)_\mathrm{max}}$. (Rayner 1979a with permission of the Company of Biologists Ltd.).

decreases somewhat as the body mass increases, because the optimal lift-to-drag ratio increases further for larger birds having higher aspect ratio wings.

4.1.2 AVAILABLE POWER

All flying creatures are powered by an aerobic engine that can deliver useful power or "available power" P_a up to the order of 10^1–10^2 watts per kg of muscle, irrespective of the size of the animal. Fig. 4.1-6 shows the mass of the flight muscles of insects and birds versus their total mass, as obtained from investigations by Hartman (1961) and Greenewalt (1962). It can be seen that the flight muscles of many insects and birds, specifically the large pectoral muscle, account for up to 10%–30% of the total mass, whereas small birds like hummingbirds, pigeons, and swallows, which are remarkably good at flying and taking off, have powerful flight musculature reaching up to about 40% of total mass.

Specific power. Weis-Fogh (1975a, 1977) reports that the specific available power of a flying animal, $K = P_a/m_m$, which is the available power generated per unit mass of muscle, varies over a wide range, as shown in Table 4.1-2. The level of muscle power is almost independent of absolute size and systematic group, and is usually considered in the range of 50–260 W/kg (or 0.07–0.35 HP/kg). However, Gallinaceae, which are small-winged and poor cruising birds, can achieve swift escape but are severely limited in their flight range because their flight musculature is composed largely of white fibers and is capable of an immediate output of appreciable quantities of power with a consequent oxygen debt (Kokshaysky 1977). The specific powers of man-made engines are also presented in Table 4.1-2 for comparison.

Metabolic rate. The available power is equal to the power converted from fuel, minus that lost in the maintainance of the drive system. Such internally consumed power during locomotion includes the internal metabolic rate for the maintenance of metabolism, power for the circulation of blood, and power for the ventilation of the respiratory system (Tucker 1973, 1975, Goldspink 1977b). The available power P_a can thus be determined from the "total power" input or "metabolic rate" P_t which can be evaluated by measuring oxygen consumption, minus the above wasted powers lumped together under the term "basal metabolic rate" P_B:

TABLE 4.1-2. Mechanical power output of wing muscle P_a/m_m during continuous flight. (Data from Weis-Fogh 1977)

Items	Specific available power $K = P_a/m_m$ (W/kg)	Mass m_m (mg)
Bats: Chiroptera		
Pteropus gouldii	780	140
Phyllostomus hastatus	93	240
Birds: Aves		
Columba livia	400	220
Larus atricilla	322	70
Corvus ossifragus	275	80
Melopsittacus undulatus	35	170
Melopsittacus undulatus	35	95
Amazilia fimbriata	5	150
Insects: Insecta		
Schistocerca gregaria	2	60–90
Drosophila virilis	2×10^{-3}	160
Large range of insects	1×10^{-3} to 13	range: 120–360 average: 140
Reciprocating engine	1,500–1,800	—
Gas turbine	3,700–7,400	—

$$P_a = (P_t - P_B)\eta_E = P_{ab} \cdot \eta_E \qquad (4.1\text{-}25)$$

where η_E is the mechanical efficiency of the flight muscles, and $P_{ab} = P_t - P_B$ is the "aerobic metabolic power." Efficiency can be expressed in many ways: η_E is identical to the "aerobic efficiency" described by Brett (1963) and Webb (1971b); and it is probably equivalent to the "contraction-coupling efficiency" of Whipp and Wasserman (1969) for human muscles. Hill (1950) gave the maximum value of $\eta_E = 0.45$, and Tucker (1977) suggested setting the value of $\eta_E = 0.2$.

Thus, the total power can be rewritten in two ways:

$$P_t = P_a/\{1 - (P_B P_t)\}\eta_E = P_a/\eta_F = P_{ab}/(\eta_F/\eta_E) \qquad (4.1\text{-}26a)$$

$$= P_a/\eta_E + P_B \qquad (4.1\text{-}26b)$$

where η_F is the efficiency of the power-producing system or the ratio of power output to power input, and is given by

$$\eta_F = P_a/P_t = \{1 - (P_B/P_t)\}\eta_E \qquad (4.1\text{-}27)$$

It has been evaluated by Tucker (1970, 1972, 1973, 1974), Bernstein et al. (1973), and Pennycuick (1972, 1975) as

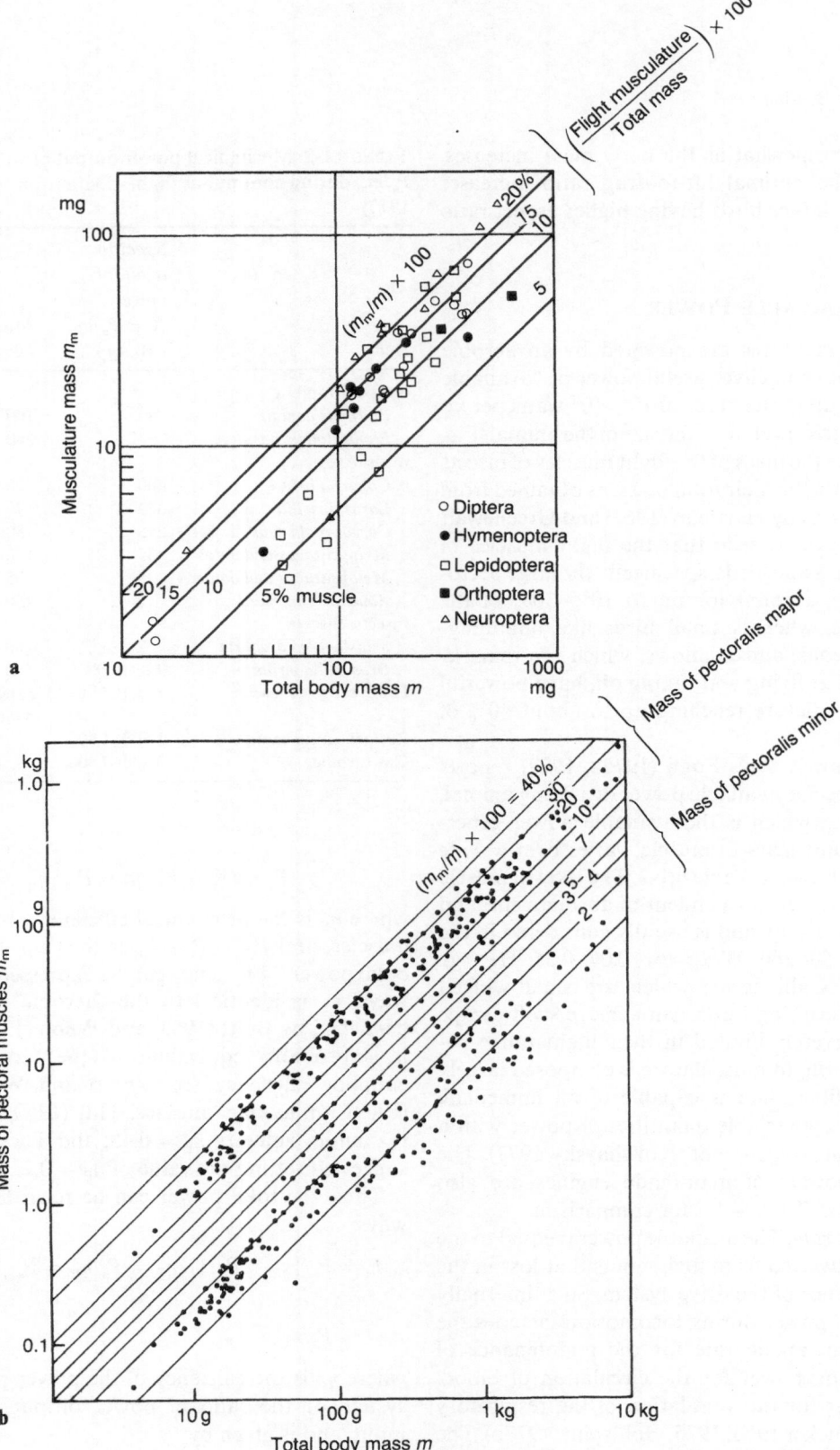

FIG. 4.1-6. **a, b.** Mass of flight muscles. (Redrawn from Greenewalt 1962). **a** Insects. **b** Birds.

$$\eta_F = 0.20-0.26 \simeq 0.23 \qquad (4.1\text{-}28)$$

In the case of human power for rowing, the efficiency η_F is appoximated at 0.18–0.23 (Prampero et al. 1971).

The mean basal metabolic rate P_B can be considered in two ways:

1. It is limited by the body surface area, because oxygen comes in and heat goes out through surfaces that are proportional to the total body surface area (Rubner 1883). By assuming that the body is governed by elastic similarity and by introducing the relationships given by Eq. (1.2-15c) in Sect. 1.2.2, the surface area S_b thus the basal metabolic rate are proportional to the product of body length and lateral dimension or body diameter ld, or

$$P_B \propto S_b \propto ld \propto m^{5/8} = m^{0.63} \quad (4.1\text{-}29)$$

2. Alternatively, if the frontal area of the body, rather than the body-surface area, provides the limitation (McMahon and Bonner 1983), then the frontal area S_f and thus the basal metabolic rate P_B are proportional to d^2 or

$$P_B \propto S_f \propto d^2 \propto m^{3/4} = m^{0.75} \quad (4.1\text{-}30)$$

Actually, however, from Brody (1945), Lasiewski and Dawson (1967), Tucker (1972, 1973, 1975), and Schmidt-Nielsen (1975, 1977), P_B can be written statistically as follows:

$$P_B = \begin{cases} (3.73-3.79)m^{0.723}: \text{ for nonpasserine} \\ \qquad\qquad\qquad \text{birds} \\ (6.15-6.25)m^{0.724}: \text{ for passerine birds} \\ (3.00-4.30)m^{0.73-0.75}: \text{ for mammals} \end{cases}$$
$$(4.1\text{-}31)$$

where coefficients increase with the body temperature (Wilkie 1977).

Wilkie (1959) gave the total power output P_t of animals as a function of the body mass m as follows:

$$\begin{aligned} P_t &= 3.43m^{0.734} = P_B \quad \text{(at rest)} \\ &= 5.52m^{0.734} \qquad \text{(maintained all day)} \\ &= 45.7m^{0.734} \qquad \text{(maintained 5–30 min)} \\ &= 84.3m^{0.734} \qquad \text{(brief effort or maintained} \\ &\qquad\qquad\qquad\quad \text{a few seconds)} \end{aligned}$$
$$(4.1\text{-}32)$$

in which the number of power of mass, 0.734, may be replaced with 0.75.

Muscular force and power. Alexander (1973) revealed that contracting muscles can develop a maximum stress of 200–400 kN/m^2, and the maximum power output occurs when they shorten against one third of this stress, i.e., against 70–130 kN/m^2. This means that the force F generated by the contraction of a striated muscle of a vertebrate is determined by the cross-sectional area of the muscle S_c. Since the stroke of locomotion is considered proportional to the length of the muscle l, the contracting speed of the muscle $\dot{l}$ is proportional to the length l times the stroke frequency f. The available power $P_a = F\dot{l}$ is proportional to the product of the cross-sectional area S_c and the length l of the muscle, and to the mass of the muscle times the frequency f, or $P_a \propto S_c lf \propto m^{7/8}f \simeq mf$. Thus, the available power P_a can be written as

$$P_a = K(m_m/m)m \qquad (4.1\text{-}33a)$$

or the available specific power, which is the available power per unit weight, can be given by

$$P_a/W = (K/g)(m_m/m) \qquad (4.1\text{-}33b)$$

where the magnitude of proportional constant K is $[K] = $ W/kg and the mass ratio m_m/m may be considered essentially free from size-dependent variations in view of biological similarity.

Respiratory system. The avian respiratory system has, as a distinctive feature, a number of air sacs that are connected to the bronchial passages and the lungs. Most of the air inhaled on a given breath goes directly into the posterior air sacs. As the respiratory cycle continues, the air passes through the lungs and into the anterior air sacs, from which it is exhaled to the outside in the next cycle (Schmidt-Nielsen 1971). The most essential feature is a single stream or unidirectional flow of air through the lung in the anterior direction on both inspiration and expiration (Schmidt-Nielsen 1975; Kuethe 1988). In other words, the air sacs act as bellows, with the help of vanes of "fluidics," switching devices utilizing the "Coanda effect," [the tendency of a jet of fluid to attach itself to a convex solid body (Massey 1979)]. The vanes assist in passing the air through the lung unidirectionally by not only pushing the air in the expiration phase but also sucking it in the inspiration phase.

This respiratory system can deliver enough oxygen for the birds to fly at high altitude. It is also

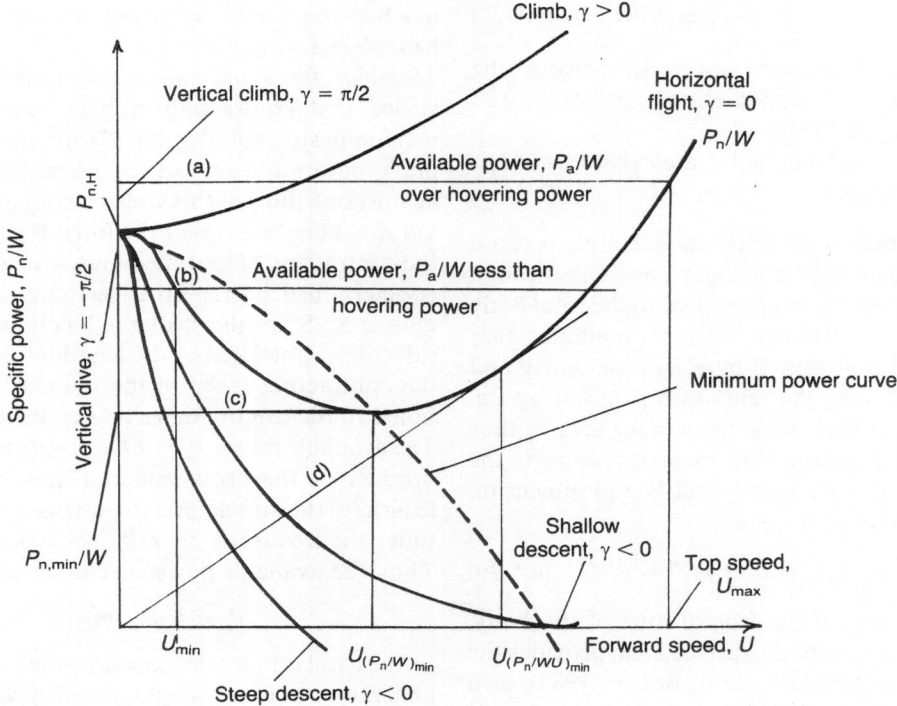

FIG. 4.1-7. Power balance for steady flights.

interesting to note that in pigeons, the blood perfusing the evaporative surfaces not only cools arterial blood flowing counter-current to it toward the brain, but also exchanges oxygen and carbon dioxide with air, thus maintaining the brain's oxygen supply during the high systematic demand of exercise and at the reduced oxygen availability of high altitude (Bernstein et al. 1984).

4.1.3 POWER BALANCE

Forward flight. From Eq. (4.1-20), the specific power or the power-weight ratio for horizontal flight ($\gamma = 0$) can be written as

$$
\begin{aligned}
P_n/W = P_0/W &+ \left\{\frac{C_{D_0}{}^2}{\pi \mathcal{R}_e} + C_{D_0}\right\} \frac{\rho/2}{W/S} U^3 \\
&+ \frac{1}{\pi \mathcal{R}_e}\left\{1 + \frac{2C_{D_0}}{\pi \mathcal{R}_e}\right\} \frac{W/S}{\rho/2} \frac{1}{U} \\
&+ \frac{1}{(\pi \mathcal{R}_e)^3} \frac{(W/S)^3}{(\rho/2)^3} \frac{1}{U^5}
\end{aligned}
$$

$$(4.1\text{-}34)$$

where the profile power P_0 is assumed constant for the change of forward speed.

Shown in Fig. 4.1-7 is the necessary specific power versus flight speed. The minimum specific

power-over-speed, which is simply given by an intersection of a straight line through the origin and tangent to the P_n/W curve, as shown by (d) in Fig. 4.1-7, can be derived from

$$(\partial/\partial U)\{(P_n/W)/U\} = 0 \qquad (4.1\text{-}35)$$

This also gives the flight of maximum range for a special case of constant weight during flight. The flight of minimum specific power, $P_{n,\,min}/W$ which is given by an intersection of a tangent-horizontal line as shown by (c) in Fig. 4.1-7, is obtained by solving the following equation:

$$(\partial/\partial U)(P_n/W) = 0 \qquad (4.1\text{-}36)$$

This gives the flight of maximum endurance for a special case of constant weight.

Shown in Fig. 4.1-8 are statistically observed data for various birds indicating (a) the speed for the probable maximum range and (b) the speed for minimum power. It is interesting to note from Fig. 4.1-8 that (1) the lift coefficients are scattered in a range of very low values of $C_L \lesssim 0.5$, specifically for the fast flyers (the actual data agree with the theory shown by solid lines in general tendency, but are not always reliable because of wind effect)

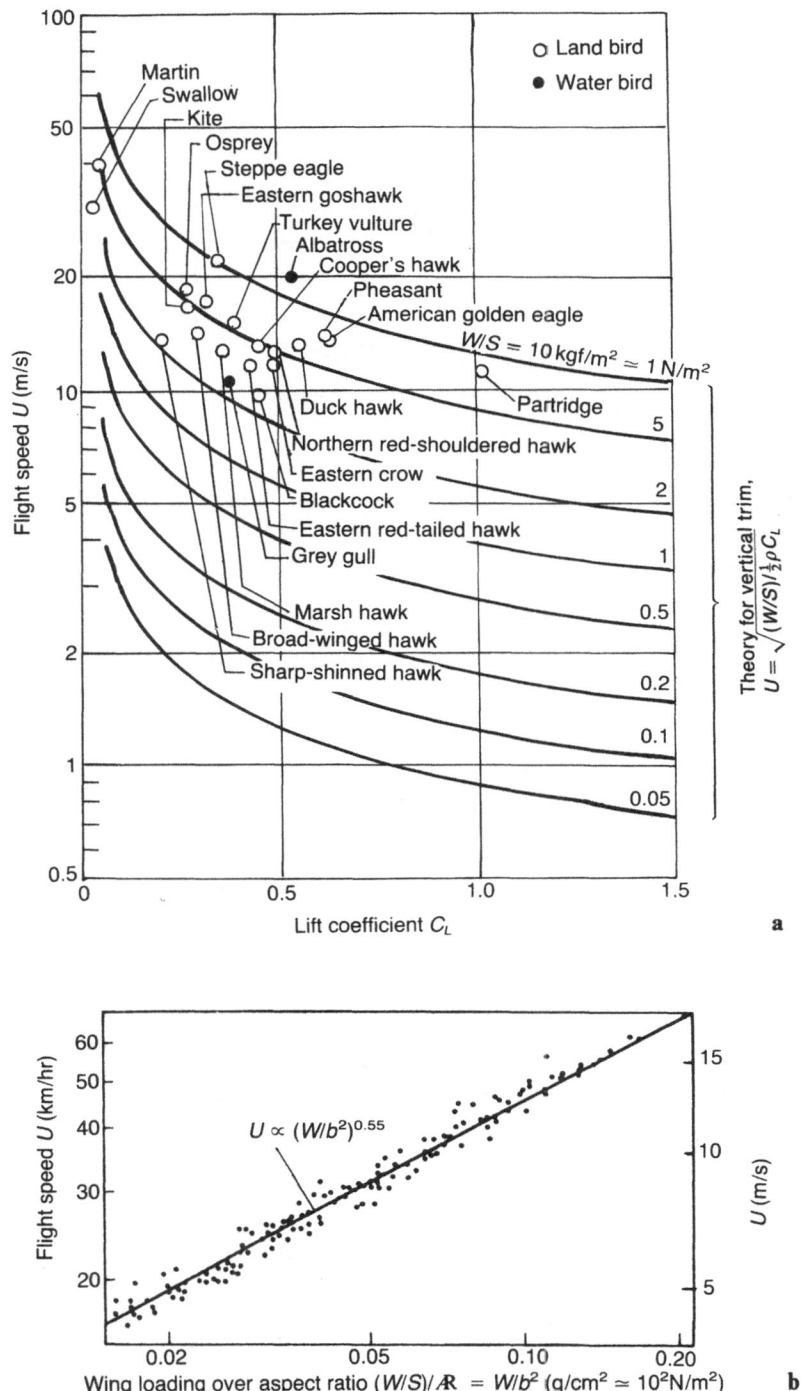

FIG. 4.1-8. **a, b.** Flight speed for probable maximum range. (Data from Pennycuick 1968b; Greenewalt 1962, 1975). **a** Speed versus lift coefficient. **b** Speed versus wing loading over aspect ratio for minimum power. (Redrawn from McMahon and Bonner 1983).

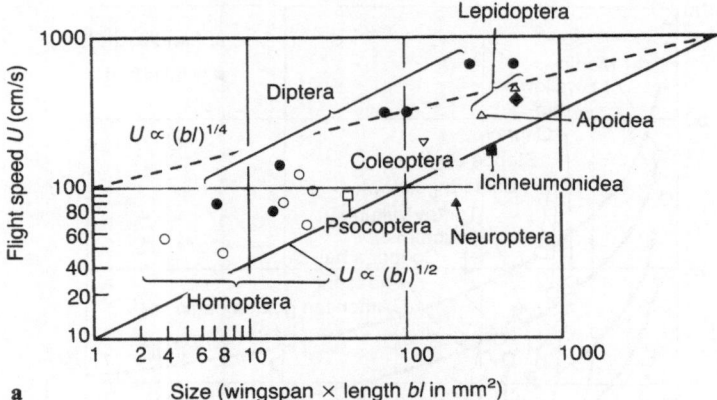

a

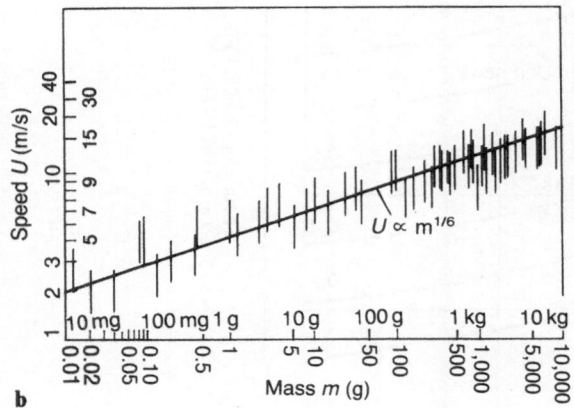

b

FIG. 4.1-9. **a, b.** Airspeed versus size for insects and mass for birds. **a** Insects. (Redrawn from Lewis and Taylor 1967 with permission; Lighthill 1977). **b** Birds. (Redrawn from Pennycuick 1969).

and (2) the flight speed for minimum power is well correlated with $(W/S)/Æ$. However, as given by equation in Table 3.2-1, the flight speed is theoretically related to $(W/S)/\sqrt{Æ}$. Rearrangement is desirable.

Shown in Fig. 4.1-9 are statistical data on the speed-size and speed-mass relationships in insects and birds respectively. Although a wide scatter is observed, a general statistical tendency can also be recognized. From the vertical force balance, the airspeed can be given by

$$U = \sqrt{\frac{(W/S)/Æ}{\frac{1}{2}\rho C_L/Æ}} \propto \sqrt{l} \propto m^{1/6} \quad (4.1\text{-}37)$$

Thus, it can be said that the airspeed is proportional to either $m^{1/6} \propto (S^{3/2})^{1/6} = S^{1/4}$ for mechanical similarity, or $m^{1/6} \propto S^{4/15} \simeq S^{1/4}$ for elastic similarity. In Fig. 4.1-9a, the broken line has a slope of $\frac{1}{4}$, while the solid line has a slope of $\frac{1}{2}$, for the logarithmic scales of U and $S = bl$. In either case, in order to achieve the maximum range, large animals must fly faster than small ones.

Hovering flight. For hovering flight, the specific necessary power, $P_{n,H}/W$, can be given by

$$\left.\begin{array}{l} P_{n,H}/W = P_0/W + Tv/W \simeq P_0/W + v \\ = P_0/W + \sqrt{(W/S)(S/S_e)/2\rho} = (P_{i,H}/W)/\eta_H \end{array}\right\} \quad (4.1\text{-}38)$$

where v and η_H are the induced velocity assumed to be constant in the stroke plane and the hovering efficiency respectively.

It can be seen from Eq. (4.1-38) that the specific power required to hover decreases with the disc loading, W/S, which is the ratio of the weight-to-disc area. Here, the nonhomogeneous induced velocity distribution does not come into consideration, so that η_H must be small ($\eta_H = 0.6$). It is believed that hovering is possible only for small birds with a mass of less than, say, 0.02 kg. As shown in Fig. 3.2-5, the wing loading of a bird with a mass of 0.02 kg is scattered between 1.5 and 30 N/m². Thus, the specific necessary power P_n/W is, assuming $\eta_H = 0.6$, in the range of 5–7 W/N (or 50–70 W/kgf), as given by Eq. (4.1-38). This requires a specific power of 20–27 W/N (or 200–280 W/kgf) for a 25% muscular weight ratio ($m_m/m = 0.25$). It can thus be said that the wing loading should be less than around 20 N/m² to allow continuous hovering flight.

Since the induced power can be obtained from Eq. (4.1-17), the total power P_t at a trimmed flight, $P_a = P_{n,H}$, is, from Eqs. (4.1-26) and (4.1-38), given by

$$P_t = P_i/\eta_H\eta_E + P_B \qquad (4.1\text{-}39)$$

Table 4.1-3 shows the flight data and the power components for hovering birds and insects presented by Weis-Fogh (1973) and Rayner (1979a).

It will be noted that the small insects have low disc loadings; thus, their metabolic rate is larger than the induced power, $P_B > P_i/\eta_H\eta_E$. On the other hand, the heavy birds such as the mallard and pigeon have high disc loading and have larger induced power than the metabolic rate, $P_i/\eta_H\eta_E > P_B$. Those in between have almost equal power components. In very small insects such as the chalcid wasp and the fruit fly, the power-mass ratio, which is the ratio between the power and the total mass of the body, is abnormally large. This probably results from the failure to take the drag component into consideration, and/or from an erroneous estimation of the coefficient of the metabolic rate.

Cruising flight. Trimmed flight speeds can be obtained by solving the following equation obtained by combining Eq. (4.1-33a) and (4.1-34):

$$
\begin{aligned}
(K/g)(m_m/m) = P_0/W &+ \{(2/\rho)(W/S/\pi \mathcal{R}_e)\}^3/U^5 \\
&+ \{(2/\rho)(W/S)/\pi \mathcal{R}_e\} \\
&\quad \cdot \{1 + 2C_{D_0}/\pi \mathcal{R}_e\}/U \\
&+ \{C_{D_0}/(2/\rho)(W/S)\} \\
&\quad \cdot \{1 + C_{D_0}/\pi \mathcal{R}_e\}U^3.
\end{aligned}
$$

$$(4.1\text{-}40)$$

The solutions are given in Fig. 4.1-7 by intersections of the specific necessary power P_n/W and the specific available power P_a/W, the latter of which is shown for: (a) $P_a/W > P_{n,H}/W$, and for (b) $P_{n,min}/W < P_a/W < P_{n,H}$,

1. If the specific available power of a bird is higher than the specific necessary power ($P_a/W > P_{n,H}/W$) it can perform not only hovering flight but also vertical takeoff and landing (VTOL). The intersection appearing in Fig. 4.1-7 gives the maximum speed in horizontal flight.
2. If the specific available power of the bird is some value between the hovering and the minimum powers, it must execute either a running takeoff from the ground (or water surface), like a conventional takeoff and landing (CTOL) aircraft, or a falling takeoff from a tree or perch until it obtains a velocity beyond which the necessary power is less than the available pow-

TABLE 4.1-3. Metabolic data and power components in hovering flight ($\eta_E = 0.9$, $\eta_H = 2/3$). (Data from Weis-Fogh 1973; Rayner 1979a)

	Body mass $m(kg)$	Body weight $W(N)$	Wing semi-span $\frac{1}{2}b(m)$	Disk loading $W/S(N/m^2)$	Stroke period $T(s) = 1/f$	Induced velocity $v(m/s)$	Induced power $P_i/\eta_H\eta_E(W)$	Metabolic rate $P_B(W)$	Total power $P_t(W)$	Power mass ratio $P_t/m(W/kg)$
Normal hovering										
Chalcid wasp, *Encarsia formosa*	2.5×10^{-8}	2.45×10^{-7}	7×10^{-4}	0.16	1/370	0.313	1.28×10^{-7}	2.18×10^{-5}	2.19×10^{-5}	875
Fruit fly, *Drosophila virilis*	2×10^{-6}	1.96×10^{-5}	0.003	0.69	1/240	0.650	2.12×10^{-5}	5.19×10^{-4}	5.40×10^{-4}	270
Crane fly, *Tipula paludosa*	2.8×10^{-5}	2.74×10^{-4}	0.0173	0.29	1/53	0.421	1.93×10^{-4}	3.51×10^{-3}	3.70×10^{-3}	132
Hover fly, *Eristalis tenax*	1.5×10^{-4}	1.47×10^{-3}	0.0127	2.90	1/182	1.54	3.26×10^{-3}	1.18×10^{-2}	1.51×10^{-2}	101
Bumblebee, *Bombus terrestris*	8.8×10^{-4}	8.62×10^{-3}	0.0173	9.18	1/156	2.41	3.41×10^{-2}	4.26×10^{-2}	7.67×10^{-2}	87.1
Moth, *Manduca sexta*	1.12×10^{-3}	1.10×10^{-2}	0.050	1.40	1/29.1	0.926	1.69×10^{-2}	5.07×10^{-2}	6.76×10^{-2}	60.4
Hummingbird, *Amazilia fimbriata*	5.1×10^{-3}	5.00×10^{-2}	0.059	4.57	1/35	1.67	0.139	0.152	0.291	57.1
Avian hovering										
Wren, *Troglodytes troglodytes*	0.01	0.098	0.085	4.32	0.084(5)	1.63	0.266	0.247	0.513	51.3
Pied flycatcher, *Ficedula hypoleuca*	0.012	0.118	0.115	2.83	0.07	1.32	0.260	0.282	0.540	45.0
Pigeon, *Columba livia*	0.033	3.26	0.316	10.41	0.15	2.52	13.7	3.13	16.9	50.6
Mallard, *Anas platyrhynchos*	1.105	10.83	0.450	17.04	0.15(4)	5.23	58.3	7.46	65.8	59.0
Long-eared bat, *Plecotus auritus*	9×10^{-3}	8.02×10^{-2}	0.115	2.13	0.08	1.14	0.168	0.229	0.397	44.1

er. Horizontal flight speed is limited within the two intersections between the available and necessary power curves. If, however, the bird takes off against a wind flowing at $U_{\min}$ or faster, it can, of course, become airborne at once without running.

3. If the specific available power is less than the specific minimum necessary power, the bird cannot maintain horizontal flight and will lose altitude even in beating flight.

TABLE 4.1-4. Optimal and maximum velocities. (Data from Houston 1986)

Items	Goldcrest	Starling	Pigeon
Body mass m (kg)	0.006	0.07	0.333
Wing span b (m)	0.147	0.3	0.632
Minimum power speed $U_{(P_n/W)_{\min}}$ (m/s)	4.91	7.83	9.08
Maximum range speed $U_{(P_n/WU)_{\min}}$ (m/s)	9.22	13.32	15.02
Maximum speed $U_{\max}$ (m/s)	14.3	17.4	21.0

Given in Table 4.1-4 are the optimal velocities, $U_{(P_n/W)_{\min}}$ and $U_{(P_n/WU)_{\min}}$, specified respectively by Eqs. (4.1-35 and 4.1-36), and the maximum speed calculated by Houston (1986) for the goldcrest (*Regulus regulus*), the starling (*Sturnus vulgaris*), and the pigeon (*Columba livia*).

Horizontal flight in wind. Regarding maximum range flight $R_{\max}$, if there is a head wind, the tangent line should be started at the head wind value, as shown in Fig. 4.1-10, so the airspeed for the minimum power-over-speed $U_{P/U_{\min}}$ or the best range $U_{R_{\max}}$ is slightly above the calm-air speed. However, since the slope of the tangent line is higher, the specific range is lower. Conversely, if there is a tail wind, the tangent line should be started at the tail wind value. This results in a slightly lower optimal airspeed but a significantly better specific range. Actually, birds and insects make good use of seasonal winds in their migrational flight (Williamson 1969; R.C. Rainey 1969).

4.1.4 ENERGY BALANCE

Wind tunnel tests of flying animals to estimate the mechanical power required for flight at various speeds were performed by Pennycuick (1968a,b) for avian flight and Weis-Fogh (1965) for insect flight. The most direct way to determine the energy

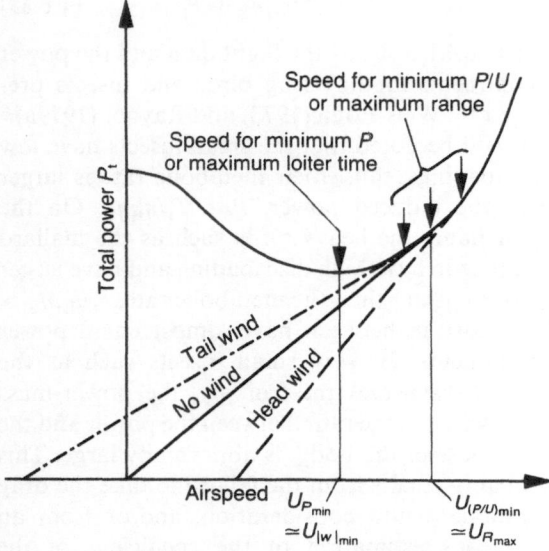

FIG. 4.1-10. Effect of wind on optimal cruising flight.

balance of a bird flying at a known speed is to measure its oxygen consumption while it is flying steadily in a wind tunnel.

Let us introduce another overall efficiency η_I that can replace the profile power P_0 and other metabolic powers by dividing the remaining powers, as in:

$$P_t = (P_p + P_{i,L})/\eta_I \qquad (4.1\text{-}41)$$

where η_I is defined as

$$\eta_I = \eta_F/\{1 + (P_0 + P_{i,T})/(P_p + P_{i,L})\} \simeq \eta_F \simeq 0.2 \qquad (4.1\text{-}42)$$

Then, by combining Eqs. (4.1-6 and 4.1-18) with Eq. (4.1-41), the flight range and endurance are approximated by

$$X = \int_0^t U\,dt = -(1/c)\int_{m_0}^m (U/P_t)\,dm$$
$$= (\eta_I/cg)(C_L/C_D)\ln(m_0/m) \qquad (4.1\text{-}43a)$$

$$E = -(1/c)\int_{m_0}^m dm/P_t$$
$$= 2(\eta_I/cg)\sqrt{\rho S/2W_0}(C_L^{3/2}/C_D)\{\sqrt{m_0/m} - 1\} \qquad (4.1\text{-}43b)$$

where c is the fuel consumption rate in the consumed mass per unit energy or $[c] = $ kg/J. The former is called the "Breguet range equation" (Perkins and Hage 1965). It is interesting to note

TABLE 4.1-5. Mass of isocaloric amounts of fuel stored by flying insects and birds. (Data from Weis-Fogh 1967; and Pennycuick 1972)

Fuel	Water content (%)	Mass per unit energy c (kg/J)
Depot fat	0	0.263×10^{-7}
Honey	20 (varies)	0.788×10^{-7}
Nectar	60 (varies)	1.601×10^{-7}
Glycogen	73	2.102×10^{-7}

that since the air density does not appear explicitly in the Breguet range equation, the range attainable is not affected by the altitude at which a bird flies, so long as the ratio of the overall efficiency to the fuel consumption, η_1/c, and the lift-to-drag ratio are constant. In practice, this conclusion is modified by the variation of η_F/c, C_L/C_D, and the effect of wind, which usually varies with altitude. Table 4.1-5 gives the mass of isocaloric amounts of fuel stored by flying insects and birds. Usually, since the specific consumption can be given by $c = 0.26 \times 10^{-7}$ kg/J for fat (Weis-Fogh 1967), the above ratio may be given by

$$\eta_1/c = 0.2/(0.26 \times 10^{-7}) \simeq 8 \times 10^6 \text{ J/kg} \quad (4.1\text{-}44)$$

By taking two typical lift-to-drag ratios of, say, $L/D = 6$ and 11, and a mass ratio of $m_0/m = 1.5$, equation (4.1-43) gives two examples of flight ranges, $X = 2,000$ km and $X = 3,600$ km, which

FIG. 4.1-11. Range of flight. (Redrawn from Tucker 1971, 1975).

respectively correspond to the distances involved in crossing the Sahara and the Aleutians-Hawaii overwater route (Moreau 1961; Pennycuick, 1972).

Shown in Fig. 4.1-11 is the predicted flight range of birds of various sizes, as presented by Tucker (1971, 1975). It can be seen that the above two examples give reasonable range estimations. In the present calculation of range, the lift-to-drag ratio and hence the aspect ratio are the key factors.

The distance traveled per unit loss of body mass, or fat consumed, which is equivalent to the distance traveled per unit loss of energy, may be called the specific range R_s and is, a simpler expression than that of Eq. (4.1-43), given by:

$$R_s = \frac{\partial X}{\partial (m_0 - m)} = -\frac{\partial X}{\partial m} = \eta_1(C_L/C_D)/Wc \,[\text{km/kg}]$$
$$(4.1\text{-}45)$$

The following facts can be noted: (1) a bird with a larger lift-to-drag ratio has a longer flight range, (2) a smaller bird has a larger value of specific range, and (3) the maximum range can be obtained at a speed of $U_{(L/D)_{min}}$, as given in Table 3.2-1.

4.1.5 BEATING FREQUENCY

The aerodynamically consumed power P_A is proportional to the product of the wing area and the cube of the tip speed, which is the wingspan times beating frequency, $P_A \propto S(bf)^3$. This is again proportional to the five-thirds power of the mass of the flying creature and to the cube of the beating frequency, or:

$$P_A \propto m^{5/3}f^3 \quad (4.1\text{-}46)$$

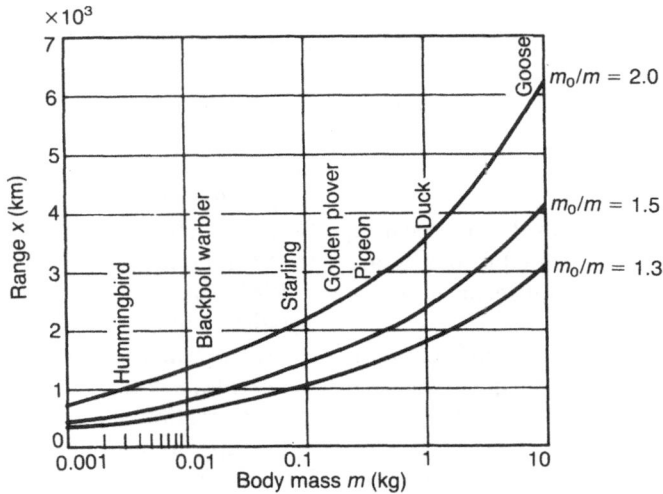

The physiologically consumed power for the inertial torque is given by the mean absolute value of the periodic change in kinetic energy and, therefore, is proportional to the product of the moment of inertia and the cube of the beating frequency. Since the moment of inertia is proportional to the fifth power of the linear dimension, i.e., either five-thirds power of the mass or square power of the mass as given by Eq. (3.1-1), the required power P_{in} for the inertial torque is also given by the following proportional relation:

$$P_{in} \propto m^{5/3} f^3 \sim m^2 f^3 \qquad (4.1\text{-}47)$$

As explained later in Sect. 4.4.4 the above inertial power can be reduced by introducing elastic constraint at the flapping hinge.

As stated before, the power generated by a muscle during one cycle of beating can be considered proportional to the product of the exponential of mass and the beating frequency:

$$P \propto (m^{2/3} \sim m) f \qquad (4.1\text{-}48)$$

By combining these proportional relations the frequency may be expressed as

$$f \propto m^{-1/3} \sim m^{-1/2} \qquad (4.1\text{-}49)$$

The first term coincides with the upper limit of the wingbeat frequency specified by Hill (1950).

However, the beating frequency is also considered to be constrained by the heat balance. Since the temperature of the body of any living creature is regulated by the surface area of the cooling system, the exothermic energy accompanying the beating motion must be limited, to keep the body temperature in the most efficient range of metabolic temperature. This imposes the constraint that the heat energy radiated from the surface area of the cooling system in one cycle of beating must be proportional to the consumed power in the beating motion, or $m^{2/3} f \propto m^{5/3} f^3$, which further yields:

$$f \propto m^{-1/2} \qquad (4.1\text{-}50)$$

The above heat balance requires that the beating frequency be inversely proportional to the square root of the body mass. This result seems to give a higher limit of relation (4.1-49).

Let us further extend the present discussion on the lower limit of wingbeat frequency. Since the thrust-and-drag equilibrium, $T = D$, leads to the relation $f \propto U/m^{1/3}$, and the lift-and-weight equilibrium, $L = W$, leads to the relation $U \propto m^{1/6}$, the frequency can be written as:

$$f \propto m^{-1/6} \qquad (4.1\text{-}51)$$

In other words, the beating frequency is inversely proportional to the square root of the linear dimension. This relation is obtained from the free balances without any consideration of power.

From the dimensional analysis, Pennycuick (1990) gave a relation among the beating frequency f, wing span b and the wing area S:

$$f = 1.08(m^{1/3} b^{-1} S^{-1/4} \rho^{-1/3} g^{1/2}) \qquad (4.1\text{-}52)$$

Then, by introducing the relation such that $b \propto m^{1/3}$ and $S \propto m^{2/3}$, the above equation supports the relation given by Eq. (4.1-51). In hovering flight, however, the lift-and-weight equilibrium gives the relation of $f \propto (m/S)^{1/2}/b^{1/2}$ which yields $f \propto m^{-1/6}$ for the geometrical similarity and $f \propto m^{-1/16}$ for the elastic similarity.

Greenewalt (1962) showed statistical data on wingbeat rate f versus wing length l_w for birds and insects, and gave the relation $f l_w^{1.15} = $ constant. In Fig. 4.1-12, data on the stroke frequency of the beating motion versus the mass of flying creatures are given. Even though the stroke frequency of insects is widely scattered, the mean slope seems to be an intermediate value between $m^{-1/2}$ given by relation (4.1-50) and $m^{-1/6}$ given by relation (4.1-51). A precise understanding of the situation will require a more exact investigation. Flying creatures, specifically small insects, that have a frequency higher than that given by $m^{-1/3}$ must find wing beating a strenuous activity because of insufficient power, whereas those, specifically large birds, that have a frequency higher than that given by $m^{-1/2}$ probably suffer from a temperature rise during wing beating. From the discussion in Sects. 1.2.1 and 1.2.2, it can be seen that the relation given by Greenewalt (1962), $f l^{1.15} = $ constant, gives a slope of either $m^{-0.38}$ for the geometric similarity, or $m^{-0.29}$ for the elastic similarity. It should, however, be noted that if a bird increases its weight, for example, by catching a prey, then the bird must slightly increase the beating frequency to maintain flight (Pennycuick et al. 1989).

Birds, like their reptilian ancestors, do not perspire. They are homoiotherms and depend on the air exhaled with each breath to regulate the temperature of the body, and, to some extent, on the evaporation of water from the internal surfaces of the lungs and air sacs. Cooling is also performed by conduction and radiation: the bills and toes are

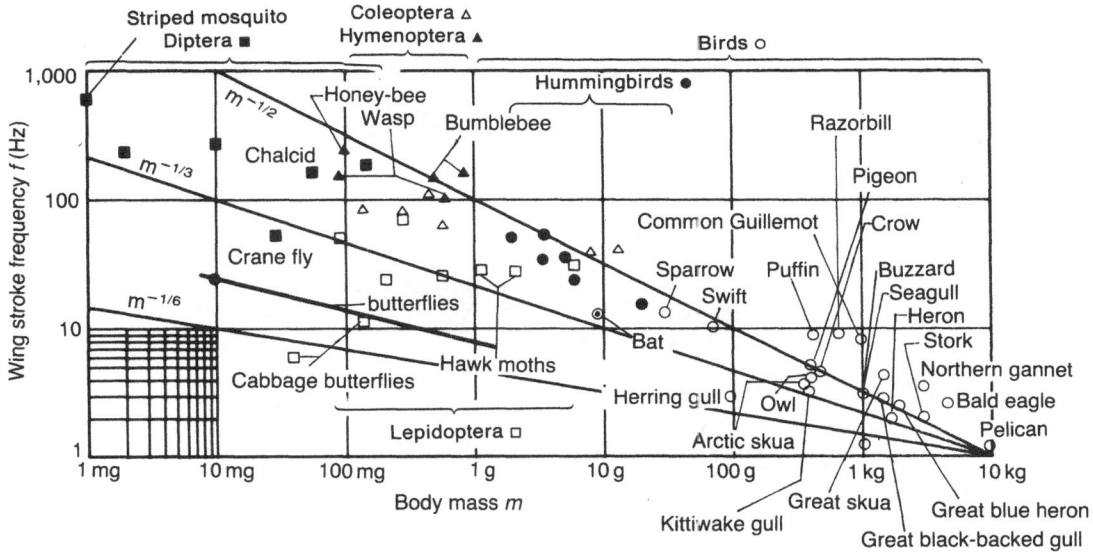

FIG. 4.1-12. Wing beat frequency.

believed to act as radiators for cooling. These cooling systems operate in a body whose temperature (40°–43°C) is considerably higher than that of human beings. During flight, the temperature may reach as high as 45°C (Bernstein et al. 1979).

It is also interesting to find that among birds, a species living in a cooler region is larger than one dwelling in a warmer region, and the former has shorter and smaller wings and bill than the latter (Higuchi 1978). Insects are, on the other hand, poikilothermal animals; thus, their body temperature is strongly dependent on that of the surroundings, as explained later.

Here let us consider again the power balance in nondimensional form, the ratio of the necessary power and the available power, P_n/P_a. As shown

in Table 4.1-6, the ratio increases with the mass of flying creatures for previously cited frequencies except one case of lower limit of the frequency. It can, thus, be confirmed that the sizes of the flying creatures are limited inevitably.

4.2 Wing Motions

In large flying creatures, the Reynolds number and the aspect ratio are so large that the creature deliberately uses its wings to sustain its weight by lift instead of drag. A wing of large span or high aspect ratio and a clean or streamlined body are all that are required for realizing long-range flight by maximizing C_L/C_D, or long duration flight by maximizing $C_L^{3/2}/C_D$.

However, in small birds and insects, the Reynolds number and the aspect ratio are small, and

TABLE 4.1-6. Power ratio P_n/P_a.

Items	Frequency	Geometric similarity		Elastic similarity	
Hovering flight	$f \propto \sqrt{m/S}/b$	$f \propto m^{-1/6}$	$P_n/P_a \propto m^{1/3}$	$f \propto m^{-1/16}$	$P_n/P_a \propto m^{1/4}$
High speed flight $U \propto \sqrt{m/S}$	$f \propto m^{-1/6}$ $f \propto m^{-1/3}$ $f \propto m^{-1/2}$	$U \propto m^{1/6}$	$P_n/P_a \propto m^0 = 1$ $P_n/P_a \propto m^{1/2}$ $P_n/P_a \propto m$	$U \propto m^{3/16}$	$P_n/P_a \propto m^{5/16}$ $P_n/P_a \propto m^{3/16}$ $P_n/P_a \propto m^{21/16}$
Fundamental elements	Wing span Body width Body length Wing area	$b \propto m^{1/3}$ $d \propto m^{1/3}$ $l \propto m^{1/3}$ $S \propto m^{2/3}$		$b \propto m^{1/4}$ $d \propto m^{3/8}$ $l \propto m^{1/4}$ $S \propto m^{5/8}$	

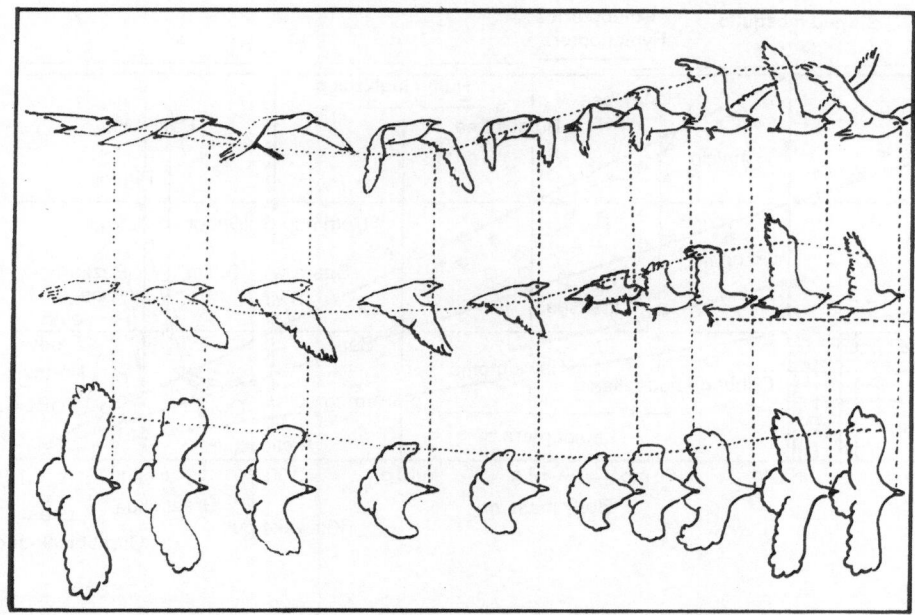

the drag specifically caused by the friction force is too large to obtain the required values for both C_L/C_D and $C_L^{3/2}/C_D$ that would enable soaring flight without power supply. They are, therefore, forced to rely mostly on the beating motion of the wings to fly. The small or low aspect ratio wings are suitable for higher beating motion and make up for the inefficient aerodynamic characteristics of the wings.

In such unsteady wing motion, the nondimensional frequency or "reduced frequency" is another important parameter which decides the aerodynamic characteristics of the wing.

Under the assumption that the induced velocity is constant over the stroke plane of beating wings, the so-called blade (or wing) element theory can be applied to obtain the aerodynamic force and moment distributions along the wingspan (Azuma et al. 1985). The analytic expression of this simple analysis is very helpful for understanding the aerodynamic characteristics of wing motions around single and double joints.

4.2.1 WING BEAT PHASES

Marey (1894) developed a time-lapse camera that can be called the first movie camera. With this, he was able to clarify and analyze a complete cycle of the beating motion of a bird. The result of his work is shown in the famous illustration of Fig. 4.2-1. Lilienthal and Lilienthal (1911) also produced, an

FIG. 4.2-1. Flight of seagull. (From Marey 1894 with permission).

illustration of beating wings from careful observation of the flight of storks.

In the mechanical sense, wing beating consists of four fundamental motions: (1) an out-plane motion called "flapping," (2) an in-plane motion called lead-lag or simply "lagging," (3) a twisting motion of the wing pitch called "feathering," and (4), an alternatively extending and contracting of the wingspan called "spanning."

Figure 4.2-2 shows two typical phases of the beating motion of a bird's wing in cruising flight. In the upstroke of the wing (or recovery stroke) (a), the downflow due to the wing beat is stronger at the outer wing than at the inner wing; thus, the lifting force, which is normal to the total inflow, is tilted backward more at the outer wing than at the inner wing. In the downstroke (or power stroke) (b), the upflow due to the wing beat is again stronger at the outer wing than at the inner wing; thus, the lifting force is tilted forward more at the outer wing than at the inner wing. The forward component of the lifting force generates a propulsive force against the drag force generated by the wings themselves and by the body. Usually, the total aerodynamic force acting on the wing is larger during the downstroke of the wing than during the upstroke. Through these beating mo-

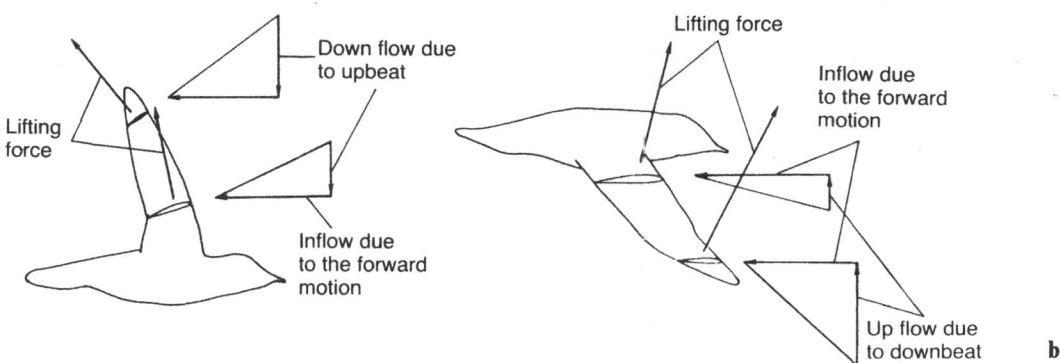

a   b

FIG. 4.2-2. **a, b.** Beating motion and resulting aerodynamic force of wing sections. **a** Upstroke. **b** Downstroke.

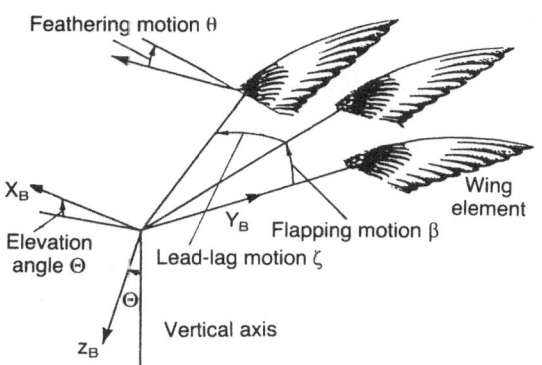

FIG. 4.2-3. Breakdown of the beating motion.

tions the wing itself must be twisted from wing root
to tip so that the airfoil section at each spanwise
station can maintain a reasonable range of angle
of attack.

Wing motion about a single hinge. The beating
motion of a right wing is shown in Fig. 4.2-3. This
is an angular motion around a universal joint,
which is either the shoulder (humeral) or the wrist
(carpal) joint for birds, and a fulcrum at the wing
root for insects, and includes a flapping motion β
(positive for flap-up), a lagging motion ζ (positive
for lead), and a feathering motion θ (positive for
pitch-up).

It must be mentioned that the universal joint
is not completely universal, but is subject to
some restriction in both amplitude and direction.
However, if the angular motions are performed
sucessively, then these angles may be called
Eulerian angles. The feathering motion is also
called "supination" for positive pitch and "pronation" for negative pitch.

Depending on the flight speed, there is a difference between bird flight and insect flight in the way
that these motions are mixed. In cruising flight,
however, there is no appreciable difference between them; the wing simply makes flapping motions at a relatively small amplitude, and to this
motion are added lead-lag and feathering motions
of still smaller amplitude. As will be explained later
in Sects. 4.2.3 , an adequate phase lag is maintained
among these motions. Therefore, as shown in
Fig. 4.2-4, the wing tip describes approximately
an elliptical orbit with respect to the body, ad-

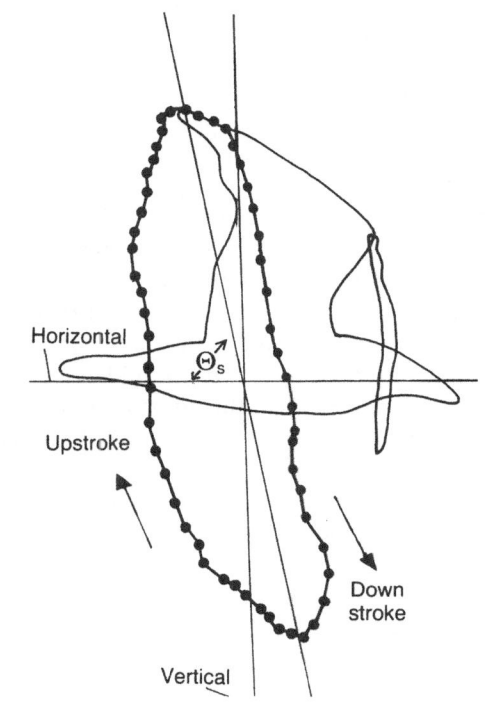

FIG. 4.2-4. Beating flight of a seagull at cruising speed.

vancing in the downstroke and retreating in the upstroke. The major axis of the ellipse is called the "stroke plane" and is tilted relative to the vertical plane with an inclination angle of Θ_s which is usually a function of forward flight speed.

The orbit of wing tip at the feathering axis can be expressed approximately by the first harmonic flapping and lead-lag angles, β and ζ. The detailed configuration of the orbit and the inclination angle of the stroke plane Θ_s are given in Table 4.2-1. The forward (horizontal) and upward (vertical) components of the aerodynamic force, which is nearly normal to the stroke plane, are respectively the propulsive force against drag and the lifting force against weight. In the case of Fig. 4.2-4 where $180° < \phi_\zeta < 270°$, since the stroke plane is that of the outer wing, the above lifting force serves to support a part of the weight, the remainder of which is sustained by the lifting force of the inner wing.

Interestingly, as seen from Figs. 4.2-1 and 4.2-4, in cruising flight, the primaries, which alone form the outer half of the wing, are partly folded backward during the upstroke and fully extended during the downstroke. These motions of the wing tip, which may be considered a local lead-lag motion, are effective for reducing unfavorable aerodynamic forces by forming a pointed wing in the upstroke, and for obtaining high aerodynamic forces without flow separation by forming a slotted wing in the downstroke. They result from the alternative change of suction force or propulsive force accompanying the flapping motion. The bird probably does not have to consume power for this lead-lag motion.

Wing motion about double hinges. Figure 4.2-5 shows the flapping pattern of a bird wing for (a) a mathematical model and (b) an actual model of an egret in slow climbing flight. The mathematical model is to show the phase difference between the flapping motions about the shoulder and wrist joints, β_1 and β_2, for the maximum mean lift coefficient. It can be seen that the two patterns are quite similar.

Aerodynamic force at a blade element. Now, let us analyze the mechanisms for generating the propul-

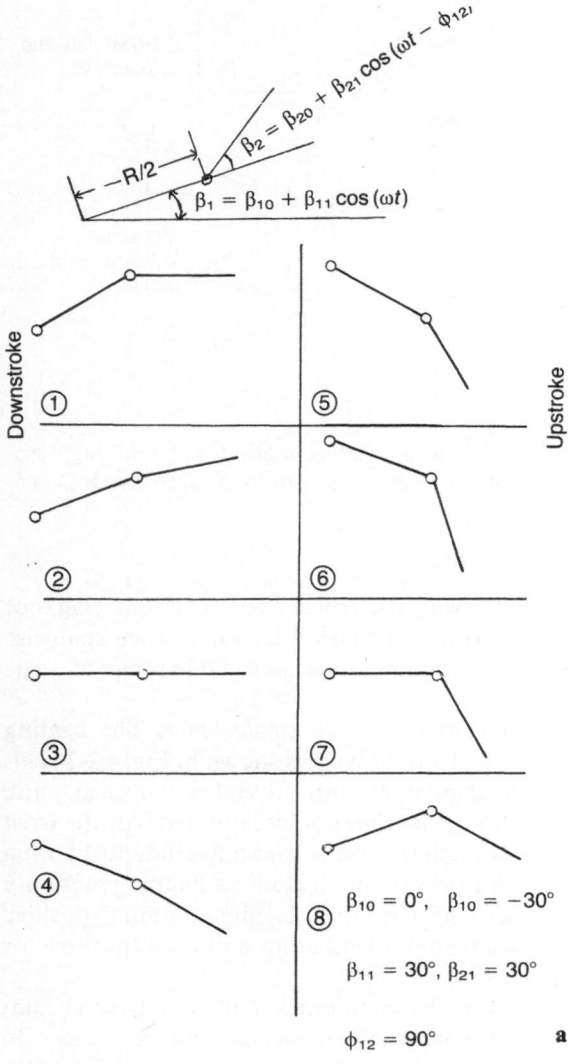

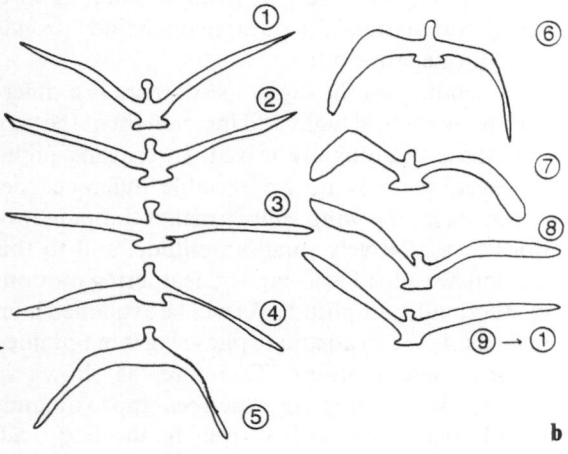

FIG. 4.2-5. **a, b.** Flapping motion of double-hinged wing. (From Sato et al. 1979; Sato 1980). **a** Mathematical model. **b** Actual flight model (Egret).

TABLE 4.2-1. Orbit of the wing tip for the first harmonic flapping and lead-lag angles

Items	Symbols	Mathematical expressions
Flapping angle	β	$\beta_1 \cos\psi$
Lead-lag angle	ζ	$\zeta_1 \cos(\psi + \phi_\zeta)$
Beating angle	γ	$\sqrt{\beta^2 + \zeta^2}$
Inclination angle of stroke plane	Θ_s	$\tan^{-1}\{\beta(\psi = \psi_{\gamma_{max}})/\zeta(\psi = \psi_{\gamma_{max}})\}$
Azimuth angle of γ_{max}	$\psi_{\gamma_{max}}$	$\frac{1}{2}\tan^{-1}[-\zeta_1{}^2 \sin(2\phi_\zeta)/\{\beta_1{}^2 + \zeta_1{}^2 \cos(2\phi_\zeta)\}]$
Maximum beating angle	γ_{max}	$\sqrt{\beta_1{}^2 \cos^2(\psi_{\gamma_{max}}) + \zeta_1{}^2 \cos^2(\psi_{\gamma_{max}} + \phi_\zeta)}$

Orbit
  configurations
  (These orbits
  are figured
  for
  $0 < \zeta_1 < \beta_1$)

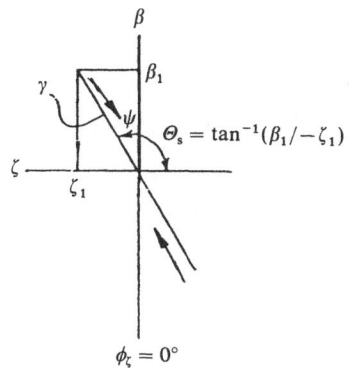

$\Theta_s = \tan^{-1}(\beta_1/-\zeta_1)$

$\phi_\zeta = 0°$

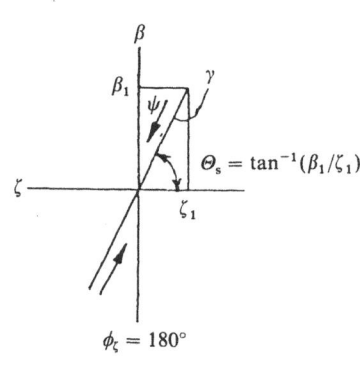

$\Theta_s = \tan^{-1}(\beta_1/\zeta_1)$

$\phi_\zeta = 180°$

⟵ Flight direction

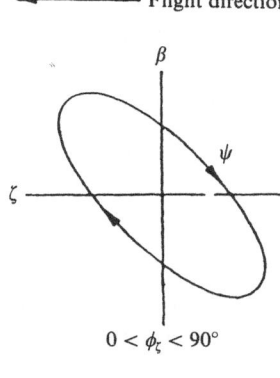

$0 < \phi_\zeta < 90°$

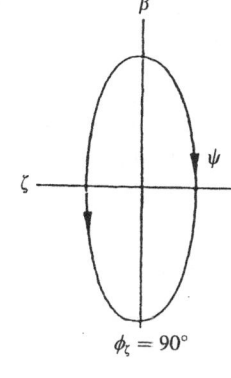

$\phi_\zeta = 90°$

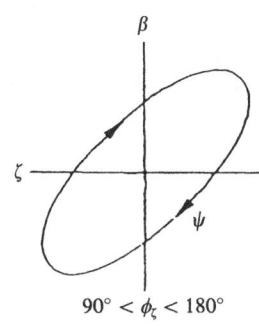

$90° < \phi_\zeta < 180°$

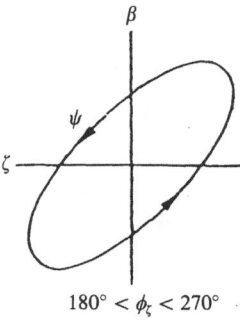

$180° < \phi_\zeta < 270°$

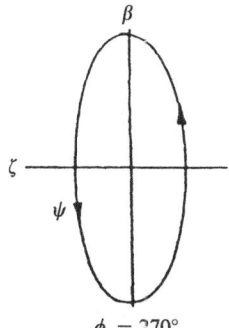

$\phi_\zeta = 270°$

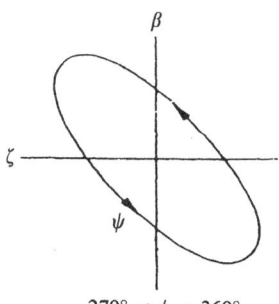

$270° < \phi_\zeta < 360°$

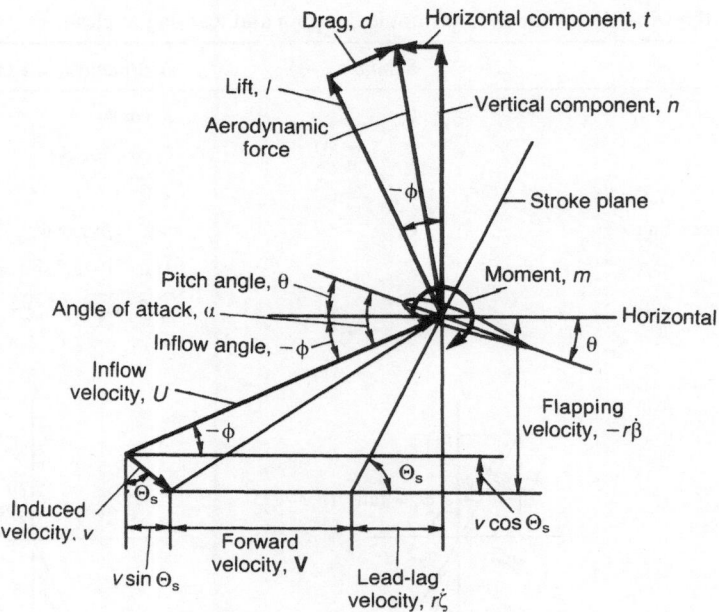

FIG. 4.2-6. Aerodynamic forces and moment at a blade element.

sive and lifting force acting on a wing section or a blade element in the beating motion of a bird in horizontal flight.

The relative or total inflow velocity $\mathbf{U}$ is, as seen from Fig. 4.2-6, given by a vectorial summation of the flapping $-r\dot{\beta}$, the lead-lag $r\dot{\zeta}$, the induced velocity $\mathbf{v}$ which is assumed normal to the stroke plane, and the forward velocity $\mathbf{V}$.

The aerodynamic force made up of two components at a spanwise station r, lift l, and drag d, which are normal and parallel components with respect to the total inflow velocity $\mathbf{U}$, is also split into two components: a lifting or vertical force

$$n = l\cos(-\phi) + d\sin(-\phi) \qquad (4.2\text{-}1a)$$

and a propulsive (horizontal) force or thrust

$$t = l\sin(-\phi) - d\cos(-\phi) \qquad (4.2\text{-}1b)$$

where

$$
\left.
\begin{aligned}
& l = \tfrac{1}{2}\rho U^2 cC_l(\alpha) \simeq \tfrac{1}{2}\rho U^2 ca(\alpha - \alpha_0) \\
& d = \tfrac{1}{2}\rho U^2 cC_d(\alpha) \simeq \tfrac{1}{2}\rho U^2 c\delta \\
& U = \sqrt{(-r\dot{\beta} - v\cos\Theta_s)^2 + (V + r\dot{\zeta} + v\sin\Theta_s)^2} \\
& -\phi = \cot^{-1}\{(V + r\dot{\zeta} + v\sin\Theta_s)/ \\
& \qquad - (r\dot{\beta} + v\cos\Theta_s)\} \\
& \alpha = \theta - \phi \\
& \qquad \simeq \theta - \{(r\dot{\beta} + v\cos\Theta_s)/(V + r\dot{\zeta} + v\sin\Theta_s)\}
\end{aligned}
\right\}
$$
$$(4.2\text{-}2)$$

where a, δ and α_0 are lift slope, mean profile drag, and "zero lift angle" respectively. In order to obtain the highest aerodynamic force, the following items must be satisfied or actually performed:

1. To get positive vertical force,

$$
\left.
\begin{aligned}
n = \tfrac{1}{2}\rho Uca\{&(\theta - \alpha_0)(V + r\dot{\zeta} + v\sin\Theta_s) \\
& - (1 + \delta/a)(r\dot{\beta} + v\cos\Theta_s)\} > 0
\end{aligned}
\right\}
$$
$$(4.2\text{-}3a)$$

the inside the braces should be positive. However in the upstroke ($\dot{\beta} > 0$), the above force is occasionally negative near the wing tip. This requires special efforts in low-speed flight.

2. To get positive thrust,

$$
\left.
\begin{aligned}
t = \tfrac{1}{2}\rho Uca\{&-(\theta - \alpha_0)(r\dot{\beta} + v\cos\Theta_s) - (\delta/a) \\
& \cdot (V + r\dot{\zeta} + v\sin\Theta_s) + (r\dot{\beta} + v\cos\Theta_s)^2 \\
& /(V + r\dot{\zeta} + v\sin\Theta_s)\} > 0
\end{aligned}
\right\}
$$
$$(4.2\text{-}3b)$$

the flapping velocity should be high and the pitch angle should be alternated so as to keep the first term in the braces positive, i.e.,

$$\theta - \alpha_0 > 0 \quad \text{for the down flap}$$

$$\theta - \alpha_0 < 0 \quad \text{for the up flap.}$$

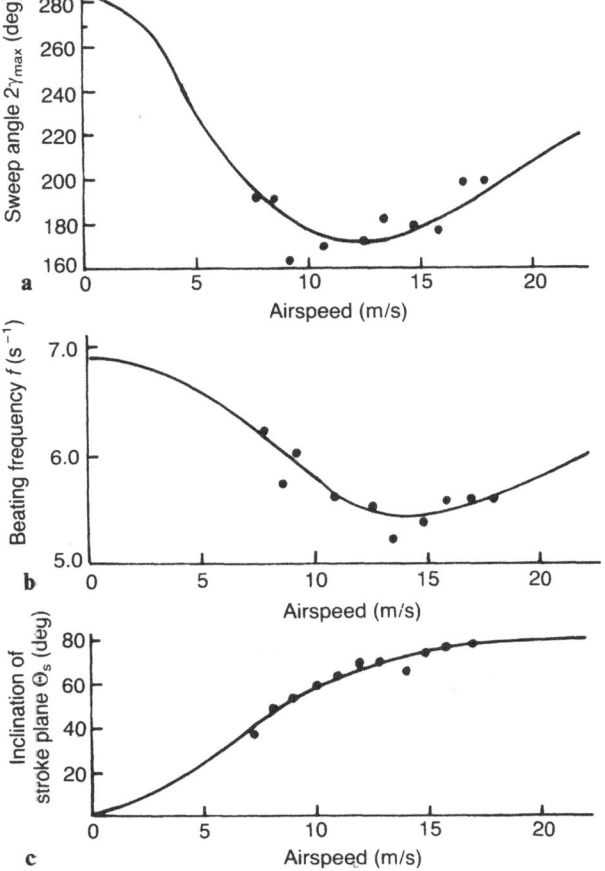

FIG. 4.2-7. **a–c.** Sweep angle, beating frequency, and inclination of the stroke plane. (From Pennycuick 1968a with permission of the Company of Biologists Ltd.). **a** Sweep angle $2\gamma_{max}$. **b** Beating frequency f. **c** Incidence of the stroke plane, H_d.

large aspect ratio selected, by which the lift-to-drag ratio increases and the induced velocity is reduced.

6. The ineffective range of beating must be reduced and the time ratio of the upstrokes and downstrokes, specifically in large birds, should be equal to about 2 : 3 (Weis-Fogh 1975a; Vogel 1966). In the upstroke, since the velocity generated by the flapping is reversed, the lifting force may be reversed, whereas the propulsive force is still positive.

Furthermore, the following items are important:

1. In the above analysis, the body attitude Θ and the flight path angle γ were assumed horizontal or $\Theta = \gamma = 0$. However this assumption can be removed simply by replacing the horizontal force and the vertical force respectively with the parallel force and the perpendicular force to the flight path ($\gamma \neq 0$).
2. The actual stroke plane is not defined by Table 4.2-1 but this unreliability is limited only for the introduction of the induced velocity through the tilt angle of the stroke plane Θ_s, $v \cos \Theta_s$ and $v \sin \Theta_s$.
3. The motion of wing element is defined by the Eulerian angles with respect to the body coordinate.
4. The kinetics on the aerodynamic and inertial forces acting on the blade element can be analyzed more accurately than other simple ways based on, for example, the stroke plane, as introduced later in Sect. 4.2-5 and Sect. 4.5-2 in which the beating angle or azimuth angle and the feathering angle are defined with respect to the stroke plane.

Observed data. Shown in Fig. 4.2-7 are (a) the sweep angle, which is twice the amplitude of the maximum beating angle in the stroke plane or $2\gamma_{max}$; (b) the beating frequency f with respect to the forward speed; and (c) the inclination of the stroke plane Θ_s observed in the beating flight of a pigeon (Pennycuick 1968b). The sweep angle, like the beating frequency, is largest at zero speed or in hovering flight, and decreases once a medium speed is attained, where the required power is minimum, because the power is proportional to the cube of the sweeping angle and the frequency. The inclination angle, on the other hand, starts from zero and approaches 90° for the highest speed.

3. The flapping and lead-lag velocities must be made high by beating with a large amplitude and high frequency, specifically in accelerated flight at takeoff and low-speed flight, in which the lifting force increases by taking a small tilt angle of the stroke plane, whereas the propulsive force increases by taking a large tilt angle of the stroke plane.
4. The bird must rely mostly on the flapping motion of the outer wing with little lead-lag motion in cruising flight, in which, as stated before, the lifting force is generated at the wing root, whereas the propulsive force is principally generated by the outer wing.
5. The wingspan must be made large and thus a

FIG. 4.2-8. Final stage of upstroke ("flick") of an osprey at low flight speed. (Courtesy of Mitsuaki Iwago 1981).

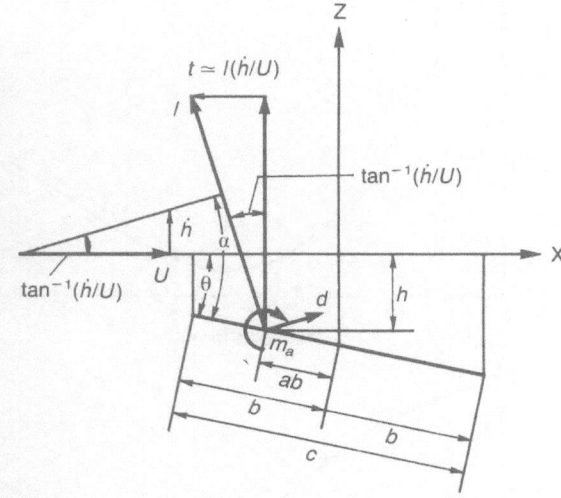

FIG. 4.2-9. Forces and moment acting on a moving flat plate in an inviscid flow.

As explained later in Sect. 4.4.5, the beating frequency of insects is locked at a specified value that is almost independent of the flight speed. In the case of bird flight, however, beating frequency is a function of flight speed. It may be supposed that (1) the power consumed by the inertial forces is not so large in comparison with the aerodynamic power because of the comparatively low beating frequency in larger birds; and (2) the moment of inertia about the shoulder hinge is varied by changing the wing spread, and therefore the resonant frequency is changed during flight.

At very low flight speed, the sweep angle and the beating frequency are so large that the relative inflow to the wing is reversed. Then, the feather arrangement is altered, typically as illustrated in Fig. 4.2-8, which shows the feather arrangement of an osprey at the final stage of the recovery stroke, called the "flick," just before the end of the recovery stroke. As mentioned in Sects. 3.1.1 and 3.1.4, the primary feathers are completely rotated around their respective rachis to allow the flow to pass through the wing from the back side, so that positive thrust as well as lift can still be generated. In addition to the above favorable deflection of the feathers, the effect of boundary layer control, probably achieved by making the wing penetrable by an appropriate arrangement of the feathers, must be remembered.

4.2.2 OPTIMAL COUPLING OF HEAVING AND FEATHERING MOTIONS

Referring to Fig. 4.2-9, let us consider a two-dimensional flat plate moving in a stationary fluid with a velocity U and a small angle of attack α. The pressure distribution C_p, lift coefficient C_l, and moment coefficient at a quarter-chord from the leading edge of the plate $C_{m,c/4}$ are given as follows:

$$\left.\begin{array}{c} C_p = -\Delta p/\tfrac{1}{2}\rho U^2 = 4\alpha\sqrt{\dfrac{1-x}{1+x}} \\[2mm] C_l = l/\tfrac{1}{2}\rho U^2 c = 2\pi\alpha \\[2mm] C_{m,c/4} = m_{c/4}/\tfrac{1}{2}\rho U^2 c^2 = 0 \end{array}\right\} \quad (4.2\text{-}4)$$

where x is the nondimensional abscissa based on the half-chord ($b = c/2$)

$$x = X/b = X/(c/2) \qquad (4.2\text{-}5)$$

The pressure distribution on the thin plate has the well-known singularity at the leading edge $X = -c/2 = -b$, which results from the simplification of the airfoil model. In actual wings, the round leading edge or smooth flow caused by the existence of a laminar separation bubble will alleviate the above singularity and generate a suction force that is parallel to the chord.

If the plate has coupled feathering and heaving motions around an axis ab with angular velocity of ω (Fig. 4.2-9) such as

$$Z = -h - \theta(X - ab)$$
$$\left.\theta = \theta_0 + \theta_1 \cos(\omega t + \phi_\theta)\right\} \quad (4.2\text{-}6)$$
$$h = h_1 \cos(\omega t + \phi_h)$$

where the amplitudes θ_1 and h_1 are assumed to be small, then, the angle of attack can be given by:

$$\alpha = \alpha_0 + \alpha_1 \cos(\omega t + \phi_\alpha) \quad (4.2\text{-}7)$$

with

$$\alpha_1 = \sqrt{\theta_1{}^2 + (h_1\omega/U)^2 + 2\theta_1(h_1\omega/U)\sin(\phi_\theta - \phi_h)}$$
$$\phi_\alpha = \tan^{-1}[\{\theta_1\sin\phi_\theta + (h_1\omega/U)\cos\phi_h\}/$$
$$\cdot \{\theta_1\cos\phi_\theta - (h_1\omega/U)\sin\phi_h\}]$$
$$(4.2\text{-}8)$$

The pressure coefficient is given by

$$C_p = A_0\sqrt{\frac{1-x}{1+x}} + 2A_1\sqrt{1-x^2} + A_2 x\sqrt{1-x^2}$$
$$(4.2\text{-}9)$$

where $x = X/b$ and the detailed expression of the parameters A_0, A_1, and A_2, are given in Table 4.2-2. Although, as stated later, the feathering axis

TABLE 4.2-2. Parameters in Eq. (4.2-9)

A_0	$\theta_0 + \theta_1\{a_c\cos(\omega t + \phi_\theta) + a_s\sin(\omega t + \phi_\theta)\}$ $+ (kh_1/c)\{-8G\cos(\omega t + \phi_h) - 8F\sin(\omega t + \phi_h)\}$
A_1	$\{-4k\sin(\omega t + \phi_\theta) + 2ak^2\cos(\omega t + \phi_\theta)\}\theta_1$ $-4k(kh_1/c)\cos(\omega t + \phi_h)$
A_2	$-2k^2\cos(\omega t + \phi_\theta)$
a_c	$4F - 2kG(1 - 2a)$
a_s	$-2kF(1 - 2a) - 4G + 2k$

can be selected at any chordwise position, it is hereafter given at a quarter-chord or $a = -1/2$. Then, the lift and moment coefficients, and the thrust of propulsive force coefficient, which is given by integrating both the suction pressure acting on the leading edge and the forward component of the normal pressure (Garrick 1936; Wu 1961, 1971a), can be obtained analytically as given in Table 4.2-3.

The two functions F and G, used in Table 4.2-3 are real and imaginary parts of the "Theodorsen

TABLE 4.2-3. Aerodynamic force and moment coefficients

C_l	$C_{l_{\theta_0}}\theta_0 + \theta_1\{C_{l_{\theta,c}}\cos(\omega t + \phi_\theta) + C_{l_{\theta,s}}\sin(\omega t + \phi_\theta)\}$ $+ (kh_1/c)\{C_{l_{h,c}}\cos(\omega t + \phi_h) + C_{l_{h,s}}\sin(\omega t + \phi_h)\}$
$C_{m_{c/4}}$	$(k\theta_1)\{C_{m_{\theta,c}}\cos(\omega t + \phi_\theta) + C_{m_{\theta,s}}\sin(\omega t + \phi_\theta)\} + C_{m_h}(kh_1/c)\cos(\omega t + \phi_h)$
C_t	$t/\tfrac{1}{2}\rho U^2 c = \theta_s{}^2\{C_{t_{\theta\theta,0}} + C_{t_{\theta\theta,c}}\cos(2\omega t + 2\phi_\theta) + C_{t_{\theta\theta,s}}\sin(2\omega + 2\phi_\theta)\}$ $+ \theta_1(kh_1/c)\{C_{t_{\theta h,0}}\cos(\phi_h - \phi_\theta - \phi_{h\theta}) + C_{t_{\theta h,c}}\cos(2\omega t + \phi_\theta + \phi_h)$ $+ C_{t_{\theta h,s}}\sin(2\omega t + \phi_h + \phi_\theta)\} + (kh_1/c)^2\{C_{t_{hh,0}} + C_{t_{hh,c}}\cos(2\omega t + 2\phi_s)$ $+ C_{t_{hh,s}}\sin(2\omega t + 2\phi_h)\} + \theta_0\theta_1\{C_{t_{\theta_0\theta,c}}\cos(\omega t + \phi_\theta) + C_{t_{\theta_0\theta,s}}\sin(\omega t + \phi_\theta)\}$ $+ \theta_0(kh_1/c)\{C_{t_{\theta_0h,c}}\cos(\omega t + \phi_h) + C_{t_{\theta_0h,s}}\sin(\omega t + \phi_h)\}$

$C_{t_{\theta\theta,0}}$	$\pi\{F^2 + G^2 - F + k^2(F^2 + G^2 - F + \tfrac{1}{2})\}$	$C_{l_{\theta_0}}$	2π
$C_{t_{\theta\theta,c}}$	$\pi\{F^2 - G^2 - F + 2kG(1 - 2F) + k^2(G^2 - F^2 + F)\}$	$C_{l_{\theta,c}}$	$2\pi\{F - kG - (k/2)^2\}$
$C_{\theta\theta,s}$	$\pi\{G(1 - 2F) + k(2G^2 - 2F^2 + 2F + \tfrac{1}{2}) + k^2G(2F - 1)\}$	$C_{l_{\theta,s}}$	$-2\pi\{G + kF + k/2\}$
$C_{t_{\theta h,0}}$	$2\pi[\{G + k(2F^2 + 2G^2 - F + \tfrac{1}{2})\}^2 + \{2F^2 + 2G^2 - F - kG\}^2]^{1/2}$	$C_{l_{h,c}}$	$-2\pi\{k + 2G\}$
$C_{t_{\theta h,c}}$	$2\pi\{G(1 - 4F) + k(2G^2 - 2F^2 + F + \tfrac{1}{2})\}$	$C_{l_{h,s}}$	$-2\pi\{2F\}$
$C_{t_{\theta h,s}}$	$2\pi\{2G^2 - 2F^2 + F + kG(4F - 1)\}$	$C_{m_{\theta,c}}$	$\tfrac{3}{16}\pi k$
$C_{t_{hh,0}}$	$4\pi(F^2 + G^2)$	$C_{m_{\theta,s}}$	$\tfrac{1}{2}\pi$
$C_{t_{hh,c}}$	$4\pi\{G^2 - F^2\}$	C_{m_h}	$\tfrac{1}{2}\pi k.$
$C_{t_{hh,s}}$	$8\pi FG$		
$C_{t_{\theta_0\theta,c}}$	$2\pi\{F - kG + (k/2)^2 - 1\}$		
$C_{t_{\theta_0\theta,s}}$	$2\pi\{-kF - G + (3k/2)\}$		
$C_{t_{\theta_0h,c}}$	$2\pi\{k - 2G\}$		
$C_{t_{\theta_0h,s}}$	$-4\pi F$		
$\phi_{h\theta}$	$\tan^{-1}\left[\dfrac{F - 2F^2 - 2G^2 + kG}{G + k(2F^2 + 2G^2 - F + \tfrac{1}{2})}\right]$		

function" (Bisplinghoff et al. 1957) given by the formula;

$$\hat{C}(k) = F(k) + iG(k) \qquad (4.2\text{-}10)$$

where

$$k = b\omega/U \qquad (4.2\text{-}11)$$

and where i is imaginary or $i = \sqrt{-1}$. It is obvious that the thrust coefficient calculated by the above method and given in Table 4.2-3 can, in a low reduced frequency range, be approximated by

$$t/\tfrac{1}{2}\rho U^2 c = C_t \simeq C_l(\dot{h}/U) \quad \text{for a small value of } k \qquad (4.2\text{-}12)$$

This shows that as stated in Sect. 4.2.1, the thrust is given by a forward component of lift inclined by the heaving motion. The mean value of the thrust coefficient through one cycle is given by

$$\bar{C}_t = C_{t_{\theta\theta,0}}\theta_1{}^2 + C_{t_{\theta h,0}}\{\theta_1(kh_1/c)\}\cos(\phi_h - \phi_\theta - \phi_{h\theta})$$
$$+ C_{t_{hh,0}}(kh_1/c)^2 \qquad (4.2\text{-}13)$$

It is interesting to find that the thrust caused by the pure heaving motion at the quarter-chord of the flat plate is proportional to $(kh_1/c)^2$, and the pure feathering motion around the quarter-chord generates a negative thrust for a low reduced frequency, or $k < 1.0$. The same result was obtained by Garrick (1936).

The maximum value of the mean thrust,

$$\bar{C}_{t_{\max}} = (kh_1/c)^2 [\pi\{F^2 + G^2 - F + k^2(F^2 + G^2 - F + \tfrac{1}{2})\}\{\theta_1/(kh_1/c)\}^2$$
$$+ 2\pi\sqrt{\{G + k(2F^2 + 2G^2 - F + \tfrac{1}{2})\}^2 + \{2F^2 + 2G^2 - F - kG\}^2}$$
$$\cdot\{\theta_1/(kh_1/c)\} + 4\pi(F^2 + G^2)] \qquad (4.2\text{-}14)$$

can be obtained at

$$(\phi_h - \phi_\theta)_{t_{\max}} = \phi_{h\theta}$$
$$= \tan^{-1}\left[\frac{F - 2F^2 - 2G^2 + kG}{G + k(2F^2 + 2G^2 - F + \tfrac{1}{2})}\right] \qquad (4.2\text{-}15)$$

Equation (4.2-14) shows that the amplitude ratio called "proportional feathering parameter" (Lighthill 1970), $\theta_1/(kh_1/c)$, has a minimum value of 2 at the reduced frequency of $k = 0.2$. In order to get the maximum thrust the proportional feathering parameter should be larger than 2 (Sato 1980; Azuma 1981b). Equation (4.2-15) requires that

$$\phi_h - \phi_\theta = -\pi/2 \quad \text{for} \quad k \to 0 \quad \text{and}$$
$$\phi_h - \phi_\theta = 0 \quad \text{for} \quad k \to \infty \qquad (4.2\text{-}16)$$

The broken line in Fig. 4.2-10 shows the phase difference between the heaving and feathering motions, $\phi_h - \phi_\theta$, for maximum thrust. Shown in Fig. 4.2-11 are the profiles of the coupling motion for two phase differences. In the case of (a) $\phi_h - \phi_\theta = -\pi/2$, the leading edge of the wing is twisted downward during the upstroke and upward during the downstroke, whereas in the case of (b) $\phi_h - \phi_\theta = 0$, it is twisted downward at the top and upward at the bottom.

FIG. 4.2-10. Phase difference between heaving and feathering motions for the maximum thrust and the maximum efficiency, or the minimum power for a given thrust. (From Sato 1980; Azuma 1980, 1981b).

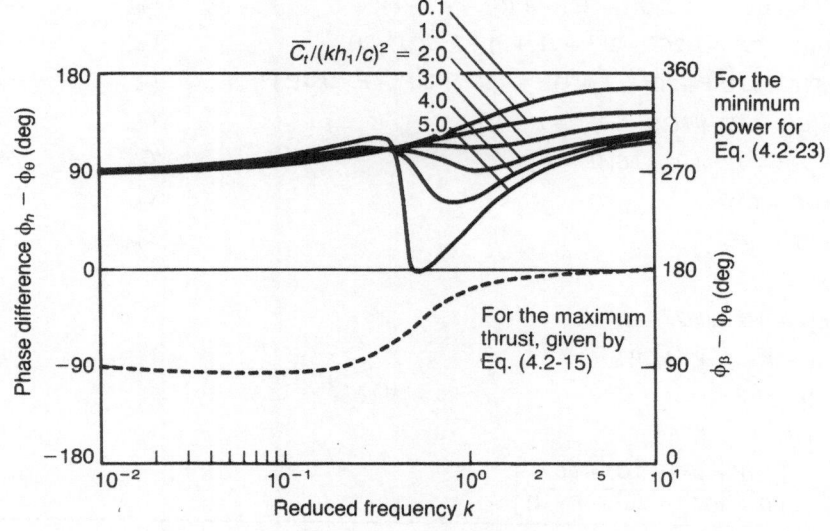

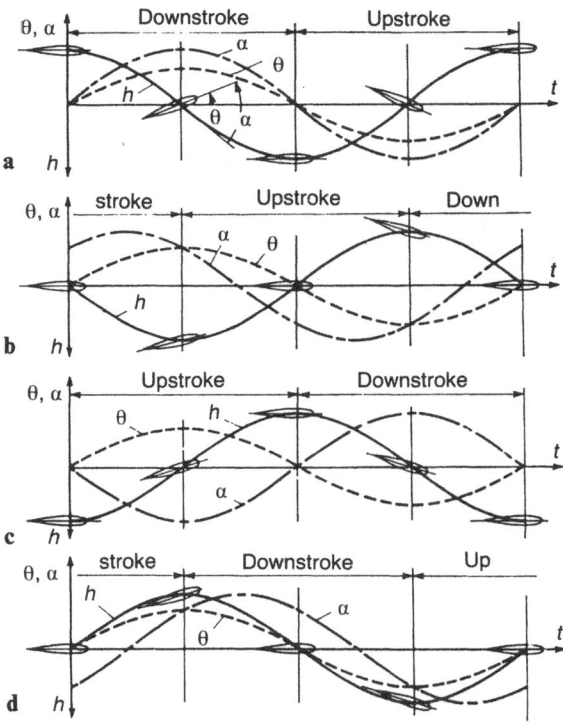

FIG. 4.2-11. **a–d.** Coupling motions for various phase differences. **a** $\phi_h - \phi_\theta = -\pi/2$. **b** $\phi_h - \phi_\theta = 0$. **c** $\phi_h - \phi_\theta = \pi/2$. **d** $\phi_h - \phi_\theta = \pi$.

Usually, since the heaving motion is actually performed by flapping (or fanning in a caudal fin as explained in Chap.6), in such a way that

$$h = -r\beta \qquad (4.2\text{-}17)$$

where r is the distance between the aerodynamic center of the blade element under consideration and the flapping or fanning hinge, and β is the flapping or fanning angle. Then, the phase difference between the flapping or fanning and the feathering is given by

$$\phi_\beta - \phi_\theta = \phi_h - \phi_\theta + \pi \qquad (4.2\text{-}18)$$

which is shown by the right ordinate in Fig. 4.2-10.

The power p consumed in this process is used for driving the heaving motion against the normal component of lift and the pitching motion against the moment. Thus, the power coefficient can be obtained analytically as given in Table 4.2-4. If any negative power in the muscle work is converted to positive power, then the power coefficients should be modified by C_p^*, also given in Table 4.2-4. Under the assumption that

$$kh_1/c \gg \theta_1, \theta_0, \quad \text{and} \quad k \qquad (4.2\text{-}19)$$

as usually established in various kinds of wing motions (beating and fanning) observed in living creatures, C_p can be approximated by

$$p/\tfrac{1}{2}\rho U^3 c = C_p \simeq 4\pi F(kh_1/c)^2$$
$$\cdot \{1 - \cos(2\omega t + 2\phi_h + \phi_1)\}$$
$$= C_p^* > 0 \qquad (4.2\text{-}20)$$

The above equation states that the physiologically consumed power is nearly equal to the mechanical power in the usual wing motions. This fact does not depend on the mode of locomotion; thus, the result can be applied equally to the beating motion of birds and insects and the fanning motion of fish.

The mean values of the above power coefficients are, then, given by:

$$\overline{C}_p = (kh_1/c)^2 [c_{p_{\theta\theta,0}}\{\theta_1/(kh_1 c)\}^2 + C_{p_{\theta h,0}}$$
$$\cdot \{\theta_1/(kh_1/c)\}\cos(\phi_h - \phi_\theta - \psi_{h\theta}) + C_{p_{hh,0}}]$$
$$= (kh_1/c)^2 [(\pi/2)k^2\{\theta_1/(kh_1/c)\}$$
$$+ 2\pi\sqrt{\{-F + kG\}^2 + \{G + k(\tfrac{1}{2} + F)\}^2}$$
$$\cdot \{\theta_1/k(h_1/c)\}\cos(\phi_h - \phi_\theta - \psi_{h\theta}) + 4\pi F]$$
$$(4.2\text{-}21)$$

The Froude efficiency can be defined by the formula

$$\eta = \bar{t}U/\bar{p} = \overline{C}_t/\overline{C}_p \qquad (4.2\text{-}22)$$

In order to obtain economical power consumption, the optimal phase difference and the maximum efficiency must be determined. This can be defined as follows: for a given slope of the thrust coefficient with respect to the square of heaving velocity, or for a given $\overline{C}_t/(kh_1/c)^2$, the minimum power $C_{p_{\min}}$ will give the maximum efficiency:

$$\eta_{\max} = \min[\{\overline{C}_p - \lambda(\overline{C}_t - \overline{C}_{t,\text{given}})\}/(kh_1/c)^2] \qquad (4.2\text{-}23)$$

where λ is Lagrange's multiplier. The reason for obtaining the maximum efficiency by keeping $\overline{C}_t/(kh_1/c)^2$ constant instead of keeping the thrust constant as Wu did (Wu 1971c) is as follows: since the mean thrust and power coefficients are, as seen from Eqs. (4.2-13 and 21), approximately proportional to $(kh_1/c)^2$, the proportional coefficients $\overline{C}_t/(kh_1/c)^2$ and $\overline{C}_p/(kh_1/c)^2$ are only a function of the reduced frequency k, the proportional feathering parameter $\theta_1/(kh_1/c)$, and the phase differences $\phi_{h\theta}$ and $\psi_{h\theta}$.

TABLE 4.2-4. Aerodynamic power coefficients

C_p	$p/\frac{1}{2}\rho U^3 c = (l\dot{h} - m_{c/4}\dot{\theta})/\frac{1}{2}\rho U^3 c \simeq C_l(\dot{h}/U) - C_{m_{c/4}}(\dot{\theta}c/U)$
	$= \theta_1{}^2\{C_{p\theta\theta,0} + C_{p\theta\theta,c}\cos(2\omega t + 2\phi_\theta) + C_{p\theta\theta,s}\sin(2\omega t + 2\phi_\theta)\}$
	$+ \theta_1(kh_1/c)\{C_{p\theta h,0}\cos(\phi_h - \phi_\theta - \psi_{\theta h})$
	$+ C_{p\theta h,c}\cos(2\omega t + \phi_h + \phi_\theta) + C_{p\theta h,s}\sin(2\omega t + \phi_h + \phi_\theta)\}$
	$+ (kh_1/c)^2\{C_{phh,0} + C_{phh,c}\cos(2\omega t + 2\phi_h) + C_{phh,s}\sin(2\omega t + 2\phi_h)\}$
	$+ \theta_0(kh_1/c)C_{p\theta_0 h,s}\sin(\omega t + \phi_h)$
$C_p{}^*$	$\{|l\dot{h}| + |-m_{c/4}\dot{\theta}|\}/\frac{1}{2}\rho U^3 c = |C_l(\dot{h}/U)| + |-C_{m_{c/4}}(\dot{\theta}c/U)|$
	$= 2\pi|(kh_1/c)^2\{2F - \sqrt{(2F)^2 + (k + 2G)^2}\cos(2\omega t + 2\phi_h + \phi_1)\}$
	$+ \theta_1(kh_1/c)\sqrt{\{F - kG - (k/2)^2\}^2 + \{G + k(F + \frac{1}{2})\}^2}\cdot\{\cos(\phi_\theta - \phi_h - \phi_2)$
	$- \cos(2\omega t + \phi_\theta + \phi_h + \phi_2)\} - 2\theta_0(kh_1/c)\sin(\omega t + \phi_h)|$
	$+ \frac{1}{2}\pi k^2|\theta_s{}^2\{1 - \sqrt{1 + (\frac{3}{8}k)^2}\cos(2\omega t + 2\phi_\theta + \phi_3)\}$
	$+ \theta_1(kh_1/c)\{\sin(\phi_\theta - \phi_h) + \sin(2\omega t + \phi_\theta + \phi_h)\}|$

$C_{p\theta\theta,0}$	$\frac{1}{2}\pi k^2$	ϕ_1	$\tan^{-1}\{(k + 2G)/2F\}$
$C_{p\theta\theta,c}$	$-\frac{1}{2}\pi k^2$	ϕ_2	$\tan^{-1}[-\{F - kG - (k/2)^2\}/\{G + k(F + \frac{1}{2})\}]$
$C_{p\theta\theta,s}$	$\frac{3}{16}\pi k^3$	ϕ_3	$\tan^{-1}\{(\frac{3}{32})k\}.$
$C_{p\theta h,0}$	$2\pi[\{-F + kG\}^2 + \{G + k(F + \frac{1}{2})\}^2]^{1/2}$		
$C_{p\theta h,c}$	$-2\pi\{G + k(F + \frac{1}{2})\}$		
$C_{p\theta h,s}$	$2\pi\{-F + kG + (k^2/2)\}$		
$C_{phh,0}$	$4\pi F$		
$C_{phh,c}$	$-4\pi F$		
$C_{phh,s}$	$2\pi\{k + 2G\}$		
$C_{p\theta h,s}$	-4π		
$\psi_{h\theta}$	$\tan^{-1}\left[\dfrac{-F + kG}{G + k(F + \frac{1}{2})}\right].$		

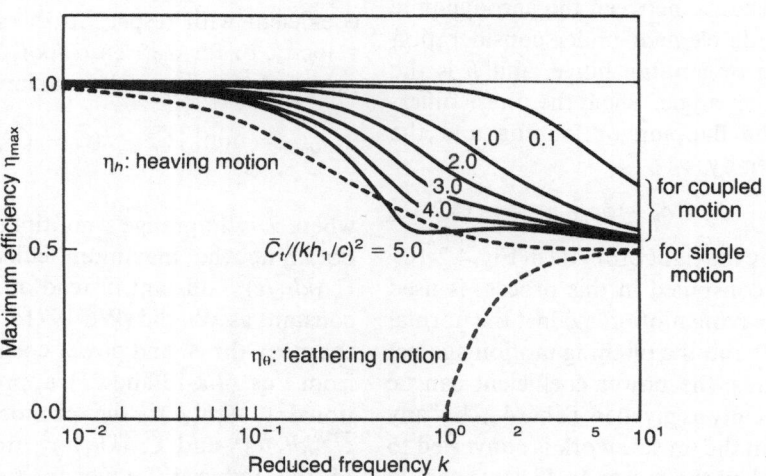

FIG. 4.2-12. Change of maximum efficiency with respect to reduced frequency. (From Azuma 1981c).

Unlike the case of the maximum thrust, the minimum power for a given proportional coefficient of thrust $\bar{C}_t/(kh_1/c)^2$ or the maximum efficiency can be obtained by taking the proportional feathering parameter less than 2 or $\theta_1/(kh_1/c) < 2$ (Azuma 1981b). As the proportional coefficient of thrust increases the proportional feathering parameter should be reduced further as Lighthill (1970) suggested. He recommended to select a value less than 1, $\theta_1/(kh_1/c) < 1$.

The solid lines shown in Fig. 4.2-10 are the phase differences $(\phi_h - \phi_\theta)_{\eta_{max}}$ for this η_{max} for various values of $\bar{C}_t/(kh_1/c)^2$. It is interesting to find that (1) for a small reduced frequency, the phase differences should be $\phi_h - \phi_\theta = -3\pi/2$ or $\pi/2$; (2) for a large reduced frequency, they should be $\phi_h - \phi_\theta = \pi/2 - \pi$; and (3) for intermediate values of the reduced frequency, $k = 0.5-2.0$ they show wide variation in their values, specifically for large values of $\bar{C}_t/(kh_1/c)^2$.

Shown in Fig. 4.2-11c,d are the profiles of the coupling motion for two phase differences. In the case of (c) $\phi_h - \phi_\theta = \pi/2$, the pitch angle is positive in the upstroke, negative in the downstroke, and flat at both the top and bottom, whereas in the case of (d) $\phi_h - \phi_\theta = \pi$, the pitch angle is maximum at the top and minimum at the bottom.

It should be remembered, however, that the above results were obtained on the assumptions of little wing perturbation to deviate from a straight and steady forward motion, and no flow separation. If the amplitude and thus the velocity of the heaving motion are large, then the straight forward motion should be replaced with a wavy motion, the general velocity of which would consist of the steady forward velocity and the heaving velocity. The perturbed motion is then limited to the feathering motion only.

The solid lines shown in Fig. 4.2-12 are the maximum efficiency η_{max} as a function of the reduced frequency for various values of $\bar{C}_t/(kh_1/c)^2$. As the reduced frequency increases, the efficiency decreases slowly for a small $\bar{C}_t/(kh_1/c)^2$ and rapidly for a large $\bar{C}_t/(kh_1/c)^2$. In contrast to the above maximum efficiency for the combined motions, the broken lines in Fig. 4.2-12 show the efficiencies for the heaving motion alone and the feathering motion alone, η_h and η_θ, which were obtained by Wu (1971c). The efficiency decreases monotonically from $\eta_h(k = 0) = 1$ to $\eta_h(k \to \infty) = 0.5$ for heaving motion, and increases monotonically from $\eta_\theta(k \cong 1) = 0$ to $\eta_\theta(k \to \infty) = 0.5$ for

pitching motion. Neither of the efficiencies for single motions can override that of the combined system.

In actual flight, birds may use a phase difference which gives the maximum thrust at takeoff, or $\phi_h - \phi_\theta < 0°$, and the minimum power for a given thrust (or the maximum efficiency) in cruising flight, or $\phi_h - \phi_\theta \gtrsim 90°$.

In the above analysis, the feathering axis was fixed at a quarter-chord or $a = -1/2$. However, if the feathering axis shifts to any other chordwise position to keep the mean square moment minimum (Yates 1986), the above optimal set of amplitude and phase differences should be modified as follows:

$$
\begin{aligned}
h^{\#}_1/b &= [\{(h_1/b)\cos\phi_h - \theta_1(a + \tfrac{1}{2})\cos\phi_\theta\}^2 \\
&\quad + \{(h_1/b)\sin\phi_h - \theta_1(a + \tfrac{1}{2})\sin\phi_\theta\}^2]^{1/2} \\
\phi^{\#}_1 &= \tan^{-1}[\{(h_1/b)\sin\phi_h - \theta_1(a + \tfrac{1}{2})\sin\phi_\theta\}/ \\
&\quad \{(h_1/b)\cos\phi_h - \theta_1(a + \tfrac{1}{2})\cos\phi_\theta\}]
\end{aligned}
\tag{4.2-24}
$$

In this section the analyses were limited to the two-dimensional flat plate in a coupled sinusoidal flapping and feathering motion. However, the results can be applied to various wing configurations of different profiles and planforms without serious corrections because the unsteady airload at any station is almost independent to the airfoil configuration and is mainly governed by the shed vortices ejected from and located near the trailing edge at that station.

4.2.3 MECHANICS IN HOVERING FLIGHT

The hovering flight of birds and insects, which may be of long or short duration, serves a different purpose from forward flight and is necessary for the precise adjustment of their absolute position with respect to the surroundings. Small size and high control characteristics are important for a good hovering ability.

Relation between reduced frequency and aspect ratio. In hovering flight, the maximum tip speed of the sinusoidally beating wing V_t is given by the formula:

$$
U \simeq \pi f b \beta_1 = V_t \tag{4.2-25}
$$

where f and b are the beating frequency and the span of the wing respectively, and β_1 is the flapping amplitude. Then, the maximum Reynolds number at the wing tip is:

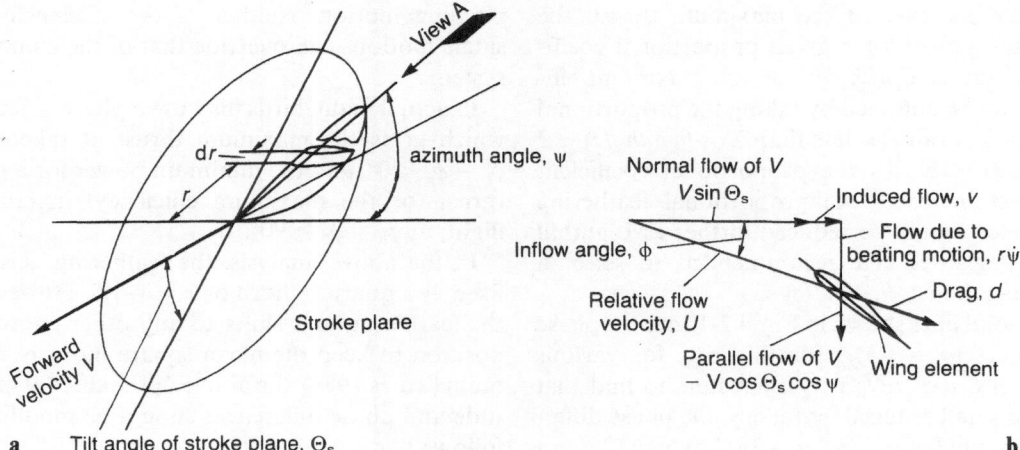

$$\mathrm{Re} = U\bar{c}/v = (\pi fb\bar{c}/v)\beta_1 = (\pi fb^2/v\mathcal{R})\beta_1 \quad (4\text{-}2.26)$$

where $\bar{c}$ is the mean chord specified by $\bar{c} = b/\mathcal{R}$, and $\mathcal{R}$ is the aspect ratio of the wing. The reduced frequency is also given by:

$$k = \pi f\bar{c}/U = 1/AR\,\beta_1 \simeq 1/\mathcal{R} \quad (4.2\text{-}27)$$

where the approximation is obtained by assuming $\beta_1 = 1.0$. This is a very important relation in the aerodynamics of hovering flight. The above equations state that any bird or insect with a wing of large aspect ratio operates at a small Reynolds number for a small span, low beating frequency, and small reduced frequency, whereas any bird or insect with a wing of small aspect ratio operates at a large reduced frequency. Thus, it can be said that a hovering bird or insect with a large aspect ratio wing relies on a quasi-steady aerodynamic force to support its weight, and that one with a small aspect ratio wing utilizes an unsteady aerodynamic force for this purpose.

4.2.4 Profile Power

Let us consider a wing moving with the velocity V and beating about a single hinge in a stroke plane. By referring to Fig. 4.2-13, a wing element beating with azimuth angle (or beating angle) ψ in the stroke plane of tilt angle Θ_s requires the profile torque dQ_0 as follows:

$$dQ_0 = \tfrac{1}{2}\rho U^2 c\delta \sin\phi r\,dr \quad (4.2\text{-}28)$$

where δ is the profile drag coefficient of the wing element ($\delta = C_{d_0}$ in Sects. 3.2 and 4.1), and U and ϕ are inflow velocity and inflow angle defined respectively by

$$U = \sqrt{(V\sin\Theta_s + v)^2 + (r\dot{\psi} - V\cos\Theta_s\cos\psi)^2} \quad (4.2\text{-}29)$$

$$\phi = \sin^{-1}\{(r\dot{\psi} - V\cos\Theta_s\cos\psi)/U\} \quad (4.2\text{-}30)$$

Integrating the above torque along the wing span, the torque of the wing becomes

$$Q_0 \simeq \frac{1}{2}\rho\delta \int_0^R Uc(r\dot{\psi} - V\cos\Theta_s\cos\psi)r\,dr \quad (4.2\text{-}31)$$

where the drag coefficient δ has been assumed constant. If the beating motion is a single harmonic motion of angular velocity ω and amplitude ψ,

$$\left.\begin{aligned} \psi &= \psi_1\sin(\omega t)\\ \dot{\psi} &= \psi_1\omega\cos(\omega t) \end{aligned}\right\} \quad (4.2\text{-}32)$$

Then the mean power of this motion against the torque Q_0 of both wings can be given as follows:

$$\left.\begin{aligned} P_0 &= \frac{\omega}{\pi}\int_{-\pi/2\omega}^{\pi/2\omega} 2Q_0\dot{\psi}\,dt\\ &= \frac{1}{2}\rho\delta\frac{2\omega}{\pi}\int_{-\pi/2\omega}^{\pi/2\omega}\int_0^R Uc\\ &\quad \cdot(r\dot{\psi} - V\cos\Theta_s\cos\psi)r\dot{\psi}\,dr\,dt \end{aligned}\right\} \quad (4.2\text{-}33)$$

Hovering flight. In hovering flight ($V = 0$), the mean profile power is given by

$$\bar{P}_{0,\mathrm{H}} = \tfrac{1}{2}\rho V_t^3 S\delta(8/3\pi)\sigma_3 \quad (4.2\text{-}34)$$

where the maximum tip speed V_t, wing area S, and σ_3 are defined by

$$\left.\begin{array}{l} V_t = R\psi_1\omega \\[2mm] S = 2\displaystyle\int_0^R c\,dr \\[2mm] \sigma_3 = \dfrac{1}{2}\displaystyle\int_0^1 (c/R)x^3\,dx \Big/ \displaystyle\int_0^1 (c/R)\,dx \end{array}\right\} \quad (4.2\text{-}35)$$

Eq. (4.2-34) states that the mean profile power in hovering flight is proportional to the product of (1) the cubic power of the maximum tip speed $V_t^3 = (R\psi_1\omega)^3$, and (2) the drag area of the wing $S\delta$ and the geometrical parameter of the wing σ_3. The profile power occupies about a quarter of the total power during hovering flight.

Forward flight. In order to obtain an analytical expression of the profile power, it is necessary to consider the high forward speed such as $V \gg r\dot\psi, v$ and $\Theta_s = \pi/2$. Under the above conditions, the mean profile power becomes

$$\bar{P}_0 = \tfrac{1}{2}\rho V V_t^2 S\delta\sigma_2 \qquad (4.2\text{-}36)$$

where

$$\sigma_2 = \frac{1}{2}\int_0^1 (c/R)x^2\,dx \Big/ \int_0^1 (c/R)\,dx \quad (4.2\text{-}37)$$

Eq. (4.2-36) states that the mean profile power is proportional to the product of the forward speed V, the square power of the maximum tip speed of beating $V_t^2 = (R\psi_1\omega)^2$, the drag area of the wing $S\delta$, and the geometrical parameter σ_2.

In medium speed range, $V \simeq R\dot\psi$, the profile power reduces from that of hovering flight as the forward speed increases, and then increases linearly with the forward speed. The profile power is, however, usually assumed constant through the whole speed range in comparison with other power components.

The drag of the wing element is also related to an additional profile power for the forward motion V, $\Delta\bar{P}_0$, as follows:

$$\left.\begin{array}{l} \Delta\bar{P}_0 = \displaystyle\int_0^R \rho U^2 c\delta \cos\phi\, V\,dr \\[3mm] \quad = \rho V^3\delta \displaystyle\int_0^R c\sqrt{1 + (r\dot\psi/V)^2}\,dr \\[3mm] \quad \simeq \tfrac{1}{2}\rho V^3 S\delta \end{array}\right\} \quad (4.2\text{-}38)$$

where the approximation is given by the same assumption of $V \gg r\dot\psi$.

The above power results from the profile drag against the main flow V and coincides with the parasite power of the wing $P_{p,w}$ given by Eq. (4.1-18), or $\Delta\bar{P}_0 = \bar{P}_{p,w}$. Thus, the above power can be included completely in the parasite power for the drag area of the body and wings.

4.2.5 WAKE SYSTEM

We have seen that the beating motion is mainly a flapping coupled with a lagging and feathering. Since the maximum and minimum lifts are, as explained in the previous section, obtained midway through the downstroke and the upstroke respectively, the vortex system of the wing in beating motion can be represented by the theoretical model shown in Fig. 4.2-14.

The passage of a wing in beating motion produces a series of periodical vortices arranged on a wavy wake surface. Here, Helmholtz's vortex law (Lamb 1930; Shapiro 1961), which states that a vortex is endless, holds for the beating wings, too. When the wing approaches a maximum lift phase following a minimum lift phase, the shed vortices γ_y assume a negative value and thus cancel the positive increment of the bound vortex, $\gamma_y = \dot\Gamma(y,t)/U$. On the other hand, when the wing leaves the maximum lift phase and approaches a minimum lift phase, the shed vortices assume a positive value and thus cancel the negative increment of the bound vortices. Therefore, in combination with the trailing vortices $\gamma_x = \partial\Gamma/\partial y$ generated by the wingspan, specifically by the wing tip and root, they make a series of vortex cells in the form of curved rectangular rings, the corners of which are actually rounded. The cells or the trailing vortices are, as shown in Fig. 4.2-14, concentrated more at the wing tip than at the wing root because of the marked spanwise change of the circulation of the wing, and, similarly, the shed vortices are concentrated near the switching points of the down- and upstrokes.

A cell of a wake vortex system produces maximum downwash at the phase of maximum lift. In the hovering flight of the hummingbird, the wavy wake surface is slightly tilted in both the forward and backward strokes; thus, the induced velocity is directed almost vertically downward in the respective strokes. However, in the forward flight of any bird, the wavy wake surface is inclined forward around the phase of maximum lift in the downstroke, and the downwash or induced veloc-

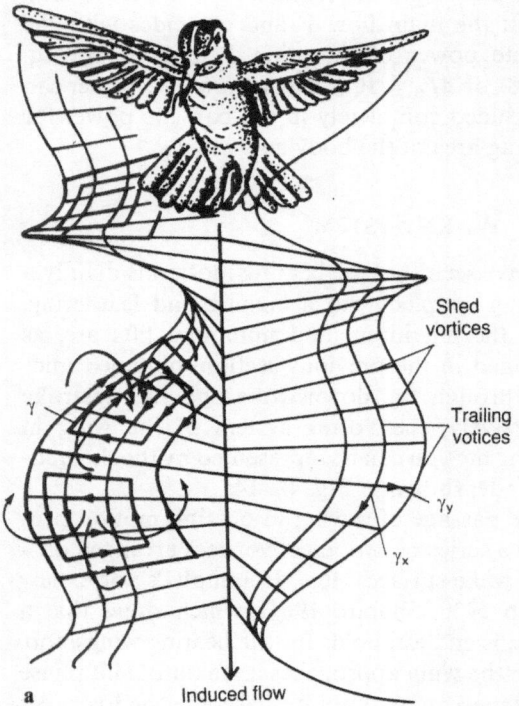

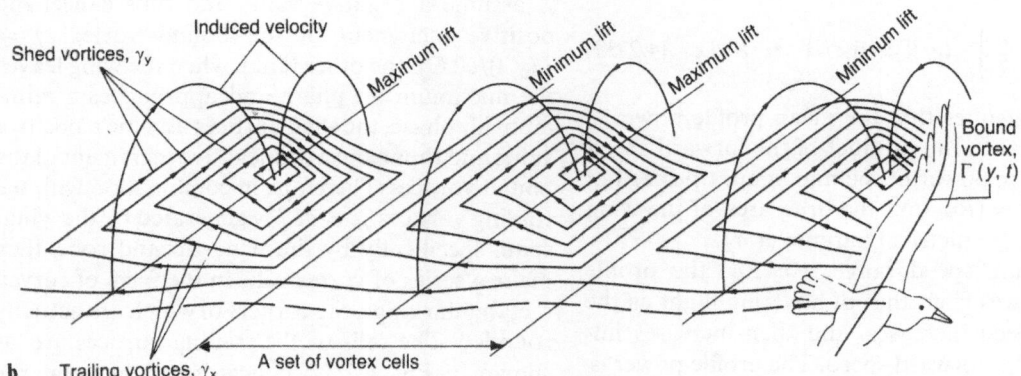

Fig. 4.2-14. **a, b.** Wake vortex systems. **a** Vortex cells generated by both wings in hovering flight. **b** Vortex cells generated by left wing in cruising flight. (The tilt angle of the vortex wake surface is much shallower than that of the stroke plane because the ridges of the vortex wake have moved down some distance by the time the valleys are completed).

ity is directed down- and backward; thus, according to the momentum theory, positive lifting force and thrust can be generated. The magnitude of the downwash is, however, very small because of the large mass of air in forward flight. As a result, the ripples observed on a water surface in hovering flight (Fig. 4.1-2) are not seen in forward flight.

Although vortex cells are formed in pairs from opposite wings, they usually combine. In an idealized schema, the vortex cells are deformed into a chain of vortex rings that are convected downward (Rayner 1979a, 1979b, 1979c). Actual observation

of the vortex loops in slow flight can be realized by using flow visualization techniques using kerosene mist for insects (as described later in Sect. 4.5.2) or helium bubbles for birds (Spedding et al. 1984, Spedding 1986). Rayner (1985a) further claims that the bound vortex strength in cruising flight is kept constant during one cycle of locomotion (power and recovery strokes) by an adequate change of feathering motion, and no transverse wake vorticity (shed vorticity) is generated. By observing the wake vortices of the beating wings of a kestrel flying at medium speed, Spedding

(1987b) also created a similar simplified wake model allowing for the change of effective span.

By knowing the downwash and, thus, the momentum change in the wake, the total aerodynamic forces acting on the wings can be determined indirectly. However, in order to calculate the airloading acting on the wing as a function of time (t) and spanwise station (y), the blade element theory must be applied by knowing the induced velocity distribution on the wing caused by the vortex system or the vorticity distribution in the wake. Thus the integral-differential equation on the elemental downwash expressed by the Biot-Savart law must be solved over the curved wake surfaces of both the left and right wings.

If the amplitude of the flapping motion is small, and thus the heaving motion at any spanwise position of the wing is small, then the wake surface may be considered flat. Since, as given by Eq. (1.2-11), the reduced frequency can be rewritten as $k = \pi c f / U = \pi / (U/C)(\lambda/c) = \pi/(\lambda/c)$ for the wake system ($U = C$), the reduced frequency increases and the unsteadiness of the fluid-dynamic force becomes important when the wavelength-to-chord ratio λ/c decreases beyond the order of $\lambda/c < 10$. Inversely, whenever the wavelength λ is larger than $10c$, the unsteady effect may be neglected. Then the blade element theory can be applied as if the wing were operating in a field of quasi-steady flow.

4.2.6 OPTIMAL LIFT DISTRIBUTION FOR A PAIR OF FLAPPING WINGS

For a given lift, the minimum induced drag of a stationary wing is obtained by making the spanwise load distribution elliptical as shown by $b/b_e = 1.0$ in Fig. 4.2-15. Here, let us consider a pair of flapping wings, the beating velocity of which is so slow compared with the forward speed that (1) the velocity of the wing element U can be assumed constant along the span or $U = V$ and (2) the effects of shed vortices can be neglected. Then, according to Jones (1950, 1980), the induced velocity distribution $v(\eta)$ and the circulation distribution $\Gamma(\eta)$ for the minimum drag under conditions of fixed total lift and fixed bending moment, are given by:

$$v(\eta) = v_0 + \omega_i(b/2)|\eta| \qquad (4.2\text{-}39)$$

$$\Gamma(\eta) = 2v_0 b \sqrt{1 - \eta^2}$$

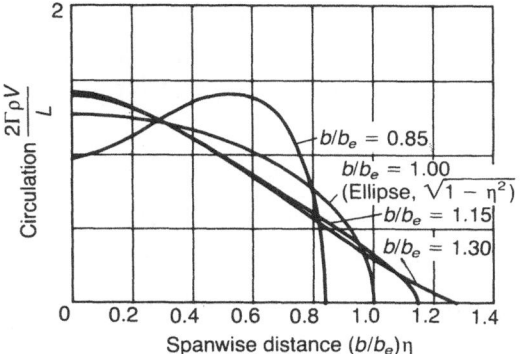

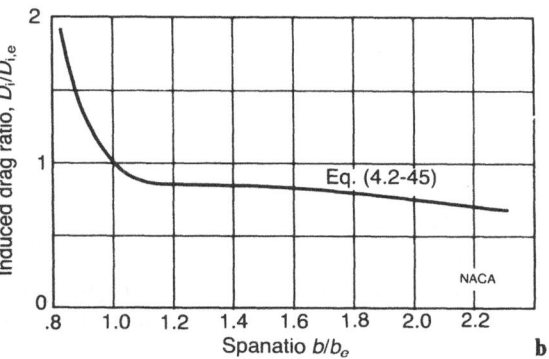

FIG. 4.2-15. **a, b.** Optimal circulation distribution and induced drag ratio. (Redrawn from Jones 1950). **a** Optimal circulation distribution. **b** Induced drag ratio.

$$+ (1/\pi)\omega_i b^2 \{\sqrt{1 - \eta^2} + \eta^2 \operatorname{sech}^{-1}(|\eta|)\} \qquad (4.2\text{-}40)$$

where

$$\eta = y/(b/2) \qquad (4.2\text{-}41)$$

It can be seen from these equations that: (1) for minimum drag in flapping motion, the induced velocity should vary linearly along the span; and (2) the circulation distribution or spanwise loading combines the elliptical distribution and an additional component that is also zero at the root and tip of the wing.

Optimal circulation distribution. Now let us consider four flat wings, three of which are the above optimal flapping wings proposed by Jones, and the remaining one is an elliptical wing, the span and thus the wing area of which are different from that of the optimal wings. If the lift L and the bending moment B of the optimal wings are equal to those of the elliptic wing (with subscript e), then the "lateral centroid" or "center of air loading" of the proposed wing η_c, is given as follows:

$$\eta_c = 2B/L(b/2) = \eta_{c,e}(b_e/b) \qquad (4.2\text{-}42)$$

where $\eta_{c,e}$ is the lateral centroid of the elliptic wing, $\eta_{c,e} = 4/3\pi$. The v_0 and ω_i in Eq. (4.2-39) are given by:

$$\left.\begin{aligned}
v_0 &= (27/2)(L/\rho U b^2)\{\eta_{c,e} - (8/9)\eta_c\} \\
&= (27/2)(L/\rho U b^2)\eta_{c,e}\{1 - (8/9)(b_e/b)\} \\
\omega_i &= (18\pi)(L/\rho U b^3)(\eta_c - \eta_{c,e}) \\
&= (18\pi)(L/\rho U b^3)\eta_{c,e}\{(b_e/b) - 1\}
\end{aligned}\right\} \qquad (4.2\text{-}43)$$

Thus, it follows that

$$\left.\begin{aligned}
v_0 &\gtrless 0 \quad \text{for } b/b_e \gtrless \tfrac{8}{9} \\
\omega_i &\gtrless 0 \quad \text{for } b/b_e \lessgtr 1
\end{aligned}\right\} \qquad (4.2\text{-}44)$$

Further, the induced drag is given in the formula:

$$D_i/D_{i,e} = (b_e/b)^2\{8(b_e/b)^2 - 16(b_e/b) + 9)\} \qquad (4.2\text{-}45)$$

Figure 4.2-15, given by Jones (1950, 1980), shows: (a) the spanwise loads or circulation distributions of the above four flat wings with span ratios of $b/b_e = 0.85, 1.00, 1.15,$ and 1.30; and (b) the induced drag ratio. Elliptic loading gives a smaller drag than any other form of loading within a restricted span. However, if the restriction on the span is removed, still lower values of induced drag can be obtained without any increase in the bending moment at the wing root. The lower values are obtained by permitting the span to increase and at the same time adopting a more tapered form of the loading curve (Jones 1950). If the restriction on the bending moment is further removed, the elliptic wing of elongated span is the best wing for minimum induced drag, assuming that the wake surface is flat.

A pointed wing tip is generally adopted by large sea birds in high-performance flight so as to reduce fatigue. Similarly, a cusped or pointed wing is chosen for airplanes for perhaps more structural considerations than aerodynamic ones (Large 1981).

Table 4.2-5 shows the general planforms of various carnivorous seabirds and land birds. It can be seen that the geometrical planform of the wing is strongly related to the birds' behavior or preying mode. The large birds have either an almost rectangular wing with slotted tips or a long and narrow wing with pointed tips and prefer glid-

ing flight, as mentioned in Sects. 3.2.1 and 3.2-2. High-performance preying birds such as swifts, swallows, and falcons have a pair of wings of either medium span with pointed tips or lunate (crescent) configuration. They can fly at high speed over long distances and maneuver skillfully to catch their prey. The aerodynamic characteristics of the lunate wing will be treated later in Sect. 6.3.5. In summary it can be said again that for a given mass of bird (1) a large wing area gives low flight speed, small rate of descent, and thus long endurance, (2) a large wing span assures long flight distance, (3) slotted tips make it possible to fly in gusty winds without loss of control, and (4) tapered wings with pointed tips enable the bird to fly using strong wing beats at high speeds, and thus to keep high maneuverability.

Sinusoidal flapping. For a slow rate of flapping, the instantaneous thrust can be given by the forward component of the air loading in the following equation (Jones 1980):

$$\left.\begin{aligned}
T &= 2\int_0^{b/2} \rho \Gamma U\{(-y\dot{\beta} - v)/U\}\mathrm{d}y \\
&= -2\int_0^{b/2} \rho \Gamma \dot{\beta} y \,\mathrm{d}y - 2\int_0^{b/2} \rho \Gamma v \,\mathrm{d}y
\end{aligned}\right\} \qquad (4.2\text{-}46)$$

where the second term is equal to the induced drag. Substituting Eqs. (4.2-39 to 41) into Eq. (4.2-46) yields;

$$\left.\begin{aligned}
T &= -\rho b^3 \dot{\beta}\left\{\tfrac{1}{3}v_0 + \tfrac{1}{4\pi}b\omega_i\right\} - D_i \\
&= -2\{(b/2)(\dot{\beta} + \omega_i)/U\}\frac{B}{(b/2)} - (v_0/U)L
\end{aligned}\right\} \qquad (4.2\text{-}47)$$

The instantaneous power is the product of the bending moment at the root B and the flapping (angular) velocity $\dot{\beta}$,

$$P = -2B\dot{\beta} \qquad (4.2\text{-}48)$$

If the flapping is a slow sinusoidal motion, or

$$\dot{\beta} = \beta_1 \omega \cos(\omega t) \qquad (4.2\text{-}49)$$

then the spanwise gradient of the induced velocity, ω_i may also be given in phase shift of π with respect to the angular velocity $\dot{\beta}$, as

$$\omega_i = -\omega_{i,1}\cos(\omega t) \qquad (4.2\text{-}50)$$

TABLE 4.2-5. Carnivorous birds. (The World Atlas of Birds; Farrand 1988, Perrins and Middleton 1985)

Diomedea exulans	Albatross family (Diomedeidae) Examples; Wandering albatross, the largest flying bird, black-browed albatross; gray-headed albatross; and royal albatross.	Their long, narrow, pointed wings make them superb gliders. They use vertical wind shear to continue soaring flight. They feed on marine organisms such as plankton, squid, and fish.
Sula leucogaster	Gannet family (Sulidae) Examples; Atlantic gannet, Australasian gannet, blue-footed booby and brown booby.	
Puffinus gravis	Shearwater family (Procellariidae) Examples; Audubon's shearwater, greater shearwater, manx shearwater, blue petrel, Bermuda petrel, and dark-rumped petrel.	
Fregata magnificens	Frigate bird family (Fregatidae) Examples; Great frigate bird, lesser frigate bird, magnificient frigate bird, and Andrew's frigate bird.	Frigate birds also have long, narrow and pointed wings with large aspect ratio. However, unlike other seabirds, they have extremely low wing loadings because of small weight and soar on thermals over the tropical sea. Frigate birds use their superb manoeuverability to harass other birds until they drop the food they are carrying.

(*continued*)

TABLE 4.2-5. (cont.)

Chaetura caudacuta	Swift family (Apodidae) Examples; Alpine swift, black swift, chimney swift, common swift, and palm swift.	
Hirundo daurica	Swallow family (Hirundinidae) Examples; American rough-winged swallow, barn swallow, blue swallow, cliff swallow, bank swallow, New World martin, purple martin, and Southern martin.	They have long, swept, and pointed wings or lunate wings, and very small legs and feet. They are, thus, entirely aerial in their way of life, extremely agile, and insectivorous.
Hemiprocne coronata	Crested swift family (Hemiprocnidae) Examples; Crested tree swift, lesser tree swift, and whiskered tree swift.	
Larus glaucescens	Gull family (Laridae) Examples; Glaucous gull, Iceland gull, ivory gull, black-headed gull, common gull, great black-backed gull, kittiwake gull, and little gull.	They are strong fliers with medium-pointed wings, often gliders. They are mainly coastal and omnivorous. Some, specifically the tern, circle high in the air over water and sometimes hover before diving for fish. The arctic tern is known to be a long-range migrant.
Sterna forsteri	Terns (Sternidae) Examples; Aleutian tern, Arctic tern, black tern, common tern, little tern, and sandwich tern.	

TABLE 4.2-5. (cont.)

Falco tinnunculus	Falcon family (Falconidae) Examples; Old World kestrel and *Falco peregrinus*. Small to medium size.	Their flight is direct and swift with less soaring than in other birds of prey. Their pointed wings enable them to scare up other birds into a long aerial chase and to strike them in midair.
Spizaetus nipalensis	Hawk family (Accipitridae) Examples; *Buteo*, golden eagle, *Butastur indicus*, *Pernis apivorus*, and kite. Strong medium sized legs and claws.	Their broad rectangular wing with slotted tips guarantees excellent gliding flight that uses upcurrents in a turbulent atmosphere. They feed on carrion, fruits and vegetables, or small animals such as birds, reptiles, insects, and fish.
Haliaeetus leucocephalus	American vulture family (Cathartidae) Examples; Black vulture, white tailed seaeagle, bald eagle, and Andean condor. Very large (span of 3 m and mass of 12 kg).	

Then the lift, the bending moment, the thrust, and the power are also in the same phase, as shown in Table 4.2-6.

The propulsive efficiency η defined by the ratio of the average propulsive work performed by the average power can be given by the formula

$$\eta = T_0 V/P_0 \simeq T_0 U/P_0$$

$$= 1 - (\omega_{i,1}/\beta_1 \omega) - \left(\frac{\pi v_0}{(b/2)\omega_{i,1}}\right)\left(\frac{\pi v_0}{(b/2)\beta_1 \omega}\right). \tag{4.2-51}$$

It can be said that the loss of efficiency results from both the ratio of the angular velocity induced by the wake to the angular velocity of the wing, and the product of the ratios of the lifting component of the induced velocity to the angular velocities of the wake and the wing. Thus, if the lifting (supporting) force or the induced velocity is zero or $v_0 = 0$, a deficiency results only from $\omega_{i,1}/\beta_1 \omega$.

4.3 Power-Saving Flight

Large birds save power in long-distance flight either by flying in tight formation in what is called "schooling flight," or by skimming just above a calm water surface to take advantage of the ground effect.

Schooling is a habit of birds that is thought to have developed for various social and genetic reasons. According to the random search theory, it may provide predatory species with advantages in feeding (Partridge 1982). In addition, it is no doubt advantageous to many species from the standpoint of self-defense, since it enlarges their visual size and makes them a more complex target to attack (Edmunds 1974). The power-saving aspect of schooling is the result of fluid-dynamic interactions. Small birds also have a way of saving power by using intermittent beatings in what is called "bounding flight."

TABLE 4.2-6. Aerodynamic characteristics of sinusoidal flapping wing, $\beta = \beta_1 \sin(\omega t)$

Lift	$L = \int_{-b/2}^{b/2} l(y)\mathrm{d}y = \rho U \int_{-b/2}^{b/2} \Gamma(y)\mathrm{d}y = L_0 + L_1 \cos(\omega t + \pi)$
Bending moment	$B = \int_0^{b/2} yl(y)\mathrm{d}y = \rho U \int_0^{b/2} y\Gamma(y)\mathrm{d}y = B_0 + B_1 \cos(\omega t + \pi)$
Thrust	$T = -2\int_0^{b/2} \rho \Gamma \dot{\beta} y\, \mathrm{d}y - 2\int_0^{b/2} \rho \Gamma v\, \mathrm{d}y = T_0 + T_1 \cos(\omega t + \pi) + T_2 \cos(2\omega t + \pi)$
Power	$P = -2B\dot{\beta} = P_0 + P_1 \cos(\omega t + \pi) + P_2 \cos(2\omega t + \pi)$
Amplitudes	$L_0 = \frac{1}{2}\pi \rho U b^2 v_0$
	$B_0 = \frac{1}{6}\rho U b^3 v_0$
	$L_1 = \frac{1}{3}\rho U b^3 \omega_{i,1}$
	$B_1 = \frac{1}{8\pi}\rho U b^4 \omega_{i,1}$
	$T_0 = -(v_0/U)L_0 - \{(b/2)(\beta_1 \omega - \omega_{i,1})/U\}\dfrac{B_1}{(b/2)}$
	$P_0 = -B_1 \beta_1 \omega$
	$T_1 = 2\{(b/2)(\beta_1 \omega - \omega_{i,1})/U\}\dfrac{B_0}{(b/2)} - (v_0/U)L_1$
	$T_2 = \{(b/2)(\beta_1 \omega - \omega_{i,1})/U\}\dfrac{B_1}{(b/2)}$
	$P_1 = 2B_0 \beta_1 \omega$
	$P_2 = B_1 \beta_1 \omega$
Coefficients	$C_{T_0} = T_0/\frac{1}{2}\rho U^2 S = \frac{1}{4\pi}A\!R(b\omega_{i,1}/U)(b\beta_1 \omega - b\omega_{i,1})/U - \pi A\!R(v_0/U)^2$
	$C_{L_0} = L_0/\frac{1}{2}\rho U^2 S = \pi A\!R(v_0/U)$

4.3.1 FORMATION FLIGHT

Trailing vortices. From the vortex theory, it is known that a vortex system trailed from an elliptic wing induces a laterally constant downwash on the wing surface and in the wake. It should be noted that far behind the wing the induced velocity is twice that on the wing surface, because the effects of the trailing vortices on the wake far behind the wing are considered to be integrated from far forward to far backward along the vortices, whereas the effects on the wing surface result only from the vortices to the rear. Outside the wing and wake, there can be observed an upwash flow in which the strongest upwash is seen just outside the wing tips.

The trailing vortex wake system generated by the ordinary lifting wing with moderate sweep and aspect ratio (Fig. 3.4-2) is unstable and tends to roll up into a pair of oppositely rotating vortices, "tip vortices" (Rossow 1977; Bilanin et al. 1977; Raj and Iversen 1978). A set of bound vortex and tip vortices is called a "horseshoe vortex." Detailed information on the rolling-up behavior, age of the wake as it dissipates with time, and the induced velocity generated by a "vortex core" can be found in many references, including Lamb (1930), Betz (1932), Westwater (1935), C.A. Cone (1960), Hancock (1970), Clements and Maull (1973), C.E. Brown (1973), Williams (1974), Ciffone (1974), and Ciffone and Orloff (1975).

Lateral and diagonal formation. Let us consider the horseshoe vortex system of a hypothetical bird flying with a constant circulation, as shown in Fig. 4.3-1. The induced velocity of this system w, positive for the upwash, at a point P outside or

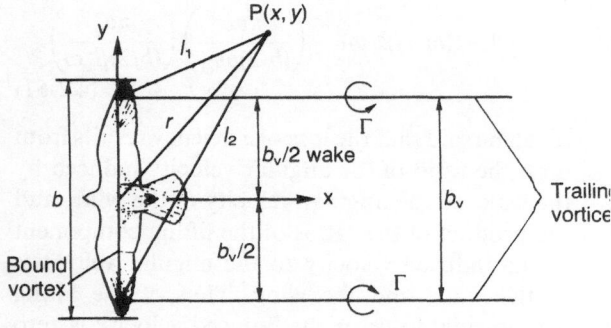

FIG. 4.3-1. Mathematical model of horseshoe vortex of a bird in stationary flight.

TABLE 4.3-1. Induced velocity w caused by a pair of trailing vortices (positive upward)

Position	Expressions
$\|y\| > b_v/2$	$\dfrac{\Gamma}{4\pi}\left\{-\dfrac{1}{x}\left(\dfrac{y+b_v/2}{l_2}-\dfrac{y-b_v/2}{l_1}\right)+\dfrac{1}{y-b_v/2}\left(\dfrac{x}{l_1}+1\right)-\dfrac{1}{y+b_v/2}\left(1+\dfrac{x}{l_2}\right)\right\}$
	$\quad =\dfrac{\Gamma}{4\pi}\left[\dfrac{1}{\sqrt{(y-b_v/2)^2+x^2}}\left(\dfrac{y-b_v/2}{x}+\dfrac{x}{y-b_v/2}\right)-\dfrac{1}{\sqrt{(y+b_v/2)^2+x^2}}\left(\dfrac{y+b_v/2}{x}+\dfrac{x}{y+b_v/2}\right)\right.$
	$\quad\quad \left.+\dfrac{1}{y-b_v/2}-\dfrac{1}{y+b_v/2}\right], \quad \|y\| > b_v/2.$
$\|y\| < b_v/2$	$\dfrac{\Gamma}{4\pi}\left\{-\dfrac{1}{x}\left(\dfrac{y+b_v/2}{l_2}-\dfrac{b_v/2-y}{l_1}\right)-\dfrac{1}{-y+b_v/2}\left(\dfrac{x}{l_1}+1\right)-\dfrac{1}{y+b_v/2}\left(1+\dfrac{x}{l_2}\right)\right\}$
	$\quad =\dfrac{\Gamma}{4\pi}\left[\dfrac{1}{\sqrt{(-y+b_v/2)^2+x^2}}\left(\dfrac{-y+b_v/2}{x}+\dfrac{x}{y-b_v/2}\right)-\dfrac{1}{\sqrt{(y+b_v/2)^2+x^2}}\left(\dfrac{y+b_v/2}{x}+\dfrac{x}{y+b_v/2}\right)\right.$
	$\quad\quad \left.-\dfrac{1}{-y+b_v/2}-\dfrac{1}{y+b_v/2}\right), \quad \|y\| < b_v/2.$

inside the wake, is given in Table 4.3-1, where the circulation Γ of each of a pair vortices generated by the wing is given by

$$\Gamma = \tfrac{1}{2}C_L c_0 U = (2b/\pi\mathcal{R})C_L U = L/\rho b_v U \quad (4.3\text{-}1)$$

and where c_0 and b_v are the center-section chord and the distance between the pair of vortices respectively, the latter of which is given by

$$b_v = (\pi/4)b \simeq 0.785 \quad (4.3\text{-}2)$$

The table shows that the upwash increases as the point P approaches the trailing vortices of $|y| \to b_v/2$ far downstream (or $x \to \infty$), if the dissipation is not considered.

Vortex systems consisting of n birds in formation make their effective aspect ratio $\mathcal{R}_e$ large. Figure 4.3-2 shows (a) the increment of effective aspect ratio $\mathcal{R}_e/\mathcal{R}$ as a function of the gap between two rectangular wings of $\mathcal{R} = 3$ arranged

FIG. 4.3-2. a–c. Effective aspect ratio and power reduction for elliptic wings in formation flight. **a** A pair of wings flying side-by-side. (Data from Schlichting 1942; Hoerner 1965). **b** A line-abreast formation. (Redrawn from Lissaman and Shollenberger 1970). **c** Power reduction of equal wings. (From Hummel 1978 with permission).

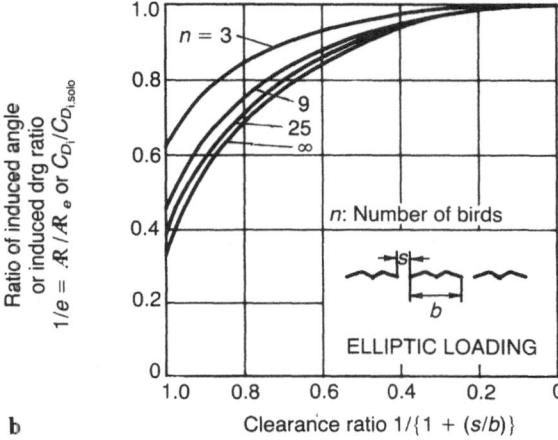

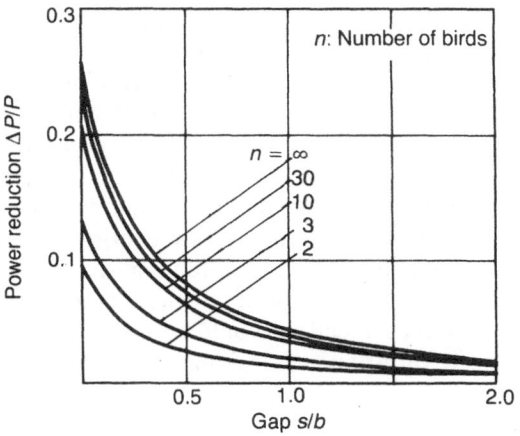

side by side, (b) the change in the effective aspect ratio, $1/e = Æ\!R/Æ\!R_e$, which is nearly equal to the induced drag ratio $C_{D_i}/C_{D_{i,solo}} \simeq 1/(Æ\!R_e/Æ\!R)$ and (c) the total power reduction $\Delta P/P$ in formation flight of birds with equal wings. The maximum saving in the induced power for an infinite number of birds will be up to 71% for a precise optimum positioning (Lissaman and Shollenberger 1970). For the V-shaped formation Hummel (1978) gave useful charts of the power reduction. The sweep angle of the V-shaped formation has almost no effect on the power reduction of the formation in the fully developed wake. These figures show essential data relating to the superior performance of formation flight. It is interesting to note that, in actuality, formation flight is not executed by the gliding birds but by the beating birds.

Shown in Fig. 4.3-3 is an example of formation flight for the swan (*Cygnus cygnus*). It is interesting to note that (1) the lateral distance between the opposite wing tips of neighboring birds is negative, which means the wing tip of each trailing bird is strongly influenced by the bird ahead of it, and (2) the swans spread their webbed feet and use them as a pair of elevators for longitudinal stability and control. Hainsworth (1987, 1988) found that brown pelicans save locomotive energy by using the ground effect and formation flight, and that a few birds have wingbeat frequencies synchronous with the bird ahead as would be needed to track vertical variation in trailing wing tip vortex positions.

FIG. 4.3-3. Formation flight of swan (*Cygnus cygnus*). (Courtesy of Minoru Takedazu).

4.3.2 GROUND EFFECT

Forward flight. The interference effect between an individual bird and a ground surface has been studied, for example, by Withers and Timko (1977), R.W. Blake (1983a), and Hainsworth (1988). A wing of circulation Γ flying close to the ground can be regarded as a combination of the wing and its image on the ground because the boundary condition on the ground can be simulated by an image wing of same size b, same height H, and opposite circulation $-\Gamma$, as shown in Fig. 4.3-4. In such a case, the wing is considered to be subject to the "ground effect."

Shown in Fig. 4.3-5 is the increase of section lift coefficient (Tomotika et al. 1933) of a two-dimensional flat plate near the ground in comparison with that in free air. The lift coefficient depends not only on the height from the ground H/c, but also on the angle of attack. It should be noted that in some conditions, the lift is less than the free-air value, in others it is higher. With increasing proximity to the ground, the absolute value of pressure is reduced over the upper surface and increased over the lower surface, so that the overall lift depends on how this balance works out (Bagley 1956). Thus, the sectional lift slope is a highly nonlinear function of the height H/c and the angle of attack α (Küchemann 1978).

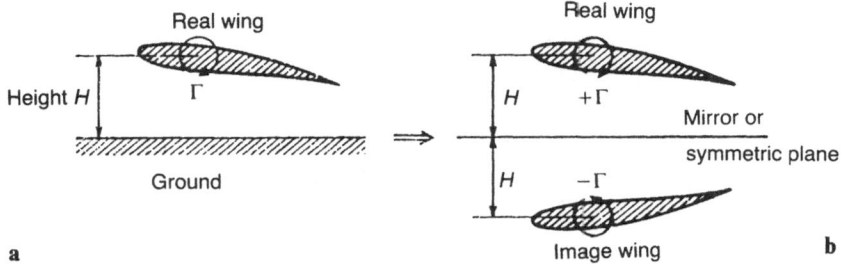

FIG. 4.3-4. **a, b.** Mathematical simulation of the ground effect. **a** Real arrangement. **b** Mathematical model.

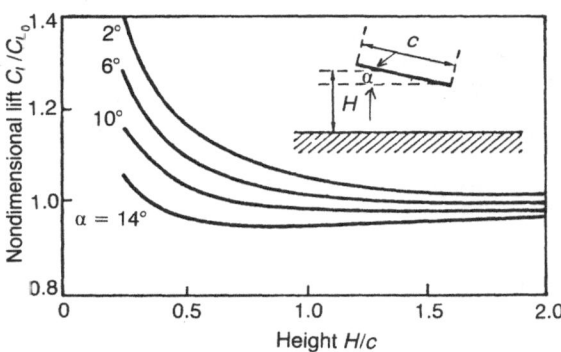

FIG. 4.3-5. Ground effect on the lift and pressure of a two-dimensional flat plate. (Data from Tomotika et al. 1933) (From Küchemann 1978 with permission).

The ground effect on a three-dimensional wing near the ground was discussed earlier in Sect. 3.5.4. The ground effect acts to increase the "effective aspect ratio" (see Eq. (3.5-27) as follows:

$$\mathcal{R}_{e,H}/\mathcal{R}_e = \{1 + 33(H/b)^{3/2}\}/33(H/b)^{3/2}$$

(Hoerner and Borst 1975)

$$= \{1 + 16(H/b)\}^2/16(H/b)^2$$

(Spedding 1987a, 1987b)

$$(4.3-3)$$

Hovering flight. As in forward flight, there is also a reduction in the induced-power penalty in hovering flight near the ground. This ground effect in hovering flight is well known in helicopter operation. Since the induced velocity in hovering flight is stronger than that in forward flight, the ground effect is pronounced. The simple analysis presented by Heyson (1959, 1960) and Lighthill (1979) for an axisymmetric actuator-disc model. The latter shows the ratio of induced power with and without the ground effect, $P_{i,in}/P_{i,out}$ as a function of the ratio H/R of height to radius of actuator disc, which is shown in Fig. 4.3-6, in comparison with experimental results (Zbrozek 1950; Knite and Hefner 1941; Miller et al. 1968). Here the radius of the actuator disk may be considered as the half-span of the hovering bird, $R \simeq b/2$.

4.3.3 INTERMITTENT FLIGHT

In small birds, such as starlings, wagtails, woodpeckers, sparrows, and others, almost all of the wing area is occupied by primary feathers, and the wing beating is performed substantially at a single joint. Thus, it seems that they do not have enough upper and lower arm length to adjust the wingspan and area in correspondence with the flight speed.

FIG. 4.3-6. The induced power ratio in ground effect.

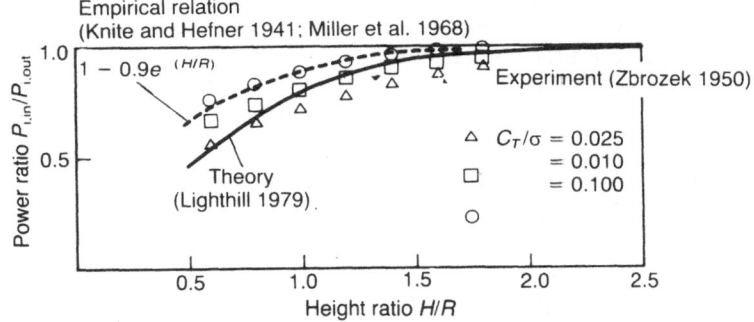

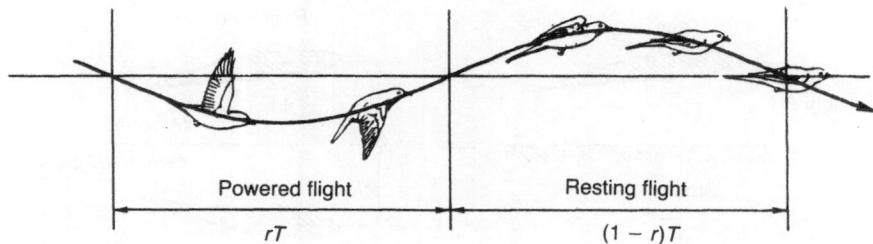

Powered flight | Resting flight

rT | $(1-r)T$

FIG. 4.3-7. Bounding flight.

They are also restricted to almost only one wing-beat frequency. To obtain the desired flight velocity (speed and path), they can use intermittent beating flight (bounding flight), in which the wings are intermittently folded. This causes them to fly like an arrow, first losing height, and then swooping up again, as shown in Fig. 4.3-7. The flight path is, then, a series of valleys (downward convex) during a time fraction r, and hills (upward convex) the rest of the time $1 - r$. The valley and hill in one cycle period T correspond to the powered flight and rest flight phases respectively. A theoretical study of this type of flight was made by Rayner (1977) and extended by himself (1985b) However, he made clear that no promising results for reducing energy consumption were obtained. *Simple analysis.* By assuming a constant induced velocity distribution over the stroke plane of a pair of beating wings, a constant profile drag δ, and a small maximum tip speed of beating V_t in comparison with the flight speed U_r, or $V_t \ll U_r$, the energy consumed for this curved flight is given by

$$\left.\begin{aligned} P = E_r/T &= \tfrac{1}{2}\rho S U_r{}^3(f/S + \delta r) \\ &\quad + \tfrac{1}{2}\sigma_2 \rho S V_t{}^2 \delta r U_r \\ &\quad + (n^2 r)\{2W(W/S)/\rho\pi\mathcal{R}_e U_r\} \\ &\quad \cdot \{1 + (C_D/C_L)^2\} \end{aligned}\right\}$$

(4.3-4)

where n is "load factor" showing an increased lift $n = L/W$, and V_t and σ_2 are already given by Eqs. (4.2-35) and (4.2-37) respectively. In the right hand side of the above equation, the first term results from the parasite drag including the profile drag of both wings extended during beating flight, Eq. (4.1-18), the second term results from the profile power caused by the beating motion, Eq. 4.2-36, and the third term also results from the induced power, Eq. (4.1-13), required for the increased lift of $L = nW$ during the beating flight for compensating the height loss during the rest flight.

The vertical velocity at the center of gravity of the body can be given by

$$\left.\begin{aligned} \dot{Z}(t) &= \dot{Z}_0 + (n-1)gt & 0 < t < rT \\ &= \dot{Z}_0 + (n-1)grt - g(t - rT) & r < t < T \end{aligned}\right\}$$

(4.3-5)

These values must be connected at switching points $t = 0$ or T and $t = rT$. Then the initial speed $\dot{Z}_0$ and the load factor n in the powered flight can be determined as follows:

$$\left.\begin{aligned} \dot{Z}_0 &= -\tfrac{1}{2}(n-1)grT = \tfrac{1}{2}(r-1)gT \\ n &= 1/r \end{aligned}\right\}$$

(4.3-6)

Thus, the minimum energy for a constant C_D/C_L and a given r, $(E_r/T)_{\min} = P_{\min}$, can be obtained at the speed of

$$\left.\begin{aligned} U_{r,P_{\min}} = \Biggl[&-\frac{1}{6}\left\{\frac{\sigma_2 V_t{}^2 \delta r}{f/S + \delta r}\right\} \\ &+ \sqrt{\left(\frac{1}{36}\right)\left\{\frac{\sigma_2 V_t{}^2 \delta r}{f/S + \delta r}\right\}^2 + \frac{2}{3}\frac{1}{\rho S(f/S + \delta r)}\frac{2W(W/S)\{1 + (C_D/C_L)^2\}}{\rho\pi\mathcal{R}_e r}} \Biggr]^{1/2} \end{aligned}\right\}$$

(4.3-7)

If the energy-speed ratio or the energy spent for a unit flight distance, $E_r/U_r T = P/U_r$, is minimum as in optimal cruising flight, then the flight speed is given by

$$\begin{aligned} U_{r,(P/U)_{\min}} = [2\{2W(W/S)/\rho\pi\mathcal{R}_e\}\{1 + (C_D/C_L)\}/ \\ \rho S(f/S + \delta r)r]^{1/4} \end{aligned}$$

(4.3-8)

Since the "minimum cost speed," which is the speed at the maximum lift-to-drag ratio, is given by

$$\begin{aligned} U_1 &= U_{(L/D)_{\max}} \\ &= \{W/\rho S\}^{1/2}[4\{1 + C_D/C_L)^2\}/\pi\mathcal{R}_e C_{D_0}]^{1/4} \end{aligned}$$

(4.3-9)

TABLE 4.3-2. Dimensions and performance of a sparrow

Items	Symbols	Units	Data
Mass	m	g	24.5
Weight	$W = mg$	N	0.24
Span	b	cm	22.6
Wing length	R	cm	9.7
Wing area	S	cm^2	104
Wing loading	W/S	N/m^2	23.0
Aspect ratio	$\mathcal{R}$	—	3.93
Body length	l	cm	14.3
Body drag coefficient	$C_{D,b}$	—	0.023
Profile drag coefficient	C_d	—	0.016
Geometrical parameter	σ_2	—	0.144
Beating frequency	f	Hz	14.0
Beating amplitude	$2\beta_1$	deg	140

the speed ratio to the above minimum cost speed is given by

$$U_{r,(P/U)_{min}}/U_1 = \{(f/S + \delta)/(f/S + \delta r)r\}^{1/4} \geq 1 \tag{4.3-10}$$

This result shows that the flight speed in bounding flight is larger than the optimal cruising speed or the minimum cost speed.

The energy difference for a unit distance is then given by the formula:

$$(E_r/U_r - E_1/U_1)/T = \frac{1}{2}\rho S U_1^2 \left[(f/S)\left\{ \sqrt{\frac{C_{D_0}}{(f/S + \delta r)r}} - 1 \right\} \right.$$
$$+ \delta\left\{ r\sqrt{\frac{C_{D_0}}{(f/S + \delta r)r}} - 1 \right\} \Bigg]$$
$$- \frac{1}{2}\sigma_2 \rho S V_t^2 \delta(1 - r)$$
$$+ \left\{ \frac{2W(W/S)}{\rho \pi \mathcal{R}_e} \right\}\{1 + (C_D/C_L)^2\}$$
$$\cdot \left\{ \frac{1}{r}\sqrt{\frac{(f/S + \delta r)r}{C_{D_0}}} - 1 \right\} \Bigg/ U_1^2 \tag{4.3-11}$$

where $E_1 = E_r(r = 1)$. The requirements for realizing bounding flight are: (1) a large profile drag of large wing $S\delta$ but small parasite drag area of the body f/S, (2) a large tip chord or large σ_2, (3) intense beating or a large V_t, and (4) a small span loading or $W(W/S)/\rho\pi\mathcal{R}_e = (W/b)^2/\rho\pi$.

Numerical analysis. An exemplified calculation was performed on a sparrow (*Passer montanus saturatus* Stejneger), the dimensions and estimated aerodynamic characteristics of which are given in Table 4.3-2. As shown in Fig. 4.3-8, the result shows that (1) the minimum power and the mini-

mum power-to-speed ratio are given by $r = 1.0$ for nonbounding flight at the speeds of $U_{1,P_{min}} = 5.5$ m/s and $U_{1,(P/U)_{min}} = 7.5$ m/s respectively, and (2) beyond the speed of $U_{1,P_{min}}$, the necessary power which is represented by an envelope for various values of r is smaller than the power for $r = 1$.

Actually, the estimated optimal speeds of nonbounding flight ($r = 1$), $U_{1,P_{min}} = 5.5$ m/s and $U_{1,(P/U)_{min}} = 7.5$ m/s, are small for the flight of the sparrow. It is supposed that the wing configuration of the sparrow is not adequate for flying at the aforesaid optimal speeds, but is better adapted for flying at higher speed or, more probably, for making frequent powerful flight (such as climbing flight). These characteristics of the sparrow's wing configuration would appear to be closely related to the bird's daily activities. Probably, the muscles are also made efficient for this powerful flight. Thus, as stated in Chap. 3, the sparrow adopts intermittent beating in level flight, but not in climbing flight.

For a given U, the time fraction r of both the minimum power and the minimum power-to-speed ratio can be given as:

$$r = \sqrt{\frac{2W(W/S)}{\rho\pi\mathcal{R}_e U}\{1 + (C_D/C_L)^2\} \bigg/ \frac{1}{2}\rho S\delta U^3\{1 + \sigma_2(V_t/U)^2\}} \tag{4.3-12}$$

Then, for $r < 1$, the flight speed must be

$$U > \left[-\frac{1}{2}\sigma_2 V_t^2 \right.$$
$$\left. + \sqrt{\left(\frac{1}{2}\sigma_2 V_t^2\right)^2 + \frac{(2W/\rho S)^2}{\pi\mathcal{R}_e\delta}\{1 + (C_D/C_L)^2\}} \right]^{1/2} \tag{4.3-13}$$

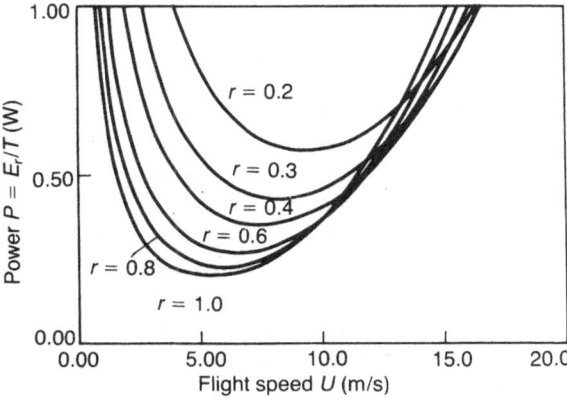

FIG. 4.3-8. Necessary power curve versus flight speed of a sparrow.

If the bird wants to fly at a speed satisfying the above condition, it would select bounding flight.

The preceding discussion relies on the power or energy balance, which is essentially an integrated form of differential equations of motion, under the assumption of a constant induced velocity during the specified period in consideration. In order to get a reliable result, a numerical calculation must be performed on the differential equations for body motion by using any computational method of time steps, and on the integral-differential equations for the aerodynamic forces acting on the beating wings by using, for example, the LCM (Azuma and Watanabe 1988). Unfortunately, at the present time, the author does not have enough data on the beating motion of the sparrow to proceed with the calculation further.

4.3.4 EFFECTS OF WIND AND HEIGHT ON MIGRATION FLIGHT

The effects of height and wind on the performance of gliding flight were presented in Chap. 3. Here, let us consider the effects of horizontal wind and height on the cruising performance of migrating birds.

The governing equations for quasi-steady flight with a small or zero climb angle γ were given earlier in Sect. 4.1.1. Considering the ground speed V as a summation of the airspeed U and the wind speed U_w,

$$V = U \pm U_w \qquad (4.3\text{-}14)$$

where the plus or minus sign denotes a tail wind or head wind respectively, the equations of motion of a flying animal and its flight range can be expressed as follows

$$T = D + W\gamma = \tfrac{1}{2}\rho U^2 S C_D + W\gamma \quad (4.3\text{-}15)$$

$$W = L = \tfrac{1}{2}\rho U^2 S C_L \qquad (4.3\text{-}16)$$

$$\frac{\mathrm{d}}{\mathrm{d}t}\begin{pmatrix} X \\ Z \end{pmatrix} = V\begin{pmatrix} 1 \\ \gamma \end{pmatrix} = (U \pm U_w)\begin{pmatrix} 1 \\ \gamma \end{pmatrix} \quad (4.3\text{-}17)$$

Here, the drag coefficient is considered to be the total drag of the flying animal in gliding flight without beating. Thus, the thrust can be generated by the beating motion without increasing the drag.

Referring to the equations given in Sect. 4.1.1, the power balance between the necessary total power and the consumed rate of mass or fuel can be expressed as

$$
\begin{aligned}
-\,\mathrm{d}m/\mathrm{d}t &= cP_t = cP_a/\eta_F \\
&= c(P_i + P_p + P_0 + P_c)/\eta_F \\
&\simeq c(\sqrt{2/\rho S}\,(W/C_L)^{3/2} \\
&\quad \cdot \{C_{D_0} + C_L{}^2/\pi A\!R_e + C_L\gamma\} \\
&\quad + P_0]/\eta_F
\end{aligned}
\right\} \qquad (4.3\text{-}18)
$$

where c is fuel consumption rate. Then, combining Eqs. (4.3-16 and 4.3-17) yields

$$\begin{pmatrix} X \\ Z \end{pmatrix} = -\frac{1}{g}\int_1^{w_f}\left(\frac{\eta_F}{c}\right) F(\gamma,\sigma,C_L;w)\begin{pmatrix} 1 \\ \gamma \end{pmatrix}\mathrm{d}w \qquad (4.3\text{-}19)$$

where

$$F(\gamma,\delta,C_L;w)$$
$$= \frac{\{1 \pm v\sqrt{\sigma}\sqrt{C_L/w}\}}{w\{(C_L/\pi A\!R_e) + (C_{D_0}/C_L) + \gamma\} + p\sqrt{\sigma C_L/w}} \qquad (4.3\text{-}20)$$

$$
\left.\begin{aligned}
w &= m/m_0 = W/W_0 \\
w_f &= W_f/W_0 \\
p &= P_0/W_0 V_0 \\
v &= U_w/V_0 \\
\sigma &= \rho/\rho_0
\end{aligned}\right\} \qquad (4.3\text{-}21)
$$

$$
\left.\begin{aligned}
U_0 &= \sqrt{2W_0/\rho_0 S C_{L_0}} \\
V_0 &= \sqrt{2W_0/\rho S} = U_0(C_{L_0} = 1)
\end{aligned}\right\} \qquad (4.3\text{-}22)
$$

and where the subscript 0 shows the values at the initial time on the ground ($Z = v = 0$). The flight path angle γ can, from Eq. (4.3-15), be rewritten as:

$$\gamma \simeq \mathrm{d}Z/\mathrm{d}X = t/w - \{(C_{D_0}/C_L) + (C_L/\pi A\!R_e)\} \qquad (4.3\text{-}23)$$

Here t is the nondimensional thrust defined by:

$$t = T/W_0 \qquad (4.3\text{-}24)$$

Then Eq. (4.3-20) can also be written as:

$$F(\sigma, C_L, t; w) = \frac{\{1 \pm v\sqrt{\sigma C_L/w}\}}{\{t + p\sqrt{\sigma C_L/w}\}} \qquad (4.3\text{-}25)$$

The wind speed, which is a function of the flying height or the air density ratio, is related to the range of flight through the <u>nondimensional wind speed</u> $v = U_w/V_0 = U_w/\sqrt{2W_0/\rho_0 S}$ and the flying height is related to the range through the ratio of efficiency and fuel consumption, η_F/c, and the air density ratio σ.

The air density ratio of Standard Atmosphere (1955) is

$$\sigma = \rho/\rho_0 = \{1 - (aZ/T_0)\}^{n-1} \quad (4.3\text{-}26)$$

where

$$\left. \begin{array}{l} n = 5.2561, \quad a = 0.065°\text{K/m} \\ T_0 = 273.15°\text{K}, \quad \rho_0 = 1.2250\,\text{kg/m}^3 \\ \qquad\qquad\qquad = 0.125\,\text{kgfs}^2/\text{m}^4 \end{array} \right\} \quad (4.3\text{-}27)$$

Eq. (4.3-26) can be inverted to

$$Z = (T_0/a)\{1 - \sigma^{1/(n-1)}\} \quad (4.3\text{-}28)$$

Since the maximum rate at which oxygen can be absorbed by the lungs decreases as the partial pressure of oxygen declines, the ratio of efficiency and fuel consumption may be assumed to be given by:

$$\eta_F/c = K_F e^{-AZ} \quad (4.3\text{-}29)$$

where K_F and A are specific available energy and fuel lapse index respectively, and their exemplified values are

$$K_F = (\eta_F/c)_{Z=0} = 0.23/(0.25 \times 10^{-7})$$
$$\simeq 10^7\,\text{J/kg (See Sect. 4.1.2)} \quad (4.3\text{-}30)$$

$$\left. \begin{array}{l} A = 6.93 \times 10^{-5}m^{-1} = A_0 \\ \quad = 3.47 \times 10^{-5}m^{-1} = 0.50A_0 \\ \quad = 2.43 \times 10^{-5}m^{-1} = 0.35A_0 \\ \quad = 1.39 \times 10^{-5}m^{-1} = 0.20A_0 \end{array} \right\} \quad (4.3\text{-}31)$$

and the first value of A_0 indicates that the ratio of efficiency is assumed to decrease exponentially with height to its half-value at $Z = 10$ km.

Combining Eqs. (4.3-28 and 4.3-29) yields:

$$\eta_F/c = K_F \exp[-A(T_0/a)\{1 - \sigma^{1/(n-1)}\}] \quad (4.3\text{-}32)$$

Then, combining Eq. (4.3-19) with Eq.(4.3-23), we get

$$X = \int_{w_f}^{1} G(\sigma, C_L, t, v; w)\,dw \quad (4.3\text{-}33)$$

$$Z = \int_{w_f}^{1} H(\sigma, C_L, t, v; w)\,dw \quad (4.3\text{-}34)$$

where the specific range $\dot{G}$ and the specific height $\dot{H}$ are respectively defined by:

$$\left. \begin{array}{l} G = (\eta_F/cg)F = G(\sigma, C_L, t, v; w) \\ \quad = \left[\dfrac{\{1 \pm v\sqrt{\sigma C_L/w}\}}{\{t + p\sqrt{\sigma C_L/w}\}g} \right] \\ \qquad \cdot K_F \exp[-A(T_0/a)\{1 - \sigma^{1/(n-1)}\}] \end{array} \right\} \quad (4.3\text{-}35)$$

and

$$\left. \begin{array}{l} H = G\gamma = H(\sigma, C_L, t, v; w) \\ \quad = \dfrac{\{1 \pm v\sqrt{\sigma C_L/w}\}K_F \exp[-A(T_0/a)\{1-\sigma^{1/(n-1)}\}]}{gw[t + p\sqrt{\sigma C_L/w}]/[t - w\{C_{D_0}/C_L + C_L/\pi A\!\!R_e\}]} \end{array} \right\} \quad (4.3\text{-}36)$$

The optimum range can be obtained by solving the variational problem of maximizing the function X (the range of the specified flying animal) in Eq. (4.3-33) under given flight conditions, such as $P_0/W_0V_0 = p$ and $U_w = U_w(Z)$, or a given $v\,(\sigma)$ constraints such as:

$$\left. \begin{array}{l} 0 \le C_L \le C_{L_{max}} \\ w_f \le w \le 1.0 \\ 0 \le \gamma \ll 1 \end{array} \right\} \quad (4.3\text{-}37)$$

and constraints imposed by Eqs. (4.3-28 and 4.3-34), or a constraint on σ,

$$(T_0/a)\{1 - \sigma^{1/(n-1)}\} = \int_{w_f}^{1} H(\sigma, C_L, t, v; w)\,dw \quad (4.3\text{-}38)$$

and terminal conditions such as $Z(w_f) = Z(1) = 0$, or

$$\sigma(w_f) = \sigma(1) = 1 \quad (4.3\text{-}39)$$

This means finding an optimal orbit in a space consisting of $\sigma(w)$, $C_L(w)$, $v(\sigma) = v(w)$ and $t(w)$ for the maximum range, in which C_L and t are control inputs.

If the wind can be assumed to be homogeneous within this space, then the flight velocity with respect to the ground is simply the sum of the wind velocity and the flight velocity relative to the air. Thus, a tail wind acts to extend the flight range, while a head wind decreases the range. Figure 4.3-9 shows the effects of head and tail winds on the achieved range (Pennycuick 1969).

Since migration flight is performed over distances of far greater magnitude than height, the height change may be discarded ($\gamma = 0$) in the equations. Then, by assuming no constraints on C_L and σ and thus discarding the constraint given by Eq. (4.3-38), the optimal height $\sigma(w)$ and lift

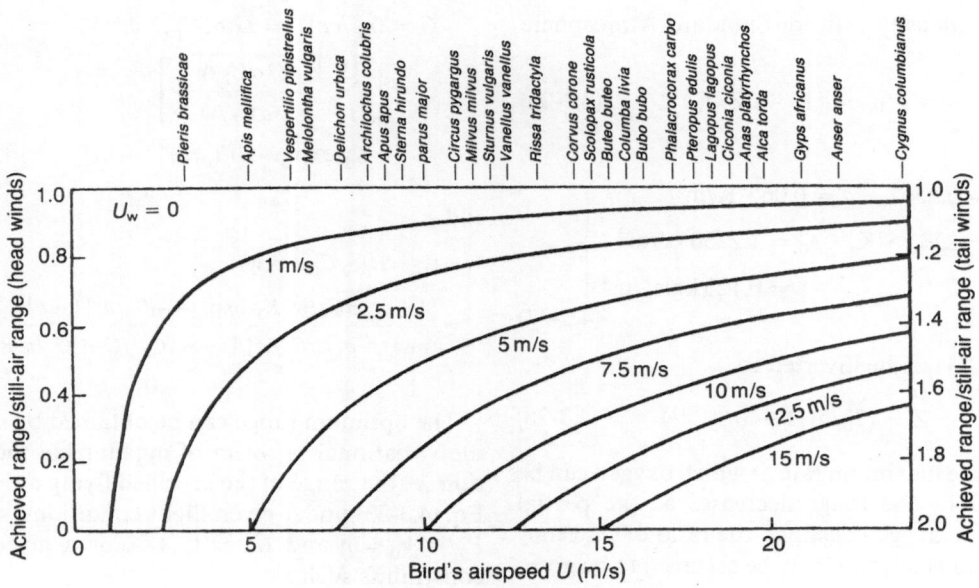

coefficient $C_L(w)$ can be determined for no wind condition, or $v = 0$, to solve the following Eulerian equations:

$$\partial G(v = \gamma = 0)/\partial \sigma = 0 \qquad (4.3\text{-}40)$$

$$\partial G(v = \gamma = 0)/\partial C_L = 0 \qquad (4.3\text{-}41)$$

where

$$G(v = \gamma = 0)$$

$$= \frac{K_F \exp[-A(T_0/a)\{1 - \sigma^{1/(n-1)}\}]}{g[w\{(C_L/\pi \mathcal{R}_e) + (C_{D_0}/C_L)\} + p\sqrt{\sigma C_L/w}]}$$
$$(4.3\text{-}42)$$

In this expression the nondimensional thrust is discarded because it has been replaced with the drag coefficient by using Eq. (4.3-15), as follows:

$$t = \{(C_{D_0}/C_L) + \{C_L/\pi \mathcal{R}_e)\}w \qquad (4.3\text{-}43)$$

The air density ratio and the lift coefficient can then be obtained for maximum range as functions of w and C_L, as follows:

$$\sigma_{X_{max}} = \sigma(w, C_L) \qquad (4.3\text{-}44)$$

$$C_{L, X_{max}} = C_L(w, \sigma_{X_{max}}) \qquad (4.3\text{-}45)$$

Now, let us apply the above analysis to the migration flight of a juvenile crane (*Grus japonensis*), the pertinent dimensions of which are given in Table 4.3-3. By solving Eqs. (4.3-40 and 4.3-41), the optimal value of the specific range is

FIG. 4.3-9. Effect of head and tail winds on specific range for eight windspeeds, from 0 to 15 m/s; the effect on the specific range for a bird flying at a given airspeed is read off the scale at left for a head wind, or off that at right for a tail wind. Calculated cruising speeds (*U*) for various animals are maked. (From Pennycuick 1969).

obtained as a function of the altitude. It should be remembered that the bird can get sufficient oxygen for the flight at high altitudes, as stated in Sect. 4.1.2. Figure 4.3-10 shows the result for various values of the fuel lapse index A, and an assumed profile power of $P_0 = V_0 p = 13.9 \times 0.0382 = 40$ W. It can be seen that for the highest value of the fuel lapse index, A_0, the flight altitude should be low, but as the fuel lapse index decreases, $A < A_0$, the optimal value of G increases with the

TABLE 4.3-3. Dimensions of a juvenile crane. (*Grus japonensis*)

Items	Symbols	Units	Values
Wingspan	b	m	2.21
Wing area	S	m²	0.640
Aspect ratio	$\mathcal{R}$	—	7.6
Mass at takeoff	m_0	kg	7.70
Minimum drag coefficient	C_{D_0}	—	0.010
Profile power	P_0	W	0 and 40
Specific available energy	K_F	J/kg	1.0×10^7

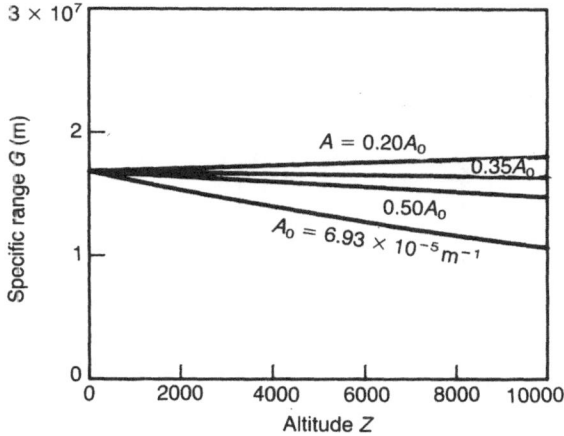

FIG. 4.3-10. Optimal specific range G.

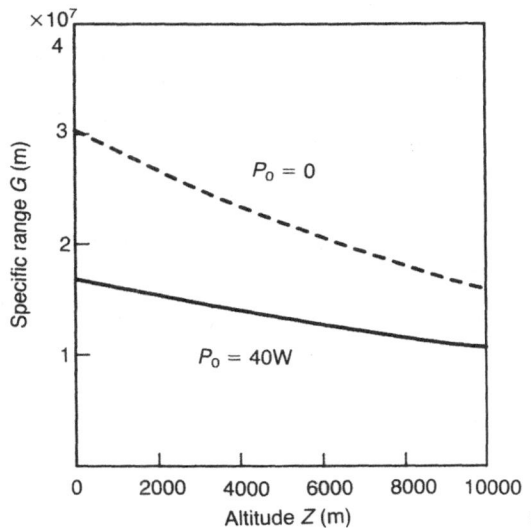

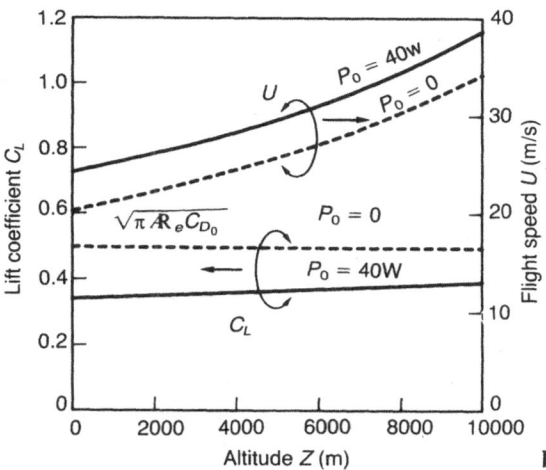

FIG. 4.3-11. **a, b.** Effect of profile power on the migration flight of a juvenile crane. **a** Specific range. **b** Flight speed and lift coefficient.

flight altitude. The actual flight range for every 10% consumption of body mass is given by Eq. (4.3-33) as 1,700 km in the present example of a crane flying at sea level, and is independent of the $A = 0.35A_0 = 2.43 \times 10^{-5}$ m^{-1}.

Figure 4.3-11 shows the effect of profile power on (a) the specific range G and (b) the flight speed U, or the lift coefficient C_L for the case of $A = A_0$. As the profile power, which was assumed constant, decreases, the specific range increases appreciably by reducing the flight speed or increasing the lift coefficient. For $P_0 = 0$, the lift coefficient coincides with that of the maximum range of $C_{L,(L/D)_{max}}$. For a positive value of the profile power, $P_0 > 0$, the crane must fly with higher speed than the above optimal value of $U_{(L/D)_{max}}$ to reduce the flight time and thus to save profile power. It is impressive to hear that after the monsoon, migrational flights of cranes over the Himalayas are sometimes observed by climbers.

4.4 Flight in Insects

Insects are basically terrestrial animals, but some live in water, and many are capable of flight. One of the interesting characteristics of insects is their small size in comparison with other flying animals, such as birds and mammals. Most present-day insects are less than 10^{-1} m in length. The smallest winged insects are wasps of the family Mymaridae, known as fairyflies. Some species are about 0.25 mm in body length and wingspread (E.W. Smith 1985). Such small insects can easily find living space nearly anywhere and have evolved a great number of species. Small size also means a shorter life span (See Fig 1.1-2). This gives these insects more opportunity for mutation, hence a greater potential for evolutionary change (Borror et al. 1976). They have been able to survive by adaptation in response to severe environmental conditions. Today there are more than seven hundred thousand recorded species.

4.4.1 GENERAL CHARACTERISTICS

Structure and wing motion. One important aspect of insect life is metamorphosis, through which insects change not only their shape but also the food they eat to adapt to the seasonal changes in their surroundings. Adult insects are divided into three main sections: head, thorax, and abdomen, each with its own particular function. The head has two kinds of eyes: two or three small simple eyes and a pair of large compound eyes. The thorax acts as an exoskeleton. The abdomen serves digestive functions. Attached to the three segments are three pairs of legs and usually one or two pairs of wings for locomotion (Pringle 1957, 1975).

A pair of completely developed wings grow out from the body wall at the mesothorax and metathorax. Because of its small size and consequent low Reynolds number, an insect's wing must operate in stickier air than a bird's wing. The wing is not usually a streamlined airfoil (Buckholz 1986), and the beating is performed intensely at high frequency. Being membranous and simple in structure, the wing is easy to fold away. These characteristics provide a wing structure and locomotion mode well adapted to various ways of life. Wings increase an insect's chances of survival by enabling it to evade unsuitable living environments (Nachtigall 1974, 1980b).

The presence of one or more supporting "veins" near the leading edge of the wing allows it to take its optimum angle of attack during beating motion, usually by an active twisting action at the joint working against the aerodynamic and inertial moments and the torsional elasticity of the wing base, rather than by, or in addition to, a passive twisting action generated by the force of air. Some wings bear tiny hairs, spines or scales which are probably effective for aerodynamic improvement of wings operating with small Reynolds number. Some front wings are thick, leather-like, or hard and sheathlike.

Some mechanical characteristics common to many insects are presented in Table 4.4-1, which also includes the wing length and area data shown earlier in Sect. 3.2.2, and the flight muscle mass and beating frequency data shown in Figs. 4.1-4 and 4.1-10. The aspect ratio, the beating frequency, and the flight speed are shown in Fig. 4.4-1 as a function of either the total mass m or the wing length l_w.

Wing drive mechanisms. Although the active wing of insects has an almost single universal joint at the shoulder instead of the three joints of a bird's wing, the movement of the wings during beating flight is an extremely complicated action driven by

TABLE 4.4-1. Dimensions and performance data on insects. (Data from Osborne 1951; Weis-Fogh and Jensen 1956; Byrne, et al. 1988)

Name of insect	Mass m (mg.)	Total wing area S (mm.^{3})	Wing loading W/S (N/m^2) or mg/mm^2	Body frontal area S_f (mm.^{2})	Length of longer wing (mm.) $l_w \simeq b/2$	Frequency of beating f(sec.$^{-1}$)	Velocity of flight V(m./s.)	Time ratio of lowering to raising $\bar{a}$	Amplitude $2\psi_1$ (rad)	Length of body l(mm.)	Aspect ratio $\mathcal{R}$ b^2/S	Reduced frequency k
Diptera:												
Tabanus borinus	276	184	14.7	63	15.5	96	4	1.5	0.785	23	5.32	0.244
Sarcophaga carnaria L.	45	36	12.3	12	7.0	160	2	1.5	0.654	12	5.44	0.281
Musca domestica	12	20	5.89	4.5	5.5	190	2	1.7	0.785	6.5	6.05	0.210
Volucella pellucens Meig.	73	78	9.18	39	12	120	3.5	1.2	0.654	13.5	7.38	0.207
Tabanus affioris (horsefly)	180	57.4	30.8	—	14.3	120	2.05	—	1.00	—	5	0.2
Aedes nearcticus (mosquito)	3.5	3.6	9.54	—	3.8	320	1.0	—	1.00	—	3	0.2

Species												
Hymenoptera:												
Xylocopa violacea	614	172	35.0	47	18	130	4	1.3		22	7.53	
Bombus terrestris Fabr.	388	142	26.8	74	16	130	3	1.1		19.5	7.21	
Vespa germanica	187	98	18.7	29	14	110	2.5	1.3	0.785	18	8.00	0.16
Vespa crabro L.	567	260	21.4	100	22.5	100	6	1.8	0.871	34	7.79	0.15
Apis mellifica L.	78	42	18.2	27	8.5	250	2.5	1.3		13	6.88	
Amonophila sabulosa V. del	45	42	10.5	82	9.0	120	1.5	1.2	0.610	18	7.71	0.21
Lepidoptera:												
Papilio podalirius	300	3600	0.814	52	37	10	3.5	—	1.22	25	1.52	0.54
Vanessa atalanta L.	134	1080	1.22	31	27	10	4	—	1.31	18	2.70	0.28
Pieris brassica L.	127	1840	0.677	35	31	12	2.5	—		23	2.09	
Macroglossa stellatorum L.	345	400	8.46	68	20	85	5	1.3	0.697	28	4.00	0.37
Plusia gamma L.	144	440	3.53	36	18	48	1.5	1.1		19.5	2.95	
Coleoptera:		mem. m. + ely.										
Melolontha vulgaris Fabr. (beetle)	961	402 642	14.7	100	28	46	2.5	1.5		28	7.80, 4.88	
Cetonia aurata	537	260 370	14.2	68	20	86	3	1.3		19	6.15, 4.32	
Lucanus corvus	2600	800 1220	20.9	250	36	33	1.5	1.0	1.048	54	6.48, 4.25	
Telephorus fuscus	109	116 166	6.44	20	12.5	72	0.8	1.4	1.263	16	5.39, 3.77	
Neuroptera:					anterior							
Brachytron pratense Mull.	557	1200	4.55	36	36.5	33	5	1.4	0.654	55	4.44	0.34
Calopteryx splendens Harr.	120	850	1.38	13	30	16	1.5	1.6	1.00	47	4.24	0.24
Pyrosoma minium Harr.	38	355	1.05	8	25	27	0.6	1.2	0.870	32	7.04	0.16
Panorpa communis L.	30	176	1.67	8	14.5	28	0.5	1.6	1.31	17	4.78	0.16
Orthetrum caerulescens Fabr.	248	1080	2.26	22	32.5	20	4	—		42	3.91	
Aeschna mixtra Latr.	530	1380	3.77	30	39.5	38	7	1.7	0.61	63.5	4.52	0.36
Orthoptera												
Schistocerca gregaria Locust	2,000	1,300	1.50	—	60.0	20	3.50	—	1000, 871	—	5.71	0.20
Odonata												
Anax parthenope	790	1,000; 1,200	3.50	75	50.0	29	7.2	—	0.63	75	10.0, 7.8	0.20
Homoptera												
Bemisia tabaci	3.3×10^{-2}	1.34	0.245	—	0.84	169						
Aleurothrixus floccosus	6.5×10^{-2}	19.4	0.336	—	1.52	166						
Aphis gossypii	0.114	1.03	1.11	—	2.18	123						
Acyrthosiphon kondoi	0.702	11.06	0.633	—	3.39	81						

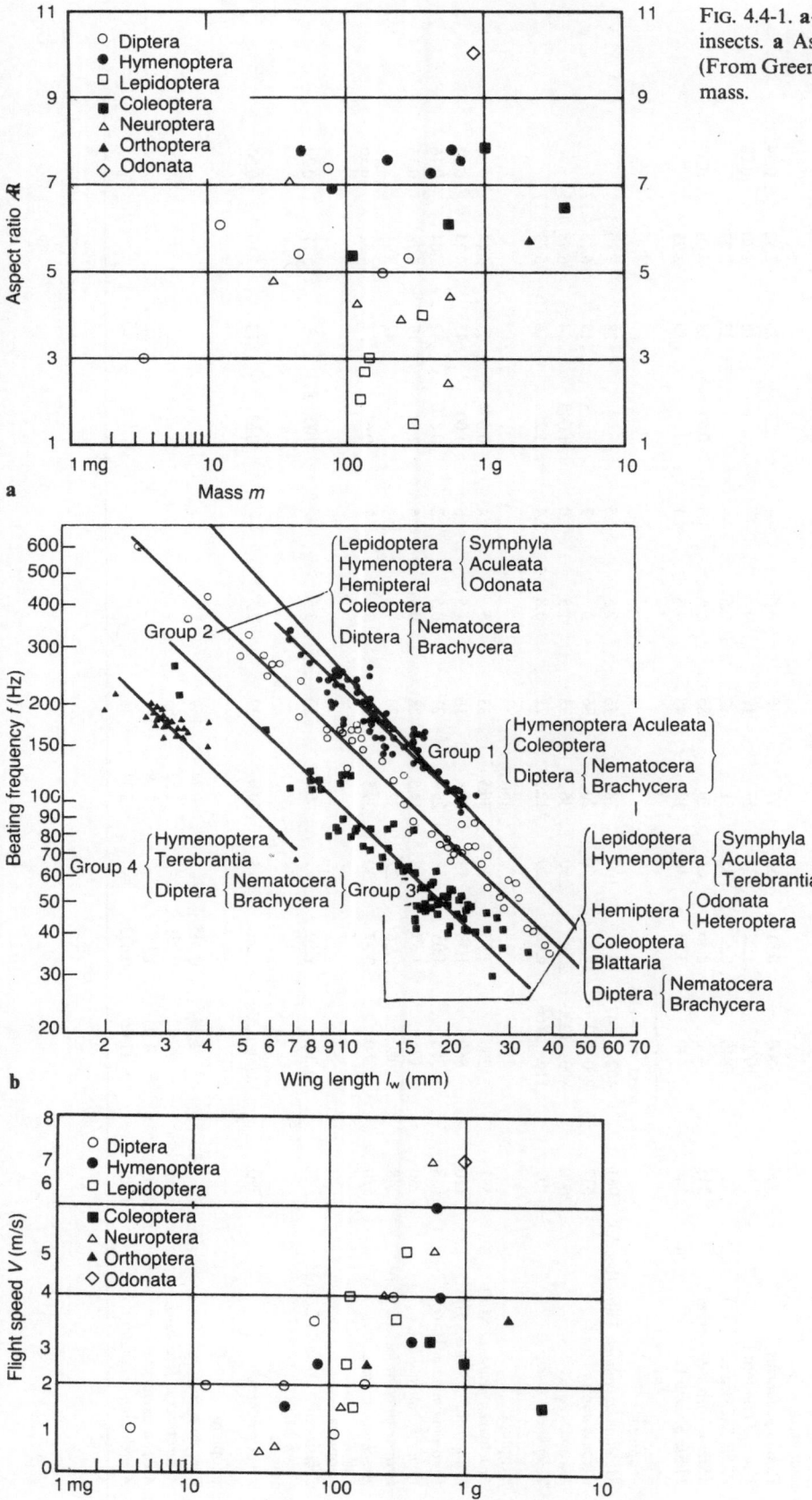

FIG. 4.4-1. **a–c.** Mechanical characteristics of insects. **a** Aspect ratio. **b** Beating frequency (From Greenewalt 1962). **c** Flight speed versus mass.

126

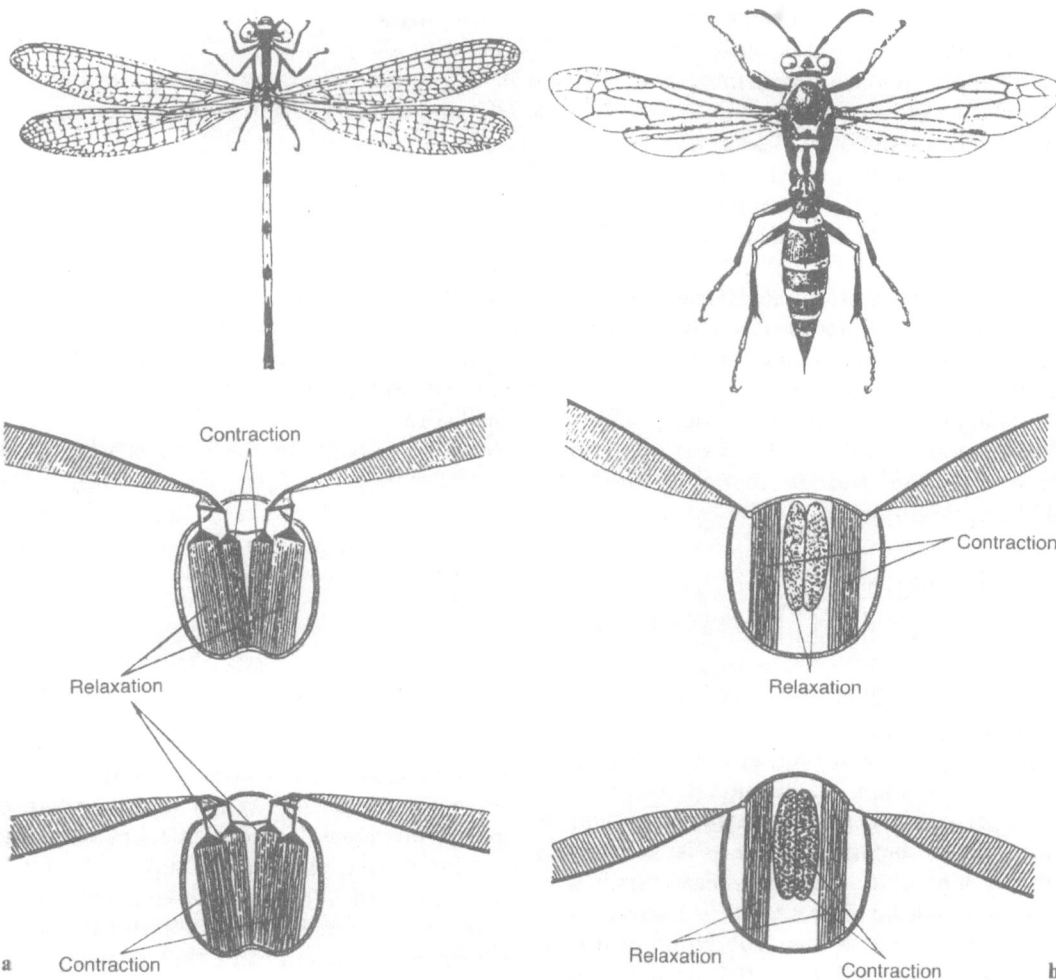

Fig. 4.4-2. **a–b.** Mechanical models of flight muscles. (From D.S. Smith 1965 with permission). **a** Indirect driving. **b** Direct driving.

the elongation and contraction of the respective muscles through either elastic deformation of the thorax or direct driving of the wing itself, as shown functionally in Fig. 4.4-2.

In the indirect driving system, wing movement is driven by muscles that distort the sclerotic cuticle constituting the thorax. The alternate contraction and relaxation of outer-vertical bundles and inner-horizontal bundles of muscles causes an oscillatory deformation of the thorax known as the "click mechanism" (Boettiger and Furshpan 1952), which in turn produces an up-and-down motion of the wing.

For a feathering motion of the wings in the indirect driving system, there are two possible mechanisms that may operate to cause cyclic wing twisting: (1) passive effects due to the inertial and aerodynamic forces acting upon wings with pliant bases and flexible veins, and (2) automatic cyclic changes brought about by the relative movements (during the wingbeat) of sclerites found at the base of the wing (Miyan and Ewing 1988).

In the direct driving system, on the other hand, the contraction of an inner pair of muscles raises the wings by leverage action, whereupon a more robust outer pair of muscles contracts to provide the downward power stroke. The muscles are said to be synchronous, because each contraction is in response to a separate stimulus from the central nervous system. At the base of the respective wings of the dragonfly are two joints for wing beating, instead of a single joint. Through the beating motion of the above two joints, with some phase lapse, both flapping and feathering motions can be attained.

Wing tissues. The meshwork of "cells" observed in the wings of many insects can be understood by considering the force balance among the tensions acting on the veins and veinlets of the wing (Thompson 1942).

These tension forces, except at the leading edge where the compression force is vital, are necessary for preventing the vertical deformation of the cells caused by the aerodynamic pressure and preserving the planform of the wing. Hence, if, for example, the wing of a dragonfly is cut at some distance from the root, the deformation at the cut edge will be appreciable.

4.4.2 FLIGHT MODES

This section discusses the beating flight of insects by means of "entomopter," a process similar to "ornithopter" in bird flight that is effected mainly by a flapping of the wings.

It must be remembered that, at less than 10 g and more often less than 1 g, the mass (or the wing and body size) of insects is much smaller than that of birds, while the beating frequency is, at more than 10 Hz, higher than that of birds. Such a small size and high beating frequency result in a low wing loading of less than 10^1 N/m^2, a low Reynolds number of less than 10^3, and a high reduced frequency of more than 10^{-1}. The flight mode of insects relies strongly on these characteristics. Specifically, the low wing loading in Lepidoptera and Neuroptera guarantees the slow flight speed that enables these insects to be easily carried by the wind.

Based on many observations of insect flight (Weis-Fogh 1956; Baker and Cooper 1979; Burton and Sademan 1961; Betts and Wooton 1988; Ennos and Wooton 1989), typical wing configurations and flight modes are summarized in Table 4.4-2.

Generally, insect flight is characterized by the ability to steer accurately, hover for long periods, and fly in any direction, even backward. Only hummingbirds approach insects in their ability to maneuver.

Recent research on the kinetics of flight in insects includes experimental and analytical studies on the locust by Weis-Fogh and Jensen (1956), on the

dragonfly by Azuma et al (1985) and Azuma and Watanabe (1988), on butterflies by Sunada et al. (1989a), on bumblebees by Dudley and Ellington (1990), and more generally on the hovering flight of insects by Ellington (1984a).

Reduced frequency. Now let us once again summarize the beating frequencies of insects. The statistical data on the frequency of the beating stroke in insects were given earlier in Fig. 4.1-12 and 4.4-1b. Weis-Fogh (1977) demonstrated the rule of $f \propto m^{-1/3}$ by carefully analyzing well-defined insect families and groups like Apoidea, Vespidae, Noctuidae, Sphingidae, Culicidae, Tipulidae, and Lamellicornes.

Here, let us consider the ratio of wing stroke to chord length, $\lambda = R\beta_{max}/c$ or $\lambda = R\zeta_{max}/c$. As stated in Sect. 4.2.4, this ratio can also be related to the aspect ratio $\mathcal{R} = 2R/c$ and the reduced frequency k at hovering flight as a reciprocal relation,

$$\left.\begin{aligned} \lambda &= R\zeta_{max}/c = \tfrac{1}{2}\mathcal{R}\zeta_{max} = \tfrac{1}{2}R\omega\zeta_{max}/\pi c f \\ &= \tfrac{1}{2}U/\pi c f = \tfrac{1}{2}/k \simeq \tfrac{1}{2}\mathcal{R} \end{aligned}\right\}$$

(4.4-1)

The above reciprocal relation for insects can be observed in Table 4.4-1. When an ordinary insect or bird wing starts to move, it takes one chord length for circulation Γ to build up, i.e., for full lift to be produced. The ratio λ is usually larger than 2.5 and varies between 3 and 5 in most of the groups, which indicates that ordinary aerodynamics applies. However, in some butterflies (Papilionidae) and in true hover flies (Syrphidae), λ is often as small as 1.4 (or $k \simeq 0.35$). As in the discussion in Sect. 4.2.3, the above condition again suggests that nonstationary flow is of major importance in these insects.

4.4.3 GENERAL PERFORMANCE OF INSECTS

The mean necessary power $\bar{P}_n$ for the intense beating motion of insects flying at a velocity V and with a mean thrust $\bar{T}$, mean wing drag $\bar{D}_w$, and mean lifting force $\bar{N}$ can, in parallel to the analysis in Sect. 4.1.1, be given by the formula

$$\left.\begin{aligned} \bar{P}_n &= \bar{T}V + \bar{D}_w U + \bar{N}v + \bar{T}u \\ &= \tfrac{1}{2}\rho V^3 f + \tfrac{1}{2}\rho U^3 S C_{D,w} \\ &\quad + \left\{\tfrac{1}{2}\rho V^2 f(C_D/C_L) + W\right\}v \end{aligned}\right\}$$

(4.4-2)

where the relative velocity U of the wing with respect to the air is approximately represented by

TABLE 4.4-2. Typical configurations and flight modes of insects. (Photographs courtesy of Satoshi Kuribayashi, 1981)

Locusta migratoria	Locusta	Orthoptera	The front of fore and hind wings has longitudinal reinforcements running approximately parallel to the leading edge and constituting the main span of the membranous wing. The fore and hind wings are neither overlapped at the base nor hooked and can move independently with equal importance in flight. Zygoptera have a smaller wingbeat frequency than Anisoptera.	By selecting an appropriate phase difference between two pairs of wings, they can fly either at high speeds of about 5–10 m/s, or at low speeds. In the case of Odonata, they can make hovering flight and have great maneuverability. In the case of *Locusta*, they can make "phase variation" to change from short wings ("brachypterous form") to long wings ("macropterous form") for long-distance flight. Unlike dragonflies, damselflies beat their wings horizontally (lead-lag stroke) even in the forward flight for utilizing the drag effectively as well as the lift.
Anax parthenope julius Brauer	Dragonflies (Anisoptera)	Odonata		
Ceylonolestes gracilis peregrinus	Damselflies (Zygoptera)			
Rhomborrhina unicolor	Drone beetle	Coleoptera	Most beetles have two pairs of wings, the front pair of which are thickened, leathery, and are, therefore, called "elytra." The hind wings are membranous and used for beating. When at rest the hind wings are folded and protected by the elytra.	Fore wings are either extended or retracted during flight and act, when extended, as a lifting surface without flapping. It is the hind wings that are largely responsible for propulsion. In the May beetle (*Melolontha japonica*), the elytra develop lift to support about 5%–10% of the body weight in rapid horizontal flight at 2.25 m/s.
Coccinella septempunctata	Ladybird beetle			
Melolontha japonica	May beetle			

(continued)

TABLE 4.4-2. (cont.)

Colias erate poliographus	Butterfly family	Lepidoptera	Wings of Lepidoptera are characterized by two pairs of wide wings, the fore and hind wings over-lapping. Each wing is framed by several ribs for supporting the airloads. In some moths there are frenula of bristles projecting from the humeral angle of the hind wing and fitting under a group of scales near the costal margin of the front wing, to synchronize the locomotion.	Flight is generally characterized by a slow flapping motion of the wings, which are moved almost synchronously. This slow-frequency beat enalbes take-off at relatively low temperature without warm-up. Many butterflies such as the cabbage white and peacock butterflies can glide for only a split second, while others like the swallowtail and searce swallow-tail can glide for many seconds.
Miltochrista striata	Moth family			
Oncotympana maculaticollis Motschulsky	Cicada family	Homoptera	Front wings have a uniform structure throughout, either membranous or slightly thickened, and the hind wings are membranous. They are thus very rigid and usually held rooflike over the body at rest.	Cicada fly by beating their wings strongly together, hooked in tandem. However, they can fly only for a short distance.
Nacaura matsumurae	Lacewings (*Chrysopa*)	Neuroptera	As the name implies, lacewings have two pairs of long and soft membranous wings, each of which consists of lacelike veins and a thin film.	Lacewings fly at night with a gentle beating of wings without phase difference.

the three-quarter-radius $(x = 3/4)$ point as follows:

$$U = \sqrt{(V + u)^2 + (\tfrac{3}{4}R\dot{\beta} + v)^2} \qquad (4.4\text{-}3)$$

Here, it should be noted that, contrary to Eqs. (4.1-18 and 4.1-20), the parasite power is given by $P_p = TV$, whereas the profile power is given by $P_0 = D_w U$. In the expression for U, $V + u \gg \tfrac{3}{4}R\dot{\beta} + v$ is not assumed for insect flight because, unlike in the cruising flight of birds, the ratio of the beating speed of the wing element to its forward speed, $\tfrac{3}{4}R\dot{\beta}/V$, is always of great significance. The induced velocities may be approximated by Eq. (4.1-5), or

$$\binom{u}{v} \simeq \left[\left\{\binom{T}{W}\middle/\tfrac{1}{2}\rho V^2 S\right\}\middle/\pi A\!\!R_e\right]V \qquad (4.4\text{-}4)$$

Shown in Fig. 4.4-3 is again a typical example of the relationship between specific aerodynamic

TABLE 4.4-2. (cont.)

 Apis mellifer	Honeybee	Hymenoptera	Hymenoptera have four membranous wings; the hind pair are slightly smaller and have a row of tiny hooks (hamuli) on the anterior margin of each wing, by which the hind wing attaches to front wing (This mechanism is also present in Hemiptera).	Hymenoptera are so small that drag mainly results from skin friction; thus, the drag coefficient based on the wing area is very large ($C_{D_0} > 1.0$). This compels the insect to perform beating flight instead of gliding. The honey-bee can change the inclination of the stroke plane, which is almost horizontal when the insect is hovering over a flower. The movement of the wing tip during active flight generally a figure eight.
 Vespa mandarinia	Wasp			
 Chironomus dorsalis	Mosquito Ortho-rrhapha	Diptera	They are relatively small and possess only one pair of wings. Their hind wings, are reduced to small appendages called "halteres." It is very interesting that the trailing edge of the wings of mosquitoes is fringed with many small, fine hairs.	Diptera can spring into the air as soon as their feet lose contact with the ground by means of a tarsal reflex. Their legs are also utilized as efficient shock absorbers. The wings of mosquitoes fringed with fine hairs, are thought to be effective in generating lift in the range of a small Reynolds number.
 Bibio tenebrosus	Fly Cyclorr-hapha			
 Heliothrips haemorrhoidalis	Thrips family	Thysanoptera	They are tiny insects, 0.5–1.0 mm in length, and have four net wings fringed with long bristles.	The flight of thrips depends to a large extent on the viscous force generated by the wing motions, rather than the inertial force; thus, they can fly in calm air.

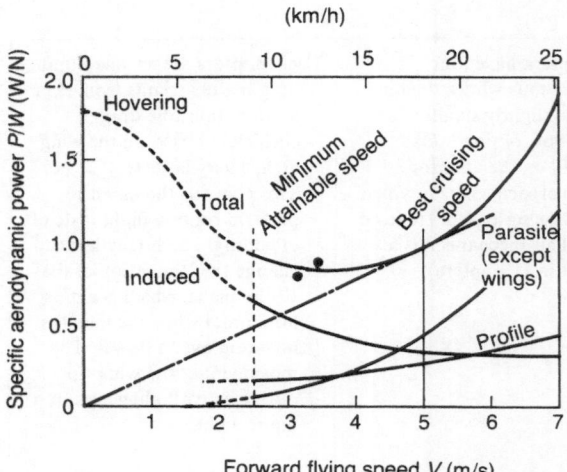

FIG. 4.4-3. Specific aerodynamic power versus forward flying speed in desert locust. (Redrawn from Weis-Fogh 1975a).

It is interesting that a real locust cannot lift its body weight at speeds lower than $V \simeq 2.5$ m/s, indicated by the dotted vertical line (Weis-Fogh 1956), because of wing stall, and that the power at this stalling speed is nearly equal to the power at the best cruising speed, or the speed of minimum cost of transport $V \simeq 5.0$ m/s. From this, it can be supposed that the available specific power of the locust for long-distance flight must be $P_a/W \simeq 1.0$ W/N for saving energy.

4.4.4 RESONANT MECHANISM

Let us consider the torque Q of flapping motion,

$$Q = Q_A + Q_I + Q_E \qquad (4.4\text{-}5)$$

where subscripts A, I, and E show aerodynamic, inertial, and elastic components respectively.

The aerodynamic torque Q_A (positive for flap up direction) has a larger value in the down flap, or $R\dot{\beta} < 0$, than in the up flap process, or $R\dot{\beta} > 0$, and is thus drawn as shown in Fig. 4.4-4. The motion proceeds along an orbit of Q_A given by ① → ② → ③ for the down flap, and ③ → ④ → ⑤ → ① for up flap. In the down flap process, $\dot{\beta} < 0$, since the driving motion is performed against positive torque by a muscle for down motion, the mechanical work must be positive. On the other hand, in the up flap process, $\dot{\beta} > 0$, since the processes ③ → ④ and ⑤ → ① (defined by the shaded areas with horizontal lines) are performed against positive torque (because of the positive lift near the wing root opposing the gravitational

power (the aerodynamic power per unit of vertical force produced) and horizontal flying speed for a desert locust, estimated by Weis-Fogh (1975a) and slightly modified by the author. It should be noted here that the parasite power does not include that of the wing, which is transferred to the profile power of the wing. The necessary power curve itself is very similar to that in bird flight. The two solid circles indicate values found independently in experiments by Jensen (1956). In insects, it is not necessary to incorporate the increased work done by the circulatory and respiratory systems during flight because these components are negligible, compared with work done against both the air and inertial forces (Weis-Fogh 1975a).

FIG. 4.4-4. **a, b.** The work performed against the aerodynamic torque. **a** Mechanically consumed energy. **b** Physiologically consumed energy.

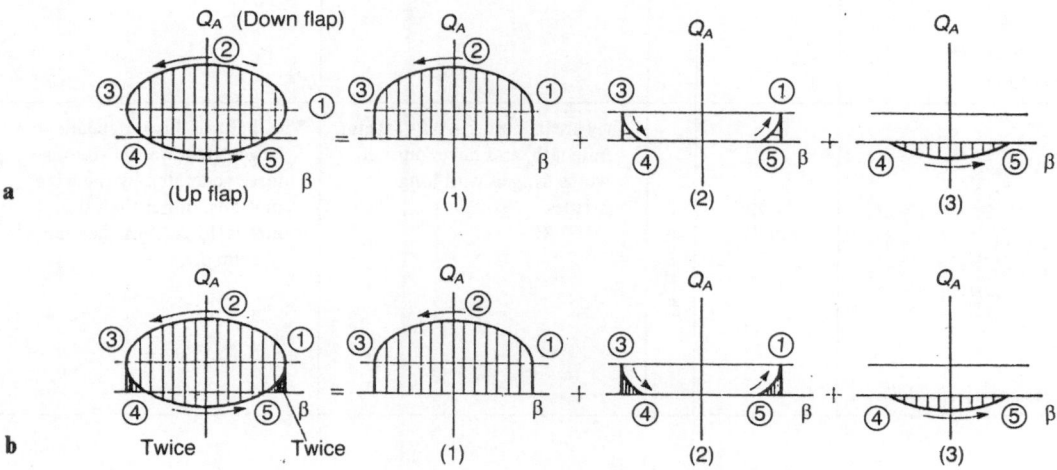

force) and for a positive $\dot\beta$, the mechanical work must be negative, and since the process ④ → ⑤ is performed by another muscle against negative torque and for a positive $\dot\beta$, the mechanical work must be positive. Thus the mechanically consumed energy must be equal to an area enclosed by the orbit of Q_A and shaded by vertical lines (Fig. 4.4-4a).

From physiological considerations, however, the muscle must support even the negative work by consuming energy if the motion is confined to the given orbit, so that the power must pay for the absolute value and is thus formulated as

$$P = (\omega/2\pi)\int_0^{2\pi/\omega} 2\{|Q_{\beta\neq0}\dot\beta| + k|Q_{\dot\beta=0}|\}\,dt$$
(4.4-6)

where the second term in the bracket { } is introduced to express the consumed energy with a proportional constant k (equivalent angular velocity), without any mechanical work, such as to support the torque caused by the gravitational force included in the inertial force. The physiologically consumed energy is given by the enclosed area shaded by the vertical lines shown in Fig. 4.4-4b, which is equal to the mechanically consumed energy, plus twice the two doubly shaded areas in phase (2) of this process. There are, however, other considerations: In the upstroke the muscle does not trace the same orbit but lets the wing move freely, and thus the motion is accelerated and the inertial torque opposed to Q_A. Even if the upstroke is performed by the same orbit, the negative work uses considerably less metabolic energy than an equivalent amount of positive work for vertebrate striated muscle (Abbott et al. 1952; Asmussen 1952; Hill and Howarth 1959; Margaria 1968), so the metabolic cost of negative work is often disregarded in animal locomotion studies (Ellington 1984a).

In a summation of the inertial and gravitational torques for the angular acceleration $\ddot\beta$ and the gravity acceleration g,

$$Q_I = I\ddot\beta - r_{cg}m_w g \qquad (4.4-7)$$

and the elastic torque for the anticipated spring of stiffness k_β

$$Q_E = k_\beta\beta \qquad (4.4-8)$$

there is no enclosed area for the mechanically consumed energy. However, if the positive work must be done for physiological reasons, and if the damping and gravitational forces may be

neglected, then, by equalizing the inertial torque with the elastic torque, or

$$I\ddot\beta + k_\beta\beta = 0 \qquad (4.4-9)$$

the torques Q_I and Q_E almost cancel each other out in the physiological process. This means that insects with the "tuned" wing-thorax system, where the angular velocity ω equals the undamped natural frequency ($\omega = \sqrt{k_\beta/I}$), consume the least energy and therefore have increased reproductive fitness. However it should be mentioned that the muscle must overcome the elastic torque initially. If the elastic constraint is absent in the present system, the additional areas, which represent the energy consumed physiologically but not mechanically in the muscles, must be larger than those given in a system with elastic constraint.

It is believed that the above elastic constraint is used by insects. As explained earlier, the major contribution to the elastic torque is from the sclerotic cuticle of the thoracic box and from the elastomer or rubber-like ligaments whose main component is the protein resilin (Weis-Fogh 1965). In addition, there is an elastic component in the myofibrils which is still unidentified chemically and structurally. A more precise description of the structural details of the click mechanism at the wing base can be found in studies by Pringle (1957), Weis-Fogh (1964), and Chapman (1969).

Small birds flying with high beating frequency must also have both neurogenic and myogenic fibers within their wing muscles (Weis-Fogh 1972; Buchthal and Weis-Fogh 1956). Furthermore, in a bird's wing, as we have seen, the elastic deformation of the wing including the primary feathers can store and release energy and thus, as suggested by Pennycuick and Lock (1976), the feathers act as elastic constraints.

4.4.5 MECHANICAL CHARACTERISTICS OF RESILIENT MATERIALS

The mechanical characteristics of resilient materials found in many animals are given in Table 4.4-3.

These resilient materials have two fundamental functions (Currey 1980): (1) To amplify muscle power output, and (2) to store energy in one stage of a locomotor cycle and release it at some other stage.

The former function can be observed in the jumping of a flea. The flea can store the necessary energy relatively slowly in elastic elements (pads of

TABLE 4.4-3. Mechanical properties of various organic materials. (Data from Alexander 1968; Bennet-Clark 1977)

Items	Young's modulus E dyn/cm²	N/m² (N/mm²)	Tensile strength σ dyn/cm²	N/m² (N/mm²)	Specific energy storage Π J/g	Notes
Resilin	1.8×10^7	1.8×10^6	3×10^7	3×10^6 (2)	2.1	*Resilin* is a protein found in arthropods, serving various functions (Anderson and Weis-Fogh 1964). It is hard when dry, but in the natural state it contains 50%–60% water and is soft and rubbery. One of the well-known elastic tendons is the apodeme in dragonflies, which is tough and inextensible for most of its length, but its middle section is almost pure resilin (Weis-Fogh 1961).
Elastin	6×10^6	6×10^5				*Elastin* is an elastic protein found in some vertebrates as thin strands in connective tissue. It forms quite a large proportion of the material in the walls of arteries, especially near the heart (Markness et al. 1957). The ligamental muscle that runs along the top of the neck in vertebrate is almost pure elastin.
Abductin	$(1–4) \times 10^7$	$(1–4) \times 10^6$				*Abductin* is an elastic protein found in scallops, where it forms an inner hinge ligament that serves as the antagonist of the adductor, opening the shell when the adductor relaxes (Alexander 1968).
Collagen	10^{10}	10^9 (1,000)	$(5–10) \times 10^8$	$(5–10) \times 10^7$ (50–100)	5	
Bone	$(1–1.4) \times 10^{11}$	$(1–1.4) \times 10^{10}$	$(1–1.8) \times 10^9$	$(1–1.8) \times 10^8$		
Locust cuticle	10^{11}	10^{10}	10^9	10^8		
Lightly vulcanized rubber	1.4×10^7	1.4×10^6				
Oak	10^{11}	10^{10}	10^9	10^8		
Mild steel	2×10^{12}	2×10^{11}	5×10^9	5×10^8 (450–2,700)	0.125–1.4	

1 dyn/cm² = 1×10^{-1} N/m² = 1.01972×10^{-2} kgf/m²; 1 dyn = 10^{-5} N

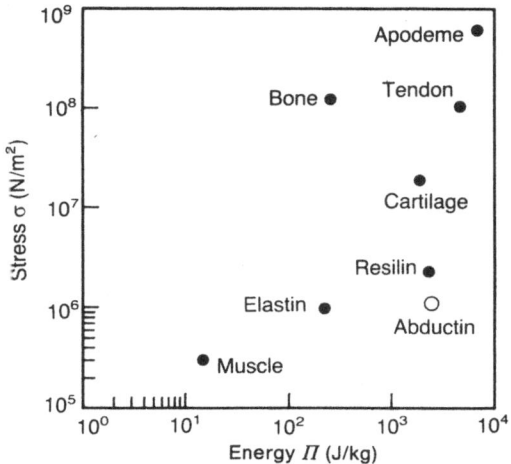

FIG. 4.4-5. Energy-storage ability and stress of various tissues. (From Currey 1980, Elder and Trueman 1980 with permission).

resilin) and then, by means of a catch mechanism, release it suddenly.

A small variation of potential energy stored along a fiber of unit area and unit length can be given by

$$\int \sigma \, d\varepsilon \simeq \frac{1}{2} E\varepsilon^2 = \frac{1}{2} \sigma^2 / E \qquad (4.4\text{-}10)$$

where σ and ε are respectively tensile stress and strain of the fiber, and E is Young's modulus, which has been assumed constant. By dividing the above potential energy by the density of muscle ρ_m, the elastic energy of the resilient material for unit mass, Π, can be obtained as

$$\Pi = \tfrac{1}{2}\sigma^2 / E\rho_m = \tfrac{1}{2}E\varepsilon^2 / \rho_m \qquad (4.4\text{-}11)$$

Figure 4.4-5 shows the energy-storage ability and stress of various tissues. It is interesting to note that (1) a muscle is one order of magnitude less effective than a tendon; and (2) the various high-energy-storing materials have great differences in stress, and thus in strain. The second statement is clearly explained by Eq. (4.4-11). When the product of the elasticity and the density $E\rho_m$ is high, as in an apodeme or tendon, the stress is great for a given stored energy, whereas when the product $E\rho_m$ is low, as for resilin and abductin, the stress is low and the strain is far greater.

In the foregoing discussion, the elasticity has been assumed constant. However, it must be mentioned that, a hysteresis can be observed in the loading-unloading stress-strain curve. (Currey 1980).

4.5 Flight Mechanics in Insects

The thin wings of insects are very suitable for flight, not only in terms of good aerodynamic characteristics at low Reynolds number flow, but also in terms of mechanical power-saving in high-frequency beating. It should be mentioned, however, that the aerodynamic characteristics are highly nonlinear; thus, determination of the correct values of the aerodynamic coefficients is very difficult. In flight at such a low Reynolds number, insects not only use the lift but also take good advantage of the drag and the unsteady aerodynamic force. As a typical example, the flight of the dragonfly, which has large aspect ratio wings, is analyzed in depth. For the low aspect ratio wings of butterflies, however, the added mass of the air bonded to the wing must be included in the calculation of the inertial force, as well as the drag.

4.5.1 WINGS AT A LOW REYNOLDS NUMBER

As pointed out earlier, at a Reynolds number of less than one million (Re < 10^5), airfoil characteristics are slightly different from those at a high Reynolds number (Re > 10^6). Typical examples are shown in Fig. 4.5-1 based on McMasters and Henderson (1979). In aeronautical engineering, the design of airfoils for high-lift capability at low flight speed has attracted great interest. Wortmann (1961, 1972), Miley (1974), and Liebeck (1978) presented new airfoils and disclosed a lot of their data. Pope and Harper (1966), Rees (1975a, 1975b), and McMasters (1989) showed the effect of the roughness and sharpness of the leading edge on the polar curves of airfoils. The effect of grit roughness was also studied experimentally by Myall and Berger (1970). A leading conclusion from these studies shows that for the best performance at a low Reynolds number, the thickness ratio should be low, the camber high, and the leading edge sharp.

Wind tunnel tests on fruit fly (*Drosophila virilis*) wings at a Reynolds number of around 200 in both flat and cambered configuration were performed by Vogel (1966, 1967a, 1967b). He revealed that, as

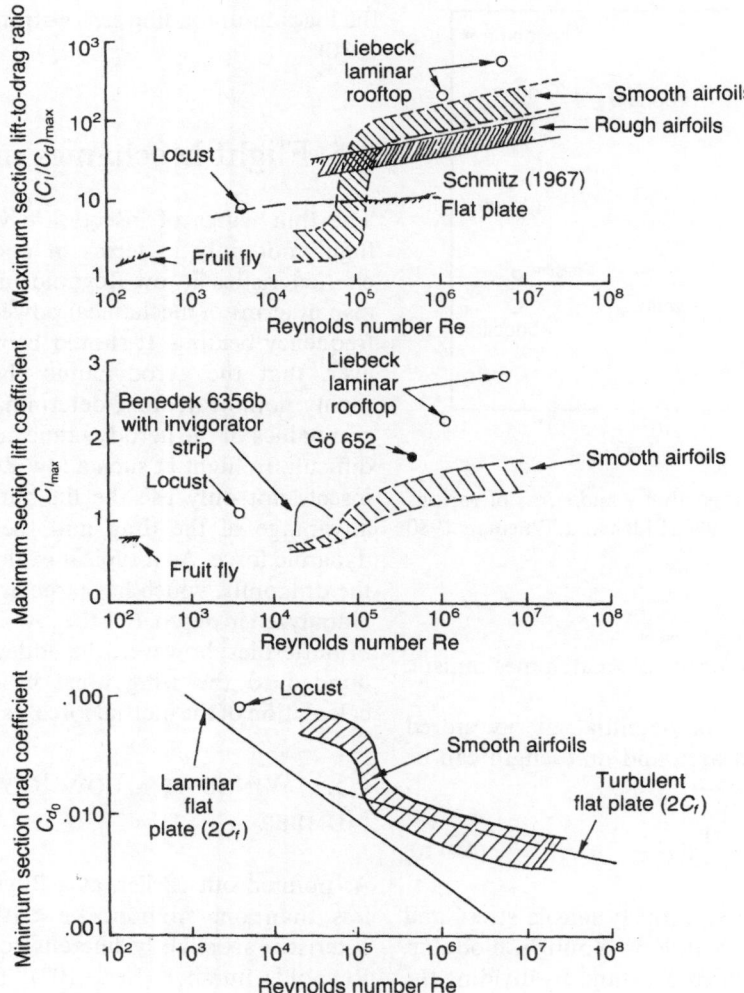

FIG. 4.5-1. Empirical survey of low-speed single element airfoil data. (From McMasters and Henderson 1979; Pressnell 1987).

shown in Fig. 4.5-2: (1) camber increased the maximum lift obtainable, with but a small increase in drag; (2) uncambered fly wings performed better at negative than at positive angles of attack, and the absolute values of their lift-to-drag ratios were higher for negative angles than for positive angles of the same magnitude at almost all values of velocity and angle of attack; and (3) the maximum lift was substantially maintained from angle of attack $\alpha = 20°$ to at least $\alpha = 50°$.

Azuma and Watanabe (1988) gave the polar curve for a dragonfly wing, which was obtained by testing a glider made of two dragonfly wings. The results are also shown in Fig. 4.5-2, along with data on other insects, such as the locust and other species of dragonfly.

Nachtigall (1976) showed that the performance of butterfly and moth wings is enhanced by scales

on the wing surface, probably because of their action on the airflow in the boundary layer. He reported that the removal of the scales from a moth (*Agrotis*) flying at a Reynolds number of about Re = 3 × 10³ reduces the lift by an average of 15%, compared with the scaled wing, without affecting the drag. Tani (1988) also verified a drag reduction obtained by roughening a smooth surface.

However, it should be remembered that the aerodynamic characteristics in low Reynolds number flow are highly nonlinear; thus, the lift and moment coefficients are not uniquely determined, but are a function not only of the angle of attack

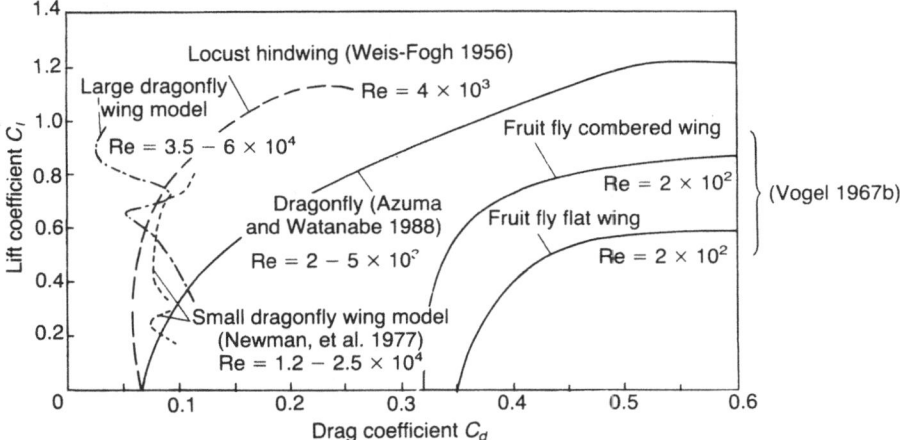

FIG. 4.5-2. Polar curves for insect wings.

at the time of measurement but also of its past values. It may therefore be considered that the polar curves of insects shown in Fig. 4.5-2 are roughly estimated mean values from widely scattered data in which hysteresis is also included.

In the range of a very low Reynolds number, such as $Re < 10^2$ in the flight of Hymenoptera, Coleoptera, and Trichoptera, the drag tends to be larger than the lift, as pointed out by Horridge (1956), and normal flight would be difficult and expensive. The highest lift-to-drag ratio is achieved at high angles of attack (Thom and Swart 1940).

Some small insects like moths and thrips have wings made of quills and hairs of very small diameter and length. This fringe construction appears to be aimed at attenuating the reflection of sonar sound waves emitted by the attack systems of predators. Also, in the case of thrips, they use the drag for flight instead of the lift, at a very low Reynolds number ($Re < 10$). Moreover, the thrips usually fly among flowers and grass, where they are protected from violent winds, predators, and the like.

Unsteady lift and drag. The effects of unsteady or transient motions of an airfoil on the lift and drag forces at low Reynolds number flow are important. Through tests on an elliptic cylinder and a double circular-arc airfoil in the Reynolds number range of $(6–12) \times 10^3$, Izumi and Kuwahara (1983) found that after the impulsive start, the related sharp peak of the lift and drag was caused by the acceleration of an added mass of fluid; then, the periodical variations of lift and drag were gener-

ated by the periodic vortex separation from the leading edge. Osborne (1951) and Bennett (1966) also pointed out the importance of unsteady effects of the flow on the flight performance of insects.

Clap and fling. The chalcid wasp, *Encarsia formosa*, is very small (1 mm) and has small aspect ratio wings with a beating frequency of 400 Hz. This yields a very small Reynolds number ($Re = 10^2$), but a fairly high reduced frequency ($k = 1/2$), so that the aerodynamic properties must be analyzed by considering the flow unsteadiness at a low Reynolds number.

Weis-Fogh (1973, 1975b) elucidated a mechanism for lift generation called "clap-and-fling." As shown in Fig. 4.5-3, by tracing from a slow-motion film, he found that a hovering wasp claps its wings straight over its back with a frequency of about 400 Hz. The wings are temporarily at rest in the clapped position and then, in the fling phase, beating occurs with a rapid rotation (or "pronation") of the wings about a common axis along their trailing edge, until the angle between them is about $2\theta = 20°$, at which time they begin to move apart. The rest of the motion is as in normal hovering until the next clap occurs.

Although this process was originally discussed for the small wasp *Encarsia formosa*, it now appears, as pointed out by Maxworthy (1979), that large insects also use the same process, which has been observed for example in the hind wings of the locust in climbing flight (Cooter and Baker 1977). Therefore, the clap-and-fling phenomenon encompasses a wide range of Reynolds numbers, from values as low as $Re = 20$ to values at least as large as 200.

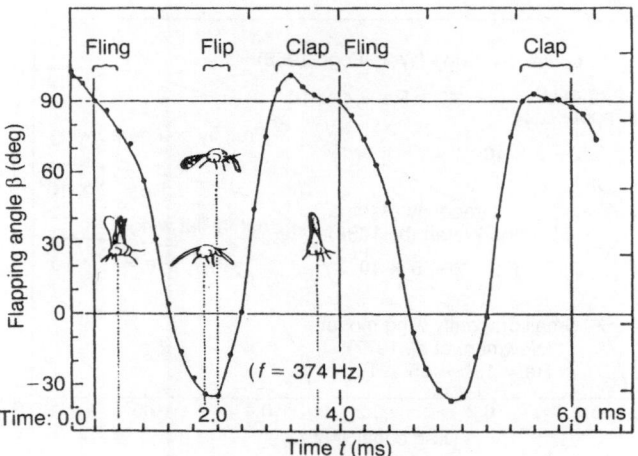

FIG. 4.5-3. Motion of wing in chalcid wasp, *Encarsia formosa.* (From Weis-Fogh 1973, 1975b with permission of the Company of Biologists Ltd.).

At the final stage of the downstroke, the wing must be stopped and supinated for the subsequent upstroke phase. This motion is called "flip" (Weis-Fogh 1973). Because of the flexibility of the wing, the trailing edge and thus the air near the trailing edge are at rest in the twisting phase (or "supination"). Then, during the rapid supination, the direction of wing circulation reverses, giving a favorable circulation for, the following upstroke.

Ellington (1975) calculated the transient circulation of a bound vortex generated in the fling and flip phases of an airfoil. Lighthill (1973, 1975b), Maxworthy (1979, 1981) and Edwards and Cheng (1982) analyzed the mechanical aspect of clapping for a two-dimensional inviscid model of a wing without separated vortices. Haussling (1979) extended the analysis of the "Weis-Fogh mechanism" to a viscous fluid by solving the Navier-Stokes equations numerically for the case of a Reynolds number of 30, and got similar results to those of Lighthill.

Inertial effects of added mass. Since the membranous wings of insects are very thin and extremely light, the "added mass" of air bound to the wing must be included in the inertial force, in addition to the mass of the wing itself. If the wing has a long span, so that the wing spanwise element becomes a two-dimensional airfoil of chord c and mean thickness $\bar{t}$, then the spanwise added mass distribution or the added mass per unit span perpendicular to the wing surface is approximated by $\rho\pi(c/2)^2$, whereas the wing mass per unit span is given by $\rho_w \bar{t} c$, in which ρ and ρ_w are the densities of air and of the wing respectively. Thus, it can be said that in a wing with a relatively large chord (or a small

aspect ratio wing), the added mass, which is approximately proportional to the square of the chord, cannot be neglected in comparison with the wing mass itself, which is proportional to the chord. Ennos (1989a) pointed out the effect of the added mass on the inertial torque of Diptera in flight.

4.5.2 DRAGONFLIES

The Japanese Islands provide a very suitable habitat for dragonflies (Anisoptera) and damselflies (Zygoptera) because of their warm weather, mountains covered with deciduous trees, and heavy precipitation that produces streams and rivers flowing from the mountains into reedy marshes and swamps. With their high aspect ratio wings independently driven by separate strong direct-driving muscles, dragonflies are swift fliers and adept at maneuvering, making it easy for them to catch other insects on the fly in open fields. Damselflies, although they have a similar configuration, are more gentle flyers with lower wingbeat frequency (Rüppell, 1989) that inhabit brushes and swamps.

By applying the local circulation method (LCM), Azuma et al. (1985) showed that *Sympetrum frequens*, can do a steady climb without using an abnormally large lift coefficient (Norberg 1975) or relying totally on unsteady aerodynamic forces (Savage et al. 1979). This result was obtained

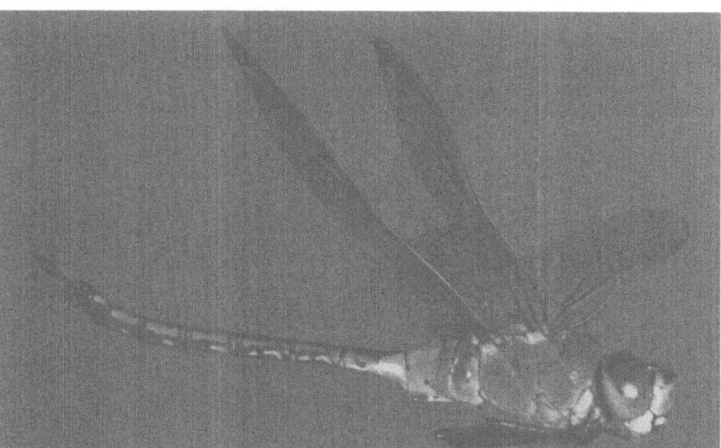

FIG. 4.5-4. *Anax parthenope julius* Brauer. (Courtesy of Satoshi Kuribayashi).

by considering the nonuniform and unsteady induced velocity distribution over a pair of stroke planes. Then by applying the same method of calculation, Azuma and Watanabe (1988) further studied the flight mechanics of *Anax parthenope julius* Brauer, shown in Fig. 4.5-4, and described its performance.

Geometrical configuration. The geometrical characteristics of two specimens of *Anax parthenope* are given in Table 4.5-1. This species is considered one of the "high-performance" species. It has excellent maneuverability and is a most active predator. It has a large wing load and high beating frequency, enabling fast and skillful flight.

The beating can be represented both as a flapping (or heaving) motion of an elastic axis assumed to be a straight line roughly passing through a quarter-chord of the airfoil section at any spanwise

TABLE 4.5-1. Geometric characteristics of two dragonflies. (Azuma and Watanabe 1988)

	Symbols and Units	Dragonfly A	Dragonfly B
Body length	l(m)	7.5×10^{-2}	8.0×10^{-2}
Mass	m(kg)	7.9×10^{-4}	7.9×10^{-4}
Wingspan			
fore wing	b^f(m)	1.0×10^{-1}	1.2×10^{-1}
hind wing	b^h(m)	9.7×10^{-2}	1.1×10^{-1}
Wing area			
fore wing	S^f(m^2)	1.0×10^{-3}	9.3×10^{-4}
hind wing	S^h(m^2)	1.2×10^{-3}	1.3×10^{-3}
Aspect ratio			
fore wing	$Æ^f = (b^f)^2/S^f$	10	15.7
hind wing	$Æ^h = (b^h)^2/S^h$	7.8	9.3
Wing loading	$W/(S^f + S^h)$(Nm^{-2})	3.5	3.5
Center of gravity	x_{cg}(m)	3.0×10^{-3}	3.0×10^{-3}
	z_{cg}(m)	6.0×10^{-3}	5.0×10^{-3}
Distance between the first joints of fore wing and hind wing	x_j(m)	8.0×10^{-3}	8.0×10^{-3}
Estimated drag area	$f = S_f C_{D,f}$(m^2)	9.4×10^{-5}	9.4×10^{-5}
Drag coefficient $C_{D,f}$		1.25	1.25
Frontal area	S_f(m^2)	7.5×10^{-5}	7.5×10^{-5}

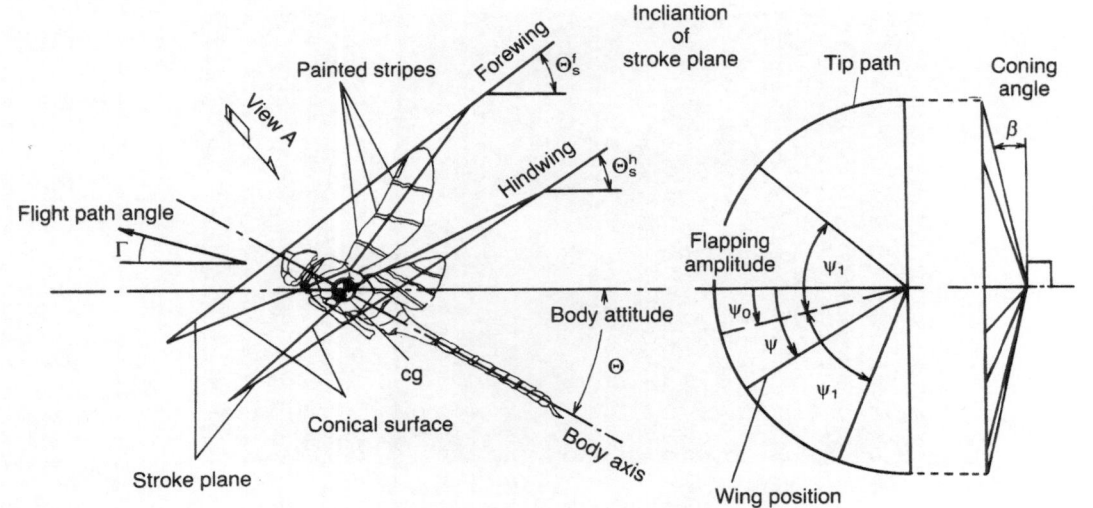

a

b

station, and as a feathering (or pitching) motion about the elastic wing axis. These motions are performed actively through two (front and rear) joints at the respective wing roots. As shown in Fig. 4.5-5, the flapping is assumed to be confined within a conical plane, the apex of which coincides with the front joint and the coning angle of which is defined by β, although the actual flapping motion deviates slightly from the conical plane during the stroke. Here the orbit of the three-quarter-radius station of the elastic (or feathering) axis is called the stroke plane, which is normal to the cone axis.

In dragonflies, the cones related to the fore wings are open toward the front, whereas the cones related to the hind wings are open mainly toward the back. Therefore, the stroke planes of the fore wings are located forward of the joints of the fore wings, and the stroke planes of the hind wings are located backward of the joints of the hind wings. The tilt angles of the two stroke planes are defined by γ^f and γ^h.

Wing motions. Free dragonfly flight can be observed in a wind tunnel and filmed, using a 16-mm high-speed movie camera. By changing the wind speed in the tunnel, the flight speed is altered. Observed modes of flight for two *Anax parthenopes* (A, B) are given in Table 4.5-2. Two examples of orbits of the wing tips with respect to the body, which roughly show the stroke planes of the respective wings, and the motions of the airfoil at a three-quarter-spanwise station in inertial space are shown in Fig. 4.5-6 for a flight speed of (a) $V = 0.7$ m/s and (b) $V = 3.2$ m/s. The traces of the azimuth angles, ψ^f and ψ^h, which are defined

FIG. 4.5-5. **a, b.** Schematic configuration of a dragonfly and its stroke planes. (From Azuma and Watanabe 1988 with permission of the Company of Biologists Ltd.). **a** Side view of dragonfly in flight. **b** Projection of stroke plane (View A).

by the flapping angles of the elastic axis, projected onto the stroke plane and measured from the horizontal line, can be expressed by the first harmonic of a Fourier series:

$$\psi = \psi_0 + \psi_1 \cos(\omega t + \delta_\psi) \qquad (4.5\text{-}1)$$

The values in the above and the following equations refer to either the fore wing (with superscript f) or the hind wing (with superscript h).

However, the feathering angles at three span positions $x = 0.25, 0.5,$ and 0.75, are expressed by the Fourier expansion series including higher harmonics as follows

$$\theta = \theta_0 + \sum_{n=1}^{4} \theta_n \cos(n\omega t + \phi_n) \qquad (4.5\text{-}2)$$

The first harmonic of the feathering angle of the respective wings is a function of the span position x,

$$\theta_1 = (\theta_{0.75} - \theta_r)(x/0.75) + \theta_r \qquad (4.5\text{-}3)$$

where the coefficients, $\theta_{0.75} - \theta_r$ and θ_r are also functions of the flight speed.

From the film analysis of the experimental date (Azuma and Watanabe, 1988), it can be seen that: (1) the beating frequency of the wings is almost unaltered ($f = 29$–32 Hz) by changes in flight speed; (2) the tilt angle of the stroke planes with

TABLE 4.5-2. Flight kinematics. Dragonflies A and B; experiments 1–4. (Azuma and Wabanabe 1988)

Dragonfly		A	A	A	B
Experiment		1	2	3	4
Velocity	V (ms^{-1})	0.7	1.5	2.3	3.2
Flight path angle	Γ (deg)	−12	−1.1	4.8	0
Body attitude	Θ (deg)	20	12	4	2.0
Beating frequency	f (Hz)	26.5	28.1	29.0	27.0
Stroke plane inclination measured from horizontal line fore wing	Θ_s^f (deg)	40	55	58	63
hind wing	Θ_s^h (deg)	38	48	52	68
measured from body axis fore wing	$\Theta_s^f + \Theta$ (deg)	60	67	62	65
hind wing	$\Theta_s^h + \Theta$ (deg)	58	60	56	70
Azimuth angle fore wing	ψ^f (deg)	36	25	25	38
hind wing	ψ^h (deg)	26	26	26	34
Phase difference of beating motion between fore wing and hind wing	$\delta_\psi^h - \delta_\psi^f$ (deg)	51	61	61	93
Phase difference between flapping and feathering fore wing	$\phi_{1,0.75R}^f - \delta_\psi^f$ (deg)	89	93	91	89
hind wing	$\phi_{1,0.75R}^h - \delta_\psi^h$	102	92	95	81
Calculated load factor	n	1.1	0.97	1.25	1.05

respect to the body axis gradually increases from about 40° to 70° as the flight speed increases; (3) the phase difference between fore and hind wings is within 60°–90° (the hind wing leading), and is not correlated with flight speed; (4) the coning angle is about 8° in the fore wing pair and about −2° in the hind wing pair, and is almost constant (within 10% deviation) throughout the beating motion; (5) the flapping amplitude ranges 25°–40° and is not correlated with flight speed; (6) the feathering amplitude ranges over 40°–60° in the fore wings and 30°–40° in the hindwings, and is not correlated with flight speed either; (7) the phase difference between the flapping and feathering is 90°, which is considered optimal for efficiency at low beating frequency (Azuma 1983; See also Sect. 4.2.3); and (8) since the beating motion is performed through two joints at the respective wing roots, in the feathering, the wing is twisted linearly for a given time—"wash out" (twisted negatively towards the tip) in the downstroke, and "wash in" (twisted positively towards the tip) in the upstroke. The motion is stabilized for wing flutter by "stigma" (Norberg 1972) which shifts the center of mass of the wing forward.

The wing motions were also observed in a smoke tunnel using either a high-speed movie camera or a stroboscopic flash, both of which make it possible to visualize not only the wing motions but also the wake vortices, as shown in Fig. 4.5-7 by the paraffin mist method (Watanabe et al. 1986). A series of trailing and shed vortices generated respectively by the span and time (or azimuthal) change of bound vortices are clearly observed in wavy wake sheets of the respective wings.

Analysis by means of the LCM. The spanwise and timewise variation of the air loading and the total aerodynamic forces and moments acting on the wings are calculated by the local circulation method ("LCM"). A detailed description of the method is given by Azuma et. al (1985). The LCM is based on the blade element analysis, but differs from other previous studies as follows: (1) the aerodynamic coefficients are, as shown in Fig. 4.5-2, used in nonlinear forms as functions of angles of attack that include the stalled range; (2) the induced velocity is not homogeneous on the stroke planes and is obtained by the LCM, which also includes the unsteady effect; and (3) the control

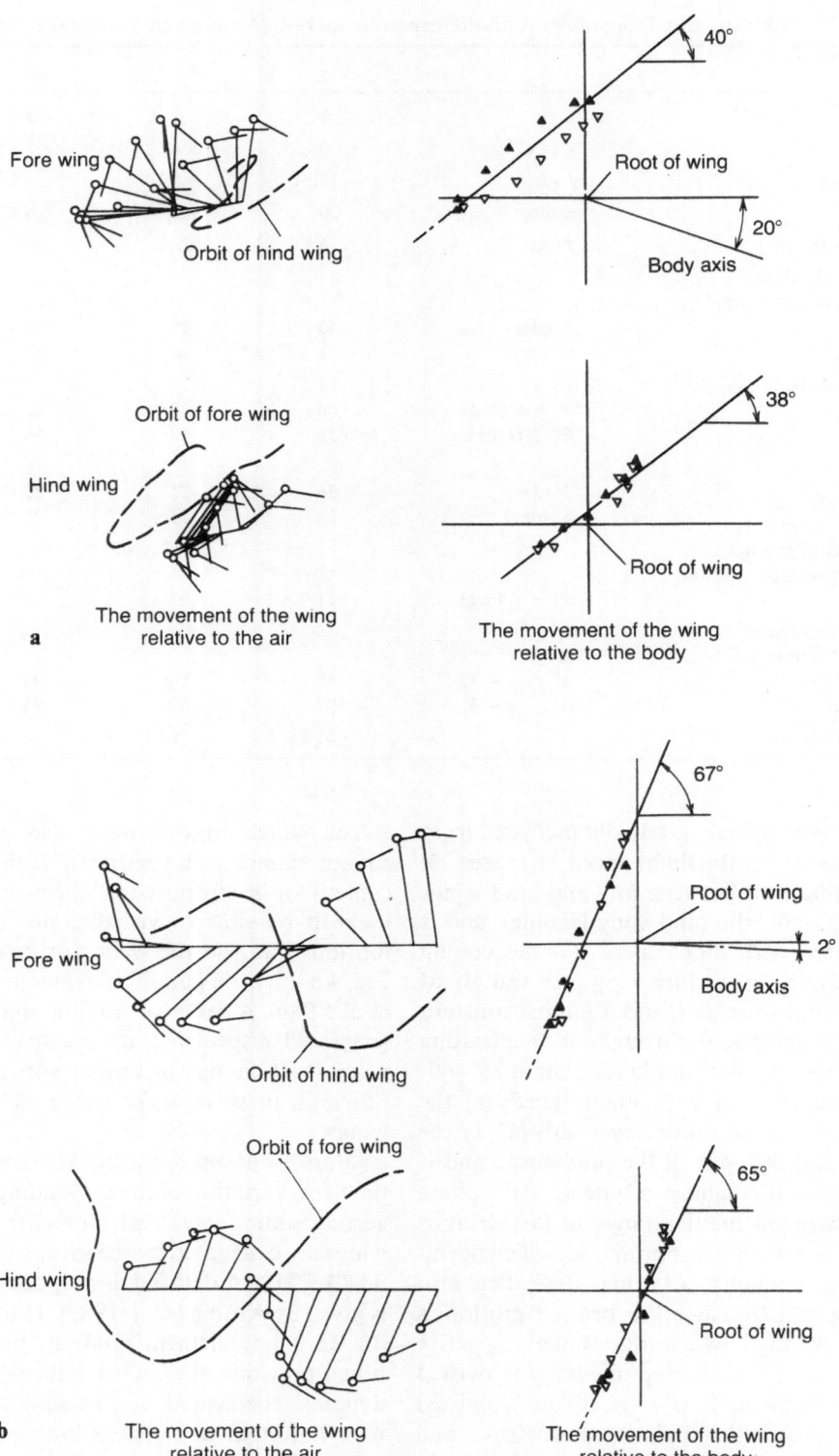

FIG. 4.5-6. **a, b.** Wing tip orbit and airfoil motion at three-quarter-spanwise station (Azuma and Watanabe 1988 with permission of the Company of Biologists Ltd.). *Open triangle*, downstroke; *Solid triangle*, upstroke. **a** $V = 0.7$ m/s, dragonfly A. **b** $V = 3.2$ m/s, dragonfly B.

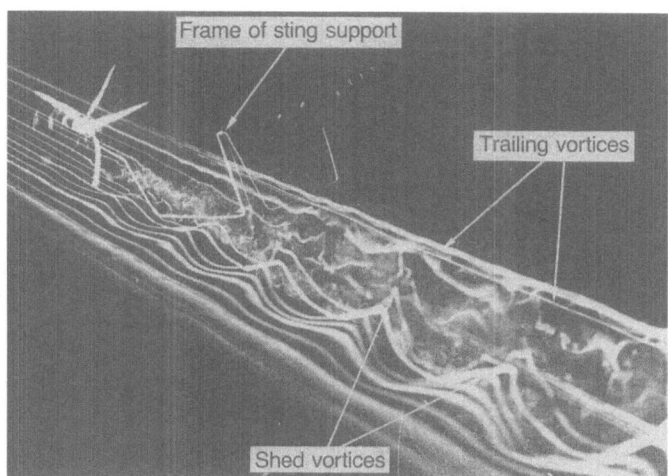

FIG. 4.5-7. Wake pattern and embedded vortices. (From Azuma and Watanabe 1988 with permission of the Company of Biologists Ltd.).

inputs for free flight in computer simulation are the feathering angles along the wingspan, the flapping angles, the tilt angles of the stroke planes, and the beating frequency for given flight speeds.

When the flight conditions, the wing motions, and the geometrical configuration of the wake sheet are known, the air loading, and thus the total aerodynamic forces and moments as well as the inertial forces and moments, can be calculated by the LCM as detailed in the paper by Azuma and Watanabe (1988). The body lift is neglected as stated in Sect. 3.5-3.

Performance. From the data given in Tables 4.5-2 and others (Azuma and Watanabe 1988) for free flights, which are not exactly trimmed flight (load factor $n > 1$), the necessary (mechanical) power is obtained from Eq. (4.4-6) under the assumption of $k = 0$ shown by the circles in Fig. 4.5-8. By referring to these data and by selecting modified feathering angles the necessary power curve of horizontal flight versus flight speed can be calculated and plotted as in Fig. 4.5-8. The curve passes through lower values than those obtained from the free flight tests, because the calculated power is based on almost completely trimmed flights ($n \simeq 1$).

By assuming an available power to musculature mass ratio of $P_a/m_m = 260$ W/kg (Weis-Fogh 1975a, 1977) and a musculature mass to total mass ratio of $m_m/m = 1/4$ (Greenewalt 1962), the avail-

able power can be estimated as $P_a = 5.75 \times 10^{-2}$ W. Then, the top speed V_{max} of this dragonfly is 7.2 m/s. This value can be increased either by reducing the estimated drag area or by increasing the available power. For example, if the drag coefficient based on frontal area is given by $C_{D,f} = 0.5$ instead of 1.25, then the top speed would be about 9 m/s. The cruising flight speed V_0 or the maximum range speed, which is found at the point at which the necessary power curve is tangent to a line drawn through the origin or $(dP/dV)_{min}$, is 3.5 m/s. The minimum power speed $V_{p_{min}}$ or maximum

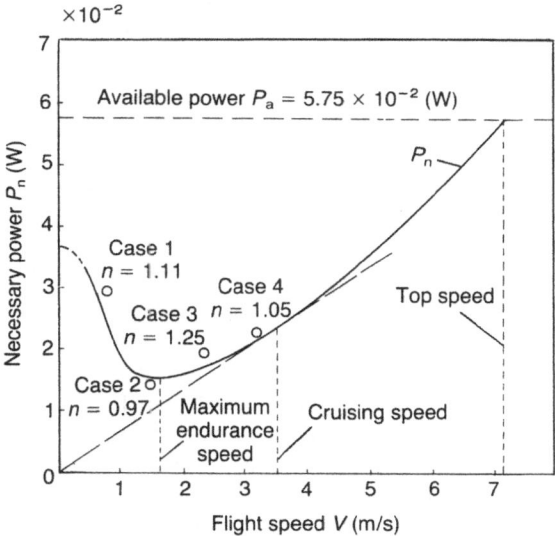

FIG. 4.5-8. Necessary power curve of a dragonfly. (From Azuma and Watanabe 1988 with permission of the Company of Biologists Ltd.). *Open circle*, experimental data.

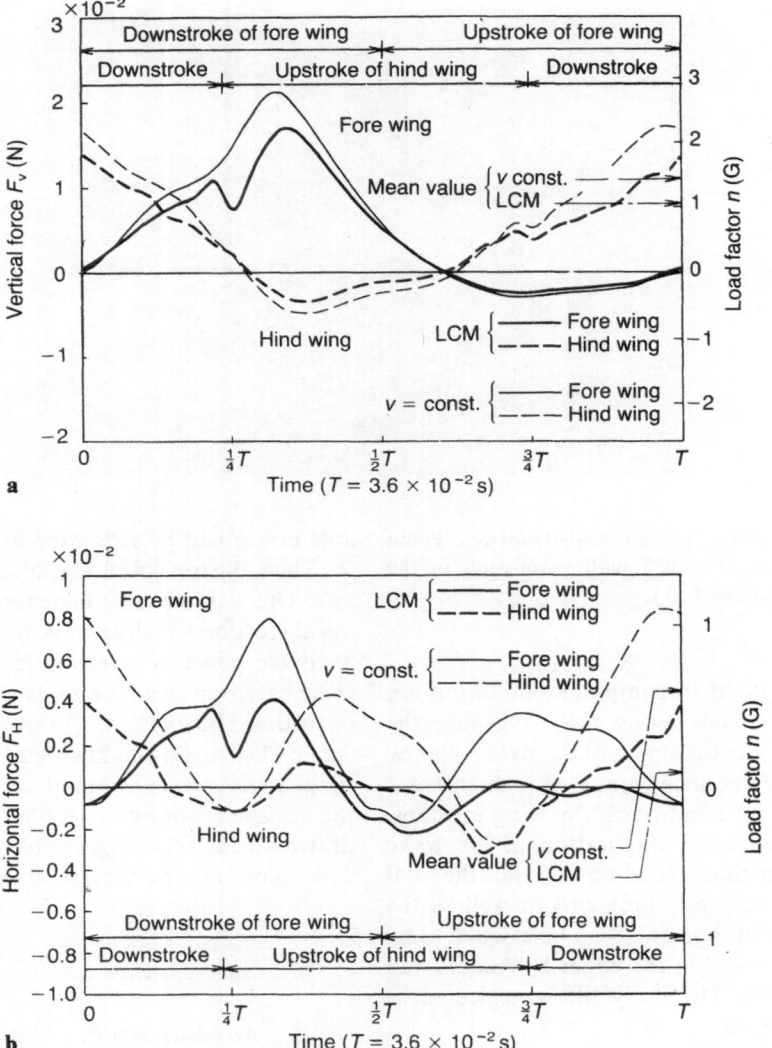

FIG. 4.5-9. **a–d.** Aerodynamic forces and torque (Case 4, $V = 3, 2\,\text{m/s}$, dragonfly B). (From Azuma and Watanabe 1988 with permission of the Company of Biologists Ltd.). **a** Vertical force. **b** Horizontal force. **c** Torque. **d** Pitching moment.

endurance flight speed is 1.7 m/s. The power required for hovering flight $P_{n, H}$ is estimated to be 3.6×10^{-2} W.

Figure 4.5-9 shows, as an example, the time variations of the vertical and horizontal components of the aerodynamic force F_V and F_H, the torque about the flapping axis Q, and the moment about the center of gravity (M_{cg}) at $V = 3.2$ m/s. They were calculated in two ways: using the LCM based on nonhomogeneous induced velocity distribution (thick lines), and using the blade element theory based on the constant induced velocity distribution (thin lines). For the calculation of the moment, the feathering moment around the elastic axis was assumed to be zero $(C_m = 0)$, because of ambiguous aerodynamic characteristics due to

complex airfoil configuration, and of the presumption that flying creatures would not have an unfavorable moment.

In the constant induced velocity distribution, some errors are introduced in the above time variations. However, as shown by the right-hand side ordinate, their mean values are close to each other in the vertical component of the force. Although the forces are in a good balanced state $F_V^f + F_V^h \simeq W$, and $F_H^f + F_H^h \simeq 0$, the mean moment

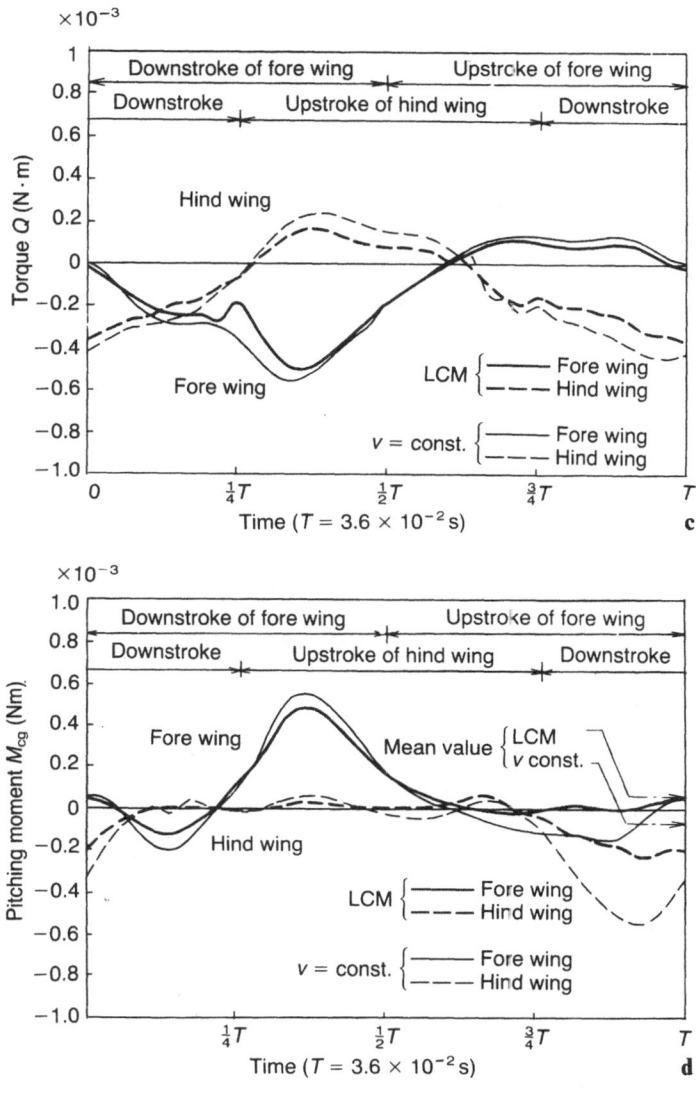

FIG. 4.5-9

about the center of gravity is not completely balanced but leaves a positive (pitch-up) value under the assumption of zero feathering moment, $C_m = 0$. Still, this value is not so large, and is within a range where adjustment is possible by taking into account the feathering moment, which is negative for a positive camber (upward convex) in the airfoil.

Figure 4.5-10 shows the inertial torque of the respective wings and the vertical component of the inertial force of the four wings in comparison with the aerodynamic components and the total resultants. Although the respective inertial and aerodynamic components are of a comparable order of magnitude, the total torque and force are not large. The variation in the total vertical force, which is given by the acceleration normalyzed by the gravity acceleration, is within $0\ G$ and $+3\ G$ in each beating cycle. It is important to recognize that, unlike in the results obtained by Somps and Luttges (1985) and Reavis and Luttges (1988), the maximum value does not exceed $n = 3G$ at the flight speed of 3.2 m/s (Azuma and Watanabel 1988).

Figure 4.5-11 details the spanwise load distribution of vertical and horizontal components of the total force acting on the fore and hind wings. Figure 4.5-12 shows the time variation of the angle

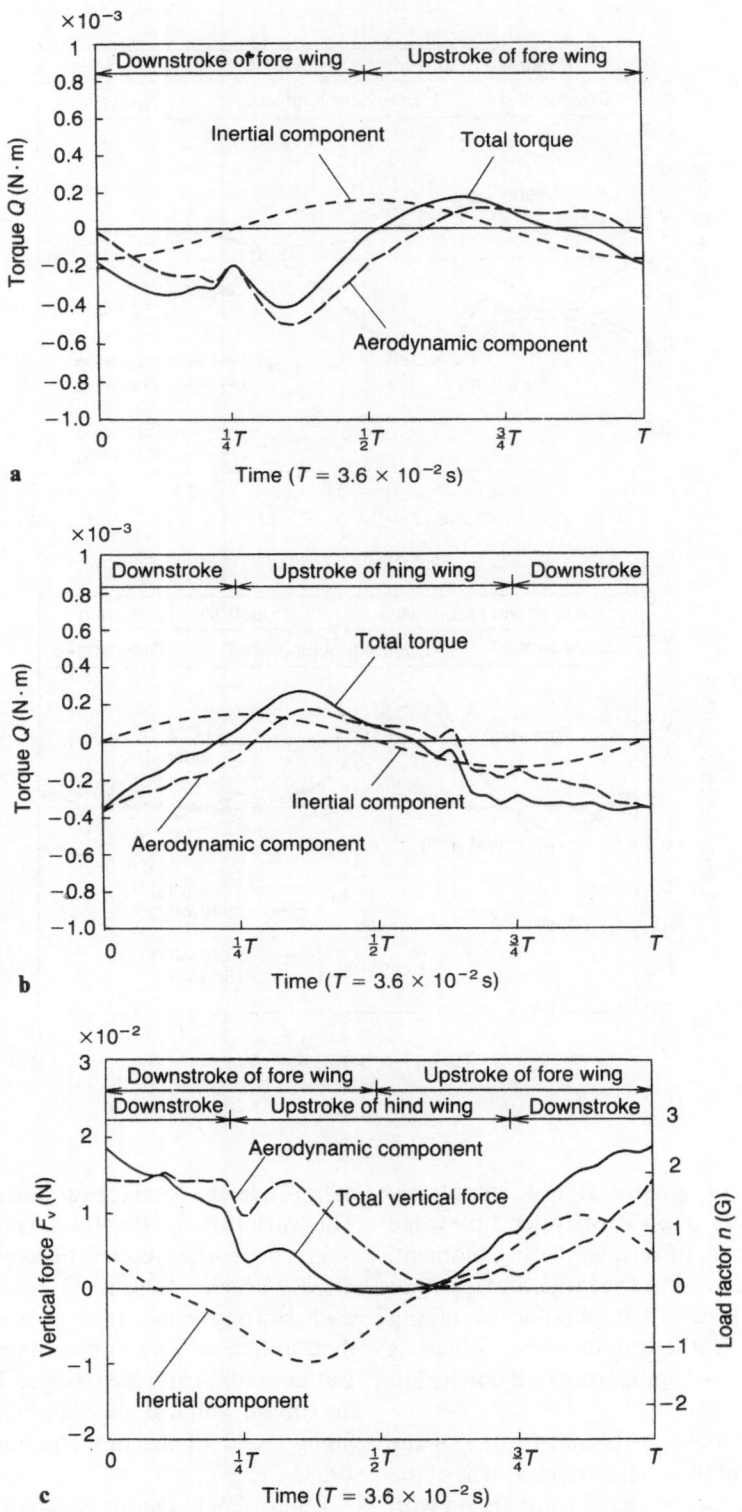

FIG. 4.5-10. **a–c.** Inertial and aerodynamic torque and force (Case 4, $V = 3.2$ m/s, dragonfly B). (From Azuma and Watanabe 1988 with permission of the Company of Biologists Ltd.). **a** Torque of fore wing. **b** Torque of hind wing. **c** Vertical force.

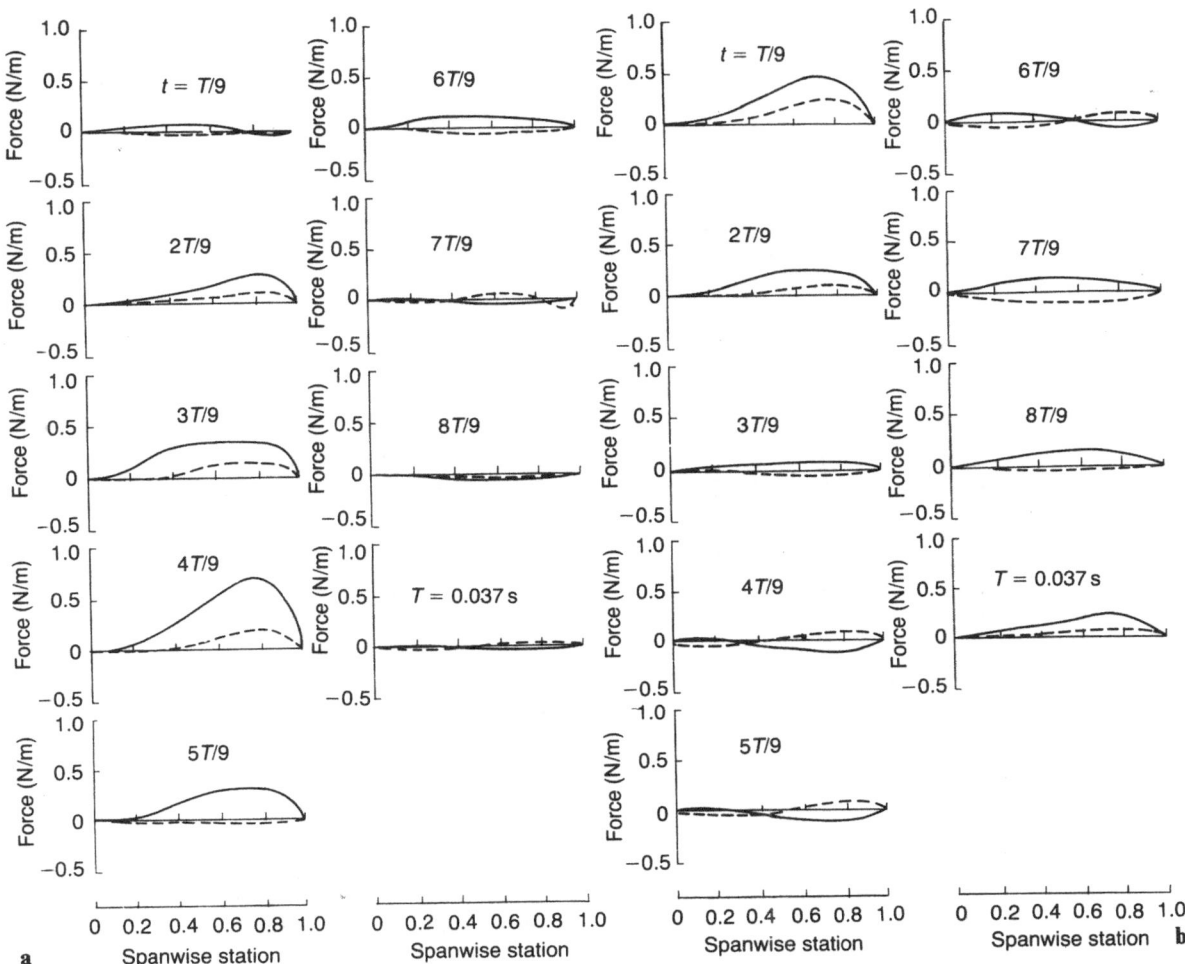

FIG. 4.5-11. **a, b.** Spanwise airload distribution (case 4, $V = 3.2$ m/s, dragonfly B, one period $T = 0.037$s). (From Azuma and Watanabe 1988 with permission of the Company of Biologists Ltd.). *Solid line*, vertical force; *dashed line*, horizontal force. **a** Fore wing. **b** Hind wing.

of attack on the stroke planes of the wings in cruising flight. It is interesting that a region of large positive angle of attack is observed in the final stage of the downstroke near the wing tip of the fore wing, and a region of large negative angle of attack is observed in the early stage of the upstroke near the midspan of the fore wing.

Figure 4.5-13 indicates the share of the lift and drag components in the mean vertical force, $\overline{F}_V$. At very low speed, including hovering, the contribution of the drag component to the vertical force cannot be neglected. This is not unconnected with the fact that the dragonfly flies with its stroke plane

tilted from the horizontal even when hovering, although this increases the induced power required. This makes the transition from hovering to other flight modes easier.

All of the above results were obtained by the LCM based on the unsteady potential theory and the conventional aerodynamic characteristics already shown in Fig. 4.5-2. It is possible to introduce higher lift coefficient generated by the unsteady separated flow near the outermost positions in the stroke plane, where supination and pronation are performed (Izumi and Kuwahara 1983; Freymuth 1988). However, the contribution in this example will be small because of the low dynamic pressure at these extreme positions.

Long range flight. If the dragonfly described above has a stored energy of 1.0×10^7 J/kg (mass per unit energy in Table 4.1-4) and continues an optimal cruising flight in various wind speeds for 12 h (within the daytime), then the ground speed V,

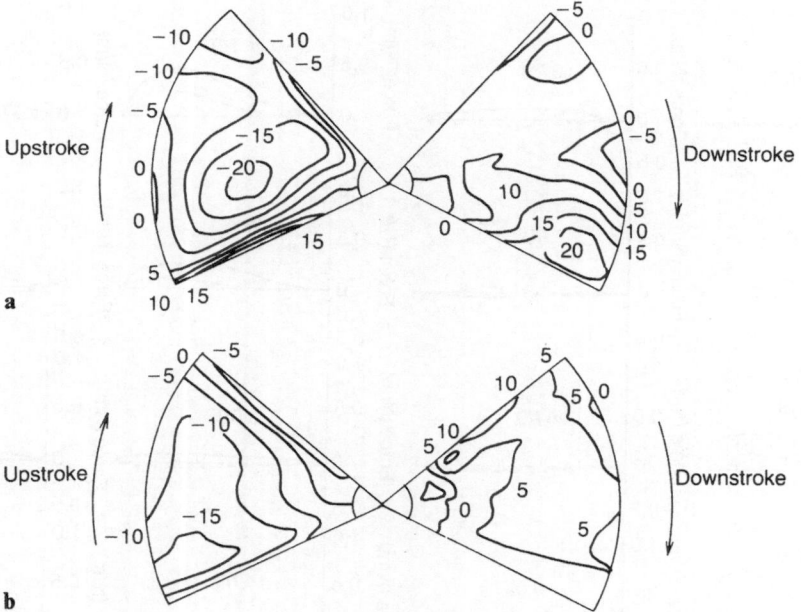

FIG. 4.5-12. **a, b.** Timewise and spanwise variation of angle of attack ($V = 3.2$ m/s, dragonfly B). (From Azuma and Watanabe 1988 with permission of the Company of Biologists Ltd.). **a** Fore wing. **b** Hind wing.

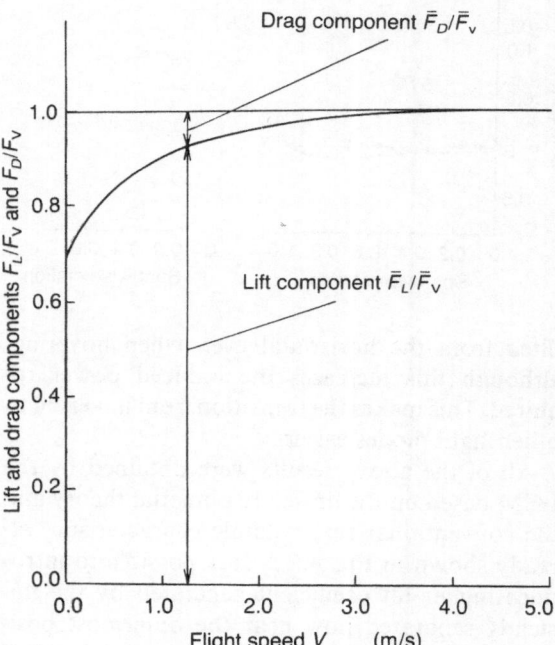

FIG. 4.5-13. Share of lift and drag components in the mean vertical force. (From Azuma and Watanabe 1988 with permission of the Company of Biologists Ltd.).

range of flight R, consumed energy E, and consumed fuel amount as a percent of the total mass $(m_f/m) \times 100$ are given as functions of wind speed U_w as shown in Fig. 4.5-14.

It can be seen that (1) in head wind and low-speed tail wind, the fuel consumption is large and long range flight is actually impossible, and (2) long range flight is acceptable in a high-speed tail wind. As an example, for cruising flight in a tail wind of speed 10 m/s, a flight range of 500 km can be attained after 12 h and with fuel of about 8% of the total mass.

4.5.3 TAKEOFF MECHANISM IN BUTTERFLIES

By considering insect flight from the viewpoint of an unsteady three-dimensional wing, it can be said that insects with large aspect ratio wings, such as dragonflies, rely principally on the steady aerodynamic force obtained by beating at a relatively low reduced frequency. On the other hand, insects with small aspect ratio wings, such as wasps and butterflies, use essentially the unsteady aerodynamic force generated by the separated vortex

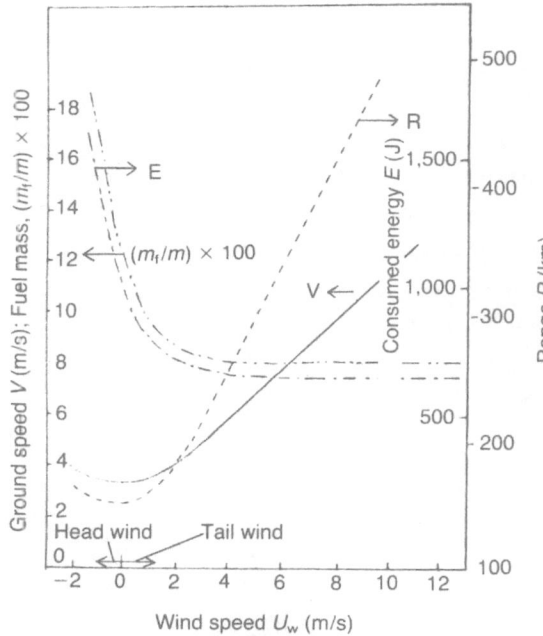

FIG. 4.5-14. Long range flight.

flow obtained by beating wings at a high reduced frequency, especially in low-speed flight (including hovering flight).

In this section, let us examine the wingbeats during takeoff of a butterfly, *Danaus plexippus* (see also Sunada et al. 1989a). Further theoretical and numerical analysis based on the observed mode of flight are left for future study.

Beating modes. Figure 4.5-15 gives a plan view of a butterfly with fully extended wings. At rest, the wings are closed, overlapping on the dorsal side of the body (see broken lines in Fig. 4.5-15b). At the instant just before flapping down, the leading edges of both wings of the respective pairs are shifted forward (solid lines in Fig. 4.5-15b) and the trailing edge of the fore wing is connected with the leading edge of the hind wing. In the ensuing flight, both pairs move in unison.

The flapping is performed around a flapping axis, a line connecting the hinges of the fore and hind wings. The flapping axis is tilted, so that the

FIG. 4.5-15. **a–c.** An observed and filmed butterfly, *Danaus plexippus. Dashed line,* in resting state; *solid line,* just before flapping down. **a** Planform for the maximum wingspan. **b** Profile change in *ii* side view. **c** Side view.

a

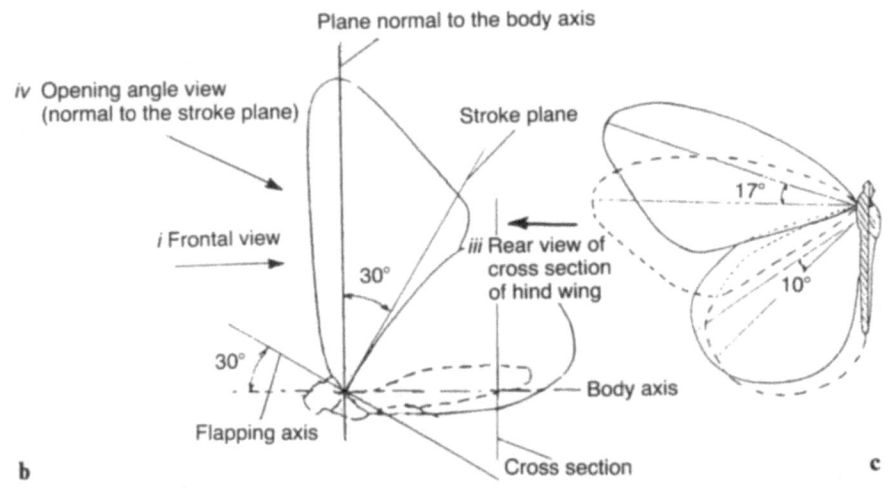

b

c

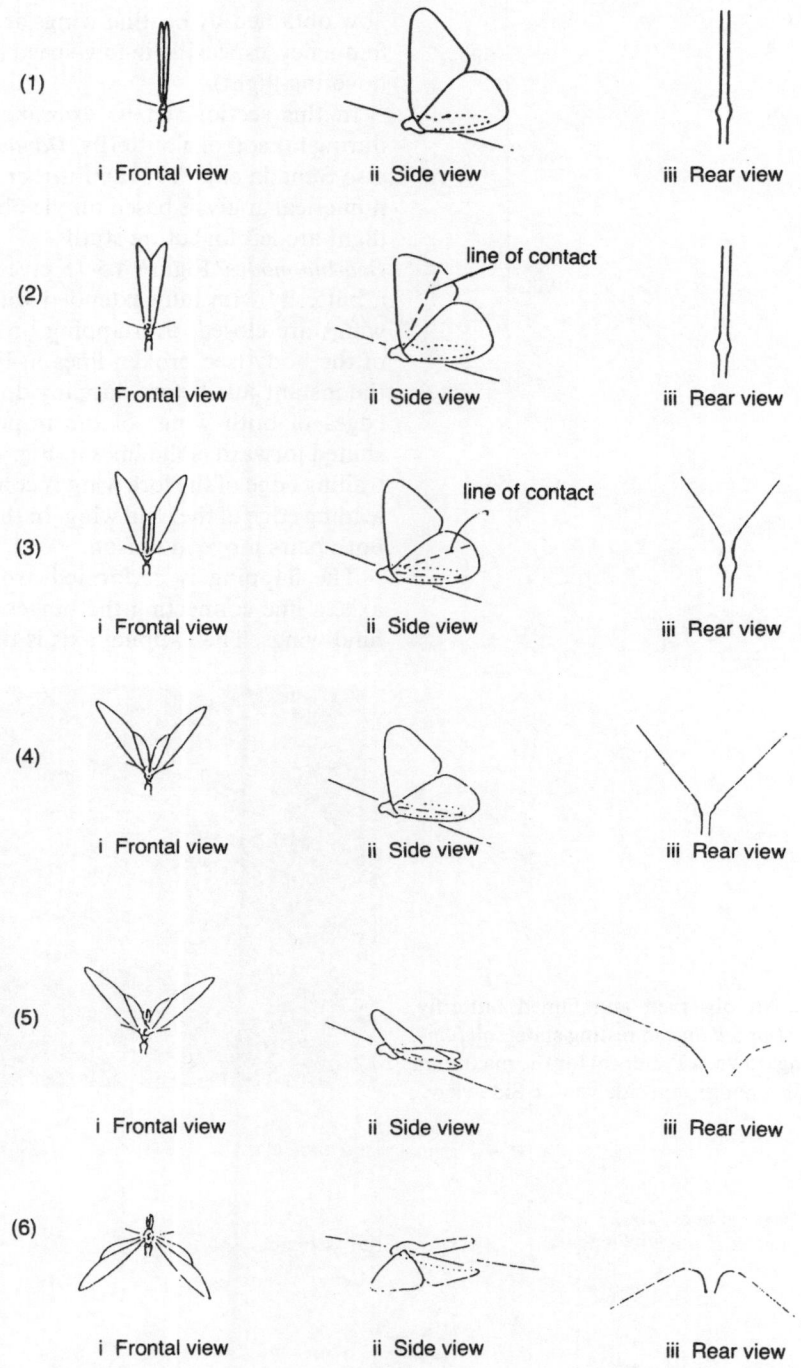

FIG. 4.5-16. Mode of beating.

stroke plane becomes shifted diagonally backward by an angle of about 30° from the normal plane, which is perpendicular to the body axis, as shown in Fig. 4.5-15c.

The wing motion of a butterfly recorded with a video camera (EKTAPRO 1,000 Motion Ana-

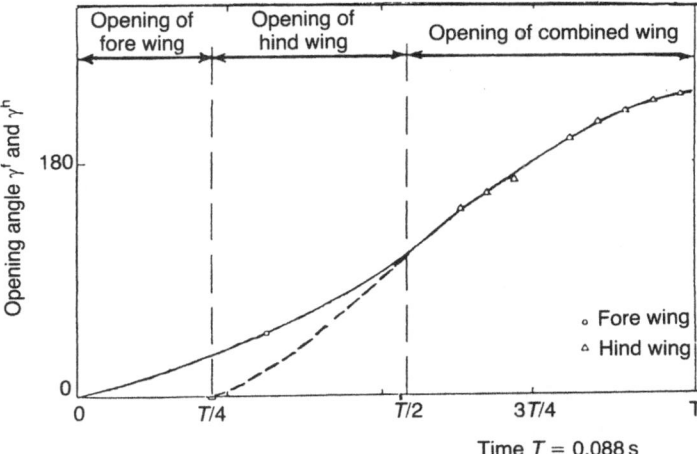

FIG. 4.5-17. Time sequence of flapping at takeoff. *Circle*, fore wing; *triangle*, hind wing.

lyzer, Kodak, Rochester, NY) is shown in Fig. 4.5-16, which presents three views of the wings corresponding to the views indicated in Fig. 4.5-15 for each of six phases of the downstroke. The frontal views of the motion are almost the same as those described by Kingsolver (1985) and listed below:

1. After the wings are shifted forward, they are pressed firmly together.
2. As the flapping down proceeds, only the veins at the leading edges of the fore wings and a part of the membranes near the veins separate from each other, the rest of the wings maintaining contact due to their flexibility. This process corresponds to the "peel" phase (Ellington 1984a), as opposed to the fling phase of the wasp mentioned in Sect. 4.5 1.
3. Next, the separation of the membranes extends to the hind wings. Thus, the tip side edges and the trailing edges of the combined wings also separate from each other. During these processes, the leading edge of each wing acts like a singular line in a mathematical sense and generates a local flow separation with a strong vortex flow, called a separation bubble, that clings to and moves with the leading edge. On the other hand, the tip side edges and the trailing edges of the separated parts of the combined wings seem to satisfy the "Kutta condition" that ensures a smooth

flow at the trailing edge of the wing. The edges leave a shed vortex wake in the fluid, the strength of which is counterbalanced by the change of circulation around the wing.

4, 5. The trailing edges of the front wings connected with the leading edges of the hind wings also separate, and the opening angle of the combined wings becomes large.
6. When the flapping down continues beyond the horizontal plane, almost all parts of the wings separate, with only the root parts of the hind wings still in contact.

The time sequence of this flapping motion viewed from the direction indicated in Fig. 4.5-15c as the "opening angle" view is shown in Fig. 4.5-17. The maximum opening angle γ exceeds 200°. During this downstroke phase the attitude of the body elevates or pitches up and the center of gravity shifts upward and backward. In the following upstroke phase, the aerodynamic force generates forward and downward acceleration and the body attitude gets pitch-down motion.

The above mode of flight suggests that the butterfly can take off with an acceleration of several G by using the following aerodynamic forces: (1) the inertial force generated by the virtual mass of the wing and of the surrounding air during the downward acceleration phase of the wing ($0 < t < T/2$); (2) the aerodynamic force, mainly drag, resulting from the circulation of the wing induced by the fully developed vortex at the leading edge and the vortices shed from the trailing edge and left in the fluid; and (3) although the feathering of the wings are small, which passively results from an elastic twist of the wings, the cyclic

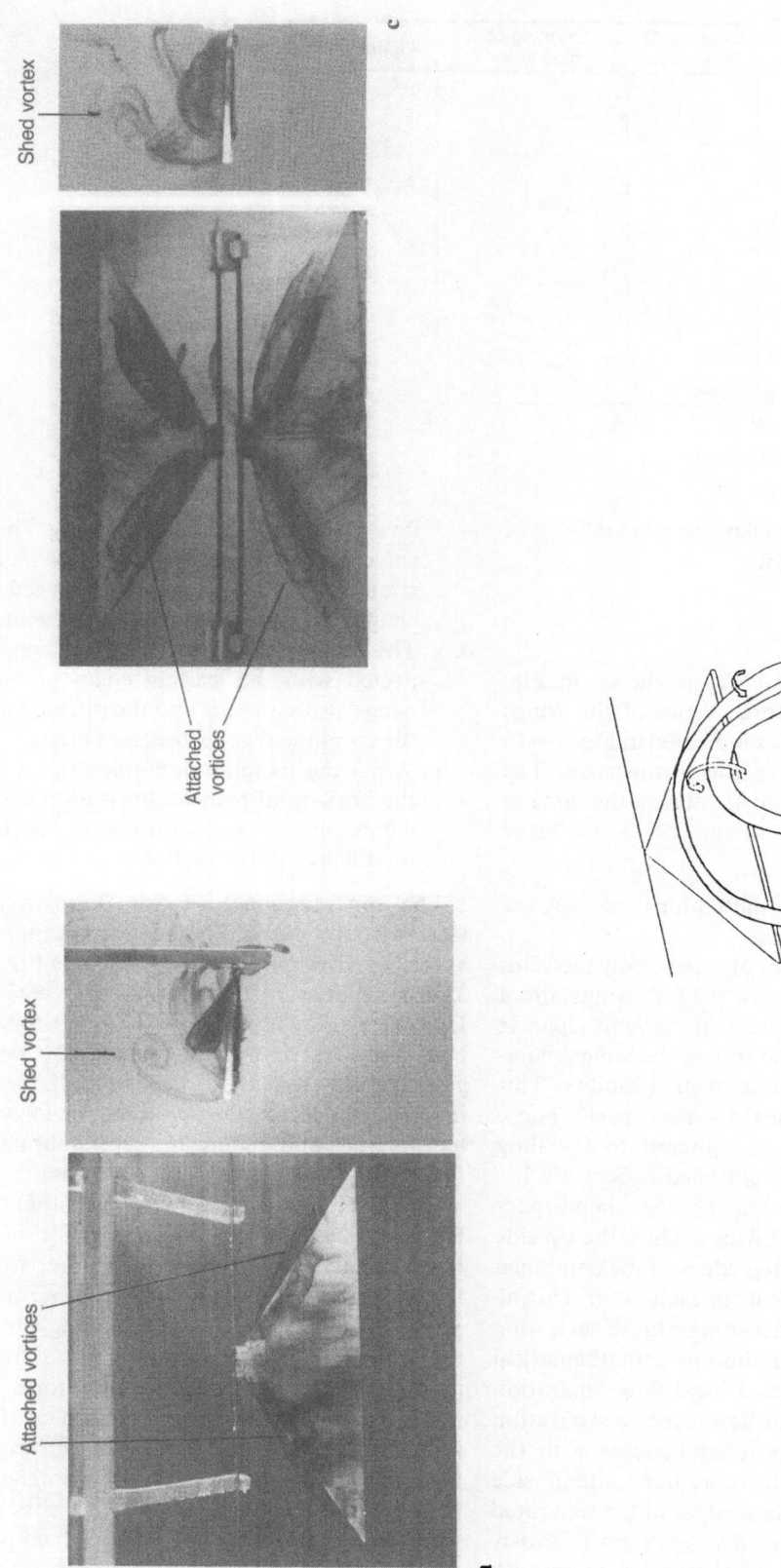

FIG. 4.5-18. **a–c.** Development of the vortices generated by flapping motion of triangular wings. (Courtesy of S. Sunada). **a** Single wing. **b** Vortex ring. **c** A pair of wings.

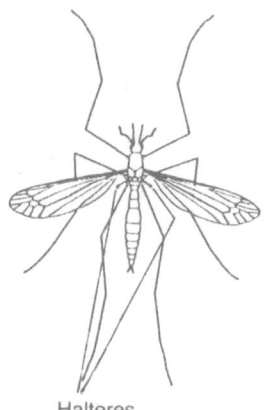

Halteres

FIG. 4.5-19. A typical example of diptera, Crane fly (*Tipula* sp.) (From Borror et al., 1976).

change of the pitching motion of the body induces a large cyclic change of the stroke plane and thus the alternation of the force direction.

Flow visualization. The aerodynamic force of a wing, either the linear "potential force" resulting from the unseparated flow or the nonlinear "vortex induced force" generated by the separated flow, is directly related to the flow circulation around the wing or the growth of attached vortices generated by the wing motion, whereas the induced velocity on the wing surface is strongly influenced by the trailing and shed vortices (Ellington 1984b; Wu and Hu-Chen 1984; Freymuth et al. 1985a; Freymuth et al. 1985b; Freymuth et al. 1989).

Figure 4.5-18a shows a pair of growing attached vortices at the leading edges of a single triangular wing moved in water, and a series of trailing and shed vortices ejected from the trailing edge, that remain in the water as a dyed vortex sheet. As the angular motion of the wing proceeds, the attached vortices at the leading edges grow stronger, so that the shed and trailing vortices also increase in strength and roll up into a kind of horseshoe vortex (Sunada and Azuma 1988; Sunada et al. 1989b). Finally, after the wing motion has stopped, all the vortices coalesce into a single ring vortex that separates from the wing and moves downward with an induced velocity generated by the vortex ring itself. The spanwise length of the rolled-up vortex sheet (the horseshoe vortex) is, as seen in Fig. 4.5-18b, smaller than the wingspan at the trailing edge.

All of the processes described above are replicated by a mirror image of the wing. Shown in

Fig. 4.5-18c is a series of vortices generated from a pair of wings made up of a real wing and its image in a mirror positioned at the hinge line of the real wing. As can be seen, the growth of both the attached vortices at the leading edges and the shed vortex generated by the trailing edge is much stronger than in the case of a single wing, because of the existence of a negative pressure generated at the backside of the space between the two wings. In the final stage, a pair of vortex rings (real and image) descend without combining into a single ring vortex (Sunada et al. 1989b). However, if the apex angle of the triangular plate increases and the shape of the plate approaches a rectangle, then the two vortex rings unite and become a single ring.

4.5.4 FLIGHT OF DIPTERA

The Diptera constitute one of the largest orders of insects and can be distinguished from other insects by the fact that they possess only one pair of wings (front wings) driven by a click mechanism. The beating kinematics and aerodynamics of the front wings were well observed and analyzed by Ennos (1987, 1989b).

The hind wings are reduced to small knobbed structures called halteres, as shown in Fig. 4.5-19 and are oscillated rapidly in the vertical plane during flight. They are used as rate gyros. By detecting the angular rate of every axis and feeding it back automatically to the wing beating mechanism, the Diptera can adjust the angular deviation from an original attitude, increase the damping to the angular motion, and thus easily maintain stationary flight.

A description of the function of halteres was given by Pringle (1948, 1976). Since the halteres are arranged diagonally with a sweep angle Λ (neither parallel nor orthogonal) on both sides of the thorax, an accidental rotation of the body in the beating plane of one haltere never coincides with the beating plane of another haltere on the opposite side. Probably by summing and subtracting the outputs of the bending moment of the right and left halters, M_B^R and M_B^L, outputs of the nervous system can detect the angular rates, rolling Ω_{X_B}, pitching Ω_{Y_B} and yawing Ω_{Z_B}, as shown in Fig. 4.5-20.

The sensitivities of the rolling and pitching angular rates are proportional to the vibration rate of the halteres and dependent on the sweep

TABLE 4.5-3. Dimensions of the halteres of fly and mosquito

	Symbols	Units	Fly (Pringle 1948)	Mosquito
Moment of inertia	I	kg·m²	10^{-15}	10^{-16}
Amplitude of oscillation	θ_1	rad (deg)	1.3 (75°)	1.3 (75°)
Frequency of oscillation	f	Hz	150	500
Angle of sweepback	Λ	deg	30	30
Primary torque	Q	N·m	10^{-9}	10^{-10}
Gyroscopic torque (to be detected)	Q_g	N·m	10^{-12}	10^{-13}

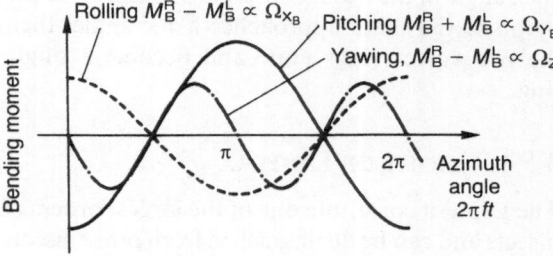

FIG. 4.5-20. Components of the summation and subtraction of the bending moments for haltere oscillation of $\theta = \theta_1 \sin(2\pi ft)$.

angle, Λ. When the sweep angle is equal to $\pi/4$ (or $\Lambda = \pi/4$), the two sensitivities coincide but differ in phase by π. If, however, the sweep angle is smaller (or larger) than $\pi/4$, then the rolling sensitivity is smaller (or larger) than the pitching sensitivity. Yawing sensitivity is a higher order small quantity because it is proportional to the square of the vibration amplitude and may be difficult to detect through the bending moment. However, since the output for the yawing oscillates at twice the frequency of the others, it can be detected if the output is filtered at twice the frequency. It is also interesting to consider that in the pitching motion, the contribution of the wings on the damping is smaller than other motions. However, in the rolling motion, the contribution of the smallest moment of inertia of the body on the response is much more appreciable than others.

Typical examples of dimensions of the halteres are shown in Table 4.5-3, for the fly and mosquito.

CHAPTER 5

Swimming by Snaking

Snaking is the most primitive way of swimming among animals. The young of most fishes swim in this mode, even though their adult locomotion may be very different. A sinusoidal deflection of either the whole body or a part of the body generates a lateral wave in the longitudinal direction. The mode of snaking used is strongly dependent on body size and body shape, because the viscous and inertial effects of the fluid surrounding the body are essentially related to the Reynolds number. The body length of microscopic organisms ranges from 1–5 µm for bacteria, to 300 µm for paramecia and spermatozoa, and their diameter is 1%–5% of the length or less. Thus, for microscopic organisms swimming in water or a slightly viscous fluid, the related Reynolds number and the reduced frequency are so small that the inertial effects of the fluid may be completely neglected. In such cases, the flow surrounding the body is said to be in the "Stokesian realm." However, for large, elongate animals a few centimeters or meters in length swimming in fresh or salt water, the Reynolds number is large enough to introduce the inertial effects of the fluid into the hydrodynamic forces and moments. Streamlined bodies may coast for a considerable distance after having stopped their locomotion. The hydrodynamic aspects of a large, elongated body in unsteady motion can be treated either by the "slender body theory" or by the "two-dimensional flexible-wing (ribbon) theory." In this chapter various snaking motions of different animals are presented, along with their modes of life.

5.1 Micro- and Medium-sized Elongate Organisms

This section will deal with the swimming behavior or locomotion in water of microscopic organisms. Such organisms have a small body compared with other animals and plants. They generally have a large number of filamentlike cilia, a whiplike flagellum, or a bundle of flagella with an almost circular cross section, the diameter of which is around 1 µm or less. Locomotion by means of cilia is based on the undulatory motion of filaments as a group, whereas locomotion by means of a flagellum or a bundle of flagella is driven by either an undulatory or spiral motion of the individual flagella. Another type of locomotion, "amoeboid locomotion," in which a pseudopodium is pushed out and the nucleus moves into it, is not considered here.

Taylor (1951, 1952a, 1953), Hancock (1953), Lighthill (1975b, 1976), and Wu (1977) approached this problem from the purely analytical side by solving the Stokes and Navier-Stokes equations, under the assumption that the fluid is highly viscous. In this section, however, a semi-empirical approach will be used to study the normal and tangential forces of a circular cylinder. The method is simpler than the above analytical methods, but the results are very close to those of Hancock (1953) and the experimental data. Most of the discussion will be devoted to the propulsion of the class of organisms known as spermatozoa and nematodes, which propagate progressive plane waves of lateral displacement along the length of the organism. The squirming or wormlike motions in which longitudinal waves are propagated will not be treated mathematically but only mentioned in passing.

TABLE 5.1-1. Wave parameters for microorganisms. (Data from Gray and Hancock 1955; Lighthill 1975b; Holwill 1966, 1977; Gray 1955)

Species	Length $l(\mu m)$[a]	Diameter $d(\mu m)$	Speed $U(\mu m/s)$	Beating frequency $f(Hz)$	Reynolds number Re[c]	Reduced frequency k[c]	Speed ratio U/C	Amplitude ratio $\pi a/\lambda$	Number of waves n
Bacterial flagella	1–5	0.02	50		10^{-6}				
Eukaryotic flagella	10–100	0.2	10–100		10^{-6}–10^{-5}				
Sperm tails									
Psammechinus miliaris sperm			191.4	35			0.220	0.523	
Lytechinus pictus sperm	10–10^3	0.2–1.0	158	30			0.233	0.639	
Ciona intestinalis sperm			165	35	10^{-5}	10^{-1}	0.214	0.614	3–5
Chaetopterus variopedatus sperm			105	26.5			0.203	0.612	
Bull sperm			97	22			0.134	0.309	
Bull sperm (pathological)			91	9			0.253	0.589	
Mastigophora									
Strigomonus oncopelti			79	24			0.053	0.339	
Blastocrithidia leptocoridis[b]			25.7	9.4			0.176	0.639	
Leptomonas oncopelti[b]			10.2	5.3	10^{-6}	10^{-1}–1	0.143	0.675	
Trypanosoma ranarum[b]			19.0	9.7			0.126	0.685	
Trypanosoma vivax[b]			9.1	10.3			0.076	0.779	
Actinomonas mirabilis[b]			18.0	16.8			0.141	0.703	
Nematoda (Thread worms)							0.4	0.9	1.5
Annelida							0.3		
Small worm (Polychaete worm)	1,000	20–40	100–1,000		10^{-1}		0.2	1.0	4

[a] Length is not total length but length in wavy form.
[b] Wave propagates proximally.
[c] Re and k in sperm tails and Mastigophora are calculated by assuming $d = 0.2$ μm.

5.1.1 PLANE UNDULATORY MOTION OF FLAGELLUM

Undulatory propulsion at low Reynolds numbers is observed in the motion of bacteria, eukaryotic cells, Mastigophora, Dinophiceae, spermatozoa, and small worms. In the first three groups, the cell body has one or more appendages called "flagella" and propagates waves by actively undulating the flagella (Brooker 1965, Brokaw 1972, Satir 1974, Warner and Satir 1974, Blum and Hines 1979). The flagella of many spermatozoa are of almost identical diameter ($d = 0.2$ μm) and length l (several tens of μm) and propel the spermatozoan at a speed of the order of 10 μm/s by oscillating at a frequency of the order of 10 Hz in a viscous fluid with a kinematic viscosity of the order of 10^{-6} m^2/s. The Reynolds number based on the body length is generally less than 10^{-3}, and the reduced frequency of the body motion based on the diameter of the flagellum is about 10^{-1}. Thus the inertial force can be completely disregarded, and the viscous force presumed to be the only one relevant to the locomotion of microorganisms. The flow field associated with the moving boundary can therefore be treated as being in a quasi-steady state. Parameters associated with the undulatory propulsion system are given in Table 5.1-1.

Although a certain amount of three-dimensional movement can be observed throughout the length of the tail of some spermatozoa like those of the squid, the bull, and man (Ritchile 1950; Bishop 1958; Zorgniotti et al. 1958), the majority of spermatozoa and other microorganisms that swim by means of flagella execute plane lateral displacement (Berg 1976).

Let us consider a microorganism with a head and a flagellum of circular cross section, that is swimming in a fluid by moving the flagellum in the manner of a plane wave with a small undulatory amplitude as shown in Fig. 5.1-1a. It is very interesting to find that in an actual organism the undulatory motion is of almost constant amplitude along the body. The waveform is assumed to be produced only by a vertical deformation of a segment or element of the flagellum given by

$$Y = a \sin\{(2\pi/\lambda)(X \mp Ct)\} \qquad (5.1-1)$$

where a, λ, and C are amplitude, wavelength, and wave propagation speed respectively. The plus-minus sign shows the direction of the wave propagation, i.e., minus for backward and plus for forward respectively. The elongation of the flagellum has been neglected under the assumption of a small amplitude in comparison with the wavelength, or $a \ll \lambda$. The wave given by Eq. (5.1-1) is called a harmonic wave. Actual bending waves of the flagella are, not harmonic (Brokaw 1965; Hiramoto and Baba 1978), but more like a series of straight segments and arcs of circles. However the first harmonic component in the Fourier expansion series of the actual waveform has been taken here.

Now, let us consider a coordinate frame (X, Y) moving with a uniform velocity with components of U along the $-$X-axis and V along the $-$Y-axis, as shown in Fig. 5.1-1b. Except for the neighborhood of the head, the velocity components of a substantial element of the flagellum in the frame of (X, Y) can be given as:

$$\dot{X} = 0 \quad \text{and} \quad \dot{Y} = \mp C(2\pi/\lambda)a \cos\{(2\pi/\lambda)(X \mp Ct)\} \qquad (5.1-2)$$

whereas the horizontal and vertical velocity components of an apparent element of the wave itself at Y can be given by

$$\dot{X}_w = \pm C \quad \text{and} \quad \dot{Y}_w = 0 \qquad (5.1-3)$$

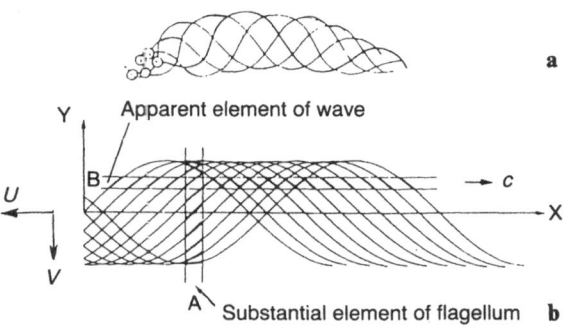

Fig. 5.1-1. **a, b.** Wave propagation. **a** Swimming of spermatozoon of starfish. (Hiramoto and Baba 1978). **b** Mathematical wave model.

This means that if an observer moving with the frame watches the snaking motion of an elongated body through a vertical slit defined by A in Fig. 5.1-1b, he can see the actual motion of the substantial element ds with the vertical speed of $\dot{Y}$, whereas if the observer watches the snaking motion through a horizontal slit defined by B, he can see the motion of an apparent element of the wave which moves along the elongated body from head to tail with the speed of C.

Hydrodynamic forces. The relative fluid velocity components normal and tangential to the element of the flagellum can be given respectively by

$$\left.\begin{array}{l} U_N = (\dot{Y} - V)\cos\theta + (U - \dot{X})\sin\theta \\ U_T = -(\dot{Y} - V)\sin\theta + (U - \dot{X})\cos\theta \end{array}\right\} \qquad (5.1-4)$$

where θ is the pitch angle of the element, or

$$\sin\theta = dY/ds \quad \text{and} \quad \cos\theta = dX/ds \qquad (5.1-5)$$

Here, let us introduce essential assumptions, namely that (1) the induced velocity generated by the flagellar motion is negligibly small, (2) the instantaneous force between the fluid and the element of the flagellum is the same as when the element moves steadily through the water at the same relative velocity (Lighthill 1975b), and (3) the end effect at the terminal ends of the body and the neighboring effect can be neglected (Chwang and Wu 1971; 1974). These assumptions are reasonable, specifically at flow fields of low Reynolds number for an elongated body.

Then, by considering the fact that the drag is proportional to the velocity instead of the square of the velocity, the normal and tangential drag forces acting on an inclined cylinder may respectively be defined as

TABLE 5.1-2. Several expressions of force coefficients for an inclined cylinder

Normal component C_N	Tangential component $C_T = \gamma C_N$	References
$2C_T$	$\dfrac{2\pi\mu}{ln(4\lambda/d) - \frac{1}{2}}$	Gray and Hancock (1955)
$\dfrac{4\pi\mu}{ln(4\lambda/d) + \frac{1}{2}}$	$\dfrac{2\pi\mu}{ln(4\lambda/d) - \frac{1}{2}}$	Chwang and Wu (1974)
$\dfrac{4\pi\mu}{ln(4q/d) + \frac{1}{2}}$	$\dfrac{2\pi\mu}{ln(4q/d) - \frac{1}{2}}$	Lighthill (1976); Wu (1977), $q = 0.09\ \lambda$
$\dfrac{4\pi\mu}{ln\varepsilon + ln2 - \frac{1}{2}}$	$\dfrac{2\pi\mu}{ln\varepsilon + ln2 - \frac{3}{2}}$	Cox (1970); Tillet (1970); Batchelor (1970) ε is the slenderness ratio $= (l/d)$

$$\left.\begin{array}{l} dD_N = \frac{1}{2}\rho|U_N|U_N C_{D_N} d\,ds = C_N U_N ds \\ dD_T = \frac{1}{2}\rho|U_T|U_T C_{D_T} d\,ds = C_T U_T ds \end{array}\right\} \quad (5.1\text{-}6)$$

where d is the diameter of the cylinder (Gray and Hancock 1955; Blum and Hines 1979). There are several expressions for C_N and C_T, given in Table 5.1-2, in which $\gamma = C_T/C_N$ is the ratio of the proportional constants of the tangential and normal components of the drag and is considered constant, being sometimes replaced simply by $\gamma \simeq 1/2$ (Hancock 1953). Then the thrust and normal forces or $-X$ and $-Y$ components of the hydrodynamic force, T and N, the power P, and efficiency η are given in Table 5.1-3.

The first term in thrust T_0 results from the relative horizontal propagation of the apparent wave with respect to the fluid, whereas the second term is caused by the relative tangential motion of the flagellum. The thrust T will drive the organism against the drag of its head. If the head is absent, the thrust is zero for a steady forward motion. Similarly the normal force N results from the normal and tangential components of the transversal speed V. It can thus be concluded that, within an implied range of validity such as $\theta \simeq 2(\pi a/\lambda) \ll 1$: (1) the lateral motion V does not contribute to the thrust or longitudinal force; (2) the thrust T_0 is proportional to the viscosity of the fluid μ and the product of the number of waves and the wavelength $n\lambda$; (3) the propagation speed C must be

$$\left.\begin{array}{l} C > U\{2(\pi a/\lambda)^2 + \gamma\}/2(\pi a/\lambda)^2(1 - \gamma) \\ \quad \text{for } 0 < \gamma < 1 \text{ (backward)} \\ C > U\{2(\pi a/\lambda)^2 + \gamma\}/2(\pi a/\lambda)^2(\gamma - 1) \\ \quad \gamma > 1 \text{ (forward)} \end{array}\right\} \quad (5.1\text{-}7)$$

for positive thrust; (4) the thrust decreases as the speed ratio U/C increases; and (5) the thrust increases with the square of the amplitude-to-wavelength ratio $(\pi a/\lambda)^2$, at a higher rate for a smaller γ, takes the maximum value at zero forward speed for every amplitude-to-wavelength ratio, and decreases as the speed increases. When the forward motion of the spermatozoon is stopped by an adhesive material, it activates the flagellum with a large amplitude and small wavelength, or with a high value of $\pi a/\lambda$ in order to get higher thrust (Hiramoto and Baba 1978).

Figure 5.1-2 shows the nondimensional speed U/C versus the amplitude-to-wavelength ratio $\pi a/\lambda$ (a) for zero thrust or trimmed condition and (b) for the maximum efficiency. It can be seen that as the amplitude-to-wavelength ratio increases, the nondimensional speed of zero thrust and of the maximum efficiency also increases initially at a higher rate for a smaller drag ratio γ, but at a small rate for larger values of the amplitude. The vertical lines show observed data for nematodes such as *Haemonchus contortus, Turbatrix aceti*, and *Panagrellus silusiae* (Gray and Lissmann 1964), which do not have any drag penalty from a head or the like and thus can cruise with zero thrust.

It is very interesting to find from Eq. (5.1-7) that when the drag ratio γ is less than one, the deflection wave should be sent backward, whereas when the drag ratio is more than one, the wave should be sent forward for positive thrust. A large drag ratio of more than one is possible for a flagellum possessing bristles along its whole length.

The efficiency has been defined as the efficiency of the snaking motion in converting the kinematic energy in the body motion into effective propulsive energy, with no considerations given to the

TABLE 5.1-3. Thrust, normal force, power, and efficiency in plane motion

Items	Equations	Note
Thrust	$T = \int \{-dD_N \sin\theta - dD_T \cos\theta\}$ $= C_N Cn\lambda[2(\pi a/\lambda)^2\{\pm 1 - (U/C)\} - \gamma\{(U/C) \pm 2(\pi a/\lambda)^2\}]$ $\quad + C_N C(l - n\lambda)[2(\pi a/\lambda)^2\{\pm 1 - (U/C)\} - \gamma\{(U/C) + 2(\pi a/\lambda)^2\}]$ $\quad + \text{additional sinusoidal terms} = T_0 + \Delta T_0 + \Delta T_1$	$(\ )_0 = \int_0^{n\lambda} \{d(\ )/ds\}\,ds$ $\Delta(\ )_0 = \text{steady terms of}$ $\int_{n\lambda}^{l} \{d(\ )/ds\}\,ds$
Normal force	$N = \int \{dD_N \cos\theta - dD_T \sin\theta\}$ $= -C_N Cn\lambda(V/C)[1 + 2\gamma(\pi a/\lambda)^2].$ $\quad - C_N C(V/C)(l - n\lambda)[1 + 2\gamma(\pi a/\lambda)^2]$ $\quad + \text{additional sinusoidal terms} = N_0 + \Delta N_0 + \Delta N_1$	$\Delta(\ )_1 = \text{sinusoidal terms of}$ $\int_{n\lambda}^{l} \{d(\ )/ds\}\,ds$ where $l - n\lambda < \lambda$ n: number of waves λ: wavelength l: length of flagellum
Power	$P = \int \{dD_T U_T + dD_N U_N\}$ $= C_N C^2 n\lambda[\{(V/C)^2 + 2(1 \mp U/C)^2(\pi a/\lambda)^2\} + \gamma\{2(V/C)^2(\pi a/\lambda)^2$ $\quad + (U/C)^2 \pm 4(U/C)(\pi a/\lambda)^2 + 6(\pi a/\lambda)^4\}]$ $\quad + \gamma\{2(V/C)^2(\pi a/\lambda)^2 + (U/C)^2$ $\quad + C_N C^2(l - n\lambda)[(V/C)^2 + 2\{1 \mp (U/C)^2(\pi a/\lambda)^2\}$ $\quad \pm 4(U/C)(\pi a/\lambda)^2 + 6(\pi a/\lambda)^4\}]$ $\quad + \text{additional sinusoidal terms} = P_0 + \Delta P_0 + \Delta P_1$ $\simeq |T|C(1 \mp U/C)$	
Efficiency	$\eta = |T|U/P = |T_0|U/P_0$ $= \dfrac{(U/C)\left[1 \mp (\gamma/2)\left\{\dfrac{(U/C) + 2(\pi a/\lambda)^2}{(1 \mp U/C)(\pi a/\lambda)^2}\right\}\right]}{\left[(1 \mp U/C) + (\gamma/2)\left\{\dfrac{(U/C)^2 \pm 4(U/C)(\pi a/\lambda)^2 + 6(\pi a/\lambda)^4}{(1 \mp U/C)(\pi a/\lambda)^2}\right\}\right]}$ $\simeq (U/C)/(1 \mp U/C)$	

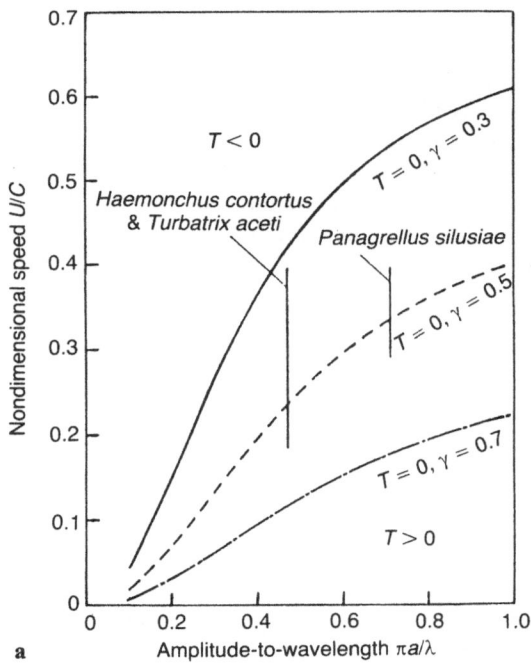

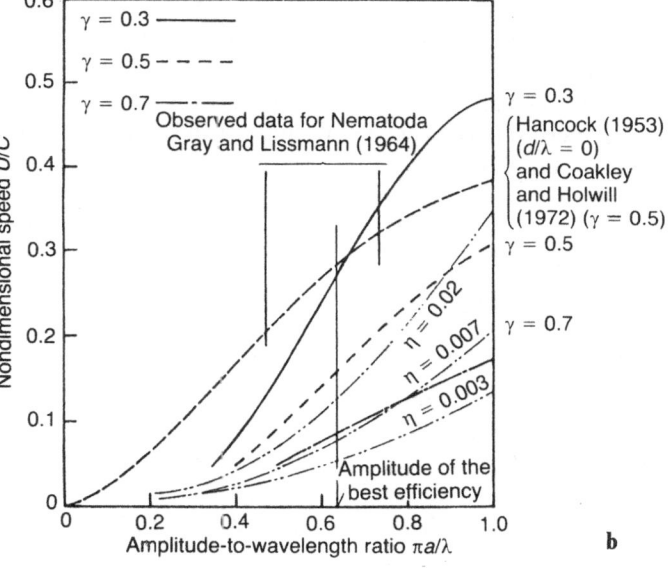

FIG. 5.1-2. **a, b.** Speed ratio-versus amplitude-to-wavelength ratio in a plane undulatory motion. **a** For zero thrust. (From Gray and Lissman 1964). **b** For the maximum efficiency.

TABLE 5.1-4. Speed ratio and reduced frequency in plane motion

Items	Equations	Note
Speed ratio	$U/C = 2(\pi a/\lambda)^2/[1 + (2\pi a/\lambda)^2 - \sqrt{1 - 2(\pi a/\lambda)^2 \{\ln(d/2\lambda) + \frac{1}{2}\}(3a_0/n\lambda)}]]$,	Taylor (1953)
	$U/C = 2(\pi a/\lambda)^2[\{K_0(\pi d/\lambda) - 1/2\}/\{K_0(\pi a/\lambda) + 1/2\}]$	Gray and Hancock (1955) K_0 is the modified Bessel function of order zero
	$U/C \simeq 2(\pi a/\lambda)^2(1 - \gamma)/\{2(\pi a/\lambda)^2 + \gamma\}$	for $a_0/n\lambda \ll 1$
	$\simeq \begin{cases} 0.5(1 - \gamma)/(0.5 + \gamma) & \text{for} \quad \pi a/\lambda \simeq 0.5 \\ 0 & \text{for} \quad \pi a/\lambda \simeq 0. \end{cases}$	
Reduced frequency	$k = \pi df/U = (\pi a/\lambda)(d/a)/(U/C) = \sigma/2(l/d)$	
	$\sigma = \omega l/U = 2\pi fl/U = 2\pi fd(l/d)/U = 2k(l/d)$	
	$k = 0.64(d/a)/0.16 \simeq 4(d/a)$	For maximum efficiency $\pi a/\lambda \simeq 0.64$, $U/C \simeq 0.16$, $\gamma = 0.5$.

thermal efficiency of converting food energy into mechanical energy. The approximate expression shows that the efficiency is independent of fluid viscosity, and is zero for $U/C = 0$ and increases as $U/C \to 1$ for backward travel of the deflection wave.

Observation data (Holwill 1966, 1974, 1977) for smooth flagella of small organisms give the optimal value of $\pi a/\lambda \simeq 0.6$. It is surprising to find that the best value for efficiency is as small as $\eta < 0.1$ for $0.3 \leq \gamma \leq 0.7$ in this approximate calculation. *Effect of head.* By assuming that the motion is performed along the X-axis only, the drag given by an inert, spherical head of radius a_0 can be derived from Eq. (2.1-3) as

$$D_{\text{head}} = 6\pi\mu U a_0 \qquad (5.1-8)$$

Equating D_{head} with the thrust T, yields

$$\pm U/C = 2(\pi a/\lambda)^2(1 - \gamma)/\{(6\mu/C_N)(\pi a/\lambda)(a_0/na) + 2(\pi a/\lambda)^2 + \gamma\} \qquad (5.1-9)$$

Various expressions of the speed ratio U/C and reduced frequency k or $\sigma = 2(l/d)k$ are given in Table 5.1-4.

5.1.2 SPIRAL MOTION OF FLAGELLUM

It is a known fact that some microorganisms, such as certain species of bacteria and eukaryotic cells, move their flagellum spirally for propulsion in a fluid. Bacteria belong to Schizomycophyta and have simple cells, usually rod-shaped and filled with active protoplasm.

Shown in Fig. 5.1-3a is a mechanical sketch of the swimming motion of a single bacterial flagel-

lum. It can be as small as $(1-3) \times 10^{-8}$ m in diameter and a few micrometers in length, as already given in Table 5.1-1. When the basal structure embedded in the cell wall and cytoplasmic membrane is driven as a rotary motor (Berg 1975), a helical filament and a short segment called a hook rotate rigidly as a propulsive system. Helical waves are then propagated down the flagellate tail with a measurable frequency of the order of 10^1-10^2 Hz, and an inert head is counter-rotated with a frequency of one order lower. When the flagellum changes its direction of rotation briefly, in what is called a "twiddle," it may cause the body to stop moving. When many flagella operate together in a bundle, the bundle separates during the twiddle. The drag of the head is balanced with the thrust generated by the flagellum at a specific speed, e.g., about 50–100 µm/s (Holwill and Burge 1963; Berg 1975).

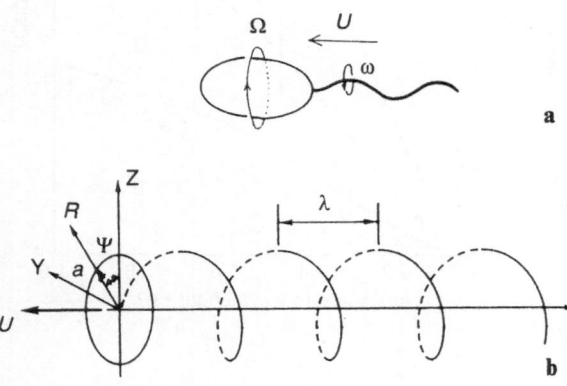

FIG. 5.1-3. **a, b.** Swimming behavior of a bacterium. **a** Body and flagellum motion. **b** A series of spiral waves.

TABLE 5.1-5. Thrust, power, efficiency, speed ratio, and reduced frequency in spiral motion

Items	Equations	Note
Thrust	$T = \int \{-dD_N \sin\Theta - dD_T \cos\Theta\}$	$\Lambda = \lambda\sqrt{1 + (2\pi a/\lambda)^2} \simeq \lambda$
	$= C_N C n \Lambda [\{\pm 1 - U/C\}(2\pi a/\lambda)^2 - \gamma\{U/C \pm (2\pi a/\lambda)^2\}]/[1 + (2\pi a/\lambda)^2]$	
Power	$P = \int \{dD_N U_N + dD_T U_T\}$	
	$= C_N C^2 n \Lambda [\{1 \mp U/C\}^2 (2\pi a/\lambda)^2 + \gamma\{U/C + (2\pi a/\lambda)^2\}^2]/[1 + (2\pi a/\lambda)^2]$	
Efficiency	$\eta = P_{\text{head}}/P = (D_{\text{head}} U + Q_{\text{head}}\Omega)/P = TU/P + \Omega/\omega$	$\omega = \omega_b - \Omega = (2\pi/\lambda)C$
		$\Omega/\omega = P/8\pi\mu a_0^3 \omega^2$
		$= P/8\pi\mu a_0^3 (2\pi C/\lambda)^2$
Speed ratio	$U/C = \pm(2\pi a/\lambda)^2(1 - \gamma)/[1 + (2\pi a/\lambda)^2](6\mu/C_N)(\pi a/\Lambda)(a_0/na) + \gamma + (2\pi a/\lambda)^2]$	Identical to Hancock (1953)
	$U/C \simeq \pm(2\pi a/\lambda)^2(1 - \gamma)/\{(2\pi a/\lambda)^2 + \gamma\}$	for $a_0/na \ll 1$
	$U/C = (2\pi a/\lambda)^2 [\{K_0(\pi d/\lambda) - \tfrac{1}{2}\}/\{K_0(\pi d/\lambda) + \tfrac{1}{2}\}]$	Taylor (1953)
	$= (2\pi a/\lambda)^2 [1 + 2(2\pi a/\lambda)^2 - \sqrt{1 + (2\pi a/\lambda)^2}\{\ln(d/2\lambda) + \tfrac{1}{2}\}(3a_0/n\lambda)]$	Holwill and Burge (1963)
	$= (2\pi a/\lambda)^3 \left[\dfrac{2\{1 + (2\pi a/\lambda)^2\}}{\{3(2\pi a/\lambda) + 2(K_1/K_0) + (2\pi a/\lambda)(K_0/K_2)\}} - \dfrac{1}{2}(2\pi a/\lambda) \right]$	Chwang and Wu (1971, 1974)
	$\simeq (2\pi a/\lambda)^4 \{\log(\lambda/\pi a) - 1.08$	
	$U/C \simeq \begin{cases} (1 - \gamma)/(1 + \gamma) & \text{for } \pi a/\lambda \simeq 0.5 \\ 0 & \text{for } \pi a/\lambda \simeq 0 \end{cases}$	
Reduced frequency	$k = 0.5(d/a)/0.18 = 2.8(d/a)$	for $\eta_{\max}$ at $\pi a/\lambda = 0.5$, $U/C \simeq 0.18$, and $\gamma = 0.5$

As shown in Fig. 5.1-3b, in a cylindrical coordinate system (X, R, Ψ), the waveform may be represented simply by

$$R = a, \quad \text{and} \quad \Psi = (2\pi/\lambda)(X \mp Ct) \quad (5.1\text{-}10)$$

In this section it is not necessary to assume small amplitude in oscillation but the radius a is assumed constant. Thus, the position at the (X, Y) and (X, Z) planes can respectively be given by

$$Y = a \sin\{(2\pi/\lambda)(X \mp Ct)\} \quad \text{and}$$
$$Z = a \cos\{(2\pi/\lambda)(X \mp Ct)\} \quad (5.1\text{-}11)$$

The velocity components at an element on the flagellum in spiral form are as follows:

$$\dot{X} = 0 \quad \dot{R} = 0 \quad \text{and} \quad \dot{\Psi} = \mp(2\pi/\lambda)C \quad (5.1\text{-}12)$$

Then, the relative velocities, the tangential U_T and the normal U_N, of the element with respect to the surrounding fluid can be given as

$$\left. \begin{aligned} U_T &= (U - \dot{X})\cos\Theta - a\dot{\Psi}\sin\Theta \\ &= U\cos\Theta \pm a(2\pi/\lambda)C\sin\Theta \\ U_N &= (U - \dot{X})\sin\Theta + a\dot{\Psi}\cos\Theta \\ &= U\sin\Theta \mp a(2\pi/\lambda)C\cos\Theta \end{aligned} \right\} \quad (5.1\text{-}13)$$

where the length of the element ds and its inclination Θ are

$$\left. \begin{aligned} \sin\Theta &= a(d\Psi/ds) = a(2\pi/\lambda)(dX/ds) \\ \cos\Theta &= dX/ds = 1/\sqrt{1 + (2\pi a/\lambda)^2} \\ ds &= \sqrt{dX^2 + (ad\Psi)^2} = dX\sqrt{1 + (2\pi a/\lambda)^2} \end{aligned} \right\} \quad (5.1\text{-}14)$$

In the above equations, the induced velocity in the slip stream generated by the spiral motion of the system has been neglected because, as stated before, the inertial effect may be neglected in this low range of Reynolds number.

By introducing here again the quasi-steady theory, the elemental thrust and power can be expressed in Table 5.1-5 in which other expressions are presented for comparison. It is interesting to note that in this spiral motion both the thrust and the power have similar expressions to those of a plane wave, and both the thrust and the power are roughly twice those of the plane wave for a small amplitude-to-wavelength ratio and a small γ.

The thrust and the power generated by a flagellum of a length of more than n wavelengths can be obtained similarly to those of the plane wave given in Sect. 5.1.1.

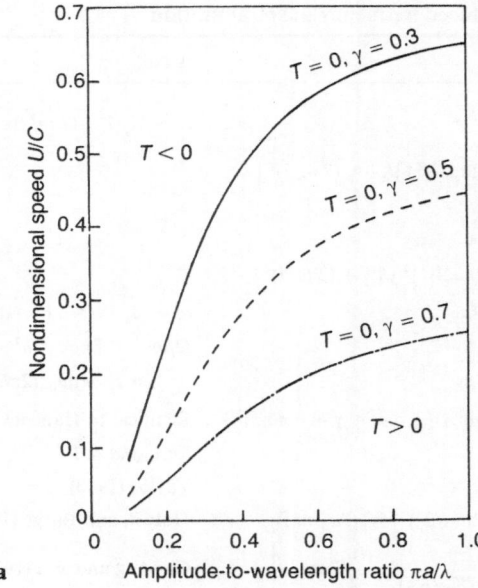

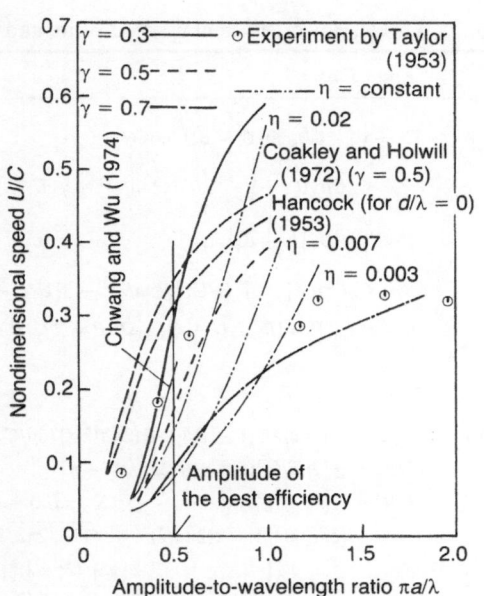

FIG. 5.1-4. **a, b.** Speed ratio versus amplitude-to-wavelength ratio in a spiral motion. **a** For zero thrust $T = 0$. **b** For maximum efficiency.

If an inert head is driven by this thrust, then the trimmed velocity can be given by equating the above thrust with the drag given by Eq. 5.1-8. It is found that for the same amplitude and wavelength, a spiral wave propels the flagellum at higher speed than a plane wave.

Unlike the case for plane undulatory motion, the rate of increase of the nondimensional thrust levels off as the amplitude-to-wavelength ratio increases, whereas the power continues to increase rapidly with the amplitude. The thrust at zero speed is maximum for other speeds at every amplitude-to-wavelength ratio and decreases as the speed increases. However, the effect of the slip condition is not large for the power.

Figure 5.1-4 shows the nondimensional speed versus the amplitude-to-wavelength ratio (a) for zero thrust and (b) for the maximum efficiency. The configuration of the curves is similar to that of plane undulatory motion. Here the optimal amplitude-to-wavelength ratio providing the best efficiency is given approximately by $\pi a/\lambda \simeq 0.5$, which is slightly lower, whereas the efficiencies themselves are slightly higher than in plane undulatory motion. Liefson (1960) and Shimada et al. (1974) gave observation data of $\pi a/\lambda = 0.4$–0.8 and $k = 2.8$ (d/a) for bacterial flagellation. Holwill and Burge (1963) also gave the maximum efficiency as about 4%, whereas Chwang and Wu (1971) reported a much better efficiency because their analysis included the useful power for the rotation of the head as follows:

$$\eta = P_{\text{head}}/P = (D_{\text{head}}U + Q_{\text{head}}\Omega)/P$$
$$= TU/P + \Omega/\omega \qquad (5.1\text{-}15)$$

where ω is the absolute spiral or spin rate of the flagellum, and is given by the rotational speed of the flagellum with respect to the head ω_b, minus the rotational speed of the head Ω, which is opposite to the direction of the flagellar motion:

$$\omega = \omega_b - \Omega = (2\pi/\lambda)C \qquad (5.1\text{-}16)$$

According to R.A. Anderson (1974), since the torque required to maintain an angular velocity ω of a sphere is given by

$$Q_{\text{head}} = 8U\mu a_0^3\Omega, \qquad (5.1\text{-}17)$$

the spin ratio Ω/ω is then given by

$$\Omega/\omega = P/8\pi\mu a_0^3\omega^2 = P/8\pi\mu a_0^3(2\pi C/\lambda)^2 \qquad (5.1\text{-}18)$$

Figure 5.1-4 also shows other theoretical and experimental values (Hancock 1953; Coakley and Holwill 1972) for comparison. Taylor's (1953) experiments were carried out on the movement of a piece of wire (0.061 cm in diameter) bent into the shape of a spiral (approximately 2.5 cm in diameter) that was rotated in viscous fluid. His results also suggest that a lower value of γ such as $\gamma \simeq 0.5$ should be chosen.

A more general discussion of the hydrodynamic effects of nonuniformities in cross section and wavelength of three-dimensional flagellar waveforms was presented by Coakley and Holwill (1972). Although the flagella of bacteria sometimes appear in bundles (Jarosch 1967; Lowy and Spencer 1968 R.A. Anderson 1974), a comprehensive study of the flagellar bundle is not as yet feasible since little is known about its hydrodynamic characteristics.

5.1.3 SNAKING IN ANNELIDA

Some animals with bodies of medium length (several centimeters) swim in water with a medium Reynolds number, such as Re = 10^1–10^2, based on their diameter. The drag coefficient in the range of these Reynolds numbers behaves with characteristics falling between those at lower and higher Reynolds number. The frictional drag coefficient is still high but the pressure drag is also important.

Marine worms. Annelida like the marine worms (*Nereis*), typical of which are Polychaeta such as the lobworms (*Nereimorpha*), have a very rough surface with numerous long projections called parapodia pointing equally in all directions (Gray 1939). Since the diameter of the projections is much smaller than the body diameter, the Reynolds number based on it is smaller than that of the body by one order. Thus, the friction force, though dependent on the exact nature of the roughness, is important, and the presence of many laterally extended parapodia on the marine worms favors the longitudinal or tangential drag over the normal drag.

For a slender body of large friction drag ($\gamma > 1$), the analysis in Sect. 5.1.1 shows that the value of U/C should be negative for optimal performance. This means that the waves of displacement should be sent forward, i.e., in the direction of locomotion (Gray 1953). In the swimming of the *Nereis diversicolor*, the deformation wave is sent more quickly than in crawling. Gray got $U/C = -0.23$ and $\pi a/\lambda = 0.57$ by assuming that the longitudinal force coefficient is more than twice the transverse coefficient: $\gamma = C_T/C_N = 2$.

5.1.4 CILIARY AND WHIPLIKE LOCOMOTIONS

Ciliary locomotion. Many microorganisms are also propelled by large numbers of cilia on their body surface. The cilia are located in rows along and across the surface, and the organism is propelled in the opposite direction to that of the effective beat of the cilia (J.R. Blake 1971).

Some species of Protozoa and Mesozoa are known to be propelled by a series of ciliary whipping motions. The cilia lie within a distinct size range with lengths usually between 20 µm and 2 mm. The swimming speed of these species is usually about 1 mm/s, faster than that of organisms propelled by flagella. However, the speed relative to body length decreases with increasing size of the organism (Sleigh and Blake 1977).

The cilia usually occur in large numbers and have the same pattern and frequency in beating or whipping, which consists of two strokes, the active and recovery strokes. Because of phase differences in the beating, the coordinated waves of beating activity are always propagated in the direction toward which the cilia move in their recovery stroke (Baba 1974). This coordination of the cilia is achieved through the intrinsically rhythmical activity of the component cilia, using a timing signal communicated among neighboring cilia, (Kinoshita and Murakami 1967; Sleigh and Barlow 1980).

Figure 5.1-5 gives a sketch of ciliary motion and the wave of the envelope of ciliar tips in a normal beat. It shows the wave crest during the "active (or power) stroke" and the wave trough during the

FIG. 5.1-5. Waving motion of cilia in normal beat.

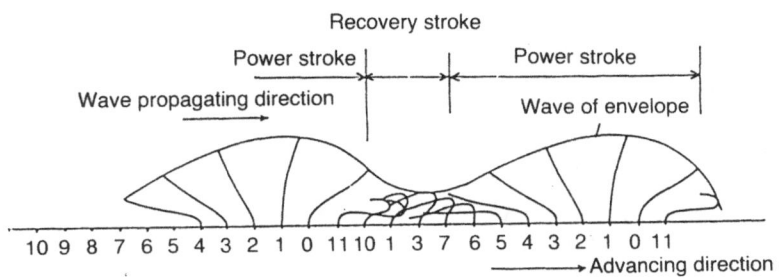

"recovery stroke". Although the active strokes occupy a large space over the surface of the organism, the recovery strokes generally take far more time. The wave propagates in the same direction as the body propulsion for *Paramecium*, and in the opposite direction for *Opalina* (J.R. Blake 1971). In the case of *Opalina*, an efficiency of 20% can be obtained at a speed ratio of $U/C = 0.19$, where the wavespeed is $C = 200-400$ µm/s. When the body is stopped, the surrounding fluid is shifted to the direction of the effective stroke (Fulford and Blake 1986).

5.2 Undulatory Motion of a Large Elongate Body

Slender creatures with long cylindrical bodies of lengths greater than a few centimeters can swim with an undulatory motion known as "anguilliform" motion. Here, anguilliform motion is defined as an undulatory motion in which a transverse wave of increasing amplitude passes along the body from head to tail. The anguilliform mode of locomotion is very similar to that of the plane flagellar motion of microorganisms, but the driving mechanism is slightly different because the inertial forces of the fluid related to the body motion cannot be discarded in this flow field at higher Reynolds numbers and larger reduced frequencies. Thus, as suggested by Lighthill (1970), the anguilliform propulsion of elongate fish uses a combination of resistive and reactive force. In this section this is clearly verified with exemplified calculations for two species of fish and a snake in comparison with actual swimming data.

With their dorsal, ventral, and anal fins and especially with their thin or flat body, fish can make better use of reactive forces. They can execute an undulatory motion by activating either the body and the dorsal, ventral, and caudal fins, or only the fins without assistance of body motion. When locomotion is performed only by extended fins, the amplitude of the motion is almost constant along the fin length. The word "elongate" implies that the body is very long in the longitudinal direction compared with the lateral dimensions; thus, the dynamic characteristics of the surrounding fluid can be treated as a two-dimensional flow at any section (Jones 1946).

Other modes of swimming, such as (1) by the oscillation of either several fins (as in the sturgeon) or a single tail (called "carangiform" or fanning motion here), (2) by the labriform motion of pectoral fins (as in the trunkfish; see Sect. 6.1.3), and (3) by the sweeping of pectoral fins or of a long tail (like in the shark), will be treated later.

5.2.1 GEOMETRICAL CONFIGURATIONS AND SWIMMING MODES

Many animals prefer to swim by anguilliform or snaking motion of the whole body: eels, loaches, morays, sea snakes, congers and gobies. This mode of locomotion of the elongate animal is characterized by lateral undulatory motion of either the entire cylindrical body or a part of the body such as dorsal and/or ventral fins, and is used over a wide range of velocities, even in backward motion. If the body length is large (more than about 1 m), and the body is smooth, without bristles, then a series of transverse waves of increasing amplitude pass along the body from head to tail.

As discussed in the previous section, many kinds of microorganisms swim without snaking their head. Similar swimming can be observed in the body stroke of eels. Since the motion of the head is substantially limited to that accompanying the longitudinal body motion, it may be called "fixed head" motion. Snakes, on the other hand, apply a snaking motion to the longitudinal axis of the body including the head. When eels and morays move slowly, however, the undulatory motion is sometimes restricted mainly to the dorsal and ventral fins themselves.

Snaking motion was carefully observed by Gray (1933a) and analyzed theoretically by Gray (1933b, 1933c), Taylor (1952b), Lighthill (1960b, 1970), Wu (1971a, 1971b), and Hess and Videler (1984). However, theoretical treatments in these studies disregard one or the other of the inertial and viscous effects. Only Vlymen (1974) analyzed the swimming energy by combining the reactive and resistive forces acting on the body of elongate fishes.

The animals studied in this section are the sea snake (*Pelamis platurus*), eel (*Anguilla japonica*) and butterfish (*Pholis gunnellus*). Their geometrical configurations are shown in Table 5.2-1, in which some of the data on the sea snake and butterfish are taken from papers written by Graham et al. (1987) and Gray (1933a). Either the body profile or the cross section, and related quantities estimated for the calculations are shown in Fig. 5.2-1 with

TABLE 5.2-1. Geometrical configuration and mode of locomotion

| Items | Species | | Sea snake[a] (Pelamis platurus) | Eel (Anguilla japonica) | Butterfish[b] (Pholis gunnellus) |
	Symbols	Units			
Body length	l	m	0.510	0.557	0.120
Height (average)	c	m	0.013	0.031	0.010
Height (maximum)	c_{max}	m	0.015	0.035	0.014
Mass	m	kg	0.037	0.225	—
Head amplitude	a	m $\times 10^{-3}$	7.4	6.5	1.7
Amplitude rise in e^{bs}	b	m^{-1}	4.7	5.2	21.5
Wavelength	λ	m	0.257	0.365	0.096
Number of wave	l/λ	—	1.70	1.59	1.39
Speed	U	m/s	0.320	0.240	0.123
Wavespeed	C	m/s	0.437	0.302	0.192
Speed ratio	U/C	—	0.732	0.796	0.640
Reduced	$k = (\pi c/\lambda)/(U/C)$	—	0.22	0.34	0.51
frequency	$\sigma = 2(\pi l/\lambda)(U/C)$	—	17	12	12

[a] Data from Graham et al (1987)
[b] Data from Gray (1933a)

the eel as an example. The animals are characteristically smooth cylinders with or without elongate ribbon-shaped fins on the upper (dorsal) and/or lower (ventral) sides of the body. The sharp edges of these fins are, as explained later, effective for generating flow separation, and thus increase the pressure drag for crossflow.

FIG. 5.2-1. Body configuration and related quantities of eel.

The normalized profile form is shown in Fig. 5.2-2, in comparison with other animals. It can be seen from this figure that in the sea snake and the eel, the body height c is almost constant along the body length, whereas the height of the butterfish and the trout varies along the body length.

The modes of locomotion analyzed from moving pictures and represented in the form of Eq. (5.2-4) given later are shown by related parameters in Table 5.2-1. An example of (a) the real mode and

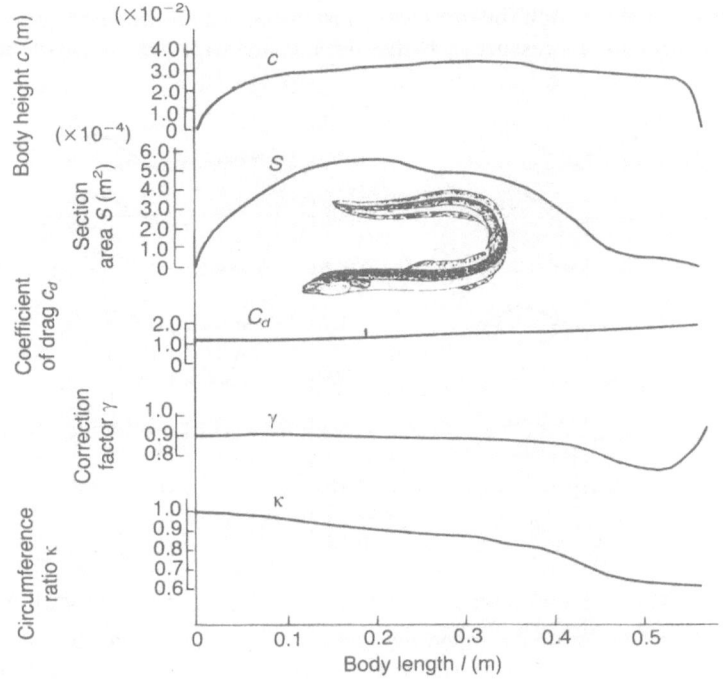

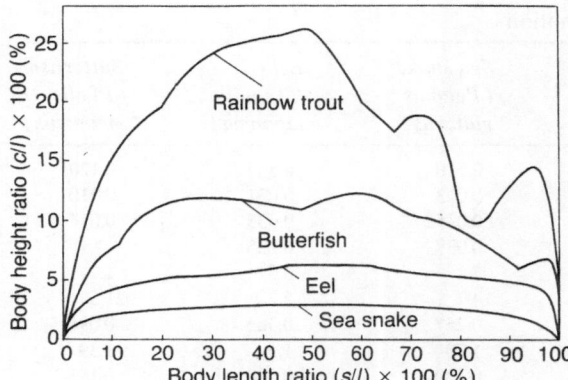

FIG. 5.2-2. Normalized form of body profiles.

(b) the mathematical mode of swimming in the eel is shown in Figs. 5.2-3. It may be supposed that the performance can be closely simulated by mathematical analysis based on the above representations.

5.2.2 FUNDAMENTAL HYDRODYNAMICS

In addition to being influenced by the friction on the body surface, the hydrodynamic forces and moments of an immersed body in swimming motion are influenced by the acceleration of the fluid surrounding the body. For a flexible or deformable body, the configuration is determined in the balance of internal stress applied through the muscles and the external hydrodynamic pressure, which

is a function of the history of the change of configuration and motion of the body. Thus, the hydrodynamic forces and moments acting on the flexible body must be obtained by simultaneously solving the elastic equation and equations of motion at every part of the body. Usually, by assuming a non-deformable cross section, the elastic equations can be discarded. Thus, in order to introduce the effects of change of body configuration into the hydrodynamic forces and moments, it is necessary to take account of the velocity and acceleration of the fluid around every part of the body due to its lateral deflection. This can be done by dividing the body into n segments or elements along the longitudinal axis of the elongate body, as shown in Fig. 5.2-4. By summarizing or integrating the friction and reaction forces acting on the body elements during linear and angular acceleration of the respective body elements themselves and of the "added masses" of the fluid surrounding them[1], the total (inertial and hydrodynamic) forces of the whole body can be obtained.

For a body element located at a distance s from the head, let us consider a local coordinate system

---

[1] The detailed expression for the added masses are shown later in Tables 6.1-1 and 6.1-2.

FIG. 5.2-3. **a, b.** Swimming mode of eel. **a** Real mode **b** Mathematical mode (based on Eq. (5.2-4)).

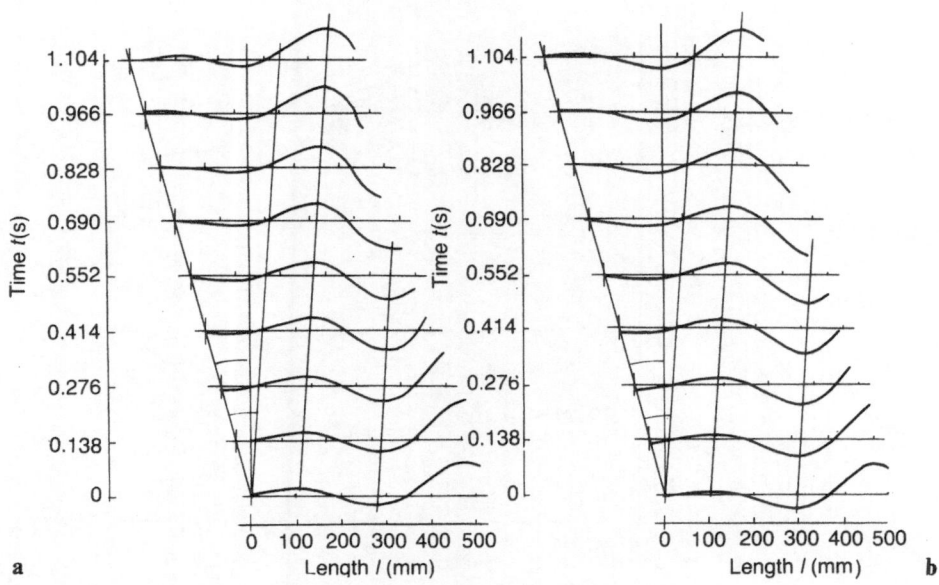

a   b

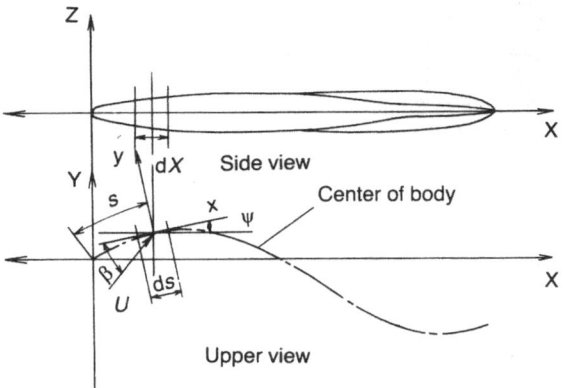

FIG. 5.2-4. Body segment and its coordinate system.

(x, y, z), the origin of which is located at the centroid of the body element, and the x-axis of which is aligned with the longitudinal body axis in deflected state, as shown in Fig. 5.2-4. The body is assumed to be elongate on the x-axis and symmetric about the (x, z) plane, and is deflected in the (x, y) plane in such a way that the velocity components of the body element along the x- and y-axes

are given by u and v respectively, and the angular velocity is given by ω.

The detailed expression of the inertial and hydrodynamic forces under the assumption of small deflection are presented in Table 5.2-2. The inertial force consists of the body force resulting from pure acceleration and of fluid force derived from both pure acceleration and height change along the body or x-axis. As stated later in Sect. 6.2.1, the force is a kind of lift. The hydrodynamic drag force consists of skin friction drag and pressure drag (coefficients C_f and C_p) based on the projected area (Eames 1967). A parameter κ is the ratio of circumferential length of an ellipse to that of a circle, the diameter of which is the reference length or the height of the body c (Bursnall and Loftin 1951; Taylor 1952b; and Casarella and Parsons 1970). Thus κ takes a value within $2/\pi \leq \kappa \leq 1$, the lower and upper limits being those for a very thin plate and a circular cylinder, respectively. In actual elongate creatures, the body surface area is important in estimating the friction drag. As an example, the relation between surface area and body mass of a swamp eel, *Synbranchus marmoratus*, (Graham

TABLE 5.2-2. Inertial and hydrodynamic forces and related quantities

Items	Symbols	Expression	Note
Inertial forces of body element	$f_{I,x}^{B}$	$-\rho_b S \dot{u}$	Cross-sectional area[a] S and density ρ_b. Inertial or body forces result from pure acceleratiion. These are approximate expressions by neglecting the effects of angular motion ω.
	$f_{I,y}^{B}$	$-\rho_b S \dot{v}$	
Inertial forces of fluid element	$f_{I,x}^{W}$	$-\{\rho_{11} S \dot{u} - u(\partial/\partial s)(\rho_{11} S u)\}$	Density of added masses ρ_{11} and ρ_{22} for longitudinal (x) and lateral (y) axes.[b] The first and second terms are related to pure acceleration and height change, respectively.
	$f_{I,y}^{W}$	$-\{\rho_{22} S \dot{v} - u(\partial/\partial s)(\rho_{22} S v)\}$	
Hydrodynamic forces	$f_{D,x}$	$\frac{1}{2}\rho U^2 c \kappa \pi C_f \cos\beta$	Skin friction drag caused by tangential flow.
	$f_{D,y}$	$\frac{1}{2}\rho U^2 c(\kappa \pi C_f \sin\beta + C_p|\sin\beta|\sin\beta)$	The first and second terms are derived from skin friction and pressure drag respectively caused by normal flow.
Body motion	U	$\sqrt{u^2 + v^2} \simeq U$	$Y = Y(X(t),t) = Y(s,t) \ll l$
	β	$\tan^{-1}(-v/-u) \simeq -v/U$	See Fig. 5.2-4
	$\cos\beta$	$-u/U \simeq 1$	
	$\sin\beta$	$-v/U,$	
	u	$\dot{Y}\sin\psi - U\cos\psi \simeq -U$	$(\ )' = \partial(\ )/\partial s$
	v	$\dot{Y}\cos\psi + U\sin\psi \simeq \dot{Y} + UY'$	See Fig. 5.2-4
	ψ	$\tan^{-1}(\partial Y/\partial X) \simeq Y'$	
	ω	$\dot{\psi} = \dot{Y}'$	

[a] Here in this section, the cross-sectional area S_c is simply expressed by S

[b] The volume of added mass of a unit length is considered to be approximately equal to the cross sectional area of the body S

et al. 1987) is not different from that of vertebrate shown later in Fig. 6.1-4.

The pressure drag coefficient C_p due to the normal flow component may be approximately equal to that for a flat plate instead of an elliptical cylinder if the eccentricity of the ellipse is large or if the thickness to the chord is small. If an elongate body has a sharp corner at its upper and/or lower edge, as in the case where dorsal or ventral fins are present, the transversal flow is easily separated from the edge at a high Reynolds number, and the resulting pressure drag, which is equivalent to the "vortex induced force" previously given in Sect. 4.5.4, is appreciable. However, if the body has round corners at the respective edges and make an undulatory motion with low amplitude or small lateral deflection, then the flow separation is not significant and the pressure drag terms, $C_p|\sin \beta|$ $\sin \beta$ in Table 5.2-2, may be neglected. The actual values of C_f and C_p can be found in the related materials (such as Hoerner 1965).

The elemental thrust and side force, dT and dH, which are negatively directed along the $-X$ and $-Y$-axes, and power dP are generated by the inertial force or "reactive force" and "resistive force" as follows:

$$
\left.
\begin{aligned}
dT/ds &= -dF_X/ds \\
&= -(dF_x/ds)\cos\psi + (dF_y/ds)\sin\psi \\
&= \{dT_{I,1} + dT_{I,2} + dT_{D,1} + dT_{D,2}\}/ds
\end{aligned}
\right\}
$$
$$(5.2\text{-}1a)$$

$$
\left.
\begin{aligned}
dH/ds &= -dF_Y/ds \\
&= -(dF_x/ds)\cos\psi - (dF_y/ds)\sin\psi \\
&= \{dH_{I,1} + dH_{I,2} + dH_{D,1} + dH_{D,2}\}/ds
\end{aligned}
\right\}
$$
$$(5.2\text{-}1b)$$

$$
\begin{aligned}
dP/ds &= -(dT/ds)U + (dH/ds)\dot{Y} \\
&= \{dP_{I,1} + dP_{I,2} + dP_{D,1} + dP_{D,2}\}/ds
\end{aligned}
$$
$$(5.2\text{-}1c)$$

where

$$
\left.
\begin{aligned}
dF_x/ds &= -f_x = -(f_{I,x}{}^B + f_{I,x}{}^W + f_{D,x}) \\
dF_y/ds &= -f_y - F_x dy/ds \simeq (f_{I,y}{}^B + f_{I,y}{}^W + f_{D,y})
\end{aligned}
\right\}
$$
$$(5.2\text{-}2)$$

and where 5.2-2, subscripts I, 1 and I, 2 denote reactive components related to the pure acceleration and the variable height respectively, and subscripts D,1 and D,2 denote resistive components related to the friction and pressure drag

respectively. Furthermore, superscripts B and W show body and fluid elements respectively. The validity of these superposed expressions of the forces has been accepted in the hydrodynamics of towing cables (Eames 1967; Casarella and Parsons 1970; Pao 1970) and in the aerodynamics of slender wings (Polhamus 1966, 1968; Lamer 1976). In the case of slender fish, as stated in the subsequent chapter, since the height change of the body is more appreciable than that of the elongate body, the pressure drag caused by normal flow is discarded in comparison with the lifting force caused by the increment of the added mass of tangential flow along the body.

The thrust and power for the total body length l can be obtained by integrating the elemental thrust and power as follows:

$$
\begin{aligned}
T &= \int_0^l (dT/ds)\,ds = T_{I,1} + T_{I,2} + T_{D,1} + T_{D,2} \\
&= T_I + T_D
\end{aligned}
$$
$$(5.2\text{-}3a)$$

$$
\begin{aligned}
H &= \int_0^l (dH/ds) = H_{I,1} + H_{I,2} + H_{D,1} + H_{D,2} \\
&= H_I + H_D
\end{aligned}
$$
$$(5.2\text{-}3b)$$

$$
\begin{aligned}
P &= \int_0^l (dP/ds)\,ds = P_{I,1} + P_{I,2} + P_{D,1} + P_{D,2} \\
&= P_I + P_D
\end{aligned}
$$
$$(5.2\text{-}3c)$$

In the integration along the body, the negative gradients of the added masses, $(\rho_{11}S)'$, $(\rho_{22}S)' < 0$, may be discarded because of the existence of a vortex wake. Lighthill (1970) showed that the dynamical effect stays practically the same if the vortex sheet is replaced by a solid fin. Wu (1971b) also took into account the contribution due to the shedding of vortices. However, in the case of almost constant body height and of undulation with increasing amplitude toward the tail, the posterior part of the body moves in an almost undisturbed flow. In light of these facts, we can make use of this simplification for the calculations in this section.

5.2.3 PERFORMANCE

The role of the resistive force was stressed in papers by Taylor (1952b), Hancock (1953), and Gray and Hancock (1955), whereas the importance of the reactive force was pointed out by Lighthill (1960b), Wu (1971b), Hess and Videler (1984), and Videler and Hess (1984).

Let us consider a simple case where: (1) the height (or depth) of a swimming body c is either constant or triangular, even if its geometrical configuration varies; and (2) the swimming is a simple sinusoidal motion such as

$$Y = ae^{bs}\sin\{(2\pi/\lambda)(s - Ct)\} \qquad (5.2\text{-}4)$$

where the amplitude a at the head is assumed small, $a \ll l$, and increases exponentially at the rate of e^{bs} $(b \geq 0, 0 \leq s \leq l)$ as the waves pass from head to tail. The propriety of this mathematical expression was verified earlier in Sect. 5.2.1.

Then, the added masses of a unit volume of fluid for longitudinal and lateral directions are, respectively

$$
\left.
\begin{aligned}
\rho_{11} &= (\rho_{11}S)' = 0 \\
(\rho_{22}S)' &= \text{constant } (= 0 \text{ for constant height} \\
&\quad \text{and} > 0 \text{ for triangular height}) \\
\rho_{22} &= \rho\{\pi(c/2)^2/S\}\gamma
\end{aligned}
\right\} \quad (5.2\text{-}5)
$$

where γ is, as shown in Fig. 5.2-5, a factor for correcting the effect of fins on the elliptical cross section and of the curved cylinder (Lighthill 1970). In order to introduce the curved cylinder effect, the correction factor γ for the sea snake and eel is modified by multiplication by 0.9 because the wavelength-to-height ratio is more than 11 for both animals, although their cross sections are not circular cylinders.

Then, by substituting Eq. (5.2-4) and (5.2-5) into Eqs. (5.2-1 through 5.2-3), and then taking time

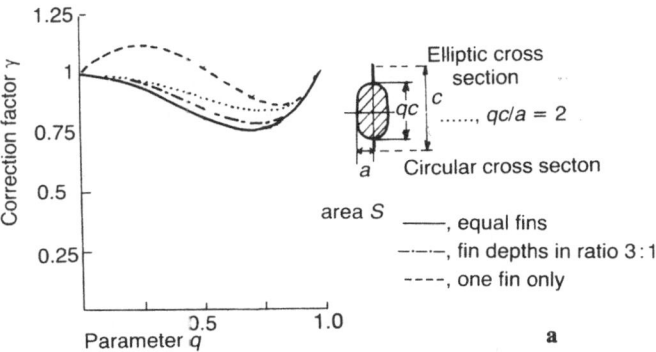

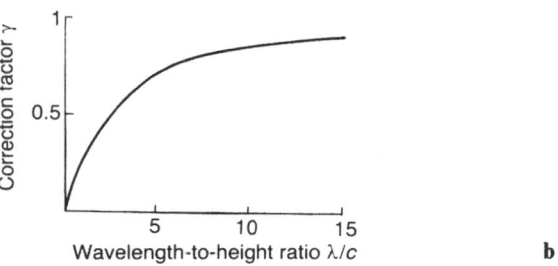

FIG. 5.2-5. **a, b.** Correction factor γ. (Based on Lighthill 1970). **a** For finned cylinder. **b** For curved cylinder.

mean $(\overline{}) = (C/\lambda)\int_0^{\lambda/c}(\quad)dt$ the inertial forces and power can be expressed as given in Table 5.2-3.

Since the analytical expressions of the inertial force have been obtained by truncating the higher harmonics of the Fourier series, the approxima-

TABLE 5.2-3. Inertial components of thrust, side force, and power of undulatory motion

Items	Expression
Inertial thrust	$\overline{T}_{I,1} \simeq \frac{1}{4}\rho\pi(c/2)^2\gamma \cdot (e^{2bl} - 1)[C^2(2\pi a/\lambda)^2\{1 - (U/C)^2\} - U^2(ab)^2]$
	$\overline{T}_{I,2} \simeq \begin{cases} 0 \text{ for constant height} \\ \frac{1}{4}\rho\pi(c/2)c'\gamma\{(e^{2bl} - 1)/b\}[UC(2\pi a/\lambda)^2\{1 - (U/C)\} - U^2(ab)^2] \\ \quad \text{for triangular height and increased amplitude } (b > 0) \\ \frac{1}{2}\rho\pi(c/2)c'\gamma \cdot lUc\{1 - (U/C)\}(2\pi a/\lambda)^2 \\ \quad \text{for triangular height and constant amplitude } (b = 0) \end{cases}$
Side force	$\overline{H}_{I,1} \simeq \overline{H}_{I,2} = \overline{H}_{D,1} = \overline{H}_{D,2} = 0$
Inertial power	$\overline{P}_{I,1} \simeq \frac{1}{4}\rho\pi(c/2)^2\gamma \cdot (e^{2bl} - 1)[UC^2\{1 - 2(U/C)^2\}(2\pi a/\lambda)^2 + U^3(ab)^2 + (2\pi a/\lambda)^2\}$
	$\overline{P}_{I,2} \simeq \begin{cases} 0 \text{ for constant height} \\ \frac{1}{4}\rho\pi(c/2)c'\gamma \cdot \{(e^{2bl} - 1)/b\}[UC^2\{1 - (U/C)\}^2(2\pi a/\lambda)^2 + U^3(ab)^2] \\ \quad \text{for triangular height and increased amplitude } (b > 0) \\ \frac{1}{2}\rho\pi(c/2)c'\gamma \cdot lC^2U(2\pi a/\lambda)^2\{1 - (U/C)\}^2 \\ \quad \text{for triangular height and constant amplitude } (b = 0) \end{cases}$

tion is only valid for low values of the amplitude rise parameter b. It is important to note that: (1) the inertial force generated by the snaking motion of an elongate fish with a constant height can be obtained only when the amplitude of the lateral deflection increases rearwardly or $b > 0$; (2) the inertial force is proportional to the square of the body height c^2, whereas the resistive force is proportional to the body height c itself, as seen in Table 5.2-2; and (3) the inertial force and power caused by the height change are proportional to the rate of change in height $c' = \partial c/\partial s$ times the height itself or the rate of change of the square the height.

It is interesting that the inertial components caused by the mass of the body itself have no impact on swimming performance because there is no longitudinal mass transfer in the undulatory motion of the cylindrical body, whereas the inertial components caused by the mass of the fluid or the added mass are important factors in the generation of thrust and power. It has thus been said that an elongate animal with constant depth (or height) can swim even in a perfect fluid (Saffman 1967) by adopting carangiform motion with increasing amplitude. Referring to Fig. 5.2-6a, this can be explained mechanically as follows: the reaction forces shown by the thick solid arrows are generated by either acceleration or deceleration of fluid around symmetrically arranged body elements. In constant amplitude swimming (a) the horizontal components indicated by thick solid arrows cancel each other out at every half-wavelength, whereas the vertical components of these forces cancel each other out at every wavelength. However, in increasing amplitude swimming (b) neither the horizontal components nor the vertical components cancel each other out because the force acting on the element at the front part of a half-amplitude is larger than that acting on it at the rear part (symmetrical phase position) of the half-amplitude; furthermore the forces and their differences (or vectorial summation) increase toward the tail, thereby generating the driving force.

It can thus be said that in the snaking motion of an elongate body with constant height and constant amplitude, the thrust can be provided solely by the hydrodynamic drag. The difference between the thrust and the drag is equivalent to the thrust of the microorganism. Actually, analytic integrations of the elemental forces related to the fluid dynamic drag over the whole length of body

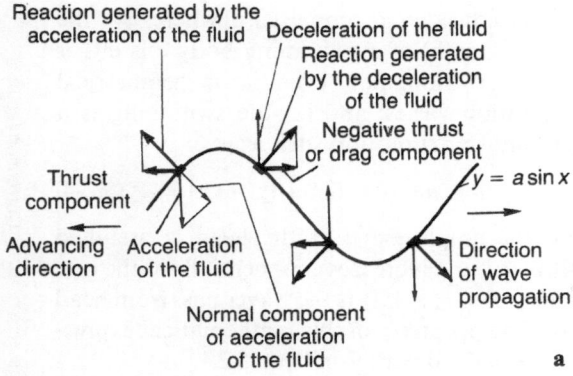

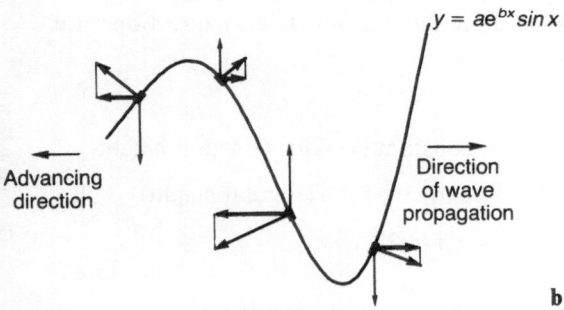

Fig. 5.2-6. **a, b.** Inertial forces generated by either accelerated or decelerated fluid. **a** constant amplitude $b = 0$. **b** Increasing amplitude $b > 0$.

cannot be obtained. Thus the numerical calculation is required to get the total forces acting on the body.

A trimmed speed U can be obtained by solving the following equation:

$$\bar{T} = \bar{H} = 0 \qquad (5.2\text{-}6a)$$

or

$$\left.\begin{aligned}\bar{T}_{I,1} + \bar{T}_{I,2} + \bar{T}_{D,2} = -T_{D,1} \equiv D_f \\ \bar{H}_{I,1} + \bar{H}_{I,2} + \bar{H}_{D,1} + \bar{H}_{D,2} = 0\end{aligned}\right\} \quad (5.2\text{-}6b)$$

where D_f is the friction drag of the whole body. At the trimmed speed, the instantaneous thrust and side force also vanish because they include the inertial components caused by the body itself,

$$T = H = 0 \qquad (5.2\text{-}7)$$

The efficiency η at this swimming speed can be expressed as

$$\eta = D_f U/P \qquad (5.2\text{-}8)$$

Numerical analysis. Numerical calculations were done by using a specified set of inputs given by the mode analysis for an eel and in the results are

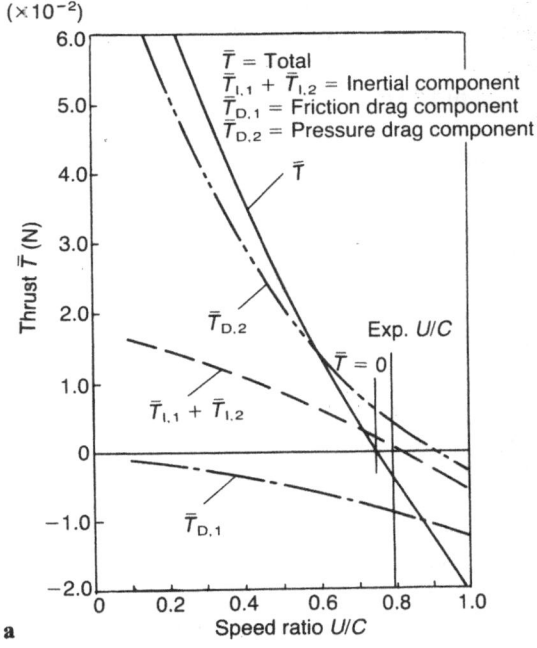
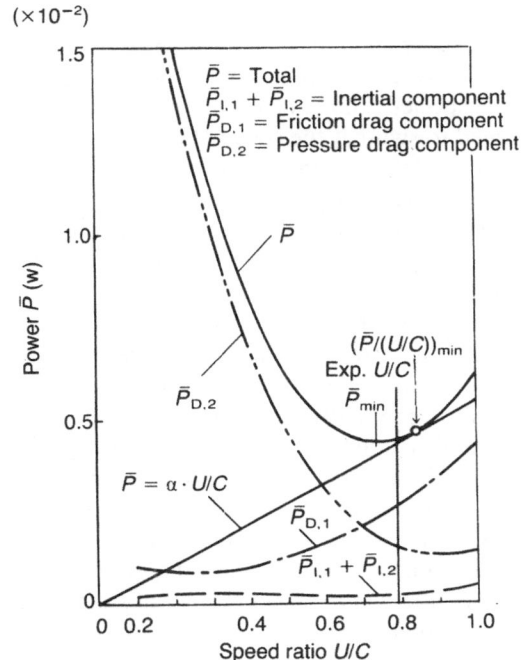

FIG. 5.2-7. **a, b.** Performance of the eel. **a** Thrust. **b** Power.

shown in Fig. 5.2-7 for (a) the thrust and (b) the power, with their respective components as a function of nondimensional speed based on the wavespeed, or "speed ratio" U/C. In general, the mean thrust $\overline{T}$ decreases as the speed ratio U/C increases, whereas the mean power $\overline{P}$ first decreases as the speed ratio increases, then increases again at a high speed ratio. A trimmed speed can be obtained at zero thrust, $\overline{T} = 0$, when the driving force is just equal to the drag force resulting from the skin friction. Table 5.2-4 gives the trimmed quantities obtained from the calculation of eel and other animals. It must be mentioned that since the locomotion is not steady except at the trimmed mean speed, the power is not that required for steady or trimmed locomotion, except at the specified speed of $\overline{T} = 0$. Thus the power required for trimmed locomotion is slightly smaller (or larger) than that given by the above calculated power under (or over) the range of the specified speed.

It is interesting that the minimum power is obtained at a speed close to the trimmed mean

TABLE 5.2-4. Results of calculation at zero thrust $\overline{T} = 0$

Items	Symbols	Units	Sea snake	Eel	Butterfish
Speed ratio	U/C	—	0.758	0.752	0.677
Inertial component of thrust	$\overline{T}_{I,1} + \overline{T}_{I,2}$	N	1.13×10^{-3}	2.09×10^{-3}	2.85×10^{-4}
Pressure drag component of thrust	$\overline{T}_{D,2}$	N	4.53×10^{-3}	6.12×10^{-3}	1.59×10^{-4}
Friction drag component of thrust	$\overline{T}_{D,1}$	N	-5.66×10^{-3}	-8.21×10^{-3}	-4.44×10^{-4}
Total thrust	$\overline{T}$	N	0	0	0
Inertial component of power	$\overline{P}_{I,1} + \overline{P}_{I,2}$	W	7.84×10^{-3}	2.19×10^{-4}	1.53×10^{-3}
Pressure drag component of power	$\overline{P}_{D,2}$	W	8.75×10^{-4}	1.75×10^{-3}	2.36×10^{-3}
Friction drag component of power	$\overline{P}_{D,1}$	W	2.47×10^{-3}	2.41×10^{-3}	7.17×10^{-3}
Total power	$\overline{P}$	W	3.42×10^{-3}	4.38×10^{-3}	1.11×10^{-4}
Efficiency	$\eta = \{U(\overline{T}_{I,1} + \overline{T}_{I,2} + \overline{T}_{D,2})/P\} \times 100$	%	54.8	42.6	52.1

speed $\bar{T} = 0$. The observed speed ratios of these animals, designated Exp. U/C in Fig. 5.2-7, are very close to the trimmed mean speed of the respective species. Thus, the exemplified species are presumed to swim at the speed of either minimum power or minimum tangential power, which is given by $\{P/(U/C)\}_{min}$ and is, as stated before, not exactly but close to optimal for cruising, because the power curve is actually close to the one obtained from the trimmed condition, $\bar{T} = 0$.

In the eel, the thrust is mostly generated by the pressure drag, especially in the low-speed range. Near the trimmed mean speed ratio, however, the inertial force is equally strong. On the other hand, in slender fish, like the butterfish, the inertial force makes an important contribution in all speed ranges. As expected from the analysis, this tendency grows stronger as the tail height increases, as discussed later in Chapter 6. The inertial component of the required power is very small for elongate fish species, and the minimum power is obtained at the speed at which the power due to the friction drag, $\bar{P}_{D,1}$, is nearly equal to the power due to the pressure drag, $\bar{P}_{D,2}$.

The efficiency is about 50% at the speed ratio of about 0.7. The reduced frequency based on the body length $\sigma = 2(l/c)k$ (see Table 5.1-4) is, so large that the above value of efficiency is similar to that estimated by Wu (1961) for the waving plate taken up in the following section.

5.2.4 WAVING PLATE

When a body is laterally thin or flat instead of round and cylindrical, the related fluid motion may be treated simply as a two-dimensional flow over a waving plate of infinite depth. In view of this fact, the Navier-Stokes equation for two extreme cases, (1) waving motion in an ideal fluid and (2) waving motion in highly viscous fluid, will be approximated by the harmonic (or Laplace's) equation and the biharmonic differential equation, respectively, and thus be solved analytically for simple forms of boundary condition.

Eulerian realm. In the first case (in the Eulerian realm), the analysis is conducted to solve the general problem of a waving plate in potential flow. This analysis was initially worked out by Lighthill (1960a, 1960b) for the swimming of elongate animals, and by Wu (1961) for the swimming of a two-dimensional flat fish.

Since Wu's analysis relates to a two-dimensional plate of infinite height, the flow going around the top and bottom sides of the plate is not considered, and the flow is specified by the boundary condition on the plate through which no flow passes. Thus, the induced flow is generated by the shed vortices that arise from the trailing edge of the plate. The force stems from by the pressure integration over the plate surface and the suction force at the leading edge of the plate, whereas the friction drag need not be considered. In Wu's analysis (1961) the optimal (maximum) efficiency can be given by some value of either the speed ratio $(U/C)_{\eta_{max}}$ or the amplitude ratio $(2\pi a/\lambda)_{\eta_{max}}$, where $(U/C)_{\eta_{max}} = 0.7$ for the grass snake and 0.4 for the eel. The former value is very close to the result obtained in the preceding section.

Stokesian realm. The highly viscous flow over a waving plate in the second case (in the Stokesian realm) was treated by Taylor (1951) and then by Reynolds (1965) and Tuck (1968). The results reveal that (1) for very small Reynolds number, trimmed speed can approximately be given by $U = 2(\pi a/\lambda)^2 C$ which is similar to the value already given in Sect. 5.1.2, and (2) as the Reynolds number increases the trimmed speed decreases.

5.2.5 ELONGATE VERTEBRATES

There are many swimming species with elongate bodies. Their motion is performed by sending a wave of lateral deflection backward or forward along all or part of the body. This undulatory motion is in accord with the theoretical results obtained in the preceding sections. Here, in this section, let us consider several species with elongate body or fins. Once one has a knowledge of the above analytical results, it is interesting to look at the body configuration and the cross section of actual elongate animals.

Squamata. As Fig. 5.2-8 shows, the tail of a sea snake (Pelamidrus) is rather flat and wide right up to the tip, while the tail of a land snake (Agkistrodon) is rounded and tapered toward the tip. Thus, the propulsive force generated by the tail part of sea snake is much larger than that of the land snake. A similar difference can be found between the tails of the marine iguana and the land iguana living on the Galapagos Islands. This suggests a way to distinguish the mode of locomotion of extinct animals of unknown ecology.

Animals with extended fins. An animal that has an extended fin or fins can easily maintain its position in flowing water relative to any obstacle such as a rock by sending a series of waves to the fin

Fig. 5.2-8. **a, b.** Tail configuration of snakes. (From Azuma 1980 with permission). **a** Sea snake. **b** Land snake.

either forwardly or backwardly. Green knifefish (*Eigenmannia virescens*), American *Mormyrus* (*Gymnarchus niloticus*), and electric eels (*Electrophorus electricus*) are characterized by long fins that are well adapted for undulatory rim motion and for swimming with the spine rigid, probably in order to keep electric generating and detecting organs aligned (Lissmann 1963).

In all forward movements, the wave motions are, as a matter of course, passed from head to tail. In general, undulatory fins unassisted by body motions are used at low speeds in situations where precise maneuvering is required. For example, the bowfin (*Amia calva*) uses only its dorsal fin for slow locomotion, although it can undulate its body for rapid movement.

Lighthill and Blake (1990) made clear the effect of a rigid body on the performance of undulatory fin motion in either "balistiform" propulsion (undulations propagated simultaneously along a posterior pair of dorsal and anal fins) or "gymnotiform" propulsion (undulations propagated along

Fig. 5.2-9. **a, b.** Coralfish. **a** Filefish. **b** Porupinefish.

a single ventral fin). They found that the fluid momentum produced by movements of these fins attached to rigid body cross section was from 3–4 times greater than the same fin movements would generate on their own.

The bodies of rays and skates are considerably flattened by a pair of enlarged, winglike pectoral fins and the positioning of the gill openings wholly on the ventral surface. In swimming, some like the thornback ray (Rajidae), eagle ray (Myliobatidae) and manta or devilfish (Mobulidae), can beat their pectoral fins like birds, but others, such as the electric ray send a series of undulatory deflections along the broad rim of the body ("rajiform" locomotion). It can be seen that the amplitude of the wave at any longitudinal section increases as the wave progresses.

Decapoda like *Sepioteuthis* and *Sepia* also swim by the undulation of extended fins surrounding the mantle. They are free to move either forward or backward.

It is also interesting to observe the locomotion of the stingray (Dasyatidae) or whip ray on the bottom of the sea. They can crawl on a flat surface by sending a series of radial furrows backward.

Flatfish. Like Rajidae, Heterosomata (e.g., sole, flounder, halibut, turbot) also move by sending vertical anguilliform motions along the whole length of the body, which is oriented horizontally. For slower motion, only the rimmed dorso-ventral fins are moved in this manner. Like other bottom-dwellers, they have no swim bladder.

Coral fish. In some coral reef dwellers, such as filefish (Monacanthidae) shown in Fig. 5.2-9a and triggerfish (Balistidae), pelvic fins are generally absent, although second or posterior dorsal and anal fins are well developed and highly flexible. These fish usually swim by sending waves along these fins either backwardly for forward motion or forwardly for backward motion. The switching of this wave-sending direction can be performed so

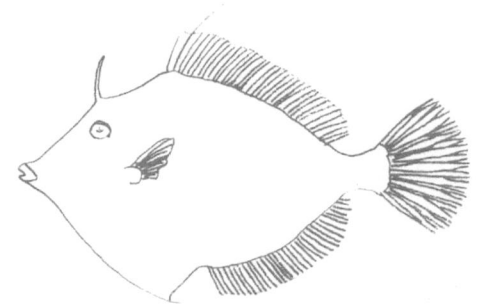

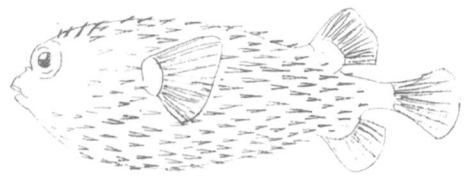

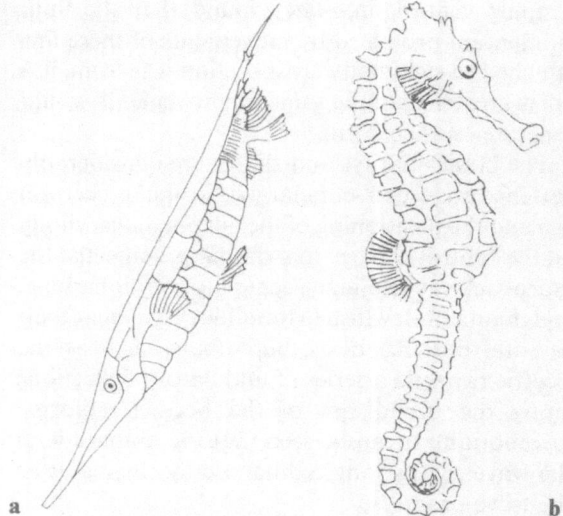

FIG. 5.2-10. **a, b.** Swimming by small fins. **a** Shrimp fish. **b** Sea horse.

rapidly that the fish can easily keep or adjust its position relative to the surrounding coral reefs and rocks, even in rough sea (Azuma 1980, 1986). The caudal fin is then used as a steering rudder.

Other coral reef dwellers such as porcupine fish (Diodontidae shown in Fig. 5.2-9b), puffers (Tetraodontidae), and trunkfish (Ostraciontidae) wear heavy armor and move by using a pair of large pectoral fins, each of which has a markedly extended base of attachment, as well as dorsal and anal fins. In slow swimming, the pectoral fins send a series of propagating waves, mostly up and down, and generate a driving force almost at the center of gravity. By tilting the surface of the pectoral fin, the direction of the driving force and thus the swimming direction can be changed.

Shrimpfish and sea horses. A similar mode of swimming appears in shrimpfish (Centriscidae), cornetfish (Fistulariidae), trumpetfish (Aulostomidae) and sea horses (*Hippocampus*). They live around reefs and rely on undulatory fins for the precise positional adjustment they need for survival.

The thin and flattened shrimpfish (*Aeoliscus strigatus*) swims upside down. In spite of having only a small fin area for swimming, as seen in Fig. 5.2-10a, the fish can swim with the good response of the first-order dynamic system. As (1) the side area of the profile or the area of the lifting surface is large in comparison with the mass of the fish, and (2) the swimming direction is on the dorsal side instead of the nose side, the radius of gyration and thus the inertial reaction moment around the longitudinal axis (nose to tail), which is almost vertical during swimming, are relatively small for the maneuvering torque or the hydrodynamic moment around this axis. Linear or first-order response can be obtained in such system with less inertia and more damping.

Certain families of Gasterosteiformes or sea horses (Hippocampus) dwell in jungles of marine plants and are thus better adapted for maneuvering at low speed than for swimming at high speed. As shown in Fig. 5.2-10b, the sea horse has small pectoral, dorsal, and anal fins, but lacks a caudal fin. It swims very slowly and gently from place to place. The driving force is obtained by undulating or sending a series of waves along the individual fins at a great variety of frequencies.

CHAPTER 6

Swimming by Fanning

In this chapter, let us look at the hydrodynamic aspects of fish and whales that swim with a "fanning" motion of the body, including pelvic, dorsal, anal, and caudal fins. These animals have developed streamlined bodies that are well adapted to their environment, mode of life, and movement. The body and the fin motion of slender fishes are accomplished by a rhythmic transverse oscillation of the entire body and can be treated as an unsteady motion of a slender body resembling the "snaking" or "anguilliform" motion discussed in the preceding chapter. On the other hand, a wide caudal fin can be better analyzed as an isolated wing performing a fanning motion consisting of a heaving and a feathering motion. In this case, the fanning motion can be divided into two types: "ostraciiform" and "carangiform" (Breder 1926; Gray 1968; R.W. Blake 1981c). In the former, the fanning motion is limited to the tail and is a rudderlike motion. It is seen typically in the Ostraciidae. The latter is, as seen in the Carangidae, performed by flexing the entire body with particular stress on the posterior part. The mechanical differences among these movements will be theoretically analyzed.

6.1 Fish

Fish usually have a swimming mode intermediate between two extremes: they can swim by sending a series of bending waves along their entire body, as seen in most slender fish, or they can swim by fanning or sweeping the tail fin from side to side or, rarely, up and down. In fanning motion, the dorsal and ventral edges of the tail fin always lead the movement. The tail follows bends developed somewhere in the middle part of the body; thus, its backward-swept tips get the greatest lateral amplitude of any movement seen on the fish's body.

Although fish have densities in the range of 1,060–1,090 kg/m^3 and are usually a little denser than either freshwater (1,000 kg/m^3) or seawater (1,026 kg/m^3), they are not limited to life at the bottom but can also live close to the water surface. Most fish that swim perpetually adjust their densities to that of the surrounding water by means of a gas-filled swimbladder. Other fish that are denser than water, like sharks and tuna, generate a lifting force when they swim by spreading their pectoral fins like a pair of wings and taking a positive angle of attack.

Furthermore, fish can steer, glide, and turn rapidly by adjusting accessory fins and the attitude of the body. As their swimming speeds can be considerable, they have Reynolds numbers that are high enough to enable them to utilize the inertial component in the fluid-dynamic forces. The frequency of motion f appears small, but, because of the large size l and slow forward speed V, the reduced frequency $k = 2\pi lf/V$ is not so small. Therefore, the apparent mass effects cannot be neglected, and the flow may be considered unsteady.

6.1.1 Form and Structure

A typical body outline of a class Pisces fish is illustrated in Fig. 6.1-1.

Body. Generally, the body of fish in this class is fusiform and oval in cross section. In free-swimming forms, the body is streamlined. Hence, the external shape is said to be slender rather than elongate, because the body length is not extremely long as in the case of the eel, but is much larger than the body height and width. Though there is

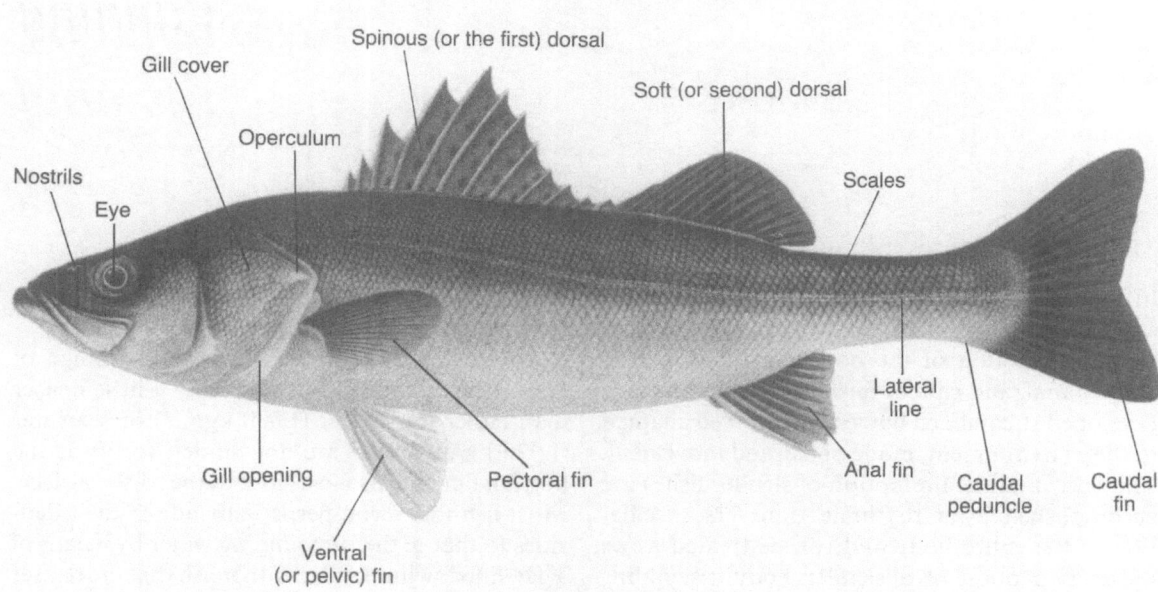

Fig. 6.1-1. Typical shape of a bony fish, sea bass. (From Anon 1967).

no neck, there is a pronounced reduction of depth (or height) called the caudal peduncle at the tail base, just anterior to the tail fin or caudal fin, where the undulation amplitude has a steeply increasing gradient.

Fins. Fish have several fins, including an anterior pair of pectoral fins, a pair of pelvic fins, one or more dorsal fins, one or more ventral or anal fins, and a caudal fin.

The caudal fin is thought to provide most of the thrust for propelling the fish (Bainbridge 1963), while the dorsal and ventral fins effectively extend the cross section to provide a large lifting surface for producing a side force. The cross section of the pectoral and caudal fins in large fish is highly streamlined and exhibits a smoothly rounded leading edge and a sharp trailing edge, similar to the airfoil shape of aircraft wing.

The dorsal and ventral fins fall roughly into two main types: the "elongated ribbon fin" and the triangular or trapezoidal "sail-shaped fin." These fins are effective in providing lateral stability and control. The pairs of fins seem to contribute little to conventional forward swimming, except in some species, and the main function is to provide stability and control in pitching and rolling motions (Harris 1936, 1938, 1953).

Gills and opercula. Fish have a pair of opercula covering a pair of gills that absorb oxygen from water into blood pumped by the heart. Water taken in through the mouth and then forced into the gill pouches is expelled through the opercula slots by rhythmical muscle contractions.

Bladder. Most (but not all) teleosts have elastic gas or swim bladders that are approximately oval and located above the abdominal cavity, beneath the vertebral column and kidney. In bony fish, the gas bladder functions as a hydrostatic organ by means of which the density of the fish is adjusted to suit the environment by regulating buoyancy. Although the shape of swim bladders varies enormously, their volume is usually about 5% of the total volume in marine fish, and 7% in freshwater fish (Thomas 1984). The swim bladder can be altered by relaxing or contracting muscles in or about the swim-bladder wall (Alexander 1959). The swim bladder can also be used to trim the pitching attitude by regulating the volume of front and rear chambers. In some fishs the gas bladder is associated with sound reception (it serves as an amplifier) and may extend into the inner ear.

Muscles. The main part of the fish is formed by many muscles supported by a spinal column and small bones arranged transversally. Behind the

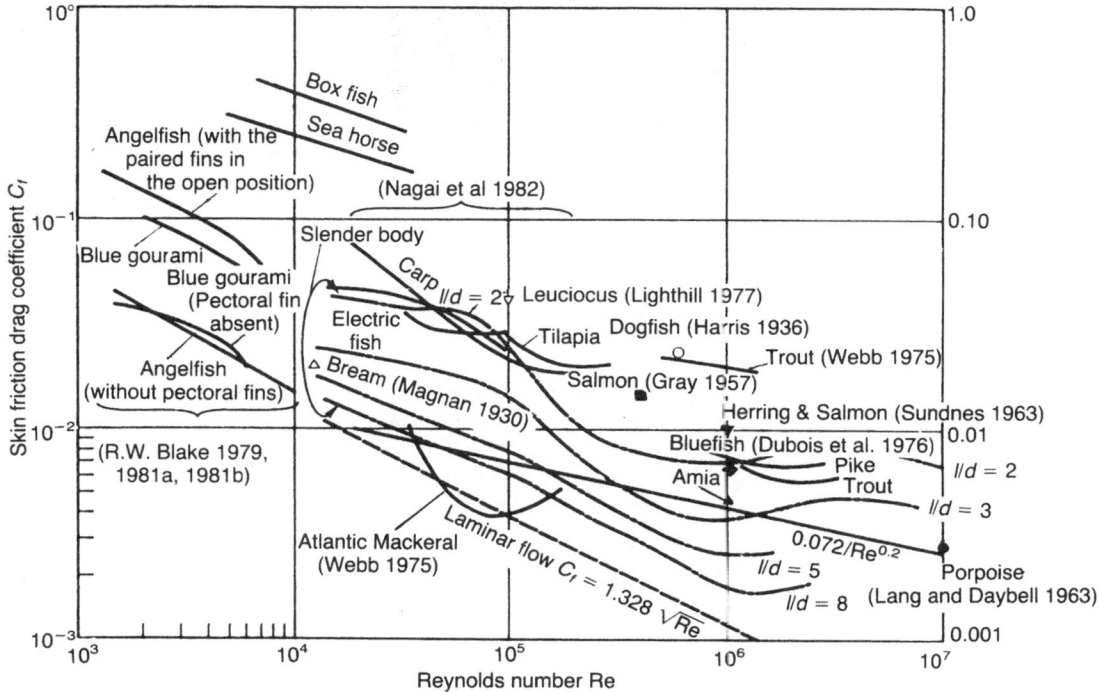

Fig. 6.1-2. Skin friction drag of fish. (Data from Bone 1975; Richardson 1936; Gero 1952; Webb 1975; R.W. Blake 1979, 1981a, 1981b, 1981c).

head, the muscles of the trunk and tail on either side of the spinal column are divided into a series of segments or myotomes. The greater the number of myotomes, the more flexible the body is likely to be (Magnan 1930; Narsall 1956; Lindsey 1978). Successive contractions and relaxations of the muscle segments from head to tail cause waves of curvature to travel down the body and culminate in transverse beats of the rear part of the body, including the caudal fin.

Fish muscles consist fundamentally of three different types of fibers, red, pink, and white, and are arranged in the longitudinal direction along both sides of the body. The red fibers, which are thin layers found on the myotomes at the outer surface and account only for a few percent of the total musculature, are responsible for "sustained swimming," defined as a slow cruising movement with low frequency fanning maintained for periods longer than 200 min (Beamish 1978). They are abundant in blood vessels and are active only when oxygen is available to the cells, and may thus be called aerobic muscle. The pink (or inter-

mediate) fibers are mobilized for prolonged swimming, which is slightly faster than slow cruising, but of shorter duration (20 s–200 min). On the other hand, the white fibers, which are arranged in a helical fashion, account for 70%–80% of the musculature and have a large store of glycogen. They are inactive in cruising and are used only for burst swimming, which is a rapid burst of activity maintained only for a short period of less than 20 s with very high energy consumption. The respiration of glycogen is known to be anaerobic and it can be converted to lactic acid without oxygen. According to Bainbridge (1961), the muscle mass of all fibers in fish generally constitute half of the body mass, and only half of it works at any time.

6.1.2 BODY PROFILE AND HYDRODYNAMIC DEVICES

The body profiles and the hydrodynamic characteristics of a range of fish are shown in Table 6.1-1. *Drag.* Several examples of the measured drag coefficient of fish, based on the wetted surface area S_w, are shown in Fig. 6.1-2. In general, the Reynolds number lies in the range of $10^3 < $ Re $ < 10^6$ for various fish, and of $10^6 < $ Re $ < 10^8$ for most cetaceans. It can be seen that fish incur larger friction drag than that of a flat plate of the

TABLE 6.1-1. Typical fish profiles. (Redrawn from Encyclopedica Zoologica 1963)

Hemiramphus sajori *Cololabis saira*	Needlefish family (Atherinidae) Examples; Hemiramph, snipefish, mackerel pike, garfish. They live near the sea surface and feed on plankton and small fish.	They have very small fins, and the dorsal and anal fins are located near the caudal peduncle and used for propulsion together with the caudal fin.
Oncorhynchus keta *Oncorhynchus masou*	Salmon family (Salmonidae) Examples; Dog salmon, red salmon, silver salmon, rainbow trout, char. They live in cold and swift rivers that are rich in oxygen.	They have slender bodies with a narrow swept caudal fin and swim in shallow rivers by sending a wavy lateral deflection with large increment of amplitude along the body axis.
Tribolodon hakonensis hakonesis	Dace family (Tribolodonidae) Examples; *Zacco platypus, Moroco steindachneri, Tribolodon hakonensis*. They dwell in midstream and downstream regions of rivers and occasionally in the sea.	
Epinephelus moara *Sebastiscus marmoratus*	Grouper family (Serranidae) Examples: *Epinephelus, Sebastiscus*, Scorpaenidae. They live at the bottom of the sea and feed on small live or dead fish and crustacea.	They have many adjustable dorsal and ventral fins that can generate lateral acceleration, and a wide fanlike caudal fin that the fish uses to produce a burst of speed from rest by extending and fanning it vigorously. Other ventral and dorsal fins arranged near the caudal peduncle may also assist in generating the sudden increase in thrust.

same surface area (shown by the broken line), except in rare cases. Alexander (1977) showed an appreciable discrepancy in the necessary power at various swimming speeds of the rainbow trout (*Salmo gairdneri*) between an expected power requirement curve calculated from the skin friction drag by assuming laminar flow, and a useful power curve obtained from oxygen consumption, the latter being 7–9 times higher than the former.

Measured data on the drag of a slender body of revolution are also shown in Fig. 6.1-2 by chain lines. Some of these coincide with those of the fish. By introducing the pressure drag correction for a three-dimensional body or body of revolution (Hoerner 1965), the drag coefficient correction term based on the wetted surface area S_w, which can be expressed as follows:

$$C_{D,w} = C_f\{1 + 1.5(d/l)^{3/2} + 7(d/l)^3\} \quad (6.1\text{-}1)$$

adds to the coefficient only a small increment, for example $C_{D,w}/C_f = 1.05$, 1.19 and 1.43 for $d/l = 0.1$, 0.2, and 0.3 respectively.

TABLE 6.1-1. (cont.)

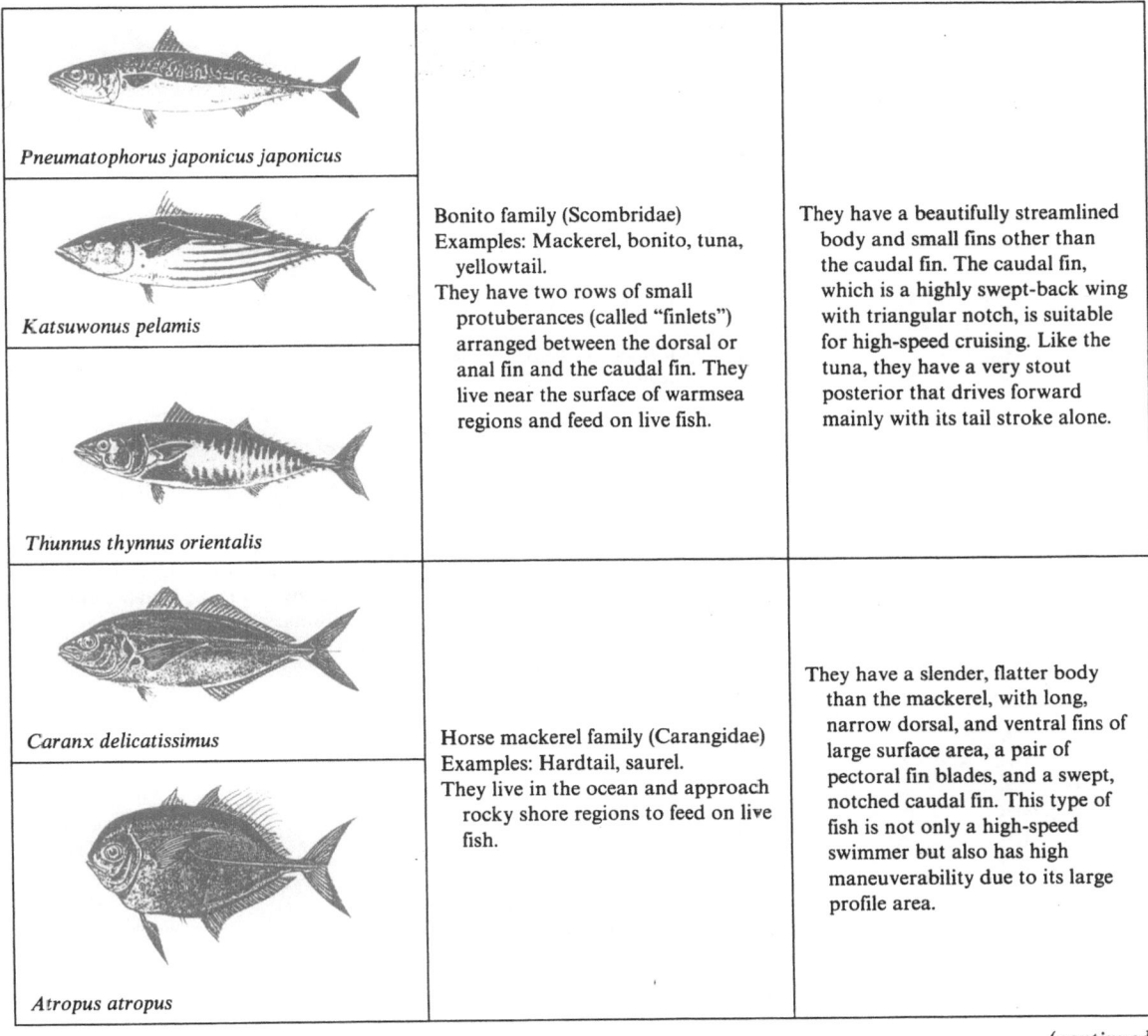

Pneumatophorus japonicus japonicus	Bonito family (Scombridae) Examples: Mackerel, bonito, tuna, yellowtail. They have two rows of small protuberances (called "finlets") arranged between the dorsal or anal fin and the caudal fin. They live near the surface of warmsea regions and feed on live fish.	They have a beautifully streamlined body and small fins other than the caudal fin. The caudal fin, which is a highly swept-back wing with triangular notch, is suitable for high-speed cruising. Like the tuna, they have a very stout posterior that drives forward mainly with its tail stroke alone.
Katsuwonus pelamis		
Thunnus thynnus orientalis		
Caranx delicatissimus	Horse mackerel family (Carangidae) Examples: Hardtail, saurel. They live in the ocean and approach rocky shore regions to feed on live fish.	They have a slender, flatter body than the mackerel, with long, narrow dorsal, and ventral fins of large surface area, a pair of pectoral fin blades, and a swept, notched caudal fin. This type of fish is not only a high-speed swimmer but also has high maneuverability due to its large profile area.
Atropus atropus		

(continued)

Possible reasons for these large values of the measured drag coefficient are the wagging of the anterior part of the fish in opposition to the tailbeat, and another associated pressure drag due to flow separation. The roughness elements such as gills and eyes are, in some cases, also responsible for increasing the drag.

In view of the uncertainty associated with direct drag measurements, it is useful to take advantage of the energy balance given by Eq. (4.2-2):

$$f = S_w C_{D,w} = P_a \eta / \tfrac{1}{2}\rho U^3 = P_{ab} \eta \eta_E / \tfrac{1}{2}\rho U^3 \tag{6.1-2}$$

The aerobic metabolic power P_{ab} can be determined by measuring the animal's oxygen con-

sumption and subtracting the standard metabolic rate. Even so, the measured data are widely scattered at far larger values than those of a flat plate (Brett 1965).

One measure of hydrodynamic shape in low drag profile for a given volume is the "slenderness ratio" or "fineness ratio," which is the ratio of length l to the body diameter in columnar shape d, or the inverse of d/l given in Eq. (6.1-1). In the case of fish, the slenderness ratio is $l/\sqrt{wh}$, where h and w are respectively the maximum height (or depth) and width of the body without fins.

Shown in Fig. 6.1-3 are the drag coefficients of a streamlined body of revolution based on (a) the wetted surface area S_w and (b) the frontal area S_f,

TABLE 6.1-1. (cont.)

Oplegnathus fasciatus	Parrot fish family (Oplegnathidae)
	Examples: *Oplegnathus fasciatus*,
	  Oplegnathus punctatus.
	They inhabit rocky shore regions
	  and feed on crustaceans, which
	  they crush with their strong teeth.
		They are characterized by a flat body with large dorsal and ventral fins, the span of which can be extended widely, and a triangular caudal fin with large area. They can swim with high maneuverability and make a rapid dash from rest.

Girella punctata	Nibbler family (Kyphosidae)
	Examples: Opaleye, *Girella mezina*,
	  Kyphosus lembus.
	They live on seaweeds and small live
	  or dead fish.	

Mylio macrocephalus
 	
Chrysophrys major
	Sea bream family (Sparidae)
	Examples: Black porgy, red sea
	  bream, snapper, surf fish.
	They live around reefs near the
	  shore or in the deep sea and feed
	  on live fish and crustaceans.	

Istiophorus orientalis
 	
Makaira mitsukurii
	Sailfish family (Istiophoridae)
	Examples: Sailfish, spearfish, blue
	  marlin.
	They range through warm sea
	  regions in pursuit of prey.
		They have a big dorsal fin with either a large surface area or extended span, a pair of long pectoral fins, and a wide lunate tail. This narrow or high aspect ratio wing must make dashing from rest difficult for them. The lunate tail is effective only at high cruising speed or in rapid acceleration.

Xiphias gladius	Broadbill family (Xiphiidae) Examples: Swordfish. They have a similar mode of life to the sailfish family.	

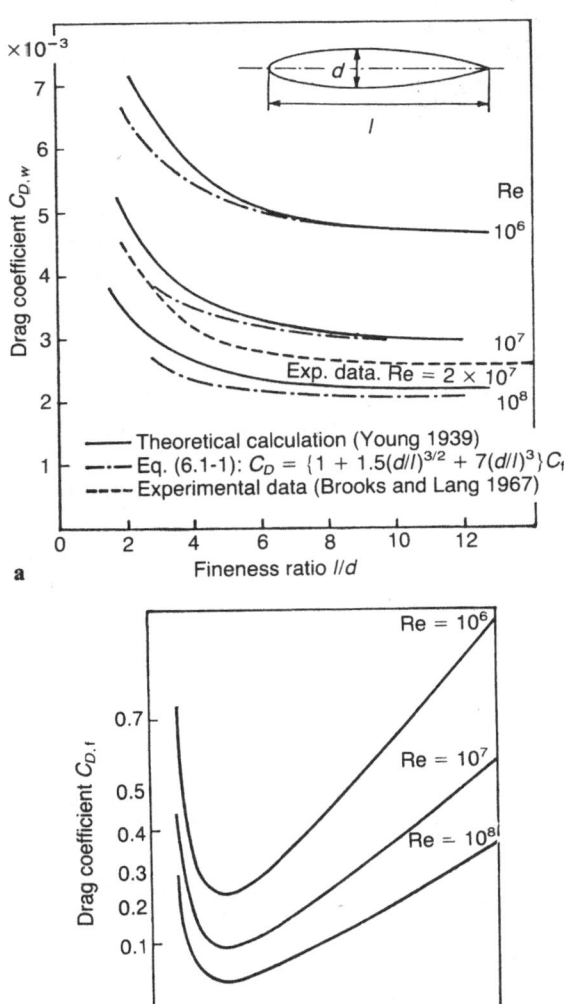

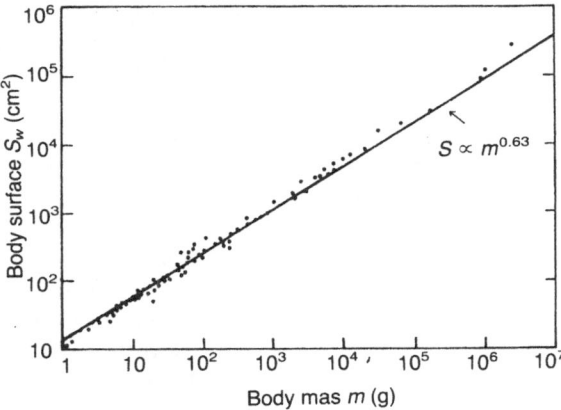

FIG. 6.1-4. Surface area of the entire body (S_w) of vertebrates. (Redrawn from McMahon and Bonner 1983).

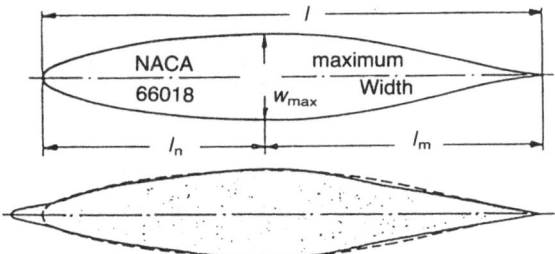

FIG. 6.1-5. Streamlined shape in planform of dolphin and NACA 66018 airfoil. (From Hertel 1966).

FIG. 6.1-3. **a, b.** Drag coefficients of streamlined body of revolution. **a** Based on wetted surface (S_w). **b** Based on frontal area (s_f).

as a function of the slenderness or fineness ratio. Figure 6.1-4 plots the surface area of the entire body in various vertebrates as a function of their mass. The slope is $m^{0.63}$.

Body shape. Let us assume that l_n and l_m are respectively the lengths of the front and rear parts of a fish or cetacean's body measured from the location of the maximum width, and l is the body length excluding the tail, or $l = l_n + l_m$. The dolphin has a slenderness ratio of about 5.6 at a

rearward position of $l_n/l = 0.45$ and a good hydrodynamic profile very close to an airfoil invented artificially for an airplane wing NACA 66018, as shown in Fig. 6.1-5. The trout has a similar correspondence with airfoils of NACA 63A016 and the laminar profile LB N-0016 proposed by Tani (1943) or Kármán (1954). The profile of a tuna coincides with the NACA 67-021 airfoil, which is one of the laminar drops of maximum volume (Hertel 1966). In this profile, however, the maximum thickness is 70% of body length, so that the pressure rise in the rear part will be quite steep and might bring about flow separation unless the skin is given some special treatment.

Another body shape factor affecting the hydrodynamic efficiency of locomotion is the "rear slimness," which is the ratio of the length of the rear part of the body l_m to the width w, or l_m/w (Hertel

1966). This ratio is used as an indication of the flexibility of the body, which is, as we shall see later, strongly related to the optimal mode of motion of the tail fin.

It is interesting to find that the position of maximum thickness generally becomes more posterior as the size of the aquatic animal increases. For large fish or mammals swimming at a high Reynolds number, the laminar flow is maintained only over the surface of the part anterior to the maximum thickness; thus, the point of maximum thickness must be as far toward the rear of the body as possible without inducing any flow separation.

Slime. It is a well-known fact that the body of most aquatic animals is covered with a slippery mucus or slime. Slime makes fish slippery and difficult for predators to grab. Slime also helps to maintain the laminar flow over the surface. The performance of hydrofoils is known to be slightly improved by the addition of dilute polymer solutions on their surfaces (Sarpkaya 1974; Sinnarwalla and Sundaram 1978). Low concentrations of slime or mucus of certain species are also known to be effective in drag reduction in turbulent flow (Rosen and Cornford 1970, 1971; Hoyt 1975). In one species, a dilute solution of its slime was found to reduce the friction of water by as much as 65.9%, and reductions of 57%–63% were frequently obtained with the slimes of other species.

Opercular slots. As explained earlier, in most species of fish, a pair of opercular slots open near the end of the anterior convex region of the head, just in front of the maximum body width (thickness). Water is taken in from the mouth into a pair of gills and, after passing through the gill chamber, is pumped out as a jet from the slots into the surrounding fluid outside of the boundary layer, at a speed either a little lower (Brown and Muir 1970) or a little higher than the surrounding fluid. The speed reduction of the internal flow generates momentum drag. The exhaled jet will energize a turbulent boundary layer if the jet speed is higher than the speed of the surrounding fluid in the boundary layer (it not always being necessary to exceed the general flow speed outside the boundary layer), and prevent the flow separation caused by unfavorable pressure conditions, specifically in a part of outward bend (Breder 1924, 1926).

Scales. Ctenoid scales can be projected slightly above the general body surface and may act as vortex generators that maintain turbulent bound-

ary layers, which are in turn, effective in preventing flow separation (Burdak 1969). In order to avoid excess generation of turbulence, the projection of ctenoid scales above the body surface must be small—for example, about 0.35 mm in the sailfish *Istiophorus* (Ovchinnikov 1966). In the case of tuna and bonito, the scales have almost completely degenerated and remain only on the part of the body anterior to the gills, probably to avoid increase of drag.

Water jet. Some marine species have an integument that allows them to eject water. Like the water pumped through the opercular slots, this jet is believed to energize the flow in the boundary layer during different phases of the swimming cycle (Bone 1975; Walters 1962; Watson and Balasubramanian 1984).

Surface roughness. The drag increment caused by surface roughness $C_{D,m}$ represented by small surface waves is given by (Belyayer and Zuyer 1969):

$$C_{D,m} = \Delta D / \tfrac{1}{2} \rho U^2 S_f = 15(h/\lambda)^2 \quad (6.1\text{-}3)$$

where S_f, h and λ are the frontal area of the waves, total wave height (trough to crest), and wavelength of the roughness, respectively. However, for the roughness of small amplitude sinusoidal wave trains with wavelengths roughly equal to the boundary layer thickness, the drag is not only a function of h/λ, but also a strong function of the Reynolds number Re* as defined by (Lin and Walsh 1984),

$$\text{Re*} = hU\sqrt{C_f/2}/\nu \quad (6.1\text{-}4)$$

thus, the drag may be reduced by the roughness.

Later, Walsh and various coworkers (1978, 1982, 1984, 1989) studied the effects of small streamwise grooves called "riblets" on the reduction of turbulent skin friction drag. Figure 6.1-6 shows that maximum drag reductions of 8% were obtained for a symmetric V-groove surface with a nondimensional height and space of $h^+ = 13$ and $s^+ = 15$, which are defined respectively by

$$h^+ = (hU/\nu)\sqrt{C_f/2} \quad \text{and} \quad s^+ = (sU/\nu)\sqrt{C_f/2}$$
$$(6.1\text{-}5a,b)$$

Here h^+ is equivalent to Re* in Eq. (6.1-4). The riblet drag reduction remained insensitive to yaw angles up to 15°. Furthermore, on the reexamination of experimental data on sand grain roughness, Tani (1988, 1989) showed that even sand grain roughness exhibited lower skin friction drag than that of a smooth surface (also see Sect. 4.5.1).

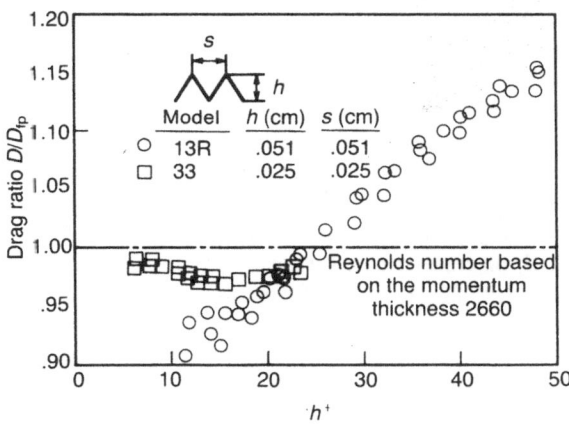

FIG. 6.1-6. The effects of riblets on the drag reduction (D_{fp}: drag of flat plate). (From Walsh 1982 with permission).

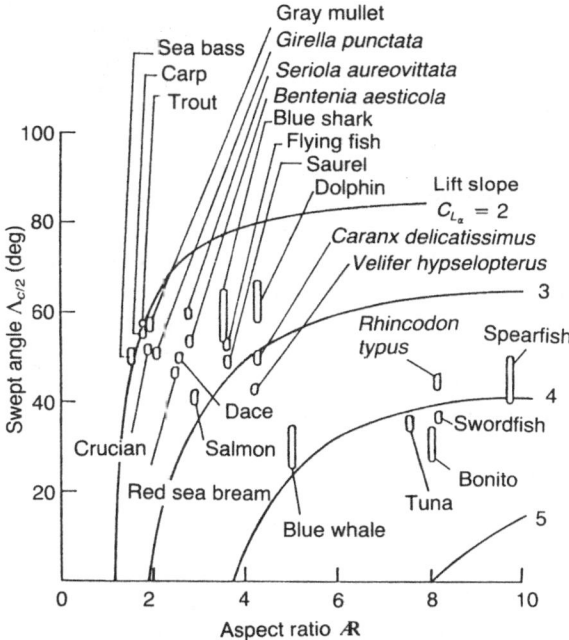

FIG. 6.1-7. Statistical relation between the swept angle and the aspect ratio. Since the data are obtained from illustrations in a book (Encyclopedia Zoologica 1963) the accuracy is not always reliable.

Compliant surface. It is supposed that a smooth and compliant skin surface is soft and indents easily, so that the surface ripples caused by disturbances in the flow tend to absorb the oscillatory energy of the boundary layer flow, and thus have a smoothing or stabilizing effect on it through resistivity and resiliency (Landahl 1961; Nonweiler 1963; Gyorgyfalvy 1967). As a result, there is a delay in the transition from a laminar to a turbulent boundary layer (Kramer 1957; Lang and Pryor 1966; Lang 1975) and accordingly a reduction of turbulent skin friction (Blick and Walter 1968; Balasubramanian 1980). Bushnell et al. (1977), Orszag (1977, 1979), and Reiss (1984) conjectured that compliant walls reduce drag in a turbulent boundary layer too. However, it was also found that while some experiments showed a substantial effect of compliant walls on friction drag (Fisher and Blick 1966; Smith and Blick 1969), others did not (Dinkelacker 1966; Lissaman and Harris 1969).

Finlets. Another device that prevents flow separation in high angle of side slipping either at turn or violent fanning at low speed is a series of caudal finlets shown in Table 6.1-1. The finlets are arranged to protrude with an alternate angle of yaw like a vortex generator in an airplane wing, so as to prevent flow separation and to get high lift (or side force) of the lunate tail of high-speed fish such as wahoo, tuna, skipjack, mackerel, and saury.

Geometrical aspects of caudal fin. Figure 6.1-7 shows the statistical relationship between the aspect ratio and the swept angle of the half-chord

of the caudal fin in various fish illustrated in Encyclopaedia Zoologica (1963), and also the lift slope in the linear range estimated by the wing theory. It is interesting to find that almost all fish have a swept-back caudal fin. In contrast, the swept angle is not so widely adopted in the wing of birds and insects. Because of the structural mechanism of the beating wing, birds can consciously superpose any desired feathering motion upon the flapping motion of the wing.

For structural reasons, a fish, particularly a fast swimmer, that propels itself with a caudal fin usually have less control of the intrinsic feathering motion than the slow swimmers (Lindsey 1978). However, as stated in Sect. 4.2.3, pitch control of the fin is critical to better performance. Thus, it is necessary to obtain pitch control or feathering motion automatically through an elastic coupling with the heaving motion. This can be done by adopting a swept angle to the caudal fin, which is very elastic. The swept fin has a large restoring moment about its elastic axis; thus, the trailing edge of the fin has a tendency to reduce the angle of side slip or to follow the flow, which increases cruising efficiency. However, as seen in Fig. 6.1-7, the lift slope, which is another measure of performance, is reduced as the swept angle increases and/or the aspect ratio decreases. Thus, high performance fish having high maneuverability and swift mobility must compromise between the above two performance factors. As we found in Sect. 3.5.5, the low aspect ratio fin has a larger maximum lift coefficient as well as nonlinear lift increment than the high aspect ratio wing. Therefore fins with low aspect ratios and high swept angles are effective for quick starts (See Sect. 4.5.4) and sudden stops, but their cruise performance is not good. On the other hand, the fast swimmers such as Scombridae, Istiophoridae, and Xiphiidae with caudal fins of a large aspect ratio exceeding 7 have good cruising efficiency.

6.1.3 SWIMMING MODE

We saw in Chap. 4 how advances in equipment and techniques have led to more precise analyses of bird and insect flights. Similarly, great progress achieved in cinematography, in experimental technology with the development of wind tunnels and water tunnels, and in hydrodynamic theory, have made it possible to gain a much better understanding of the manner and mechanism of the swimming motion of fish.

The undulatory mode of propulsion of an elongate body with constant height is called anguilliform or snaking motion. It was demonstrated in Chap. 5 that although a combination of resistive and reactive forces is applied for propulsion, animals using this mode of locomotion have developed a more exclusive use of the reactive force by magnifying the amplitude toward the tail.

On the other hand, among fish that vary in height or depth from head to tail, a great majority use an undulatory motion called fanning. In this motion, there is usually a steep increase of amplitude toward the rear and a phase lag between the anterior and posterior sections that is adequate for sending a propagating wave backward. It is difficult to observe a whole wavelength at any one time. Since the amplitude of undulation becomes significant only in the posterior half, the lateral oscillation of the caudal fin is generated by an undulatory motion of a flexible body, at one terminal end of which the leading edge of the central section of the fin is attached; thus, the motion is similar to that of a fan for producing wind. To refer to such motion as fanning seems very appropriate. As we shall see in Sect. 6.3.3,

Bainbridge (1958) found that the maximum amplitude of caudal fin movement for goldfish, dace, and trout is scattered around 20% of the total body length.

When the fish has a relatively rigid body and a very short tail that oscillates almost symmetrically about the caudal peduncle, the motion is substantially limited to the tail and is called ostraciiform, because it can be seen in such Ostraciidae as trunkfish. In ostraciiform motion, the body is essentially a rigid shaft or bar with a hinge at one end about which the tail is periodically swung, as Fig. 6.1-8a shows. Thus, the caudal fin motion is a coupled motion composed of sinusoidal heaving and feathering without any phase difference.

When the body of the fish is more flexible, the body may be represented by a rigid shaft with at least two hinges, that is swung periodically around both hinges with an appropriate phase difference, as shown in Fig. 6.1-8b. Then, the caudal fin motion is a coupled undulatory motion, called carangiform motion (Breder 1926; Lighthill 1969), composed of sinusoidal heaving and feathering with a phase difference realized by selecting an appropriate combination of the parameters of hinge location. Referring to the symbols and definitions shown in Fig. 6.1-8, the angle of incidence θ (pitch-up positive) of the fin or wing surface, considered to be two-dimensional, and the heaving distance h (downward positive), and their ratios and phase differences are respectively given in Table 6.1-2.

Thus, in the case of a single hinge, the amplitude ratio between heaving and feathering (pitching) motion is given by 1, whereas no phase difference can be obtained. In the case of two hinges, on the other hand, the amplitude ratio $h_{max}/\{(l_1 + l_2)\theta_{max}\}$ and phase difference ψ between heaving and feathering motions are given by functions of the arm length ratio (l_2/l_1), the amplitude ratio θ_{2s}/θ_{1s} and the phase difference ϕ between the motions about the first and second hinges. They are, thus, selectable.

Actually, such a coupling motion is attained by introducing the swept angle to the tail wing and by utilizing the torsional rigidity along the aerodynamic center line of the tail wing. An adequate phase difference between the heaving and feathering motions is necessary for the optimal performance of the two-dimensional wing, as seen in Sect. 4.2.2. It can thus be concluded that in order to attain optimal performance over a wide range

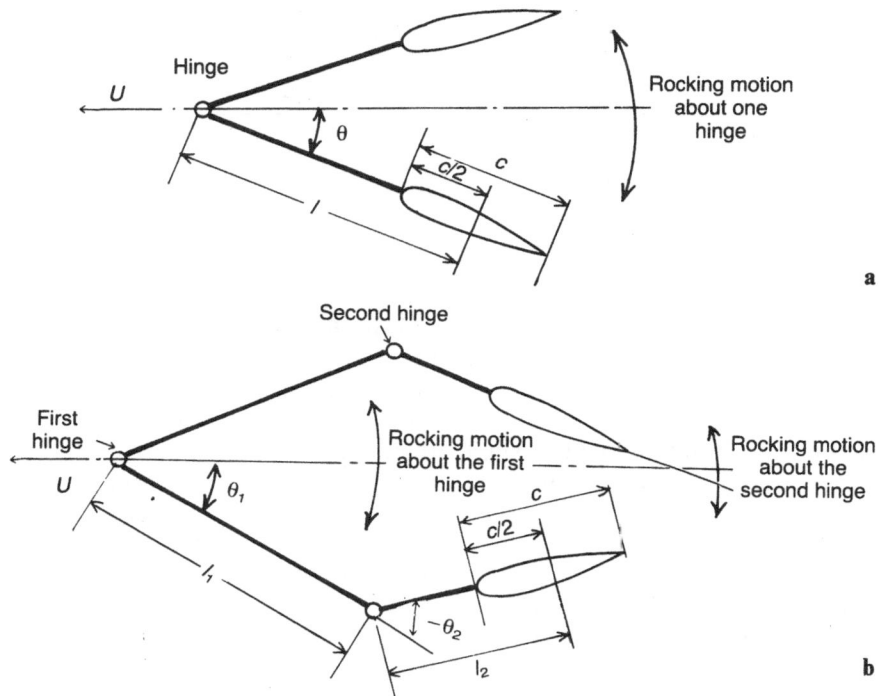

FIG. 6.1-8. **a, b.** Mechanically idealized fanning. **a** Single-hinge or Ostraciiform motion. **b** Double-hinge or Carangiform motion.

TABLE 6.1-2. Fanning about single and double hinges

Items	Expression
Single hinge	
Feathering	$\theta = \theta_s \sin(\omega t)$
Heaving	$h = l\theta_s \sin(\omega t)$
Amplitude ratio	$h/l\theta = 1$
Phase difference	0
Double hinges	
Feathering	$\theta = \theta_1(t) + \theta_2(t)$
	$= \theta_{1s} \sin(\omega t) + \theta_{2s} \sin(\omega t + \phi)$
	$= \sqrt{(\theta_{1s} + \theta_{2s} \cos \phi)^2 + (\theta_{2s} \sin \phi)^2}$
	$\cdot \sin\left\{\omega t + \tan^{-1}\left[\dfrac{\theta_{2s} \sin \phi}{(\theta_{1s} + \theta_{2s} \cos \phi)}\right]\right\}$
Heaving	$h = (l_1 + l_2)\theta_{1s} \sin(\omega t) + l_2\theta_{2s} \sin(\omega t + \phi)$
	$= \sqrt{\{(l_1 + l_2)\theta_{1s} + l_2\theta_{2s} \cos \phi\}^2 + \{l_2\theta_{2s} \sin \phi\}^2}$
	$\cdot \sin\left\{\omega t + \tan^{-1}\left[\dfrac{l_2\theta_{2s} \sin \phi}{\{(l_1 + l_2)\theta_{1s} + l_2\theta_{2s} \cos \phi\}}\right]\right\}$
Amplitude ratio	$\dfrac{h_{max}}{(l_1 + l_2)\theta_{max}} = \sqrt{\dfrac{\{(l_1 + l_2)\theta_{1s} + l_2\theta_{2s} \cos \phi\}^2 + \{l_2\theta_{2s} \sin \phi\}^2}{(\theta_{1s} + \theta_{2s} \cos \phi)^2 + (\theta_{2s} \sin \phi)^2}} \Big/ (l_1 + l_2)$
Phase difference	$\psi = \tan^{-1}\left[\dfrac{l_2\theta_{2s} \sin \phi}{(l_1 + l_2)\theta_{1s} + l_2\theta_{2s} \cos \phi}\right] - \tan^{-1}\left[\dfrac{\theta_{2s} \sin \phi}{(\theta_{1s} + \theta_{2s} \cos \phi)}\right]$

TABLE 6.1-3. Velocity, wavespeed and speed ratio. (Data from Gray 1933a)

Species	Velocity of fish $U(m/s)$	Wavespeed $C(m/s)$	Speed ratio U/C
Butterfish (*Centronotus gunnellus*)	0.117	0.175	0.67
Whiting (*Gadus merlangus*)	0.168	0.25	0.67
Dogfish (*Acanthias vulgaris*)	0.29	0.55	0.53
Mackerel (*Scomber scombrus*)	0.425	0.77	0.55
Ammodytes (*Animodytes lanceolatus*)	0.80	0.16	0.50

of reduced frequencies, the fanning motion of the fish's tail surface must be performed by a multi- or at least two-hinged body (or flexible body) to get an adequate phase difference between the angular (feathering) and the translational (heaving) motions of the surface. However, if the reduced frequency is around $k \simeq 1-3$, the phase difference may be zero for maximum efficiency, so that a single-hinged body is acceptable.

Some extreme examples of fanning motion, ostraciiform and carangiform motion, will now be considered.

The body of the trunkfish is encased in a thick bony shell that is incomplete only around the mouth, eyes, gill-openings, and where the dorsal, anal, and caudal fins protrude. This fish swims slowly, moving with gentle sculling movements of the pectoral and posterior fins, but it can also swim by fanning its caudal fin. Since the stiff body cannot bend for swimming, the fanning motion of the caudal fin is limited to a single-hinge motion around the peduncle; thus, the driving is performed at high reduced frequency, without any phase difference between the heaving and feathering motions, except for a small deflection of the fin. *Epinephelus* is also comparatively slow but capable of extremely sudden short bursts of speed by impulsive motion of a fanlike fin at the initial stage.

In the cruise swimming of mackerel, Gray (1933a) reported that in addition to very low reduced frequency, the angle of attack at the tail surface takes some phase difference with respect to the heaving motion and thus the fin surface is directed to follow the trajectory of the penducle. The motion is similar to that shown in Fig.4.1-1c for the maximum efficiency. Similarly, by analyzing serial photographs of a whale, *Thursiops truncatus*, Parry (1949) also found the mode of maximum efficiency performed by the flukes, which can be represented by the two hinged motion of a wing.

Table 6.1-3 gives examples of the velocity of the fish U, the velocity of the wave in undulation C, and their ratio U/C for several fish in the carangiform mode of swimming. It is interesting to note that, unlike the beating of bird wings, the fanning of a fin is performed uniformly along the fin span and is thus more powerful for the generation of propulsive force only, as is required in swimming where no lifting force is necessary on the fin.

Various Ways of Using Pectoral Fins. A surprising number of fish swim with fins, particularly pectoral fins, rather than with the main part of the body and the caudal fin. Pectoral fins with many different geometrical configurations and used in various styles can be observed at an aquarium.

The pectoral fin is usually composed of many fin rays separated by a highly flexible membrane and is sometimes capable of changing its shape. Such fins of low aspect ratio in salmon, pike, barracuda, cod, and carp are used for various modes of swimming as well as in maneuvers such as beating, braking, sweeping, and shoveling. Although the paired appendages in fish are not ordinarily controlled by huge muscle masses as in the tetrapods, the pectoral fins in these species are well adapted for multiform movements (R.W. Blake 1980, 1983b, 1983c, 1983d).

In quick turning motion, one pectoral fin is held out at an appropriate angle to the body, and the other is folded against the side to allow water exhalation through only one gill orifice. A horizontally extended pair of fins can be used to maneuver and to stabilize the body, and thus the path of vertical movement by using the resulting lift. When the angle of attack of the respective fins is either collectively or differentially altered, head-up or head-down moment or rolling moment can be generated. The function of pectoral fins in stabilizing in body attitude is considerable.

Labriform motion. When a pectoral fin is used as a paddle rotating about a universal hinge at the fin base, as shown in Fig. 6.1-9, the motion is a

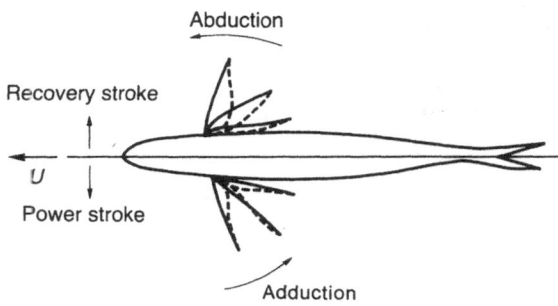

FIG. 6.1-9. Labriform locomotion (viewed from the dorsal side).

kind of paddling, although it is called "labriform" (Breder 1926; Webb 1973a; R.W. Blake, 1981a, 1981b, 1981c). In the recovery stroke "abduction" begins with the first complete fin ray rotating outwards. As the leading edge swings forward, the more flexible remaining rays lag behind to give a feathering motion, through which the fin surface is tilted obliquely to the fin motion so as to reduce the adverse hydrodynamic force. In the power stroke, after the abduction of the posterior fin rays, the leading edge swings backward. Following the adduction of the anterior fin rays, the power stroke is completed by the adduction of the posterior fin rays. During the power stroke, however, the fin surface is kept nearly vertical (i.e., normal to the direction of fin motion) in order to keep the hydrodynamic force at its maximum.

Through a complete stroke cycle, the forward velocity varies in correspondence to the power and recovery strokes (Archer and Johnston 1989). Similar but inverse locomotion can be used to give a gentle backward motion.

6.1.4 AVAILABLE POWER AND ENERGY

As mentioned in Sect. 6.1.1, it is assumed that only half of the muscle mass, which is estimated at half the body mass, is worked at any time. Bainbridge (1961) defined the relationship between the active muscle mass m_m and the body length l as

$$m_m = 0.005l^{2.9} \simeq 0.005l^3 \qquad (6.1\text{-}6)$$

The active muscle mass is determined by the formula

$$m_m = (m_m/m)m = \begin{cases} 0.2m & \text{(Gero 1952)} \\ 0.25m & \text{(Bainbridge 1961)} \end{cases}$$

$$(6.1\text{-}7)$$

A general discussion of the mechanical properties of vertebrate muscles was presented earlier in Sect. 4.1.2. The muscle power or available power of fish can be estimated as $P_a/m_m = 20$ W/kg for red muscle and 80 W/kg for white muscle (Webb 1974) in cold water. Recent studies on marine teleosts by Altringham and Johnston (1990b) showed slightly smaller values such that the maximum mechanical power output was 5–8 W/kg for slow and 25–35 W/kg for fast muscle fibres at 4°C and Johnson and Johnston (1991) also showed a seasonal variation at 15°C from 9 W/kg to 30 W/kg in the winter- and summer-acclimatized fish, respectively. These values are smaller than those for birds and insects given in Sect. 4.1.2. On the other hand, values of 40 W/kg and 170 W/kg have been suggested by Bainbridge (1961) for fish from tropical waters. Gero (1952) gave 62.5 W/kg as the power loading of the muscle of a perch (*Perca flavescens*) with a mass of 0.11 kg. The largest power loading of a fish measured and recorded by Gero was 65 W/kg, produced by a barracuda whose mass and top speed were 4.54 kg and 12 m/s, respectively. Furthermore, Weis-Fogh and Alexander (1977) estimated the maximum specific power output of vertebrate striated muscles to be 250 W/kg.

6.2 Slender Bodies

The bodies of almost all living creatures active in air or water can be considered slender (provided that extended wings are ignored), i.e., the length of the creature is larger than its width and height, and the cross-sectional area of the body changes slowly along its length. For thin or slender bodies it may be assumed that the fluid is set into motion by a longitudinally distributed body action; therefore, the longitudinal variation of the fluid motion along the forward velocity is gradual. By the "slender body theory," each vertical slice of the fluid perpendicular to the general flow is influenced primarily by the body sections close to the slice. A slender fish with a caudal fin can, by applying rhythmic transverse oscillations to its entire body, attain higher utilization of reaction force than resistive force in comparison with the elongate fish discussed in Chap. 5.

Since fish have to generate lift or side force solely when turning and otherwise need to produce only driving force or thrust, their locomotion is symmetric in both left- and rightward oscillations and thus consists of only power strokes in both directions.

6.2.1 MECHANICS OF LOCOMOTION

In this section, let us examine the unsteady flows past a slender and flat body with trailing edges, from which an oscillating vortex sheet is shed. While a steady flow past the rear part of a slender wing (the part after the maximum span section) produces no lift (Jones 1946), this is not true for an unsteady flow.

In contrast with the fish with long cylindrical bodies discussed in Chap. 5, slender fish with upstream lifting surfaces (side fins) and a downstream appendage (caudal fin) are known as strong and active swimmers. Such fish include the salmon, trout, and carp, among others, all of which are capable of swimming up rapid streams. To analyze this type of body configuration in unsteady motion, it is necessary to consider the interaction among the vortex sheet, the downstream portion of the body and the tail fins. The following analysis is fundamentally based on the results presented by Wu (1971a, 1971b, 1971c, 1971d) and Wu and Newman (1972).

Let us introduce a mathematical model of a flat slender body and relate it to a coordinate system as shown in Fig. 6.2-1. (The coordinates axes are different from those in Fig. 5.2-4). The slender body is, to simplify, assumed to be symmetrical with respect to the xz plane and the body motion is represented by a function of distance x and time t as follows:

$$y = \pm \tfrac{1}{2}b(x) \quad \text{and} \quad z = h(x, t) \quad (6.2\text{-}1)$$

The origin of the coordinate axes is at the midpoint of the maximum height, and the body extends from its nose at $x = -l_n$ to its tail at $x = l_t$. Furthermore, the body may have a tail base neck at $x = l_m$, i.e., just in front of the caudal fin. As Fig. 6.2-1a shows, a vortex sheet is continuously shed into the wake from sharp side edges in the trailing edge section, where $0 < x < l_m$.

Then, the "sectional added mass" in the z-direction A_{33} is given by

$$A_{33}(x) = \tfrac{1}{4}\rho\pi b^2(x)\gamma, \quad (6.2\text{-}2)$$

where the factor γ is, as stated in Sect. 5.2.3, a correction factor for a finned body of limited length (i.e., for any body other than one that is

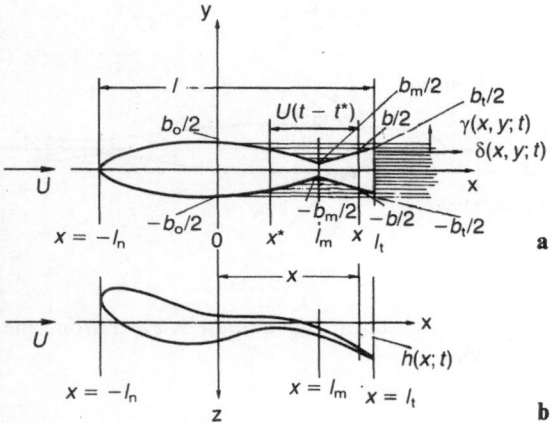

FIG. 6.2-1. **a, b.** Mathematical model of a thin slender body. **a** Profile. **b** Plan view.

either perfectly flat or round, finless, and of infinite length).

The velocity components of steady flow U and the cross-flow w due to the displacement h determine lift, moment, thrust, and other relevant quantities. That is to say, the instantaneous lift per unit length in the front or upstream section, $-l_n < x < 0$, and the tail section, $l_m < x < l_t$, in which the side edges are (usually round) leading edges, can be given as a reaction force that is equal and opposite to the rate of change of the z-component of momentum of the fluid passing the station x (Yates 1983), or

$$l(x, t) = \int_{-b/2}^{b/2} (-\Delta p)\,\mathrm{d}y = -\frac{\mathrm{d}}{\mathrm{d}t}\{A_{33}(x)w(x, t)\}$$

$$\text{for} \quad (-l_n < x < 0) \quad (6.2\text{-}3a)$$

$$= -\frac{\mathrm{d}}{\mathrm{d}t}\{A_{33}(x)w(x, t)\} + w^*\frac{\mathrm{d}}{\mathrm{d}t}\{A_{33}(x)\}$$

$$\text{for} \quad (l_m < x < l_t) \quad (6.2\text{-}3b)$$

where Δp is the pressure difference across the body, $\Delta p = p(x, y, +0, t) - p(x, y, -0, t)$; and w^*, as defined by

$$w^* = w(x^*, t^*) \quad (6.2\text{-}4)$$

represents the effects of the vortex sheet on the tail; x^* is the value x in the trailing edge region $(0 < x^* < l_m)$ where $b(x^*) = b(x)$; and t^* is the "retarded time" taken for the vorticity shed at x^*,

TABLE 6.2-1. Total forces, moment, power and energy

Lift or Side force

L

$$-\left\{\int_{-l_n}^{0}+\int_{l_m}^{l_t}\right\}\left[\frac{d}{dt}\{A_{33}(x)w(x,t)\}\right]dx-\int_{0}^{l_m}A_{33}(x)\frac{d}{dt}w(x,t)dx+\int_{l_m}^{l_t}w^*\frac{d}{dt}\{A_{33}(x)\}dx$$

$$=-\int_{-l_n}^{l_t}\frac{d}{dt}\{A_{33}(x)w(x,t)\}dx+U\int_{0}^{l_m}w(x,t)\frac{\partial}{\partial x}A_{33}(x)dx+\int_{l_m}^{l_t}w^*\frac{d}{dt}\{A_{33}(x)\}dx$$

Moment

M

$$\int_{-l_n}^{l_t}x\frac{d}{dt}\{A_{33}(x)w(x,t)\}dx-U\int_{0}^{l_m}x\left\{w(x,t)\frac{\partial}{\partial x}A_{33}(x)\right\}dx-\int_{l_m}^{l_t}xw^*\frac{d}{dt}\{A_{33}(x)\}dx$$

Power

P

$$-\int_{-l_n}^{l_t}\frac{\partial}{\partial t}h(x,t)l(x,t)dx$$

$$=\int_{-l_n}^{l_t}\left\{\frac{d}{dt}\left(A_{33}w\frac{\partial h}{\partial t}\right)-A_{33}w\frac{\partial w}{\partial t}\right\}dx-U\int_{0}^{l_m}\frac{\partial h}{\partial t}w\frac{\partial A_{33}}{\partial x}dx-U\int_{l_m}^{l_t}\frac{\partial h}{\partial t}w^*\frac{\partial A_{33}}{\partial x}dx$$

$$=\frac{\partial}{\partial t}\int_{-l_n}^{l_t}A_{33}w\left(\frac{\partial h}{\partial t}-\frac{1}{2}w\right)dx+U\left[A_{33}w\frac{\partial h}{\partial t}\right]_{x=l_t}-U\int_{0}^{l_m}\frac{\partial h}{\partial t}w\frac{\partial A_{33}}{\partial x}dx-U\int_{l_m}^{l_t}\frac{\partial h}{\partial t}w^*\frac{\partial A_{33}}{\partial x}dx$$

Rate of kinetic energy

E

$$\left\{\int_{-l_n}^{0}+\int_{0}^{l_t}\right\}\left[\frac{d}{dt}\left(\frac{1}{2}A_{33}w^2\right)\right]dx+\int_{0}^{l_m}A_{33}\frac{d}{dt}\left(\frac{1}{2}w^2\right)dx-\int_{l_m}^{l_t}\frac{1}{2}(w^*)^2\frac{d}{dt}\{A_{33}\}dx$$

$$=\int_{-l_n}^{l_t}\frac{d}{dt}\left\{\frac{1}{2}A_{33}w^2\right\}dx-U\int_{0}^{l_m}\frac{1}{2}w^2\frac{\partial}{\partial x}A_{33}dx-U\int_{l_m}^{l_t}\frac{1}{2}(w^*)^2\frac{\partial}{\partial x}A_{33}dx$$

$$=\frac{1}{2}\frac{\partial}{\partial t}\int_{-l_n}^{l_t}A_{33}w^2dx+\frac{1}{2}U[A_{33}w^2]_{x=l_t}-\frac{1}{2}U\int_{0}^{l_m}w^2\frac{\partial}{\partial x}A_{33}dx-\frac{1}{2}U\int_{l_m}^{l_t}(w^*)^2\frac{\partial}{\partial x}A_{33}dx$$

Thrust

T

$$-\frac{\partial}{\partial t}\int_{-l_n}^{l_t}A_{33}w\frac{\partial}{\partial x}h\,dx+\left[A_{33}\left(\frac{1}{2}w^2-Uw\frac{\partial h}{\partial x}\right)\right]_{x=l_t}$$

$$-\int_{0}^{l_m}\left\{\frac{1}{2}w^2-Uw\frac{\partial h}{\partial x}\right\}\left(\frac{\partial A_{33}}{\partial x}\right)dx+\int_{l_m}^{l_t}\left\{\frac{1}{2}(w-w^*)^2-\frac{1}{2}w^2+Uw^*\frac{\partial h}{\partial x}\right\}\left(\frac{\partial A_{33}}{\partial x}\right)dx$$

$$t^*=t-(x-x^*)/U \qquad (6.2\text{-}5)$$

When the effect of the trailing vortex sheet on the flow around the caudal fin is negligible, the approximations $x=x^*$, $t=t^*$ and $w=w^*$ can be established.

By assuming that the presence of the vortex sheet alongside a tail section ($0<x<l_m$) does not have any effect on the pressure distribution at that section due to the slenderness, and that the flow at the sharp trailing edge satisfies the Kutta condition, the instantaneous lift per unit length in the rear or downstream section can be given by

$$l(x,t)=-A_{33}(x)\frac{d}{dt}w(x,t) \quad \text{for} \quad 0<x<l_m$$

$$(6.2\text{-}3c)$$

in which the added mass $A_{33}(x)$ refers to the body cross section alone. Thus, the effect of vortex shedding is equivalent to treating A_{33} as if it were a constant in calculating the changes of the cross-flow momentum (Wu 1971d). By using the following relations:

$$w(x,t)=\frac{d}{dt}h(x,t)=\left(\frac{\partial}{\partial t}+U\frac{\partial}{\partial x}\right)h(x,t) \quad (6.2\text{-}6)$$

$$A_{33}(-l_n)=0 \qquad (6.2\text{-}7)$$

$$|\partial h/\partial x|\ll 1 \quad \text{and} \quad |\partial h/\partial t|\ll U \qquad (6.2\text{-}8)$$

the total lift or side force in the z-direction L, the moment about the origin of coordinate frame M, the power P, and the rate of kinetic energy E are those given in Table 6.2-1.

Since the power input P is equal to the rate of work done by the thrust TU and the kinetic energy lost to the fluid in unit time E,

$$P=TU+E \qquad (6.2\text{-}9)$$

the instantaneous thrust can be given as $T=(P-E)/U$. Its detailed expression is also presented in Table 6.2-1.

The thrust is also given by both the inclination of the pressure along the body and the suction force T_s at the leading edge as follows (Wu, 1971a):

$$T = \int_{-l_n}^{l_t} \left\{ \int_{-b/2}^{b/2} (-\Delta p) \mathrm{d}y \frac{\partial h}{\partial x} \right\} \mathrm{d}x + T_s$$

$$= \int_{l_n}^{l_t} l(x,t) \frac{\partial h}{\partial x}(x,t) \mathrm{d}x + T_s \qquad (6.2\text{-}10)$$

where T_s is obtained from the "Blasius formula"

$$T_s = \int_{-l_n}^{0} \left\{ \frac{1}{2} w^2 \frac{\partial A_{33}}{\partial x} \right\} \mathrm{d}x$$

$$+ \int_{l_m}^{l_t} \left\{ \frac{1}{2} (w - w^*)^2 \frac{\partial A_{33}}{\partial x} \right\} \mathrm{d}x \qquad (6.2\text{-}11)$$

The hydrodynamic efficiency η can again be defined by

$$\eta = U\bar{T}/\bar{P} = 1 - (\bar{E}/\bar{P}) \qquad (6.2\text{-}12)$$

where $\bar{T}, \bar{P}$, and $\bar{E}$ are, as defined later, mean values of T, P, and E respectively in one period. It is interesting to note that in the formulation of the thrust T in Table 6.2-1, the second term within square brackets involves a contribution only from the posterior end l_t, and by contrast, the first term is merely the time derivative of a fluctuating quantity, so that its time mean is zero. The third term, which is given by the integration spanning the range of reduced height, in which $\partial A_{33}/\partial x < 0$, shows the effect of the vortex sheet through which the added mass maintains its contribution even in the range of reduced height along the downstream section. That is to say, in the third term, the first part shows the contribution of the kinetic energy of the fluid per unit time, and the second part gives the energy change due to the x-component of the shedding of momentum into the vortex wake per unit time. In another interpretation, this term is composed of the resultant pressure force acting over an inspection plane at the trailing edge and the rate of momentum transported across the inspection plane (Lighthill 1970).

The third term of the power in Table 6.2-1 is, on the other hand, the work done by the lateral force, which is given by the rate of shedding of the momentum into the vortex wake per unit time. The final terms given in Table 6.2-1 and related to w^* show the effect of the wake on the tail.

Taking the time average of the thrust and the power for one period and using Eq. (6.2-6) will yield:

$$\bar{T} = \tfrac{1}{2} A_{33}(l_t) \overline{\{(\partial h/\partial t)^2 - U^2 (\partial h/\partial x)^2\}}_{x=l_t}$$

$$\left. \begin{array}{l} - \dfrac{1}{2} \displaystyle\int_0^{l_m} \overline{\{(\partial h/\partial t)^2 - U^2(\partial h/\partial x)^2\}} (\partial A_{33}/\partial x) \mathrm{d}x \\[2mm] + \displaystyle\int_{l_m}^{l_t} \overline{\{\tfrac{1}{2}(w^*)^2 - w^*(\partial h/\partial t)\}} (\partial A_{33}/\partial x) \mathrm{d}x \\[2mm] = \bar{T}_t + \bar{T}_m + \bar{T}_i \end{array} \right\}$$

$$(6.2\text{-}13\text{a})$$

$$\bar{P} = A_{33}(l_t) U \overline{[(\partial h/\partial t)\{(\partial h/\partial t) + U(\partial h/\partial x)\}]}_{x=l_t}$$

$$\left. \begin{array}{l} - U \displaystyle\int_0^{l_m} \overline{\{w(\partial h/\partial t)(\partial A_{33}/\partial x)\}} \mathrm{d}x \\[2mm] - U \displaystyle\int_{l_m}^{l_t} \overline{\{w^*(\partial h/\partial t)(\partial A_{33}/\partial x)\}} \mathrm{d}x \\[2mm] = \bar{P}_t + \bar{P}_m + \bar{P}_i \end{array} \right\}$$

$$(6.2\text{-}13\text{b})$$

in which the bar denotes the time average.

The first terms on the right-hand side of the above equations, $\bar{T}_t$ and $\bar{P}_t$, are the contribution of the tail-end section associated with the vortex sheet and are known as Lighthill's results (Newman 1977). By making the slope at the tail zero, or $(\partial h/\partial x)_{x=l_t} = 0$, the power will decrease, and the thrust and thus the efficiency defined by Eq. (6.2-12) will increase. As already stated in Sect. 6.1.2 and again in Sect. 6.3.1, this can be achieved with the swept-back fin, as the twisting moment causes the trailing edge of the fin to follow the stream in cruising. In a two-dimensional isolated wing this is, as shown in Fig. 4.2-11c, equivalent to taking a feathering angle with a phase difference of 90° between heaving and feathering for maximum efficiency.

On the other hand, according to Wu (1971d), the second terms in the form of both equations, $\bar{T}_m$ and $\bar{P}_m$, represent the contribution due to the shedding of vortices along all the trailing edges, including dorsal and ventral fins. A positive contribution to the thrust can be obtained wherever $\{\partial h/\partial t - U(\partial h/\partial x)\}$ and $w = \{\partial h/\partial t + U(\partial h/\partial x)\}$ are positively correlated, since $\mathrm{d}A_{33}/\mathrm{d}x < 0$ in the range of $0 < x < l_m$. Similarly, the second term in the power expression is positive for a positive $\{(\partial h/\partial t)(\mathrm{d}h/\mathrm{d}x)\}$ in the range of $\partial A_{33}/\partial x < 0$, $0 < x < l_m$. The third terms, $\bar{T}_i$ and $\bar{P}_i$, arise from the interaction of the vortex sheets shed by the body (including dorsal and ventral fins) with the caudal fin.

Now, let us consider a simple progressing trans-

TABLE 6.2-2. Thrust and power coefficients for a sinusoidal wave with constant amplitude

Thrust coefficient	C_T	$\overline{T}/\frac{1}{2}\rho U^2 l^2 = C_{T,\mathrm{t}}[1 + (1/\eta_e)\{1 - (b_\mathrm{m}/b_\mathrm{t})^2\}\{1 - (\sin\theta^*/\theta^*)\}]$
Power coefficient	C_P	$\overline{P}/\frac{1}{2}\rho U^3 l^2 = C_{P,\mathrm{t}}[1 + \{1 - (b_\mathrm{m}/b_\mathrm{t})^2\}\{1 - (\sin\theta^*/\theta^*)\}]$
Efficient	η_e	$\frac{1}{2}\{1 + (U/C)\} = C_{T,\mathrm{t}}/C_{P,\mathrm{t}}$
Parameters	$C_{T,\mathrm{t}}$	$\frac{1}{2}\pi\sigma^2(h_1/l)^2\{1 - (U/C)^2\}(b_\mathrm{t}/2l)^2$
	$C_{P,\mathrm{t}}$	$\pi\sigma^2(h_1/l)^2\{1 - (U/C)\}(b_\mathrm{t}/2l)^2$
	σ	$2\pi f l/U$
	θ^*	$(l_\mathrm{t}/l)\sigma\{1 - (U/C)\},$

verse wave with constant amplitude, such as

$$h(x,t) = h_1\cos(2\pi/\lambda)(x - Ct) \quad (6.2\text{-}14)$$

This form is an optimal shape (Wu, 1971d) that allows integration after substituting Eqs. (6.2-2) and (6.2-14) into Eq. (6.2-13). By assuming that the caudal fin reaches the same height as the dorsal fin and does not extend beyond this point ($b_\mathrm{t} = b_0$), the mean thrust and the power in coefficient form are given in Table 6.2-2 (Yates 1983). Then, the hydrodynamic or Froude efficiency defined by Eq. (6.2-12) becomes

$$\eta = C_T/C_P$$
$$= \frac{C_{T,\mathrm{t}}}{C_{P,\mathrm{t}}}\frac{1 + (1/\eta_e)\{1 - (b_\mathrm{m}/b_\mathrm{t})^2\}\{1 - (\sin\theta^*/\theta^*)\}}{1 + \{1 - (b_\mathrm{m}/b_\mathrm{t})^2\}\{1 - (\sin\theta^*/\theta^*)\}}$$
$$(6.2\text{-}15)$$

which is equal to η_e in Table 6.2-2 if there is no vortex sheet interaction.

The following conclusions can thus be drawn:
1 Positive thrust can be obtained by making the wave speed larger than the forward speed $C > U$.
2 The contribution of the height or wingspan at the trailing edge b_t to the thrust is obvious and increases with the square of the height, whereas the height at the tail base (caudal peduncle)$\frac{1}{2}b_\mathrm{m}$ should be as small as possible, or $b_\mathrm{t} > b_\mathrm{m}$.
3 The efficiency can become quite high when the wavespeed is close to the forward speed, or $C/U \simeq 1$, because the drag component is absent here.
4 When the ratio l_t/l and the reduced frequency, $\sigma = \omega l/U$, are small, or when the value of θ^* is small, the vortex sheet interaction is negligible, and the efficiency is high and approximated by $\eta = \eta_e$. These results were also pointed out by Lighthill (1977). In an actual case, the speed ratio U/C was reported to be 0.7 for goldfish (Bainbridge 1958).

It is also interesting to note that, as we have seen in Sect. 5.2.3, a sinusoidal wave of constant ampli-

tude does not generate any thrust in one wavelength for an elongate body of constant height (i.e., $b_\mathrm{t} = 0$ and $b_0 = b_\mathrm{m}$) swimming in nonviscous fluid, whereas a fishlike body with a point of maximum height, a narrow caudal peduncle, and a wide tail generates thrust efficiently by sending a transverse wave of constant amplitude along its body. It must be remembered that, as pointed out in Sect. 6.1.3, fish with a wide and flat body and a caudal fin with a laterally extended wing of large surface area are good swimmers with high maneuverability.

The side force and yawing moment generated in response to the fanning motion of the caudal fin are resisted by the inertial force and moment generated at other parts of the body. Since (1) the above inertial force is proportional to the added mass of the water; (2) the added mass of water for lateral motion is proportional to the square of the combined body height (dorsal fin, body, and ventral fin); and (3) the dorsal and ventral fins are arranged close to the center of mass where the body height is usually maximum, the lateral translation of the center of mass caused by a fanning motion can be significantly minimized. This is specifically true in the species with large fins near the center of mass. However, fins are usually folded into special slots during fast motion for reduced resistance.

When turning is required, a large side force and yawing moment can reduce the turning radius of the center of the body mass, because the side force is trimmed with the centrifugal force, which is inversely proportional to the turning radius. In order to increase the side force, the fish can extend their dorsal and ventral fins near the center of body mass and enlarge the side area of the body. Then, by using the maximum height of body $b_{\max}$, the height-to-body-length ratio (or simply height ratio) $b_{\max}/l$, or the aspect ratio based on the profile area $b_{\max}^2/S$, are important parameters for the turning performance of the fish. Thus, it can be said

that the ratio of the turning radius to the body length, R/l, becomes small as the height ratio increases.

Lighthill (1971) extended the above results to the problem of large-amplitude oscillation by replacing the term $w = \partial h/\partial t + U\partial h/\partial x$ with the velocity of the fluid normal to the backbone U_N and the forward velocity U with the tangential velocity U_T, as follows:

$$\left. \begin{aligned} \bar{T} &= [-A_{33}\{\tfrac{1}{2}\overline{U_N w} - \overline{U_N(\partial h/\partial t)}\}]_{x=l_t} \\ \bar{E} &= [\tfrac{1}{2}A_{33}\overline{U_T^2 U_T}]_{x=l_t} \\ \bar{P} &= \bar{T}U + \bar{E} \end{aligned} \right\} \quad (6.2\text{-}16)$$

Further extensions of the large-amplitude theory were made to include the effects of body elasticity and centerline curve by Katz and Weihs (1979). Lighthill's study was adapted by Weihs (1972, 1973a) for turning and fast-starting maneuvers.

6.2.2 RIVER FISH

Environment. Mountain streams are usually clear and cool, and flow fast through narrow valleys; they often gather in the basins of waterfalls before gushing away. Fish inhabiting such mountain streams must, above all else, be slender. They also need strong power for large acceleration and deceleration. Members of the trout family, like char or bull trout, trout, and *Oncorhynchus*, meet these requirements.

At the foot of a mountain, rivulets merge into larger streams where the flow speed decreases

FIG. 6.2-2. Freshwater fish, Sweetfish (*Plecoglossus altivelis*) (Courtesy of Kazutoshi Hieda 1982).

slightly with the decrease in slope. There are many round stones among large rocks, and as pools and shoals follow one another, the stream flows alternatively slow and fast. These larger streams are also inhabited by slender fish such as rainbow trout and dace.

Living a little further downstream are the dace, *Zacco platypus*, and, in Japan, the *ayu* or sweetfish (*Plecoglossus altivelis*) shown in Fig. 6.2-2. These fish are also active, rushing out from behind rocks into a swift current, and then back again into still water.

As a general feature, these fish have a transverse cross section with rather rounded edges anterior to the section of maximum depth, turning to a more or less lenticular shape with fairly pointed edges at the posterior part, where there may be a great variety of dorsal, ventral, pectoral, and possibly other smaller fins, and then merging into the caudal fin.

In rivers that flow slowly across flat regions and in deep ponds or lakes, the fish have developed high bodies with a large profile area. Such fish include the crucian (*Carassius carassius*) or carp, *Acheilognathus moriokae*, and sunfish.

In estuaries, the flow is very slow, except during floods. Here salt- and freshwater meet, so that two types of fauna intermingle. These waters are inhabited by pike, gray mullet, bass, and black porgy. Some are slender and well adapted for sudden acceleration and deceleration, and thus for preying on small fish, while the height or profile area of others is large, giving them superior maneuverability. It is reported that the overall mean maximum acceleration rates were 12–13G for pike and 6G for trout, where G is the nondimensional acceleration based on the gravity acceleration (Harper and Blake 1990, 1991).

6.2.3 SWEPT WING FOR MINIMUM INDUCED DRAG

The minimum induced drag on a thin flat wing of given lift is obtained when the downwash is constant owing to elliptic chord (or circulation) distribution. Küchemann (1953) extended the above result to swept wings wherein the minimal induced drag is obtained when the spanwise chord distribution is:

$$c(\eta)/(b/2) = (8/\pi A\!R)\sqrt{1 - \eta^2}\{a_0 \cos \Lambda/a(\eta)\}$$
(6.2-17)

where a_0, a, Λ and η are two-dimensional lift slope 2π, sectional lift slope C_l/α, sweep angle, and nondimensional spanwise distance $2y/b$ respectively. When the term within braces in Eq. (6.2-17,) is one or $\{a_0 \cos \Lambda/a(\eta)\} = 1$, the above equation gives the elliptical lift distribution. Shown in Fig. 6.2-3 is an example of a swept wing with minimum induced drag. This optimal wing has a planform wider at the root and narrower at the tips than the elliptical shape shown by dotted curves, and resembles the shape of the caudal fin in various fish. Similar analyses have been performed by other researchers such as Theilheimer (1943), Lobert (1981), and Ashenberg and Weihs (1984).

Slender fish dwelling in the shallow stream must send the deflection waves along the body with

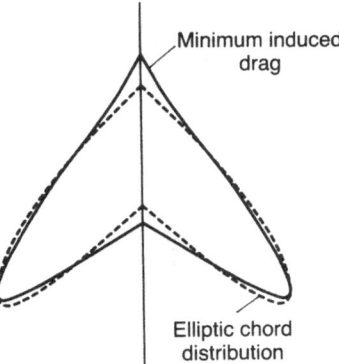

Minimum induced drag

Elliptic chord distribution

FIG. 6.2-3. Swept wing for minimum induced drag. (From Küchemann 1953 with permission).

increasing amplitude to ban or to beat the caudal fin of low-aspect-ratio and swept wings. It must also be remembered that a caudal fin shaped in this way does not lose its function even with a high angle of attack caused by the violent fanning in low swimming speed.

6.3 Fanning Mechanics

For a wide body or fin, the vertical extent of the fish body, specifically of the tail, is so large that the vertical or spanwise change of motion of the fluid is gradual. Thus, the action of the tail's lateral motion on different horizontal slices of the fluid also gradually varies and can be regarded as two-dimensional. The caudal fin, particularly the lunate tail of high-speed swimmers, is very much like an airplane wing in many aspects. Its chordwise section resembles an airfoil that has a rounded leading edge to support the suction force and a sharp trailing edge to ensure tangential flow past the edge. In the carangiform mode, to propel the whole body a fish needs to oscillate its caudal fin in a fanning motion.

As we saw in Sect. 6.1.3, the motion of the caudal fin can be divided into two coupled fundamental strokes: (1) a feathering produced by a large curvature at the caudal peduncle and (2) a heaving, with some phase difference relative to the feathering, produced by a bending deflection of small curvature at the posterior part of the body. These coupled motions function to give optimal performance for the needs of the fish concerned.

6.3.1 HYDRODYNAMICS OF AN ISOLATED FANNING WING

In the fanning propulsion of a high-speed fish, the caudal fin is extended widely in the spanwise direction and body undulation becomes significant only in the posterior part. The fin thus behaves like an isolated wing, enabling highly efficient sustained swimming at high cruising speeds.

The simplest way to analyze this motion is to apply the lifting-line theory for the spanwise loading and the two-dimensional unsteady-wing theory for the chordwise loading in sinusoidal heaving and feathering oscillation. This is equivalent to assuming that the local flow around each cross section remains two-dimensional, but the local angle of incidence is influenced by the whole

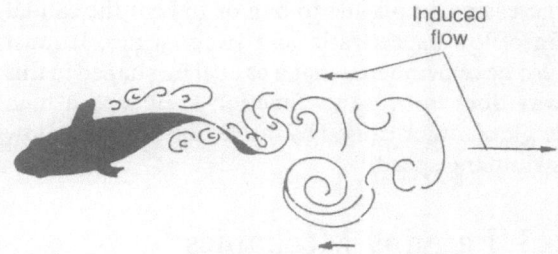

FIG. 6.3-1. Wake formation due to fanning motion. (From Abe 1981 with permission).

pattern of the time-dependent streamwise and spanwise wake vorticity (Chopra 1974).

Shed-wake formation. The actual formation of a series of wake vortices generated by the fanning motion of a fish can be seen in a two-dimensional projection of the motion in Fig. 6.3-1. The vortices, called shed vortices, are produced to satisfy the Kutta condition at the trailing edge of the wing and remain in the wake after the wing has passed. Unlike the direction of circulation of the vortices left in the wake of a blunt body, the vortices in the present case clearly produce an induced flow directed backward. As a result, the momentum can also be carried in the same direction. Thus, the wing produces thrust as a reaction force to this fluid motion. The important thing in this mechanism of thrust generation is the change of circulation caused by the undulatory heav-

ing and pitching motion. Therefore part of the power is consumed as parasite power to propel the animal through the water at some velocity *V* against the viscous resistance, and the remainder of the power is wasted in the vortex wake as induced power and profile power.

Three-dimensional wake system. The three-dimensional wake system here is very similar to that of the beating wing motion. However, since the fanning motion is assumed to be symmetrical with respect to a vertical plane for sideward fanning, the plane of maximum lift appears in both leftward (portside) and rightward (starboard) strokes. The lift is directed rightward in the leftward stroke, and leftward in the rightward stroke, and is always inclined forward to produce the thrust. Thus, due to the sinusoidal fanning motion, a sinusoidal normal force and a sinusoidal thrust with a double frequency can be generated.

The sinusoidal change of lift creates a series of vortex cells, (Fig. 6.3-2), like the beating wing shown in Fig. 4.2-14. Associated with the change of thrust, the rearward wake velocity is also induced in the wake along the x-axis, and the forward wake velocity is generated outside of the wake. The latter is, in some cases, utilized to save locomotive energy by schooling (Abrahams and Colgan 1985).

FIG. 6.3-2. Three-dimensional wake system and forces.

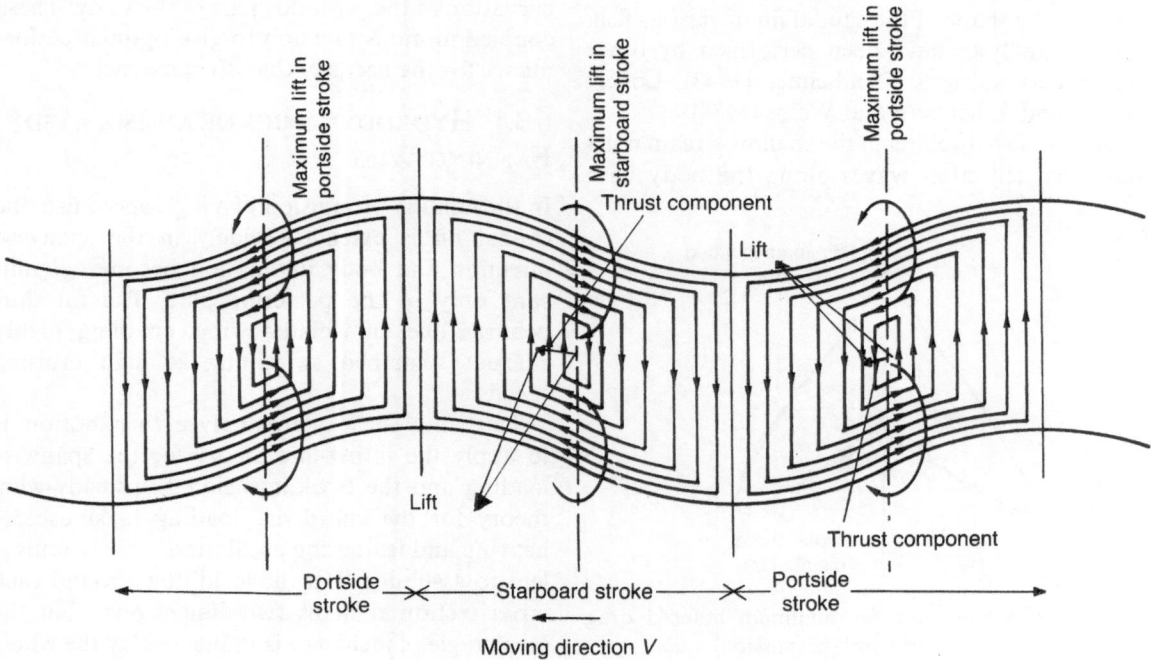

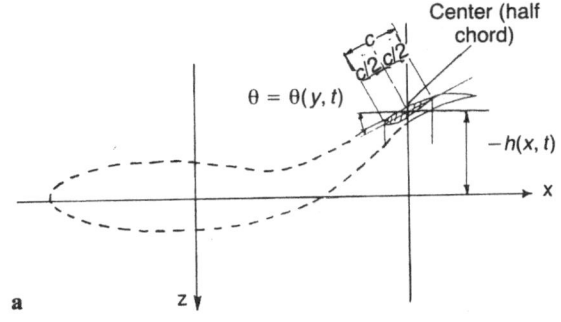

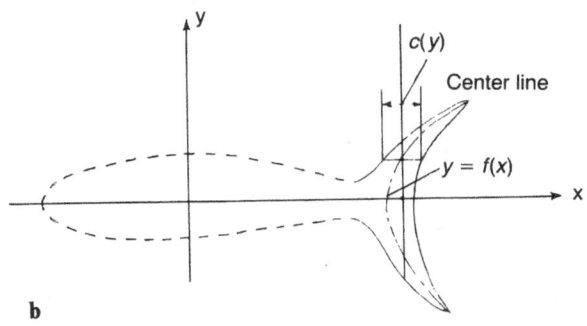

FIG. 6.3-3. **a, b.** Mechanical model of a caudal fin in fanning. **a** (x, z) plane. **b** (x, y) plane.

Hydrodynamic forces and moments. Similarly to the beat of a wing (See Sect. 4.2.1), the fanning motion of a caudal fin generates a relative inflow velocity that consists of the lateral velocity resulting from the fanning itself, the induced velocity generated by the wake vortices trailed by the wing, and the velocity caused by the forward motion of the fish as a vectorial summation.

In Fig. 6.3-3, the wing position represented by the center line of the airfoil of a caudal fin can be defined by:

$$y = f(x): \quad \text{ordinate of center line} \quad (6.3\text{-}1a)$$

$$z = -h(x, t): \quad \text{lateral displacement} \quad (6.3\text{-}1b)$$

$$\theta = \theta(y, t): \quad \text{feathering angle} \quad (6.3\text{-}1c)$$

where it is assumed that the airfoil is rigid and thus fin motion is performed without changing the airfoil configuration.

The relative velocity components of the flow around the airfoil at the point of interest on any section can be given by:

$$\left.\begin{aligned} U_x &= V + \bar{a}(c/2)\dot{\theta}\sin\theta + v\sin\phi \\ U_y &= 0 \\ U_z &= \dot{h} + \bar{a}(c/2)\dot{\theta}\cos\theta - v\cos\phi \end{aligned}\right\} \quad (6.3\text{-}2)$$

where v is the induced velocity generated by the trailing and shed vortices at the point of interest, $\bar{a}(c/2)$, and ϕ is the inflow angle at the same point,

$$v = v_t + v_s \quad (6.3\text{-}3)$$

$$\phi = \tan^{-1}(U_z/U_x) \quad (6.3\text{-}4)$$

Then, the angle of attack α at the point of interest, lift l, drag d, and moment m can be formulated as:

$$\alpha = \theta + \phi \quad (6.3\text{-}5)$$

$$l = \tfrac{1}{2}\rho U^2 c C_l \quad (6.3\text{-}6a)$$

$$d = \tfrac{1}{2}\rho U^2 c C_d \quad (6.3\text{-}6b)$$

$$m = \tfrac{1}{2}\rho U^2 c^2 C_m \quad (6.3\text{-}6c)$$

where

$$U = \sqrt{U_x{}^2 + U_z{}^2} \simeq \sqrt{V^2 + (\dot{h} - v)^2} \quad (6.3\text{-}7)$$

The lift, drag and moment coefficients may include unsteady terms. The thrust component t and the side force (or normal force) component n are respectively defined by:

$$t = l\sin\phi - d\cos\phi \quad (6.3\text{-}8a)$$

$$n = l\cos\phi + d\sin\phi \quad (6.3\text{-}8b)$$

The total hydrodynamic thrust T, total side (or normal) force N, and the power P can be expressed as:

$$\begin{aligned} T &= \int_{-b/2}^{b/2} t\,dy \\ &= \int_{-b/2}^{b/2} \{l\sin\phi - d\cos\phi\}\,dy \end{aligned} \quad (6.3\text{-}9a)$$

$$\begin{aligned} N &= \int_{-b/2}^{b/2} n\,dy \\ &= \int_{-b/2}^{b/2} \{l\cos\phi + d\sin\phi\}\,dy \end{aligned} \quad (6.3\text{-}9b)$$

$$\begin{aligned} P &= \int_{-b/2}^{b/2} (n\dot{h} - m\dot{\theta})\,dy \\ &= \int_{-b/2}^{b/2} \{\dot{h}(l\cos\phi + d\sin\phi) - m\dot{\theta}\}\,dy \end{aligned} \quad (6.3\text{-}9c)$$

The bending moments M_x and M_z, and the twisting or feathering moment M_y at the center section of the caudal fin can be given by the following formulas:

$$M_x = \int_0^{b/2} (ny)\,dy = \int_0^{b/2} \{(l\cos\phi + d\sin\phi)y\}\,dy$$

$$(6.3\text{-}10a)$$

$$M_z = \int_0^{b/2} ty\,dy = \int_0^{b/2} (l\sin\phi - d\cos\phi)y\,dy$$

$$(6.3\text{-}10b)$$

$$\left.\begin{aligned}
M_y &= \int_{-b/2}^{b/2} \{-n(x-x_t) + th + m\}\,dy \\
&= \int_{-b/2}^{b/2} \{(l\cos\phi + d\sin\phi)(x_t - x) \\
&\quad + (l\sin\phi - d\cos\phi)h + m\}\,dy
\end{aligned}\right\}$$

$$(6.3\text{-}10c)$$

In order to get a high propulsive force with less required power, the following conditions must be met: (1) a large forward velocity and transversal velocity within the "cavitation limit," because the fluid-dynamic force is proportional to the square of the inflow velocity; (2) a large feathering angle for tilting the fluid-dynamic force in the forward direction, within the "stall limit" of the fin; (3) an optimal angle of attack in order to get the largest lift-to-drag ratio, which is increased by a large aspect ratio, with the result that the induced velocity is also reduced.

6.3.2 SIMPLE ANALYSIS

Let us consider a simple analysis of the fanning motion of a wing by assuming that: (1) the wing is a rectangular plate having a homogeneous induced velocity distribution, and a small coupled heaving and feathering motion is exerted around its aerodynamic center ($\bar{a} = -1/2$) in such a way that

$$\left.\begin{aligned}
\text{heaving:} \quad & h = h_1\cos(\omega t) \\
\text{feathering:} \quad & \theta = \theta_1\cos(\omega t + \phi_\theta)
\end{aligned}\right\} \quad (6.3\text{-}11)$$

where h_1/b and θ_1 are less than one, and (2) the induced velocity is also represented by the first harmonic

$$v = v_1\cos(\omega t + \phi_v) \qquad (6.3\text{-}12)$$

whereas its spanwise distribution is homogeneous in spite of the rectangular planform.

The inflow velocity components, the inflow ratio, and the angle of attack of the wing are given in Table 6.3-1.

Then the lift, drag, thrust, normal force, and moment acting on a fin element are approximated

TABLE 6.3-1. Relative velocity and angles

Items	Expression
Velocity	$U_x \simeq V$
	$U_y = 0$
	$U_z = \dot{h} + (\bar{a}c/2)\dot{\theta} - v$
	$\quad = -\omega h_1\sin(\omega t) + (c/4)\omega\theta_1\sin(\omega t + \phi_\theta)$
	$\quad\quad - v_1\cos(\omega t + \phi_v)$
Inflow angle	$\phi = \tan^{-1}(U_z/U_x)$
	$\quad \simeq \{\phi_c k - (v_1/V)\cos\phi_v\}\cos(\omega t)$
	$\quad\quad + \{\phi_s k + (v_1/V)\sin\phi_v\}\sin(\omega t)$
Angle of attack	$\alpha = \theta + \phi$
	$\quad = \{\theta_1\cos\phi_\theta + \phi_c k - (v_1/V)\cos\phi_v\}\cos(\omega t)$
	$\quad\quad - \{-\theta_1\sin\phi_\theta + \phi_s k + (v_1/V)\sin\phi_v\}\sin(\omega t)$
Parameters	$\phi_s = \{-2(h_1/c) + \frac{1}{2}\theta_1\cos\phi_\theta\}$
	$\phi_c = \frac{1}{2}\{\theta_1\sin\phi_\theta\}$
	$k = \omega c/2U \simeq \omega c/2V$

in the following expressions:

$$\left.\begin{aligned}
l &= \tfrac{1}{2}\rho U^2 c C_l \simeq \tfrac{1}{2}\rho V^2 c a_0 \alpha \\
d &= \tfrac{1}{2}\rho U^2 c C_d \simeq \tfrac{1}{2}\rho V^2 c \delta_0 \\
t &= l\sin\phi - d\cos\phi \simeq l\phi - d \\
n &= l\cos\phi + d\sin\phi \simeq l + d\phi \\
m &= \tfrac{1}{2}\rho U^2 c^2 C_m \simeq \tfrac{1}{2}\rho V^2 c^2 C_{m_\alpha}\alpha
\end{aligned}\right\} \quad (6.3\text{-}13)$$

where a_0, δ_0, and C_{m_α} are respectively lift slope, drag coefficient (assumed constant), and the moment slope of a two-dimensional airfoil. Then, the thrust T, the normal force N, and the power P can be obtained as given in Table 6.3-2.

It is very important to note that: (1) by adopting the expression of thrust given in Eq. (6.3-13), the suction force that acts at the leading edge of the wing is taken into account in the above calculations; (2) the power is the mechanical rather than the physiological power because, as stated in Sect. 4.2.3, the two powers are nearly equal, or $C_p = C_p^*$ in the usual operation range; (3) the phase difference between the heaving and feathering motions, ϕ_θ, may be selected to satisfy the optimal coupling given in Fig. 4.2-10, as a function of the reduced frequency; (4) the thrust and the power have second harmonics, whereas the normal force has the first harmonic only; and (5) as the reduced frequency increases, the forces and the power also increase.

TABLE 6.3-2. Thrust, normal force, and power for rectangular plate

Thrust

T

$$\int_{-b/2}^{b/2} (l\phi - d)\mathrm{d}y = \tfrac{1}{2}\rho V^2 S\{a_0[\tfrac{1}{2}\{\phi_c k - (v_1/V)\cos\phi_v\}\{\theta_1\cos\phi_\theta + \phi_c k - (v_1/V)\cos\phi_v\}\{1 + \cos(2\omega t)\}$$
$$+ \tfrac{1}{2}\{\phi_s k + (v_1/V)\sin\phi_v\}\{-\phi_1\sin\phi_\theta + \phi_s k + (v_1/V)\sin\phi_v\}\{1 - \cos(2\omega t)\}$$
$$+ \tfrac{1}{2}\{\phi_s k + (v_1/V)\sin\phi_v\}\{\theta_1\cos\phi_\theta - \phi_c k - (v_1/V)\cos\phi_v\}\sin(2\omega t)$$
$$+ \tfrac{1}{2}\{\phi_c k - (v_1/V)\cos\phi_v\}\{-\theta_1\sin\phi_\theta + \phi_s k + (v_1/V)\sin\phi_v\}\sin(2\omega t)] - \delta_0\}$$

Normal force

N

$$\int_{-b/2}^{b/2} (l + d\phi)\mathrm{d}y = \tfrac{1}{2}\rho V^2 S\{a_0[\{\theta_1\cos\phi_\theta + \phi_c k - (v_1/V)\cos\phi_v\}\cos(\omega t)$$
$$+ \{-\theta_1\sin\phi_\theta + \phi_s k + (v_1/V)\sin\phi_v\}\sin(\omega t)]$$
$$+ \delta_0[\{\phi_c k - (v_1/V)\cos\phi_v\}\cos(\omega t) + \{\phi_s k + (v_1/V)\sin\phi_v\}\sin(\omega t)]$$

Power

P

$$\int_{-b/2}^{b/2} (\dot{h}l - m\dot{\theta})\mathrm{d}y = \tfrac{1}{2}\rho V^3 S\{-a_0(h_1/c)k[\{-\theta_1\sin\phi_\theta + \phi_s k + (v_1/V)\sin\phi_v\}\{1 - \cos(2\omega t)\}$$
$$+ \{\theta_1\cos\phi_\theta + \phi_s k - (v_1/V)\cos\phi_v\}\sin(2\omega t)]$$
$$+ (C_{m_a}\theta_1)k[\sin\phi_\theta\{\theta_1\cos\phi_\theta + \phi_c k - (v_1/V)\cos\phi_v\}\{1 + \cos(2\omega t)\}$$
$$+ \cos\phi_\theta\{-\theta_1\sin\phi_\theta + \phi_s k + (v_1/V)\sin\phi_v\}\{1 - \cos(2\omega t)\}$$
$$+ \cos\phi_\theta\{\theta_1\cos\phi_\theta + \phi_c k - (v_1/V)\cos\phi_v\}\sin(2\omega t)$$
$$+ \sin\phi_\theta\{-\theta_1\sin\phi_\theta + \phi_s k + (v_1/V)\sin\phi_v\}\sin(2\omega t)]\}$$

TABLE 6.3-3. Induced velocity and its components

Induced velocity	v/V	

$$L/2\rho\pi(b/2)^2 V^2 = \int_{-b/2}^{b/2} l\,\mathrm{d}y / \{\tfrac{1}{2}\rho V^2 S(\pi \cancel{R})\}$$
$$= (a/\pi \cancel{R})[\{\theta_1\cos\phi_\theta + \phi_c k - (v_1/V)\cos\phi_v\}\cos(\omega t)$$
$$+ \{-\theta_1\sin\phi_\theta + \phi_s k + (v_1/V)\sin\phi_v\}\sin(\omega t)]$$

Parameters	$(v_1/V)\cos\phi_v$	$\{(a/\pi \cancel{R})/(1 + a/\pi \cancel{R})\}\{\theta_1\cos\phi_\theta + \phi_c k\}$
	$(v_1/V)\sin\phi_v$	$\{(a/\pi \cancel{R})/(1 + a/\pi \cancel{R})\}\{\theta_1\sin\phi_\theta - \phi_s k\}$
	v_1/V	$\dfrac{(a/\pi \cancel{R})}{\{1 + (a/\pi \cancel{R})\}}\{(\theta_1\cos\phi_\theta + \phi_c k)^2 + (\theta_1\sin\phi_\theta - \phi_s k)^2\}^{1/2}$
	ϕ_v	$\tan^{-1}\left[\dfrac{\theta_1\sin\phi_\theta - \phi_s k}{\theta_1\cos\phi_\theta + \phi_c k}\right]$

The induced velocity v can be determined by the momentum theory from the total lift, as given in Table 6.3-3.

By taking the mean value of one stroke period, the following steady values can be obtained:

$$\bar{T} = \tfrac{1}{2}\rho V^2 S\{a_0[\tfrac{1}{2}\{\phi_c k - (v_1/V)\cos\phi_v\}\{\theta_1\cos\phi_\theta + \phi_c k - (v_1/V)\cos\phi_v\}$$
$$+ \tfrac{1}{2}\{\phi_s k + (v_1/V)\sin\phi_v\}\{-\theta_1\sin\phi_\theta + \phi_s k + (v_1/V)\sin\phi_v\}] - \delta_0\} \equiv \tfrac{1}{2}\rho V^2 S\bar{C}_T$$

(6.3-14a)

$$\bar{N} = 0 \tag{6.3-14b}$$

$$\bar{P} = \tfrac{1}{2}\rho V^3 S\{-a_0(h_1/c)k[\{-\theta_1\sin\phi_\theta + \phi_s k$$
$$+ (v_1/V)\sin\phi_v\}] + (C_{m_a}\theta_1)k[\sin\phi_\theta\{\theta_1\cos\phi_\theta$$
$$+ \phi_c k - (v_1/V)\cos\phi_v\} + \cos\phi_\theta\{-\theta_1\sin\phi_\theta$$
$$+ \phi_s k + (v_1/V)\sin\phi_v\}]\} \equiv \tfrac{1}{2}\rho V^3 S\bar{C}_P$$

(6.3-14c)

Since the mean thrust is equal to the drag of the body,

$$\bar{T} = D = \tfrac{1}{2}\rho V^2 f \tag{6.3-15}$$

where f is the drag area of the body, the mean power can be rewritten as:

$$\bar{P} = P_0 + P_p + P_m \qquad (6.3\text{-}16)$$

where

$$P_0 = \tfrac{1}{2}\rho V^3 S \delta_0 \qquad \text{(profile power)}$$

$$P_p = \tfrac{1}{2}\rho V^3 f = \tfrac{1}{2}\rho V^3 S C_{D_0}$$

$$ = DV \qquad \text{(parasite power)}$$

$$\begin{aligned}
P_m = \tfrac{1}{2}\rho V^3 S \{ a_0 [&\tfrac{1}{2}\theta_1 (v_1/V)\cos(\phi_\theta - \phi_v) \\
&+ \tfrac{1}{2}\theta_1 (v_1/V)k \sin(\phi_\theta - \phi_v) \\
&- \tfrac{1}{8}\theta_1^2 k^2 - \tfrac{1}{2}(v_1/V)^2 + \tfrac{1}{2}(h_1/c)\theta_1 k^2 \cos\phi_\theta \\
&- \tfrac{1}{2}(h_1/c)(v_1/V)k \sin\phi_v] \\
&+ (C_{m_\alpha}\theta_1)k[\cos\phi_\theta\{-\theta_1 \sin\phi_\theta + \phi_s k \\
&+ (v_1/V)\sin\phi_v\} + \sin\phi_\theta\{\theta_1 \cos\phi_\theta \\
&+ \phi_c k - (v_1/V)\cos\phi_v\}]\}
\end{aligned}$$

$$\text{(miscellaneous power)}$$

$$(6.3\text{-}17)$$

The above equations show that, unlike the necessary power for flying, the necessary power for swimming increases monotonically as the swimming speed increases, from its minimum value given by the profile power, which is usually very small, to infinity, in proportion to the cube of the speed. The miscellaneous power includes the induced power related to the generation of the induced flow and the twisting power for fin feathering.

Then, the mechanical (or Froude) efficiency η can be given by

$$\eta = \bar{T}V/\bar{P} = \bar{C}_T/\bar{C}_P \qquad (6.3\text{-}18)$$

Let us consider a typical example of an isolated rectangular wing in fanning motion, the aspect ratio of which is $\mathcal{R} = 8$. In order to learn the effects of the amplitudes of the heaving motion h_1/c, the feathering motion θ_1, the phase difference ϕ_θ, and the reduced frequency k on the mean thrust coefficient $\bar{C}_T$, the power coefficient $\bar{C}_P$, and the efficiency η, these are calculated under the assumption of homogeneous induced velocity distribution, even though the wing is rectangular. The results are shown in Fig. 6.3-4 only for $\phi_\theta = -90°$ or $\phi_h - \phi_\theta = 90°$.

1. The thrust and power coefficients increase monotonically as the heaving amplitude h_1/c and the fanning frequency k increase. However, as the feathering amplitude increases, they decrease for negative and zero phase advance, $\phi_\theta = -90°$ and $\phi_\theta = 0°$, and increase for positive phase advance, $\phi_\theta = 90°$. That is to say, the thrust and power increase as the angle of attack increases near the most effective strokes, $\psi = \omega t = 90°$ and $270°$, where the relative velocity with respect to the fluid is maximum.

2. On the other hand, the efficiency increases as the heaving amplitude increases, but soon saturates at a higher heaving amplitude ratio h_1/c. In the present examples, the maximum efficiency is given for the positive phase difference $\phi_h - \phi_\theta = 90°$ or $\phi_\theta = -90°$, the highest reduced frequency $k = 0.3$, and the largest feathering amplitude $\theta_1 = 0.3$. In a trimmed condition for a given fish, combining Eqs. (6.3-14a) and (6.3-15) yields the thrust coefficient $\bar{C}_T = $ constant, and thus the reduced frequency, $k = \omega c/2U \simeq (\omega/2)(c/l)(l/V) = $ constant, if the fish swims with constant amplitudes of θ_1 and h_1, and at the phase difference $\phi_\theta = -90°$. This further leads to the conclusion that the "specific speed," or the ratio of speed and body length V/l, is proportional to the fanning frequency ω. The above will be verified in the following section.

It should be remembered that the above results have been obtained in the linear range of the solution without any flow separation, and that the effect of the phase difference ϕ_θ on performance is very similar to that of a two-dimensional wing in beating motion, as will be recalled from the explanation in Sect. 4.2.2 in which the effect of the unsteady flow was taken into account.

These statements are generally confirmed by a numerical computation based on the local circulation method (LCM), which can treat the induced velocity in a more real or complex form. It is thus significant that the main information on swimming mechanics can be obtained from the foregoing simple analytic expressions (Kawachi and Azuma 1988).

6.3.3 PERFORMANCE OF FISH

Speed and relative speed. The measured speeds of various fish under natural conditions are plotted in Fig. 6.3-5 as a function of body length for sustained, burst, and long-distance swimming. Generally, as the size increases, the speed and the specific speed also increase. As a rule of thumb, a specific speed of 10, or $V/l = 10$ s^{-1}, represents a

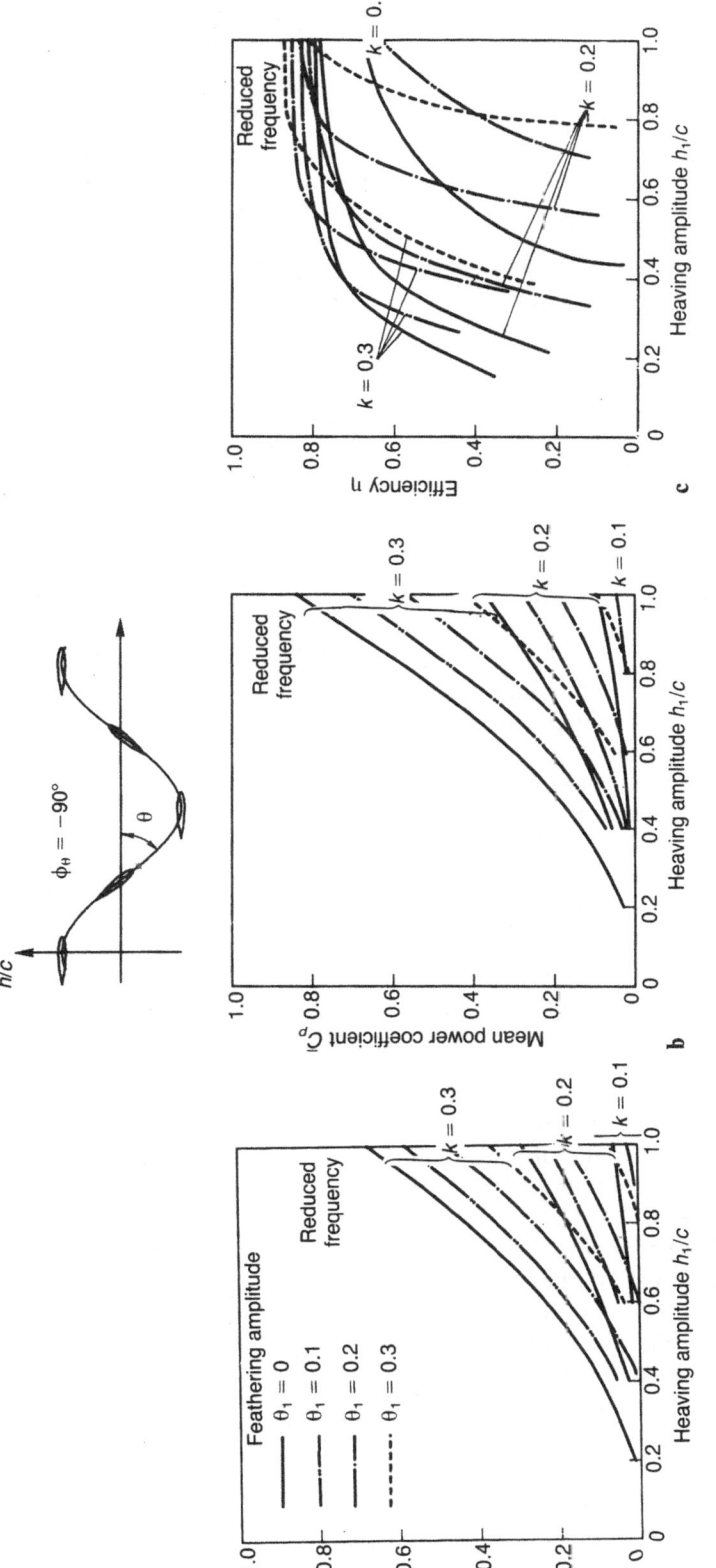

Fig. 6.3-4. **a–c.** Performance of an isolated wing ($\mathcal{R} = 8.0$) with the phase difference of $\phi_B = -90°$, under the assumption of homogeneous induced velocity distribution. **a** Thrust. **b** Power. **c** Efficiency.

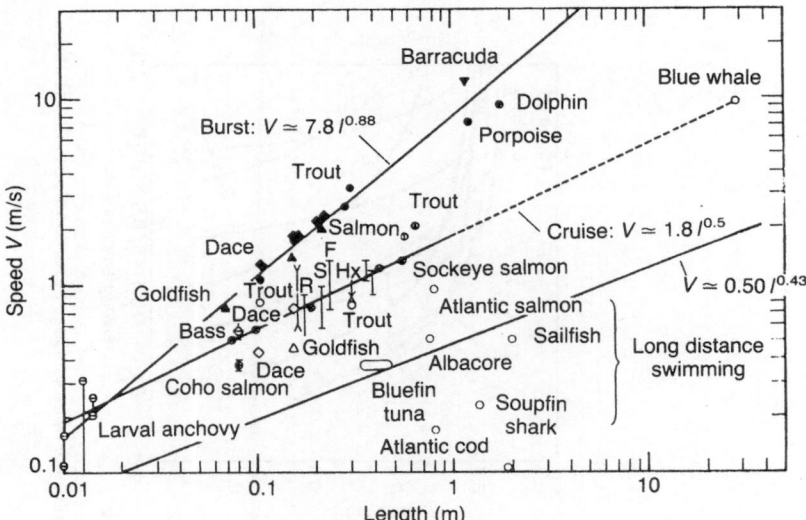

FIG. 6.3-5. Speed of fish. (Redrawn from McMahon and Bonner 1983; Bainbridge 1960; Weihs 1977; Beamish 1978; Lindsey 1978 for long-distance movement).

reasonable upper limit to the swimming performance of cruising fish, irrespective of their size. It is interesting to find that the larger fish can increase their speed in going from cruise to burst mode by a much larger factor than smaller fish, and fish smaller than about 10 mm in length appear incapable of burst mode swimming.

In the simplest case, if the basal metabolic rate P_B is proportional to the mass of the fish m and the drag area $C_{D_0}S$ is proportional to $m^{2/3}$ (geometric similarity)–$m^{5/8}$ (elastic similarity), then the cruising speed V is proportional to $m^{1/9}$–$m^{1/8}$. If, as in the general case, the specific speed V/l is considered, it is roughly proportional to $m^{-2/9}$–$m^{-1/8}$. By replacing m with l^3–l^4, these results can be rewritten for the dimension of length as follows: $V \propto l^{1/3}$–$l^{1/2}$, or $V/l \propto l^{-2/3}$–$l^{-1/2}$. This relation coincides nearly with the observed data for long-distance swimming and cruise swimming shown in Fig. 6.3-5. For example, Weihs (1977) proposed the use of $V = 0.50 \, l^{0.43}$.

On the other hand, by assuming that the available power is proportional to the mass of the propulsive musculature, which is proportional to $l^{2.9}$, and by considering the effect of the Reynolds number on the skin friction drag, Bainbridge (1958) obtained the following relations: $V \propto l^{0.56}$ or $V/l \propto l^{-0.44}$ for laminar flow, or $V \propto l^{0.39}$ or $V/l \propto l^{-0.61}$ for turbulent flow. For the maximum burst speeds, Wardle (1975) proposed the relation

$$V = \Lambda/2T = (\Lambda/l)l/2T \qquad (6.3\text{-}19)$$

where Λ is the forward distance resulting from one complete tail beat cycle (or one "stride"), and T is muscle contraction time, which is estimated to be about 0.08 s for fish about 2 m in length and decreases with an increase in body length. The value of Λ/l will be given later. The observed slope of burst swimming of 0.08 or $V \propto l^{0.88}$, shown in Fig. 6.3-5, might result from $T \propto l^{0.12}$. Because of the definition of $\Lambda = (\lambda/C)V$, the contraction time T is equivalent to $T = \frac{1}{2}(\lambda/C)$, where λ is the wavelength. Thus, the following relations are useful for a rough estimation:

$$V = 0.50 l^{0.43} \quad \text{for long-distance swimming}$$
$$= 1.8 l^{0.50} \quad \text{for cruise swimming}$$
$$= 7.8 l^{0.88} \quad \text{for brust swimming}$$
$$(6.3\text{-}20)$$

Wardle (1977) further reported that, "sprint" of a fast and large fish with a lunate tail can reach a speed of $V \simeq 20 \, l$ or more.

Reynolds number. The values of the Reynolds number based on body length are of the order of 10^8 for the most rapid cetaceans, 10^6 for migrating fish, 10^5–10^3 for a great variety of fish, and about 10^2 for tadpoles. Thus, for a large Reynolds number, swimming propulsion using the fanning motion depends primarily on the inertial effect, and

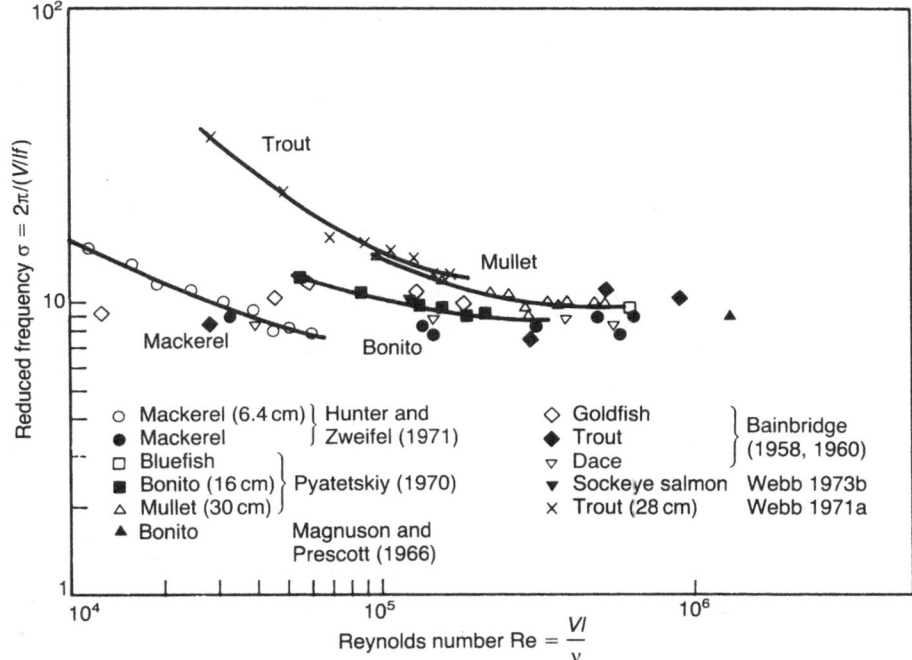

FIG. 6.3-6. Measured reduced frequencies. (Redrawn from Yates 1983).

for a medium Reynolds number, it depends on both the inertial and resistive effects.

Reduced frequency. Bainbridge (1958) found that the specific speed is related to the frequency of tail strokes. Two important facts must be noted: (1) except at very low speed, there is a linear relation expressed as

$$(V/l)_{mean} = (0.6\text{--}0.7)f \qquad (6.3\text{-}21a)$$

or

$$(V/lf)_{mean} = (V/C)(\lambda/l) = 0.6\text{--}0.7 \qquad (6.3\text{-}21b)$$

where λ and f are the wavelength and the fanning frequency; and (2) there is a minimum frequency of tailbeat, $f > 1\text{--}2$ Hz, under which the fish seems to rely on the pectoral or other fins for locomotion. As stated in Sect. 4.1.5, Eq. (6.3-21) gives the same relation obtained from the force balance of $T = D$. This also suggests a nondimensional parameter $V/lf = 0.6\text{--}0.7$, which may be called "nondimensional speed" rather than specific speed, and is a reciprocal form of the reduced frequency k or σ,

$$k = \pi \bar{c} f/V = \pi(\bar{c}/l)/(V/lf)_{mean} \simeq 0.5$$

(based on the mean half-tail chord $\bar{c}/2$, and assumed $\bar{c}/l \simeq 0.1$) (6.3-22a)

or

$$\sigma = 2\pi/(V/lf)_{mean} = 2k(l/\bar{c}) \simeq 10$$

(based on the body length l) (6.3-22b)

Since the speed V is proportional to $l^{0.5}$ in cruise or optimal swimming, Eq. (6.3-21) further states that the frequency is inversely proportional to the square root of the body length or $f \propto l^{-0.5}$. The same relation was experimentally observed by Altringham and Johnston (1990a) as $f = 1.67 l^{-0.52}$.

Yates (1983) gave some correspondences between the reduced frequency σ and the Reynolds number Re for various species of fish, as given in Fig. 6.3-6, which shows clearly that as Re increases σ approaches 10. These nondimensional relations, as stated in Sects. 1.2.1 and 1.2.2, are important for finding a general trend of performance in animal locomotion. The dimensional specific speed V/l may, in the hydrodynamic sense, be replaced by the above nondimensional speed V/lf or the reduced frequency, for a mechanical understanding of swimming. The relatively higher value of the reduced frequency based on the mean half-chord probably results from the fact that the fish considered are not cruising fish with continuous economic swimming. Thus, as seen from Fig. 4.2-10, the phase difference between heaving and feathering is nearly, $0° < \phi_h - \phi_\theta < 90°$ for a fish tail compared with 90° for bird and insect wings. Except the bursting, the coupled heaving and feathering motion of a fish's tail takes an intermediate mode between those of Figs. 4.2-11b and 4.2-11c.

In fanning, if the body lacks flexibility and thus if the mathematical model of the fanning motion is expressed by a single-joint motion, then, as explained earlier, the tail wing must, because of the body configuration, inevitably take the ostraciiform mode or coupling motion without any phase difference. This is strictly in dash motion with large amplitude at start or at very low speed. In order to take lower values of the reduced frequency in economical swimming, much more flexibility in body motion would be required. This need is actually answered in the body of the shark and the lunate tail of high-performance cruising fishes such as the marlin. By adopting a narrow caudal peduncle and lunate tail that is a swept wing of high aspect ratio, the marlin can get a large phase difference and make the reduced frequency in the cruising motion small.

Bainbridge (1958) gave the following relation between the "distance travelled per unit tailbeat" ($\Lambda = \lambda V/C = 2\pi l/\sigma$), the amplitude $h_{\max}$ and the "distance per beat over body length" Λ/l for dace, trout, and goldfish: $\Lambda/h_{\max} \simeq 3.4$ and $\Lambda/l \simeq 0.6$. More generally, from the relation (6.3-22b) and Fig. 6.3-6, the ratio Λ/l can also be approximated by

$$\Lambda/l = 2\pi/\sigma \simeq 0.6 \qquad (6.3\text{-}23)$$

in large Reynolds numbers. Bainbridge further studied speed and stamina in dace, trout, and goldfish. He showed that for more than 10 s of sustained time, the specific speed approaches $V/l \simeq 4(\text{s}^{-1})$. A similar tendency can be observed in salmon (Hartt 1966; Shepard et al. 1968).

In these considerations, it must be remembered that: (1) unlike flying creatures, swimming creatures are not required to generate a lifting force against their weight, but only a driving force for locomotion; and (2) the available power is, as stated in Sect. 4.1.2, also proportional to the contracting speed of the musculature or to the frequency of the locomotion, and thus is a function of the temperature and the oxygen concentration of the surrounding fluid.

Study of swimming performance can be further developed by investigation of a self-propelled mechanical model fish designed to allow the artificial alteration of such locomotion factors as ampli-

tude, frequency, and phase shift between heaving and feathering. A mechanical fish simulating the carp was made by Watanabe and Kimura (1976) and swam like a real carp.

Skimming or skittering. Some fish skim or skip along the water surface either by a fanning motion of the caudal fin, mostly under water, or by a vigorous beating of the caudal fin against the wave surface, during which parts of the body other than the caudal fin are above water. By this skimming or skittering the fish can reduce its hydrodynamic drag and increase its swimming speed. Mullet (*Mugil corsula*), halfbeak (*Hemiramphus*), needlefish (*Tylosurus*) and many flying fish use this technique for gaining speed before takeoff (Breder 1926; Hubbs 1933; Ganguly and Mitra 1962).

6.3.4 Effect of Dynamic Lift in Continuous Swimming

Let us consider the swimming performance of two fish with specific gravities larger than that of the water they swim in, one obtaining lift for horizontal swimming from the hydrodynamic force acting on a wing or a pair of laterally extended pectoral fins such as tuna (*Parathunnus sibi*) and shark (*Carcharodon carcharias*) and the other from the buoyant force generated by a gas in its air bladder or the lipid content of its liver as observed in many other fish.

By assuming that the only additional lift is the dynamic lift of the wing or pectoral fins swimming with speed U, the additional lifting force compensating for insufficient buoyant force is equal for both swimmers, or

$$\tfrac{1}{2}\rho U^2 S C_L = (\rho - \rho_b)g V_b \qquad (6.3\text{-}24)$$

where S and V_b are the extended fin area and additional bladder volume respectively, and ρ and ρ_b are the density of the water and contents of the bladder or liver. Then, the ratio of additional lift to the original weight of the fish is expressed by

$$(\rho - \rho_b)g V_b/\rho_B g V_B = \{(\rho - \rho_b)/\rho_B\}(V_b/V_B)$$
$$\simeq V_b/V_B. \qquad (6.3\text{-}25)$$

where ρ_B and V_B are the density and volume of the original body respectively. This approximate expression is obtained by assuming $\rho_B \simeq \rho$ and $\rho_b \ll \rho$.

The drag increments caused by the inflated part of the body and by the extended wing are respectively given by

$$D_b = \frac{1}{2}\rho U^2 C_f k\{(V_b + V_B)^{2/3} - V_B^{2/3}\}$$

$$\simeq \frac{1}{3}\rho U^2 C_f k V_B^{2/3}(V_b/V_B) \qquad (6.3\text{-}26a)$$

$$D_w = \frac{1}{2}\rho U^2 S C_{D,w} \qquad (6.3\text{-}26b)$$

where k is the ratio of the wetted surface area S_w to two-thirds of the volume, or $k = S_w/V_B^{2/3}$, and C_f and $C_{D,w}$ are the friction drag coefficient of the body surface and the drag coefficient of the wing itself. The approximate expression in Eq. (6.3-26a) is obtained by assuming a small inflation of the bladder, or $V_b/V_B \ll 1.0$. Then, by combining Eqs. (6.3-26a) and (6.3-26b), the drag ratio D_b/D_w becomes

$$
\left.\begin{array}{l}
D_b/D_w = \frac{1}{2}\{(1 + V_b/V_B)^{2/3} - 1\}(V_B/V_b) \\
\qquad \cdot \mathrm{Fr}^2 k C_L(C_f/C_{D,w})\{\rho/(\rho - \rho_b)\} \\
\qquad \simeq \frac{1}{3}\mathrm{Fr}^2 k C_L(C_f/C_{D,w})\{\rho/(\rho - \rho_b)\}
\end{array}\right\}
$$
$$(6.3\text{-}27)$$

where Fr is the Froude number based on the body length, defined in terms of the cubic root of the body volume $V_B^{1/3}$, or $\mathrm{Fr} = U/\sqrt{gV_B^{1/3}}$.

By supposing that the drag coefficient of the extended wing can be made optimal (See Sect. 3.2.3), as

$$C_{D_i} = C_L^2/\pi A\!R_e \simeq C_{D_0} + 2.04 C_f \qquad (6.3\text{-}28)$$

$$C_{D,w} = 2.04 C_f + C_{D_i} = C_L^2/\pi A\!R_e + 2.04 C_f$$

$$= C_{D_0} + 4.08 C_f \qquad (6.3\text{-}29)$$

where C_{D_0} is the parasite drag coefficient of the body based on the wing area S, we can obtain (without any flow separation)

$$C_{D_0} = k(V_B^{2/3}/S)C_f \qquad (6.3\text{-}30)$$

Thus, substituting these relations into Eq. (6.3-27)

yields

$$
\left.\begin{array}{l}
D_b/D_w = \frac{1}{2}\{(1 + V_b/V_B)^{2/3} - 1\}(V_B/V_b)\mathrm{Fr}^2 C_L k \\
\qquad \cdot \left\{\dfrac{1}{k(V_B^{2/3}/S) + 4.08}\right\}\left\{\dfrac{\rho}{\rho - \rho_b}\right\} \\
\qquad \simeq \frac{1}{3}\mathrm{Fr}^2 k[\sqrt{\pi A\!R_e}\{k(V_B^{2/3}/S) + 2.04\}C_f/ \\
\qquad \{k(V_B^{2/3}/S) + 4.08\}]/\{1 - (\rho_b/\rho)\}
\end{array}\right\}
$$
$$(6.3\text{-}31)$$

The above equation shows that the buoyant force is disadvantageous: (1) for a large Froude number Fr, (2) for a large friction drag, and (3) when the density of the bladder or liver content is close to the density of the water. Thus, performance will be improved for large fish swimming with high speed by the presence of a wing of large aspect ratio and small wing area.

In the case where buoyancy is regulated by the lipid content of the liver, extensive vertical movements are possible because the buoyant force is unaffected by changes in hydrostatic pressure (Thomas 1984). However, the total lift cannot be adjusted rapidly, and, as Alexander (1959) assessed, there would be a potential increase in additional drag.

To illustrate, let us consider a hypothetical shark—sharks have a wing but no bladder—and a hypothetical equivalent fish that has either a bladder or lipid to support the excess weight. The hydrodynamic characteristics are given in Table 6.3-4.

The volume ratio $V_B/V_b = 5$ has been selected with reference to members of the subclass Chondrichthyes identified by Marshall (1965) as having the lowest value. However, the results will not be altered greatly at larger values, or $V_B/V_b > 5$. The drag ratio exceeds one when Fr > 2.85

TABLE 6.3-4. Performance of a hypothetical shark

Items	Symbol	Unit	Value	Notes
Aspect ratio of wing	$A\!R$	—	4	
Area ratio	$kV_s^{2/3}/S$	—	6	
	$S_w/V_b^{2/3}$	—	7	
Volume ratio	V_s/V_b	—	5	
Friction drag	C_f	—	0.003	at Re $= 10^7$–10^6
Density ratio	ρ_b/ρ	—	0 and 0.9	
Drag ratio	D_b/D_w	—	0.123 Fr^2	for $\rho_b/\rho = 0$
		—	1.23 Fr^2	for $\rho_b/\rho = 0.9$

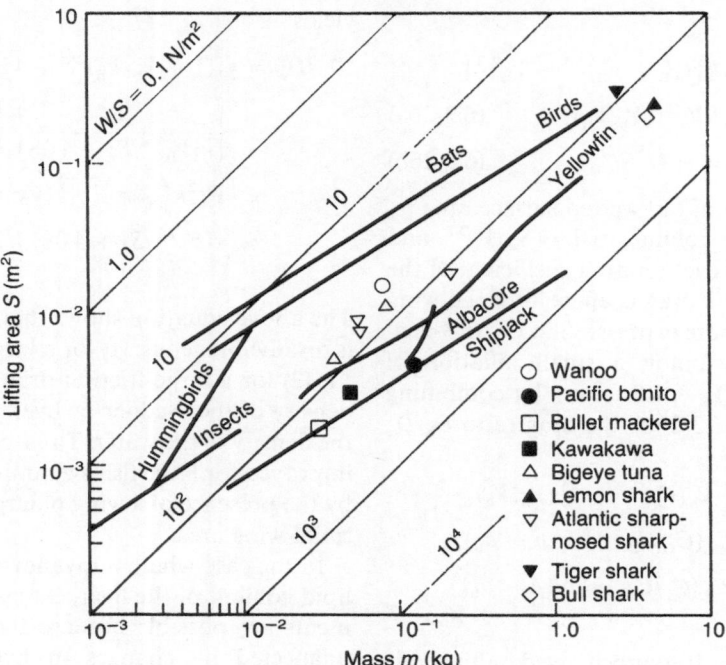

Fig. 6.3-7. Lifting area and weight of scombrids and sharks in comparison with birds and insects. (From Magnuson 1978 with permission).

for $\rho_b/\rho = 0$, and Fr > 0.90 for $\rho_b/\rho = 0.9$. In the present example, this is attained when the speed of the shark exceeds approximately 8.9 m/s ($\rho_b/\rho = 0$) or 2.8 m/s ($\rho_b/\rho = 0.9$) for $V_B^{1/3} = 1$ m, and 12.6 m/s ($\rho_b/\rho = 0$) or 4.0 m/s ($\rho_b/\rho = 0.9$) for $V_B^{1/3} = 2$ m. This exemplified calculation clarifies that (1) the bladder should be replaced with pectoral fins only in high-speed swimmers, and (2) this tendency is more pronounced in larger swimmers.

Figure 6.3-7 shows a comparison of the wing loading of lifting surfaces among scombroids, sharks, birds, insects, and bats. Since wing loading can be related to flight speed in trimmed flight by the formula $W/S = \frac{1}{2}\rho U^2 C_L$, among creatures of the same species, those that have lower wing loading can fly or swim at slower sustained speed than those with higher wing loading. Furthermore, if the lift coefficient used is not altered for flight in air and swimming in water (see Sect. 3.2.3), then the speed ratio is given by $(W/S)_W/(W/S)_A = (\rho_W/\rho_A)(U_W/U_A)^2 \simeq 10^3(U_W/U_A)^2$. Thus, if the wing loading of a fish is ten times higher than that of a bird, $(W/S)_W/(W/S)_A \simeq 10$, the speed of the fish will be roughly one-tenth that of the bird, $U_W/U_A \simeq 1/10$.

Actually, in the case of sharks, except for the fast-swimming families, the excess weight is probably supported by the hydrodynamic lifts generated by a pair of extended pectoral fins held at a positive angle of attack, and by the upper leaf of the caudal fin, which is heterocercal and considerably swept-back, in snaking or anguilliform motion, or more likely in the sweeping motion explained in Sect. 7.3.2.

In the above discussion, the only lifting device considered was the pectoral fins. Another possible device is the body proper. However, as pointed out in Sect. 3.4.2, it is less effective to allocate lift to a device of lower aspect ratio (such as the body) than to one of higher aspect ratio (pectoral fins). Actually, kawakawa, for example, swims at sustained speeds with zero angle of attack to the body (Magnuson 1970). The posterior part of the body, including the caudal peduncle and fin, can generate a small lift for moment trim.

Vestigial bladders can be seen in most members of the order Selachii (sharks), certain families of fast-moving fish such as Scombridae (mackerel and tuna), and Aciperseridae. The skipjack has no gas bladder, and 94% of its weight is offset by

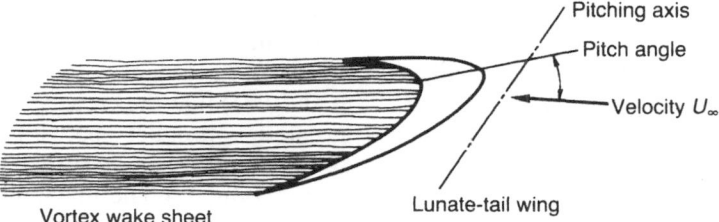

FIG. 6.3-8. Lunate wing and its vortex wake sheet.

buoyancy. Many deep-sea sharks, e.g. *Centroscymnus*, attain near-neutral buoyancy because their enormous squalene deposits (accounting for about 80% of the body weight) provide just enough lift to compensate for the weight in water of the denser parts of the body. They must be slow swimmers.

6.3.5 HIGH-PERFORMANCE FISH WITH LUNATE TAILS

A highly extended wing in the shape of a crescent moon, called a "lunate tail," is adopted in high-performance fish such as tuna, wahoo, marlin, swordfish, and others, as well as fast-swimming families of shark. Some of these were shown earlier in Table 6.1-1.

Nonplanar wake. The planform of the lunate-tail wing is characterized by a large aspect ratio, and a swept angle and taper ratio that increase along the wingspan from root to tip. Thus, as the wing's pitch angle or the angle of attack increases, as shown in Fig. 6.3-8, the trailing edge of the wing shifts downward progressively toward the wing tips, and the vortex wake sheet makes a downward concave surface. We saw in Sect. 3.3.1 that such a curved wing and wake system will bring about an improvement in performance (C.D. Cone 1962a).

Zimmer (1979, 1983) and Dam (1987a, 1987b) also demonstrated that, owing to the nonplanar rolled-up trailing vortex sheet, the crescent planform shape can increase the level of induced efficiency to a value greater than that of the flat untwisted wing of elliptical shape. Their results were obtained with a vortex-lattice method for prescribed wake, and a nonlinear surface panel method for free wake (Maskew 1982).

Dam considered the following wing shape:

$$\left.\begin{array}{l} x_{le}(\eta) = x_t(1 - \sqrt{1 - \eta^2}) \\ c(\eta) = c_r\sqrt{1 - \eta^2}. \end{array}\right\} \quad (6.3\text{-}32)$$

where x_{le}, x_t and c_r are the ordinate of leading edge, wing tip and the root chord respectively. Three exemplified shapes and calculated spanwise loads are shown in Fig. 6.3-9. It can be seen that the backward tip sweep (wing 2 and 3) produces an increase in fluid-dynamic loading outboard toward the tip, while forward sweep (wing 1) produces a reduction in tip loading. The influence of wing tip location x_t on induced drag is represented by taking the ratio of the induced drag coefficient to that of the elliptic wing, $C_{D_i}/(C_{D_i})_e$, which is also shown in Fig. 6.3-9b. Thus, it is clear that the nonplanar wake shape produces a slight induced drag reduction. This may help to explain why fast marine animals and high-performance

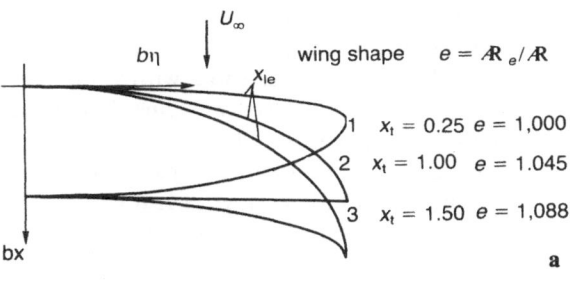

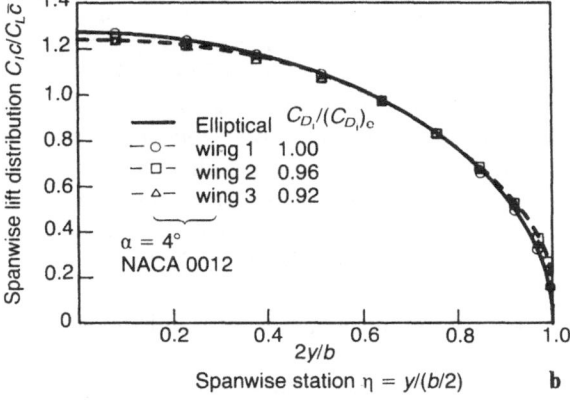

FIG. 6.3-9. **a, b.** Lunate wings and their spanwise loads. (From Dam 1987a with permission). **a** Swept tip ($C_r = 1.0$, $Æ = 7$). **b** Lift distribution.

birds have evolved long lunate wing lifting surfaces.

Induced feathering motion. On the other hand, the pitching (or feathering) moment at the root of a lunate-tail wing is much larger than that at the root of a straight wing; thus, the lunate-tail wing has a synchronized feathering motion at the thin caudal peduncle produced by the hydrodynamic moment and the elastic restoring moment generated by the lateral fanning motion of the wing. To maintain the high efficiency of the wing, the reduced frequency of the fanning motion is low, and the phase shift between the heaving and feathering motion is $\phi_n - \phi_\theta = \pi/2$, as explained in Sect. 4.2.3; therefore, the mode of fanning is similar to that shown in Fig. 4.2-11c. In this "thunniform" motion, most of the thrust is concentrated at the caudal fin while the body is held in a fairly streamlined form without large lateral deflection of the caudal peduncle (Lindsey 1978; Yates 1983).

Hydrodynamics of the lunate-tail wing. Theoretical studies of a lunate-tail wing in fanning motion were conducted by Chopra (1974), James (1975), Chopra and Kambe (1977), and Lan (1979), and then elaborated on by Karpouzian et al. (1990), who applied the "asymptotic theory" (Cheng and Murillo 1984) based on the "unsteady (or oscillatory)-lifting-surface theory" to performance analysis of high aspect ratio wings with a curvature and sweep of the planform center line. Numerical calculations were performed on lunate-tail wings for various combinations of thrust coefficient, mode shapes, and morphological features. Under the assumption of small heaving and feathering motion, the following results are made clear by Karpouzian et al. (1990): (1) For the maximum efficiency there is an optimum sweep angle which depends on the thrust coefficient and proportional feathering parameter; (2) while the peak efficiency decreases somewhat with increasing thrust, the benefit of sweep is more noticeable at a higher C_T; and (3) with the relatively large heaving amplitude of the peduncle and/or the proportional feathering parameter, the optimum sweep is so small that practically any sweep tends to reduce the efficiency.

The above results are derived from a three-dimensional investigation of the vortex wake system. They contribute further information on the fundamental concepts obtained from the two-dimensional studies made by Lighthill (1969, 1970) or presented in Sect. 4.4.2 by myself, and the three-dimensional study of straight wings presented in Sect. 6.3.2.

It is, from a wind tunnel test, reported by Dam et al. (1991) that the crescent wing, with its highly swept tips, produces better high angle-of-attack aerodynamic characteristics than the unswept elliptic wing: (1) The maximum lift coefficient is higher, (2) the poststall pitching-moment behaviour is improved, and (3) the prestall dihedral effect of the crescent wing is stronger and more linear than the elliptic wing.

Scombroid fish. As we have seen, many scombroid fish, including bonitos, mackerels, and tunas, are negatively buoyant (Beamish 1978). To overcome the negative buoyancy, they swim continuously with extended pectoral fins in low speed and generate a dynamic lift that balances their weight in the water (Magnuson 1970, 1973, 1978). The posture, including the span, area, swept angle, dihedral angle, pitch angle, and probably even the camber, of the extended pectoral fins can be varied in correspondence to the swimming conditions.

6.3.6 TWO-STAGE SWIMMING

Weihs suggested two methods by which fish with negative buoyancy can save energy with a two-stage swimming mode. In the first, swimming at a constant speed would include a phase of powerless gliding with gradual increase of depth followed by active upward swimming at an angle to the original height (Weihs 1973b). In the second, the swimming would have alternating periods of accelerated motion and powerless coasting (Weihs 1974).

Steady swimming. Referring to Fig. 6.3-10a, equations of motion for trimmed flight of a swimming animal with speed U in water can be given by

$$L = (W - B)\cos\gamma \qquad (6.3\text{-}33a)$$

$$T = (W - B)\sin\gamma + D \qquad (6.3\text{-}33b)$$

where lift L, drag D, thrust T, buoyant force B, and weight in the fluid $W - B$ can be expressed by

$$\left.\begin{array}{l} L = \frac{1}{2}\rho U^2 S C_L \\ D = \frac{1}{2}\rho U^2 S C_D \\ T = \frac{1}{2}\rho U^2 S C_T \end{array}\right\} \qquad (6.3\text{-}34)$$

$$\left.\begin{array}{l} W - B = (\rho_B - \rho)g V_B = \rho\sigma g V_B \\ \sigma = (\rho_B/\rho) - 1 \end{array}\right\} \qquad (6.3\text{-}35)$$

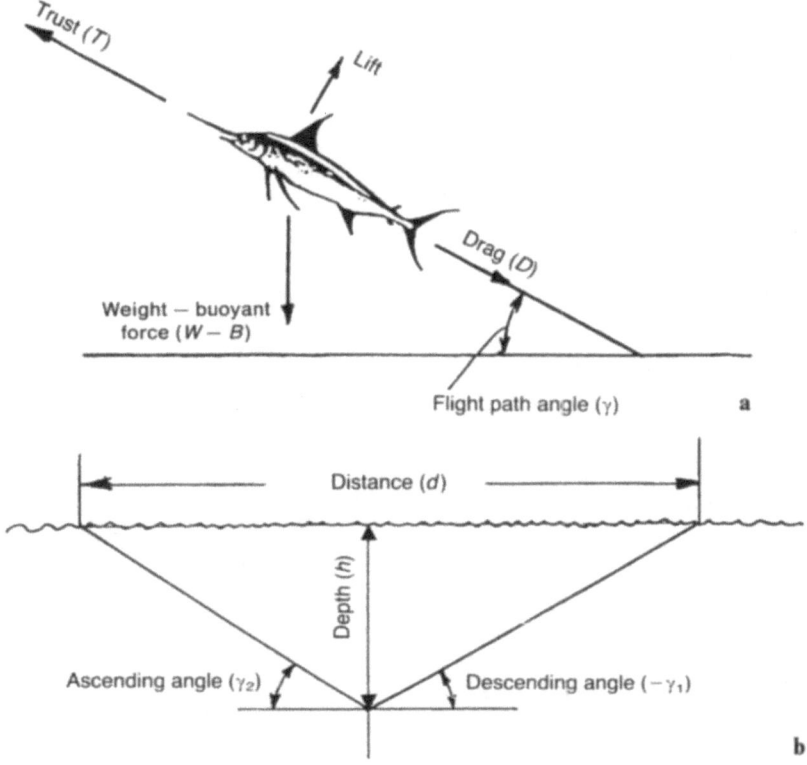

FIG. 6.3-10. **a, b.** Two-stage swimming consisting of descending and ascending phases. **a** Trimmed flight. **b** Two-stage swimming.

and

$$C_D = C_{D_0} + C_L{}^2/\pi A\!\!R \qquad (6.3\text{-}36)$$

and where $A\!\!R$ is a simple expression of the effective aspect ratio of the animal including the body itself and a pair of pectoral fins or finlike forelegs if they are extended, and the minimum drag coefficient C_{D_0} is assumed constant even though the animal is slender. Then, by combining the above equations, Eq. (6.3-33b) yields

$$C_T = \{2\sigma g(V_B/S)/U^2\}\sin\gamma + C_{D_0}$$
$$+ \{2\sigma g(V_B/S)/U^2\}^2\cos^2\gamma/\pi A\!\!R \quad (6.3\text{-}37)$$

The energy consumed by the animal in the two-stage swimming (subscripts 1 and 2 for a steady descending stage and a steady ascending stage) shown in Fig. 6.3-10b is given by

$$E_y = T_1 U_1 t_1/\eta_1 + T_2 U_2 t_2/\eta_2 \quad (6.3\text{-}38)$$

where η_1 and η_2 are the mechanical efficiencies of the thrust generator, and t_1 and t_2 are the times spent for the respective stages,

$$\left. \begin{aligned} t_1 &= (h/U_1)/\sin(-\gamma_1) \quad \text{for } -\gamma_1 > 0 \\ t_2 &= (h/U_2)/\sin\gamma_2 \qquad \text{for } \gamma_2 > 0 \end{aligned} \right\} \quad (6.3\text{-}39)$$

Here, h is the maximum depth of this flight, and $-\gamma_1$ and γ_2 are the steady descending and ascending or climbing angles.

The mechanical efficiency is given by the ratio of effective work performed by the thrust of the caudal fin or flukes to the total energy spent for this flight, the latter of which includes the work acting on the fluid. Thus the efficiency η can be expressed by the formula

$$\eta = \eta_0\{TU/T(U+u)\} = \eta_0/\{1+(u/U)\} \quad (6.3\text{-}40)$$

where η_0 is mechanical efficiency excluding the contribution of the induced power Tu and is thus constant for any flight speed. The induced velocity u is given by

$$\left. \begin{aligned} u &= T/\{2\rho(\tfrac{1}{4}\pi b^2 U)\} = 2T/\rho\pi b^2 U \\ &= C_T \cdot U/\pi A\!\!R \end{aligned} \right\} \quad (6.3\text{-}41)$$

The denominator of Eq. (6.3-40) can be formulated thus:

$$1 + u/U = 1 + C_T/\pi A R \simeq 1 \qquad (6.3\text{-}42)$$

where the approximation is valid for high-performance teleosts such as Scombridae, Istiophoridae, Xiphiidae, Selachii, and Cetacea, all of which have large aspect ratio fins.

Equation (6.3-38) can also be rewritten as

$$E_\gamma \simeq (T_1 U_1 t_1 + T_2 U_2 t_2)/\eta_0 \qquad (6.3\text{-}43)$$

When combined with Eqs. (6.3-35) through (6.3-39), this equation further yields

$$
\begin{aligned}
E_\gamma = (\tfrac{1}{2}\rho S/\eta_0)&h[C_{D_0}\{U_1{}^2/\sin(-\gamma_1) + U_2{}^2/\sin\gamma_2\} \\
&+ \{(2\sigma g V_B/S)^2/\pi A R\} \\
&\cdot \{\cos^2(\gamma_1)/U_1{}^2 \sin(-\gamma_1) \\
&+ \cos^2\gamma_2/U_2{}^2 \sin\gamma_2\}]
\end{aligned}
\qquad (6.3\text{-}44)
$$

For a horizontal flight of distance d with the speed of U,

$$\gamma_1 = \gamma_2 = h = 0 \qquad (6.3\text{-}45)$$

and

$$t = d/U \qquad (6.3\text{-}46)$$

this becomes

$$
\begin{aligned}
E_{\gamma=0} = Td/\eta_0 = (\tfrac{1}{2}\rho S/\eta_0)d \\
\cdot [C_{D_0}U^2 + \{(2\sigma g V_B/S)^2/\pi A R\}/U^2]
\end{aligned}
\qquad (6.3\text{-}47)
$$

and the following constraint exists:

$$d/h = \cot(-\gamma_1) + \cot\gamma_2. \qquad (6.3\text{-}48)$$

Now, let us define the nondimensional relative energy ratio:

$$
\begin{aligned}
(E_\gamma &- E_{\gamma=0})/E_{\gamma=0} \\
&= (h/d)[C_{D_0}\{U_1{}^2/\sin(-\gamma_1) + U_2{}^2/\sin\gamma_2 \\
&- (d/h)U^2\} + \{(2\sigma g V_B/S)^2/\pi A R\} \\
&\cdot \{\cos^2\gamma_1/U_1{}^2 \sin(-\gamma_1) + \cos^2\gamma_2/ \\
&U_2{}^2 \sin\gamma_2 - (d/h)/U^2\}]/ \\
&[C_{D_0}U^2 + \{(2\sigma g V_B/S)^2/\pi A R\}/U^2]
\end{aligned}
\qquad (6.3\text{-}49)
$$

In comparing energy consumption, either the "block time" required to travel a given distance or the speed of flight must be specified. The following two cases are interesting:

1. If the time of flight is constant, or

$$t = t_1 + t_2 \qquad (6.3\text{-}50)$$

then, when combined with Eqs. (6.3-39), (6.3-46), and (6.3-48), Eq. (6.3-50) yields

$$d/h = (U/U_1)/\sin(-\gamma_1) + (U/U_2)/\sin\gamma_2 \qquad (6.3\text{-}51a)$$

When combined with Eq. (6.3-48), this yields

$$
\begin{aligned}
\{(U/U_1) &- \cos(-\gamma_1)\}/\sin(-\gamma_1) \\
&= -\{(U/U_2) - \cos\gamma_2\}/\sin\gamma_2 \qquad (6.3\text{-}51b)
\end{aligned}
$$

Eq. (6.3-49) can then be reformulated:

$$
\begin{aligned}
(E_\gamma &- E_{\gamma=0})/E_{\gamma=0} \\
&= (h/d)[C_{D_0}\{(U_1/U)^2(1 - (U/U_1)^3)/\sin(-\gamma_1) \\
&+ (U_2/U)^2(1 - (U/U_2)^3)/\sin\gamma_2\}U^2 \\
&- \{(2\sigma g V_B/S)^2/\pi A R\}(1/U^2) \\
&\cdot \{(U/U_1)(1 - (U/U_1)\cos^2\gamma_1)/\sin(-\gamma_1) \\
&+ (U/U_2)(1 - (U/U_2)\cos^2\gamma_2)/\sin\gamma_2\}]/ \\
&[C_{D_0}U^2 + \{(2\sigma g V_B/S)^2/\pi A R\}/U^2].
\end{aligned}
\qquad (6.3\text{-}53)
$$

This equation states that if U_1 and U_2 are larger than the horizontal speed U (or U/U_1, $U/U_2 < 1$), then the "relative energy ratio" $(E_\gamma - E_{\gamma=0})/E_{\gamma=0}$ is negative under a "critical speed" U^*, $U < U^*$ which can not be given analytically. It is interesting to find that the critical speed U^* is approximately proportional to the square root of $2\sigma g V_B/S$ and inversely proportional to the quadric root of $\pi A R \, C_{D_0}$. Any swimming animal having large body size $V_B{}^{1/3}$, high density σ, well streamlined and low drag configuration C_{D_0}, and small aspect ratio $A R$ can have a large critical speed.

For the range of speed ratio in the first stage, $0 < U/U_1 < 1$, the speed ratio in the second stage can be determined from Eq. (6.3-51) as

$$
\begin{aligned}
U/U_2 = -\{\sin\gamma_2/\sin(-\gamma_1)\} \\
\cdot \{U/U_1 - \cos(-\gamma_1)\} + \cos\gamma_2.
\end{aligned}
\qquad (6.3\text{-}54)
$$

Since the above speed ratio must also satisfy the condition $0 < U/U_2 < 1$, the speed ratio U/U_1, is limited within

$$\left.\begin{array}{l} \cos(-\gamma_1) - \{(1 - \cos\gamma_2)\sin(-\gamma_1)/\sin\gamma_2\} \\ \quad < U/U_1 < \cos(-\gamma_1) \\ \quad + \{\sin(-\gamma_1)\cos\gamma_2/\sin\gamma_2\} \end{array}\right\} \quad (6.3\text{-}55)$$

Usually, the energy or mean power can be saved by performing steady two-stage swimming only at very low swimming speeds. It should be remembered that, unlike in Weihs analysis (Weihs 1973b), the steady descent swimming in the first stage is not limited to coasting (or $T = 0$), but includes any powered flight (or $T > 0$). Here also thrust is the propulsive force counteracting the constant parasite drag and the induced drag of the animal.

2. If the flight speed is made optimal by having the induced drag always equal to the parasite drag, or

$$\left.\begin{array}{l} C_{D_0} = C_L{}^2/\pi A\!R \quad \text{or} \quad C_L = \sqrt{\pi A\!R C_{D_0}} \\ \text{and} \\ \quad C_D = 2C_{D_0} \end{array}\right\} \quad (6.3\text{-}56)$$

then, the speeds in the respective phases are given by

$$\left.\begin{array}{l} U = \sqrt{2\sigma g(V_B/S)}\sqrt{\pi A\!R C_{D_0}} \\ U_1 = U\sqrt{\cos(-\gamma_1)} \\ U_2 = U\sqrt{\cos\gamma_2} \end{array}\right\} \quad (6.3\text{-}57)$$

and the relative energy ratio becomes

$$(E_\gamma - E_{\gamma=0})/E_{\gamma=0} = 0 \quad (6.3\text{-}58)$$

The above equation states that the energy is not changed by trimming the lift coefficient C_L for optimal performance; thus, energy saving cannot be attained by introducing steady two-stage swimming.

If the pectoral fins of the animal are retracted and the effective aspect ratio is very small, the expression of the drag coefficient given by Eq. (6.3-36) is not always valid. The above statement is therefore not fully supported.

Intermittent swimming. If the swimming is an intermittent and horizontal flight consisting of "bursting" (powered) and "coasting" (unpowered) phases, then equations of motion can be given by

$$\left.\begin{array}{l} (m + m_w)\dot{U} = T - D \\ L = W - B \end{array}\right\} \quad (6.3\text{-}59)$$

where $m = W/g$ and m_w are the mass of the animal and the added mass of the surrounding fluid respectively, and T is positive in the bursting phase and zero in the coasting phase.

Combining Eq. (6.3-59) with Eqs. (6.3-34–36) yields

$$\dot{U} = aU^2 - b/U^2 \quad (6.3\text{-}60)$$

where

$$a = (C_T - C_{D_0})/2(V_B/S)(1 + \sigma^\#) > 0$$

$$\text{in the bursting phase} \quad (6.3\text{-}61a)$$

$$= -C_{D_0}/2(V_B/S)(1 + \sigma^\#) < 0$$

$$\text{in the coasting phase} \quad (6.3\text{-}61b)$$

$$b = 2(\sigma g)^2(V_B/S)/\pi A\!R(1 + \sigma^\#) \quad (6.3\text{-}62)$$

and

$$\sigma^\# = \sigma + (m_w/\rho V_B) \quad (6.3\text{-}63)$$

By assuming $C_T - C_{D_0}$ and $\sigma^\#$ to be constant, Eq. (6.3-60) can be integrated analytically as preserved in Table 6.3-5, where the following initial and final conditions are specified:

TABLE 6.3-5. Equations for determining speed of intermittent flight

Bursting phase	$F_1(U) \equiv \dfrac{1}{4(b/a)^{1/4}}\left[\ln\left\|\dfrac{U - (b/a)^{1/4}}{U + (b/a)^{1/4}}\right\| + 2\tan^{-1}\left\{\dfrac{U}{(b/a)^{1/4}}\right\}\right]$
	$= at + F_1(U_0)$
Coasting phase	$F_2(U) \equiv \dfrac{1}{4\sqrt{2}(-b/a)^{1/4}}\left[\ln\left\|\dfrac{U^2 - \sqrt{2}(-b/a)^{1/4}U + (-b/a)^{1/2}}{U^2 + \sqrt{2}(-b/a)^{1/4}U + (-b/a)^{1/2}}\right\|\right.$
	$\left. + 2\tan^{-1}\dfrac{\sqrt{2}(-b/a)^{1/4}U}{(-b/a)^{1/2} - U^2}\right] = a(t - t_1) + F_2(U_1)$
Speed at $t = t_1$	$F_1(U_1) = at_1 + F_1(U_0).$

$$t = 0 \qquad U = U_0$$
$$t = t_1 \qquad U = U_1 \qquad (6.3\text{-}64)$$
$$t = t_2 \qquad U = U_0$$

Then, the speed can be determined by numerically solving each equation given in Table 6.3-5 for the respective phases, bursting and coasting.

The total energy required for intermittent and horizontal flight, E_1, can be given by

$$E_1 = \int_0^{t_1} TU\,dt/\eta_0 = \{\tfrac{1}{2}\rho V_B^{2/3} C_T/\eta_0\} \int_0^{t_1} U^3 dt \qquad (6.3\text{-}65)$$

whereas the energy required for normal cruising flight with speed of $U_c = d/t_2$ is given by

$$E_c = Td/\eta_0 = Dd/\eta_0$$
$$= \{\tfrac{1}{2}\rho S/\eta_0\} d[C_{D_0} U_c^2 + \{(2\sigma g V_B/S)^2/\pi \mathcal{R}\}/U_c^2] \qquad (6.3\text{-}66)$$

These equations can be used to determine the energy ratio E_1/E_c. Usually, there cannot be any energy saving ($E_1/E_c < 1$). Here again, the thrust is the propulsive force overcoming the drag, consisting of constant parasite drag and induced drag. If the drag increases in powered flight, as Weihs (1974) assumed, the thrust in the above analysis will include the increment of the drag as a negative component.

6.4 Cetacea and Sirenia

Cetaceans (whales and dolphins) and sirenians (dugongs and manatees) are the mammals most completely adapted to an aquatic life. A uniform layer of fat concentrated in a part of the dermis covers the whole body and serves to conserve heat and to provide an auxiliary source of energy.

The motion of the tail fin, which consists of two "flukes" and is horizontally attached instead of vertically like in fish, is kept in the dorso-ventral plane, whereas that of a fish is in the lateral plane. This is probably inherited from terrestrial ancestors with well-developed ventral and dorsal muscles for locomotion on land. It is known that the anatomical structure of the propulsive muscles of a dolphin is simpler than that of fish (Gray 1936).

Cetaceans are streamlined and swim as far as the best swimmers among fish; a maximum speed of 15 knots or more has been authenticated (Parry 1949). Almost all cetaceans swim with the tail fin. In fish, the lateral tailbeat induces a slight lateral

motion of the center of gravity, which is countered by the vertically flat-shaped body. Similarly, in cetaceans the vertical beating of the tail fin causes a vertical motion that is suppressed to some extent by the flatness of the body in the horizontal plane and, in addition, by the presence of a pair of flippers evolved from fore-limbs.

6.4.1 Habitat and Ecology

There are three large groups or orders of mammals that have separately readapted to the marine environment: Cetacea (whales, dolphins, and porpoises), Pinnipedia (seals, sea lions, and walruses), and Sirenia (dugongs and manatees). This adaptation is most extreme in the cetaceans, which have adopted a totally fishlike form.

Mysticeti, including the blue whale, right whale, rorqual, California grey whale, and humpback whale, are characterized by a large body size and a weight exceeding 100 t. These large mammals are incapable of securing their food by aggressive hunting using high-speed swimming. Thus, the large whales and whale sharks, whose weight is supported by the distributed pressure or buoyant force, feed on plankton simply by filling their mouth with krill-laden water, and trapping the krill with their fine strainers.

Odontoceti, the dolphins and porpoises, are gregarious creatures and keep in contact with each other by efficient use of sound. They have a large brain, learn quickly, and exhibit a rich vocal repertoire. They send out sounds and can determine the position of objects by analyzing echoes. Their body size and weight fall within the medium range. They can swim fast and dive deep through the vertical motion of flukes, and can feed on fish and squid by group hunting. This is especially true of killer whales, which eat penguins and seals.

Sirenia, or manatees (*Trichechus*) and dugongs (halicores) are the only plant-eating aquatic mammals living today. They have such large and bulky bodies (a few meters in length and several hundred kilograms in mass) that they cannot move about on land. They have a streamlined body and paddlelike forelimbs and flattened tail flukes. The hindlimbs have been dispensed with during their evolutionary history. Because of their feeding requirements, they are inhabitants of rivers, shallow coastal waters, and river estuaries. They have few natural enemies except man.

The tail of the manatee is wide and more like a rounded fan than the notched tail of the dugong.

This difference suggests that the dugong is adapted to swim faster, whereas the manatee is good at making a sudden start from rest.

6.4.2 PERFORMANCE

Many observations regarding the performance of dolphins and whales (Gray 1936; Gawn 1948; Johannessen and Harder 1960; Norris and Prescott 1961; Lang 1975) showed that either their power is higher than that of a human or their boundary layer is almost laminar and not chiefly turbulent, as would be expected for a high Reynolds number of greater than $Re = 10^7$. As explained earlier, it is believed that the hydrodynamic measures for drag reduction are probably the principal cause of these animals' better-than-expected performance.

Gray's paradox. Gray (1936) attempted to calculate the power required to overcome the drag force that would be acting on animals at their observed swimming speeds, and to equate this to the power available from their propulsive musculature. He considered the case of a 1.22-m porpoise and a 1.82-m dolphin (see Table 6.4-1). The results were as follows: the drag coefficient based on the wetted surface was estimated at $C_{D,w} = 1.5 \times 10^{-3}$ for the porpoise and 1.3×10^{-3} for the dolphin, by assuming 40% laminar flow on the surface. The necessary powers were then estimated to be 448 W and 1,940 W for the porpoise and dolphin swimming at 7.6 m/s and 10.1 m/s respectively. In order to supply these powers, their propulsive muscles, assumed to have a mass of 4.05 kg (17% of total mass of 24.0 kg) and 15.9 kg (17.5% of total mass of 91 kg), had to generate specific powers of 111 W/kg and 122 W/kg respectively. These numbers were considered too high in comparison with that of 17 W/kg for an oarsman (Henderson and Haggard 1925), although the figure for the oarsman is now considered too small. This, then, is "Gray's paradox" (Parry 1949), which provided the initial stimulus for research on drag reduction.

By recording a velocity of 12.2 m/s for a 1.22 m *Sphyraena* barracuda and by employing similar methods to Gray's under the assumption that 50%–75% of the surface of the fish experiences laminar flow, Gero (1952) obtained a result that supports the paradox, namely a result corresponding to a specific power value in the region of 240–330 W/kg.

An interesting analysis was conducted by Lang for the performance of cetaceans and fish of various sizes (Lang 1966). In the calculations, Lang assumed that (1) the general shape of these animals is approximated by a 6:1 ellipsoid with an added tail region that extends by 20% the ellipsoid length; (2) the weight of each body is that of the basic 6:1 ellipsoid of neutral buoyancy; (3) the surface area of each body is 20% larger than that of the basic ellipsoid to compensate for the fins and flukes; (4) the drag coefficients are those resulting from experimental tests reported by Hoerner (1965) on rigid, smooth, streamlined bodies as a function of their Reynolds numbers; (5) the available power loading is equal to those of experimental human power ratios; and (6) the propulsive efficiency is $\eta = 0.85$.

The solid lines in Fig. 6.4-1 plot the estimated top speeds, and the broken lines show the same, with the additional factor of full laminar flow, which is applicable to small fish. Despite increased turbulent flow, the expected top speed continuously increases as body size increases, and, because of the crossover from laminar to turbulent flow, the top speed is fairly constant for a body mass between 10 and 100 kg.

There is an interesting way to estimate the mass and wetted area of whales. Hill (1950) proposed the following formulas for the relations among length l, mass m, and total wetted surface S_w (including fins, flukes, and flippers) in two animals:

$$m = 10l^3 \quad \text{and} \quad S_w = 0.45l^2 \quad \text{for dolphins} \tag{6.4-1}$$

$$m = 6l^3 \quad \text{and} \quad S_w = 0.40l^2 \quad \text{for blue whales} \tag{6.4-2}$$

where l, S_w and m are expressed in m, m^2, and kg respectively.

Fig. 6.4-2 shows the metabolic rate of mammals, P_B. The available power per unit of muscle mass can be assumed to be about 20 W/kg for long-duration exertion and 80 W/kg for short-duration exertion. The muscles as a percentage of body mass have been measured at 20%–25% for porpoises, and 40%–45% for whales (Kanwisher and Sundnes 1966). It is interesting to compare the mass ratios of blubber to muscles: 40%–45% for porpoises, and 20%–25% for whales. In the calculations of Fig. 6.4-1, the power loading for the total mass was assumed to be only a function of the exertion period as depicted in the drawing.

Notty. One of three nicknamed and trained dolphins examined in Table 6.4-1, Notty was a Pacific white-sided dolphin, *Lagenorhyncus obliquidens*, trained by Ralph Penner, and tested in the Gen-

TABLE 6.4-1. Dimensions and performances of cetacea. (Data from Gray 1936; Lang and Daybell 1963; Lang 1966, 1975)

Items	Symbols	Units	Porpoise (Wu, 1971a)	Porpoise (Phocoena communis) (Gray, 1936)	Dolphin (Delphinus delphis) (Gray, 1936)	Whale (Hill, 1950)	Pacific white-sided dolphin (Lagenorhyncus obliquidens) (Lang, 1975)	Pacific bottle-nosed dolphin (Trusiops gillii) (Lang, 1975)	Spotted porpoise (Stenella attenuata)
Length	l	m	2.04	1.22	1.82	25.6	2.09	1.91	1.86
Mass	m	kg	90.0	24.04	90.9	9.00	91	89	53
Speed	V_0	m/s	5.18	7.62	10.1	7.5	—	3.08	—
	V_{max}	m/s	—	—	—	—	7.76×2	8.31×7.5	11.05×2
Duration of exertion	t	s	—	—	—	—	—	7.0×10; 6.08×50; $3.08 \times \infty$	—
Reynolds number	Re		4.2×10^6	—	—	1.25×10^8	—	—	—
Froude number	Fr		1.16	2.20	2.38	0.484	—	—	—
Mass of muscles	m_m	kg	4.05	15.9	—	—	—	—	—
$m_m/m \times 100$			17	17	—	—	—	—	—
Wetted surface	S_w	m²	1.56	0.650	1.39	210	—	—	—
Drag coefficient based on wetted surface	$C_{D,w}$		2.7×10^{-3}	1.5×10^{-3} (40% laminar)	1.3×10^{-3}	—	—	—	—
Measured drag coefficient based on $V_B^{2/3}$	$C_{D,v}$		—	—	—	—	0.031 0.028 (Estimated as fully turbulent)	0.033	0.027
Necessary power	P_n	W	—	440	1,910	—	1,120–1,420	2,800	—
Specific power	P_n/m_m	W/kg	—	111	122	—	70–90	—	—

FIG. 6.4-1. Predicted top speed of cetaceans and fish. (Lang 1966; Norris 1966 with permission).

FIG. 6.4-2. Metabolic rate of mammals. (Kanwisher and Sundnes 1966; Kanwisher and Ridgway 1983; Norris 1966 with permission).

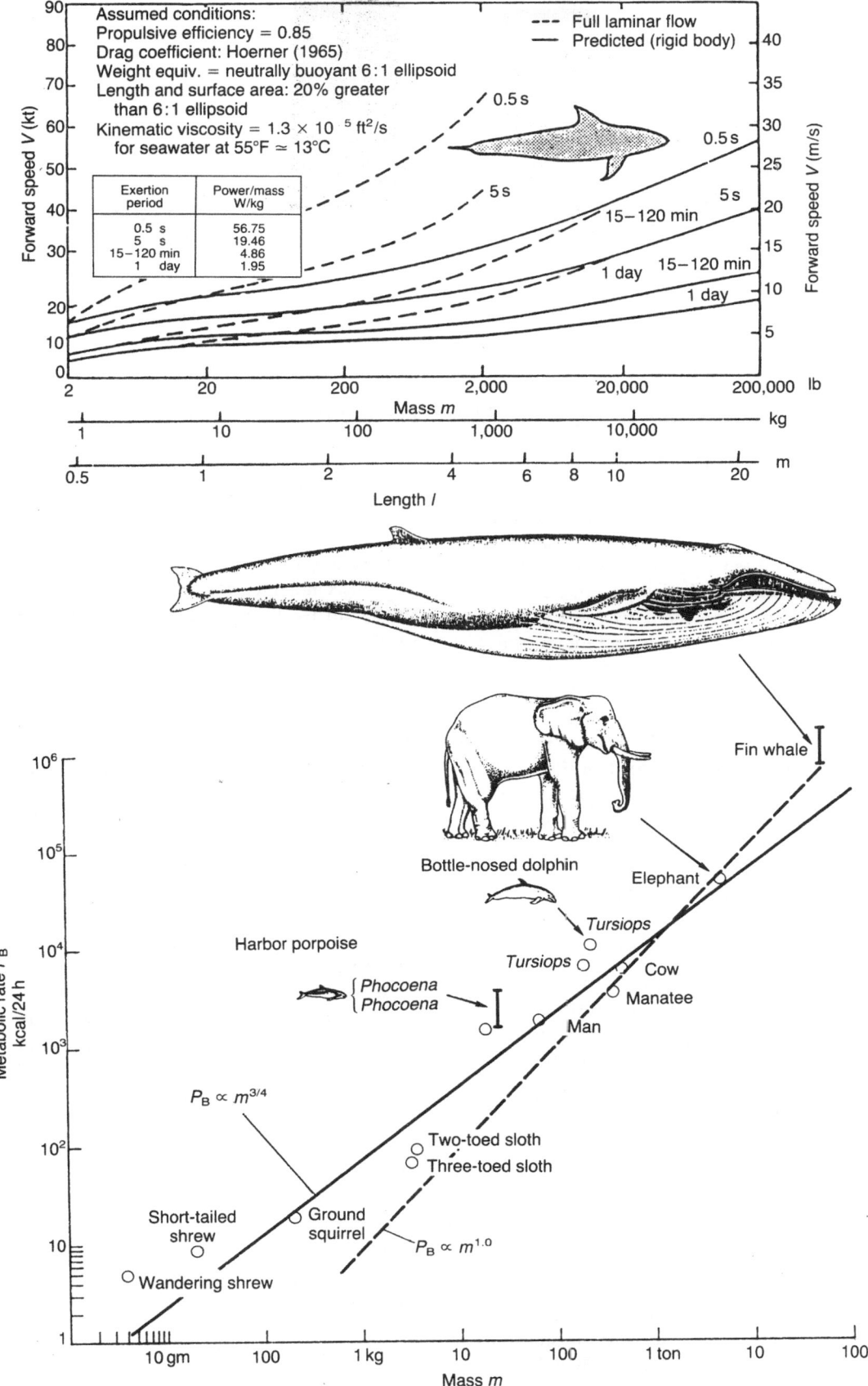

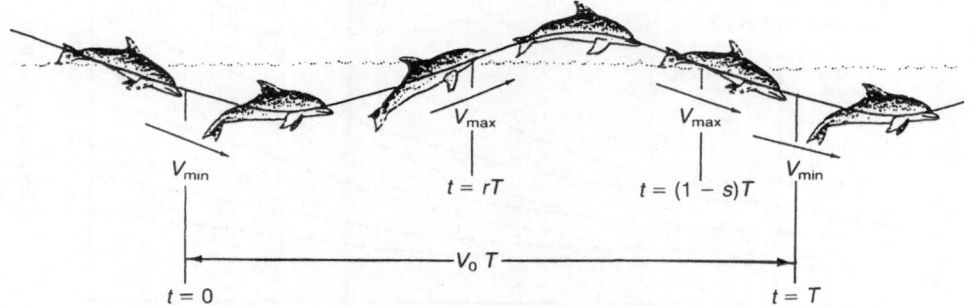

Fig. 6.4-3. Porpoising. (Redrawn from Whitehead 1985).

eral Dynamics towing tank at San Diego (Lang and Daybell 1963). Notty was trained to swim at maximum speed, glide through hoops, and swim with maximum effort while wearing collars of various thicknesses. From Table 6.4-1, it will be noted that the high-speed swimming at faster than 10 m/s was limited to short bursts of about 7 s in duration. The resulted drag coefficient based on the square-third power of the volume, $V_B{}^{2/3}$, was 0.028, which is close to the estimated value of 0.031, and the specific power was 70–80 W/kg, which is smaller than Gray's initial assessment and is within a reasonable range for short-duration swimming. Notty helped eliminate "Gray's paradox" regarding the performance of porpoises and dolphins. If during a short period of exertion, porpoises and dolphins can produce more power than humans, Gray's paradox can be cleared up even without considering the reduced drag caused by the laminar flow (Lang 1966; Hui 1987b).

6.4.3 PORPOISING AND BREACHING

Cetaceans engage in many activities that involve leaping out of the water. Two distinct ones are porpoising and breaching. The former is known as an action for breathing, and probably for power saving in dolphins, but the latter, which is observed in some whales, has not been clearly explained. *Porpoising.* As shown in Fig. 6.4-3, porpoising is a series of horizontal leaps above the water surface made while traveling fast. Theoretical analyses on the effectiveness of porpoising in dolphins have been performed by Au and Weihs (1980a) and R.W. Blake (1983b). However, as mentioned by Gordon (1980), the energy required for leaping against gravity is almost completely recovered and the energy loss is limited only to the splashing on penetration of the water (Au and Weihs 1980b).

If the swimming of a porpoise is performed in the water in a trimmed buoyant condition, then the necessary power $P_{n,0}$ can be obtained from Eq.

(6.3-21) as follows:

$$P_{n,0} = P_p + P_0 + P_m \qquad (6.4\text{-}3)$$

where again P_p, P_0, and P_m are the parasite, profile and miscellaneous power respectively. Then, the steady thrust T_0 opposing the parasite drag and the induced drag represented by the miscellaneous power can be expressed as follows:

$$T_0 = \tfrac{1}{2}\rho_w V^2{}_0 f + P_m/V_0 = (P_p + P_m)/V_0 \quad (6.4\text{-}4)$$

where the drag area f is considered to be a function of the swimming depth (Blake 1983b) but independent to the steady velocity V_0.

If, however, the porpoise swims in the water with a constant forward acceleration α during time fraction r and spends the rest of the time by jumping into and flying through the air primarily for respiration, and by penetrating the water for acceleration following deceleration, then is it possible to save the power?

As shown in Fig. 6.4-3, let us assume that the acceleration in the water initiates with the minimum speed V_{min} at time $t = 0$ and ends with the maximum speed V_{max} at time $t = rT$, where T is a time period of this process. Then the following relation results

$$V_{max} = V_{min} + \int_0^{rT} \alpha \, dt = V_0 + \Delta V = V_{min} + 2\Delta V$$

$$(6.4\text{-}5)$$

where V_0 and ΔV are a steady cruising speed in continuous swimming and a small increment of the speed in the leaping process, respectively,

$$\left.\begin{array}{l} V_0 = (V_{max} + V_{min})/2 \\ \Delta V = (V_{max} - V_{min})/2 \end{array}\right\} \qquad (6.4\text{-}6)$$

During this acceleration the thrust must be increased by a thrust factor n times that of the steady continuous swimming T_0, and thus its increment opposes the parasite and induced drags and approximates the inertial force of the body mass m and the added mass A_{11} as follows:

$$nT_0 = \tfrac{1}{2}\rho_w V_0^2 k_1 f + n^2 P_m/ \left.\begin{array}{c} V_0 + 2(m + A_{11})\Delta V/rT \\ = (k_1 P_p + n^2 P_m)/V_0 \\ + \{2(m + A_{11})V_0/rT\}\Delta V/V_0 \end{array}\right\} \quad (6.4\text{-}7)$$

where the miscellaneous power P_m is considered to be multiplied by n^2 (like the load factor given in Sect. 4.3.3) and the drag area f is modified by multiplying by a mean attenuate factor k_1 for the depth change. At the time of leaping, the added mass A_{11} will be sprayed as water particles in the air.

After leaping with maximum speed V_{max}, the porpoise makes a nearly parabolic orbit and penetrates the water again with the same maximum speed V_{max}, because the drag in the air is very small and the energy loss is negligible.

After re-entry, the kinetic energy of the dolphin $\tfrac{1}{2}mV^2_{max}$, will be reduced because of (1) the kinetic energy loss due to the added mass $\tfrac{1}{2}A_{11}V^2_{min}$ (2) the energy loss due to the mean hydrodynamic drag $\tfrac{1}{2}\rho V^2_0 k_s f = k_s T_0$, where k_s is another mean attenuate factor for drag reduction caused by the depth variation from that of continuous swimming, and (3) also because of the energy dissipated in the splash, E_s. Thus, the speed of the porpoise becomes the minimum value of V_{min} within the time sT as follows:

$$\tfrac{1}{2}mV^2_{max} = \tfrac{1}{2}(m + A_{11})V^2_{min}$$
$$+ (\tfrac{1}{2}\rho_w V^3_0 k_s f)sT + E_s \quad (6.4\text{-}8a)$$

Substituting Eq. (6.4-5) yields

$$\Delta V/V_0 = \{\tfrac{1}{2}A_{11} + \tfrac{1}{2}\rho_w V_0 k_s fsT + E_s/V^2_0\}/$$
$$(2m + A_{11}) \quad (6.4\text{-}8b)$$

The above velocity difference must be recovered in the acceleration phase. Combining the above equation with Eg. (6.4-7) further yields

$$n = 1 - \frac{\{(1 - k_1)P_p + (1 - n^2)P_m\}}{(P_p + P_m)} \left.\begin{array}{c} \\ + \left\{\dfrac{1 + (A_{11}/m)}{1 + \tfrac{1}{2}(A_{11}/m)}\right\} \\ \cdot \left\{\dfrac{(\tfrac{1}{2}A_{11}V_0^2 + E_s) + k_s P_p sT}{rT(P_p + P_m)}\right\} \end{array}\right\} \quad (6.4\text{-}9a)$$

or

$$r = \left\{\dfrac{1 + (A_{11}/m)}{1 + \tfrac{1}{2}(A_{11}/m)}\right\}$$
$$\cdot \left[\dfrac{\{(\tfrac{1}{2}A_{11}V_0^2 + E_s)/T\} + k_s s P_p}{P_p(n - k_1) + P_m n(1 - n)}\right] \quad (6.4\text{-}9b)$$

Here, the time fraction of the acceleration r is in a linear relation with the time fraction of deceleration s and in an inverse relation approximately with the thrust factor n. It can be seen that for a given body configuration and muscle of porpoise, the thrust or the acceleration is limited by a maximum value n_{max} and thus the minimum time fraction r_{min} is specified from the above equation.

The energy spending for one complete cycle in the continuous swimming E_c is given by

$$E_c = (P_0 + P_p + P_m)T. \quad (6.4\text{-}10)$$

On the other hand, the energy for porpoising E_1 is consumed only in the accelerative swimming and is thus given by

$$E_1 = P_0 rT + (nT_0 V_0)rT \left.\begin{array}{c} \\ = \{P_0 + k_1 P_p + n^2 P_m\}rT + \left\{\dfrac{1 + (A_{11}/m)}{1 + \tfrac{1}{2}(A_{11}/m)}\right\} \\ \cdot \{(\tfrac{1}{2}A_{11}V_0^2 + E_s) + P_p k_s sT\}. \end{array}\right\} \quad (6.4\text{-}11)$$

In the above equation the attenuate factors k_1 and k_s are considered to be less than one whenever the porpoising is performed at a depth deeper than that of the continuous swimming. Then, the mean power difference in one complete cycle of swimming can be given by

$$(E_c - E_1)/T \left.\begin{array}{c} \\ = P_0(1 - r) + P_p(1 - k_1 r) + P_m(1 - n^2 r) \\ - [\{1 + (A_{11}/m)/\{1 + \tfrac{1}{2}(A_{11}/m)\}] \\ \cdot [\tfrac{1}{2}A_{11}V_0^2 + E_s)/T + sk_s P_p] \end{array}\right\} \quad (6.4\text{-}12)$$

Although the thrust factor n given by Eq. (6.4-9) is not completely subtracted from the above equation, it can be said that for making $(E_c - E_1)/T$ positive or the porpoise economical, the following items are important: (1) the time fraction for acceleration r should be as small as r_{min}, (2) the body shape must be slender in order to make the added mass A_{11} and the drag area f small, and (3) the dissipated energy for the splash E_s must also be small.

By assuming small s and A_{11}/m, Eq. (6.4-12) gives a critical speed V_0^* beyond which porpoising is economical,

$$V_0^* \simeq \left[\frac{(E_s/T) - P_0(1 - r) - P_m(1 - n^2 r)}{\frac{1}{2}\rho_w f(1 - k_1 r - k_s s)} \right]^{1/3}$$

(6.4-13)

where again the thrust factor n is not subtracted but is considered a constant such as $n = n_{max}$ and the time fraction is specified by $r = r_{min}$. The above equation states that the critical speed V_0^* must be large for violent splashing with a small cycle duration and for small drag area. As will be mentioned later in Sect. 7.4.2, king penguins also perform porpoise leaps.

Gray mullet and flying fish do not use this power-saving method of swimming, presumably because of (1) physiological factors, such as an inability to breathe air efficiently, and (2) the danger of attack by airborne predators.

Breaching. Breaching is an action in which a whale emerges from and falls back into the water. First, the whale swims horizontally until it has developed enough speed. Then, it tilts its head upward and raises its flukes or tail. These actions convert the horizontal momentum into vertical momentum and the whale emerges from the water (Whitehead 1985). Most humpback whales, *Megaptera novaeangliae*, leave the surface of the water at an angle of about 35° with a speed of about 7.5 m/s or 15 knots, and twist by flailing "flippers" and land on the back. This is what Whitehead called a true breach. He also pointed out another form of breaching, called the belly flop, in which the whale remains dorsal side up throughout the breach and lands on its belly. As far as is known, the whale gains no hydrodynamic advantage from breaching.

Wave-Making resistance. As a swimming object approaches the water surface, its drag increases to many times its normal value. This drag increase results from wave making and is expressed as a function of the Froude number, $Fr = V/\sqrt{gl}$, and submergence ratio h/l. Figure 6.4-4 gives a theoretical prediction (Wigley 1953) of the magnitude of the wave drag on slender spheroidal bodies. It can be seen that: (1) the wave drag is maximum at the surface ($h/l = 0$); (2) the wave drag decreases sharply with submergence, $h/l > 0.6$ or $h/d > 5.0$; and (3) the wave drag increases at a certain Froude number, the "critical Froude number," $Fr \simeq 0.6$.

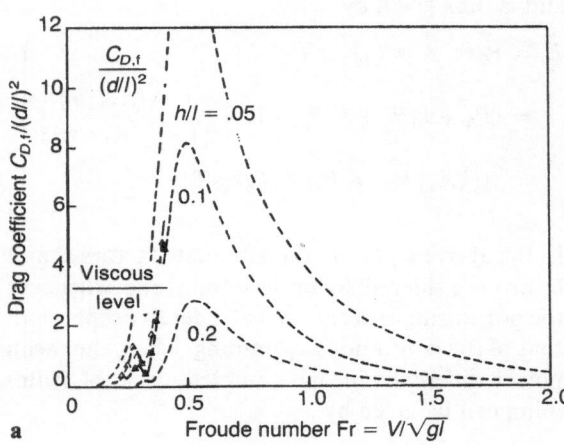

a

Fig. 6.4-4. **a, b.** Wave making drag of submerged streamlined bodies based on frontal area. (Hoerner 1965). **a** Effect of Froude number. **b** Effect of submergence.

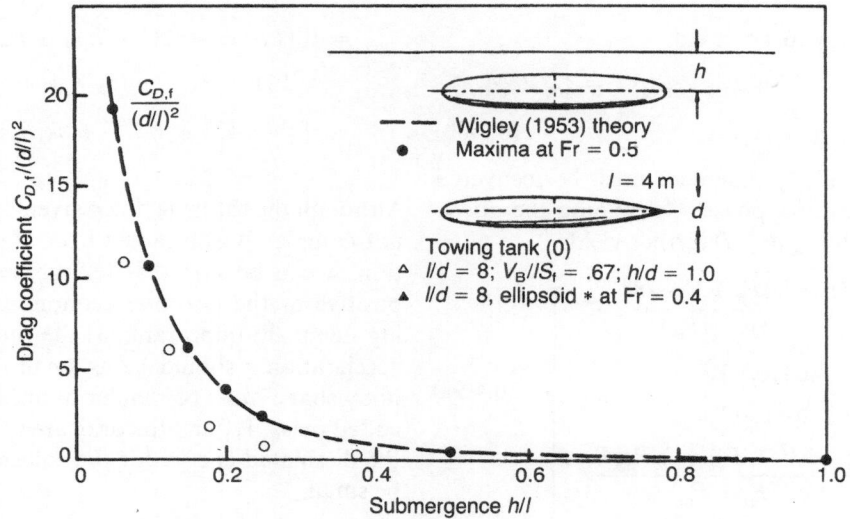

b

CHAPTER 7

Swimming by Other Methods

Eight primitive but fundamental ways of locomotion, paddling, whipping, jetting, sweeping, beating, sailing, skating, and wave riding are presented and analyzed in this chapter. These swimming methods can be performed with simple swimming devices and mechanics, which accounts for their prevalence among swimming animals. The animals studied in this chapter are limited to those considered large enough for the inertial force to be positively utilized. Like the pressure drag, the inertial force is a reaction force of fluid motion but it is accompanying the acceleration or unsteady motion of swimming devices. Thus, paddling, whipping, and jetting are more effective for sudden starts from rest than for continuous propulsion. Sweeping is a mode of locomotion in which a wing or blade is swept in its own plane, whereas beating, as described earlier in Chap. 4, is a mode of locomotion in which a wing is flapped up and down, out of its own plane.

With regard to the three other types of locomotion on the water surface also discussed in this chapter, (1) sailing gets the driving force from wind, and it is possible to travel at an angle to the wind, using some device similar to the centerboard of a sailing boat to prevent lateral drift; (2) skating gets the driving force from the action of legs that also support the weight passively with the assistance of the surface tension of the water; and (3) wave riding utilizes the horizontal component of the buoyant force acting normal to the free surface of a surface wave: the vertical component of the buoyant force supports the weight, and the direction of movement is limited within a fairly narrow range of angles centered on the wave direction.

7.1 Paddling and Whipping

Unlike snaking, which is a continuous movement, paddling and whipping are reciprocal motions consisting of a power stroke and a recovery stroke. In the power stroke, a paddle is fully extended and driven backward to generate forward thrust, while in the recovery stroke, the paddle is returned to its initial position through the water or the air, with its frontal area reduced so as to lower unwanted drag. On the other hand, whips are sometimes used evenly in order to generate thrust in the recovery stroke too.

7.1.1 PADDLING MECHANISM

Steady translation. Let us consider a boat driven by a three-dimensional paddle, as shown in Fig. 7.1-1, and moved with a constant velocity V. The boat can actually stand in for any living creature, such as waterfowl swimming in the water. To simplify the following calculations, the paddle is assumed to move backward along a straight line with a constant velocity U.

The steady driving force T_D generated by the fluid dynamic drag of the paddle, which is proportional to the dynamic pressure of the relative speed $U - V$, can be expressed in the formula

$$T_D = \tfrac{1}{2}\rho(U - V)|U - V|S_f C_{D,f} \quad (7.1\text{-}1)$$

where S_f and $C_{D,f}$ are the frontal area and the drag coefficient of the paddle. Here again the induced velocity, being small, has been neglected. For a positive thrust, the rowing speed U must be larger than the boat speed V, or $U > V$. The power P required to drive the paddle is given by

$$P = T_D U = \tfrac{1}{2}\rho U(U - V)|U - V|S_f C_{D,f}. \quad (7.1\text{-}2)$$

In the power stroke at constant boat speed V, the hull drag of the boat D, which is equal to the

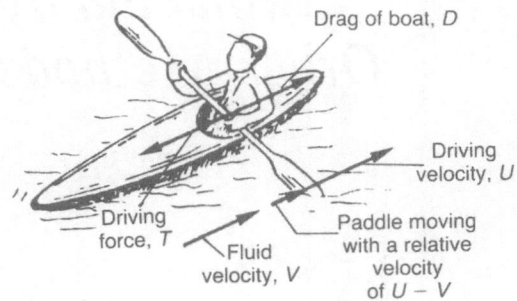

FIG. 7.1-1. Operation of linear paddle motion.

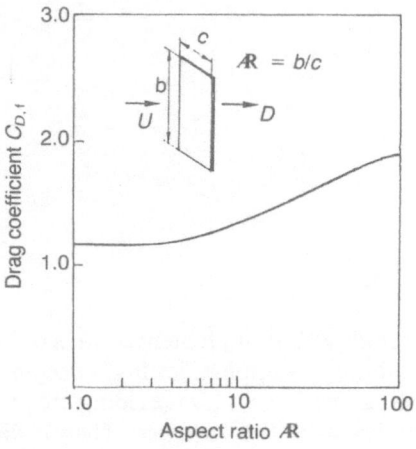

FIG. 7.1-2. Drag coefficient of rectangular, flat plate in normal flow.

driving force T_D, and the necessary power for the parasite drag, or the useful power of the boat W, are respectively given by

$$D = \tfrac{1}{2}\rho V^2 f = T_D \qquad (7.1\text{-}3)$$

$$W = DV = T_D V = \tfrac{1}{2}\rho V^3 f \qquad (7.1\text{-}4)$$

where f is the drag area of the boat.

Thus, the efficiency η, i.e., the ratio of the useful power to the necessary power for the parasite drag, can be given as follows:

$$\eta = W/P = V/U = 1/\{1 + \sqrt{f/S_f C_{D,f}}\} \quad (7.1\text{-}5)$$

This equation states that in a steady driving system, (1) the efficiency η will approach 1 as the driving speed of the fluid approaches the speed of the boat, $U \to V$; and (2) this can be attained by reducing the drag area of the boat compared with the drag area of the paddle as much as possible, $f/S_f C_{D,f} \to 0$.

The actual value of the drag coefficient of a paddle C_D is unsteady and strongly dependent on the paddle planform and concavity. The mean drag coefficient of a rectangular flat plate in normal flow is, as shown in Fig. 7.1-2, a function of the aspect ratio of the plate and the Reynolds number. It changes from $C_{D,f} = 1.2$ at $\mathcal{R} = \pi/4$ to $C_{D,f} = 2.0$ at $\mathcal{R} = 100$. However, as pointed out later, the total hydrodynamic force consisting of the lift as well as the drag is not necessarily small for a low aspect ratio plate operated obliquely. The drag coefficient of other two- and three-dimensional shapes at high Reynolds number are given in Fig. 7.1-3. As Fig. 7.1-4 shows, the drag coefficient of a concave plate increases as the height ratio h/d increases, from $C_{D,f} = 1.17$ at $h/d = 0$ (disk) to the maximum value of $C_D = 1.42$ at $h/d \simeq 0.5$ (hemisphere), and then falls off again to the theoretical minimum value of $C_{D,f} = 1.0$.

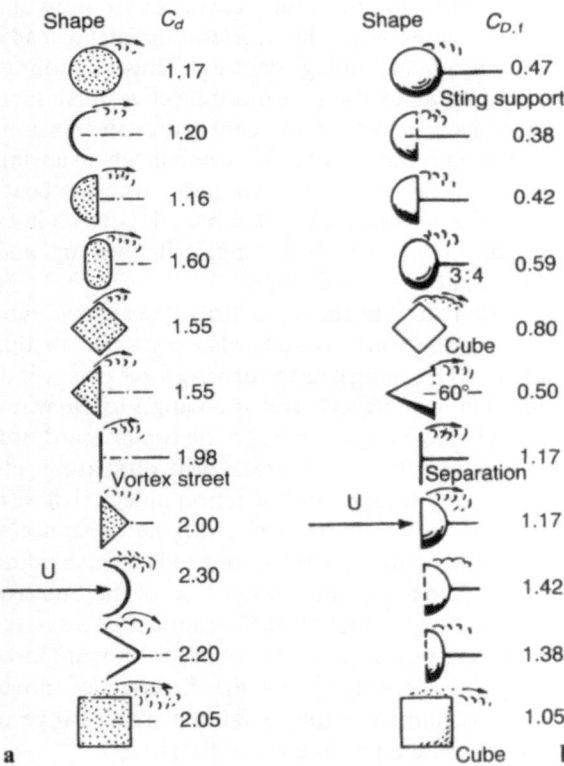

FIG. 7.1-3. **a, b.** Drag coefficients of various two- and three-dimensional bodies (Hoerner 1965). **a** Two-dimensional bodies. **b** Three-dimensional bodies.

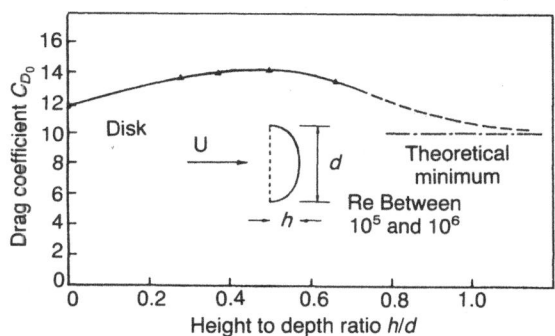

FIG. 7.1-4. Drag coefficient of sheet-metal caps as a function of their height ratio. (Hoerner 1965).

If the flat plate is replaced with a two-dimensional, extensible membrane, the drag also increases because of the cup-shaped profile. Figure 7.1-5 shows the two-dimensional drag as a function of the "excess-length ratio," $\varepsilon = (l - c)/c$, in which l and c are the length and chord of the membrane. At small values of the excess-length ratio, the drag coefficients of two membranes of different mass experimentally obtained lie on a single curve and increase with increasing concavity of the membrane as the theory proposed by Newman and Low (1981) indicates. However, for a large ε, the drag is greater for the heavier membrane.

The flow is substantially insensitive to change

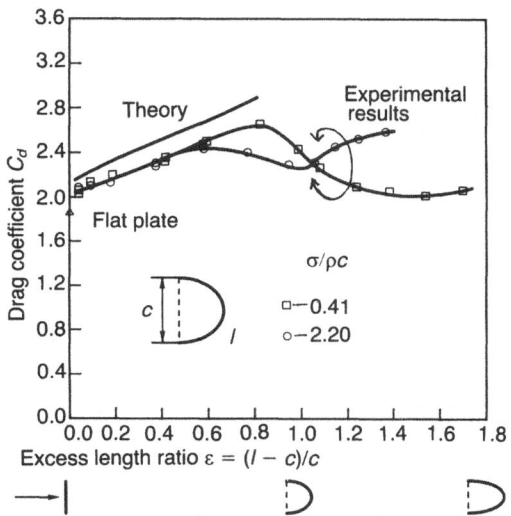

FIG. 7.1-5. Drag coefficient of a two-dimensional membrane. (From Newman and Low 1981 with permission). σ, Mass of membrane per unit area; ρ, density of fluid; c, chord length; l, length of membrane.

in Reynolds number above approximately 10^4 (Hoerner 1965; Newman and Low 1981).

Unsteady translation. Here let us consider a hypothetical creature propelled by a paddle. In its power stroke, the motion consists of three phases: (1) an accelerating phase, (2) a driving phase with nearly constant speed, and (3) a decelerating phase. In the first phase, the propulsive force is generated mostly by the inertial force, since the speed of the paddle is low and the acceleration predominates. In the second phase, which takes up most of the time of the power stroke, the effect of the drag force, which is similar to the vortex induced force specified in Sect. 4.5.4, overshadows the inertial force because the relative speed of the paddle is predominant. In the third phase, the paddle speed is reduced in preparation for the following recovery stroke, which is usually performed with a reduced drag area of the paddle; therefore the inertial force caused by the deceleration acts against the drag force. All of these forces are actually given by a summation of the horizontal components of pressure integration over the whole wetted surface of the paddle and the inertial force of the paddle itself. It should be noted that the inertial force of the added mass of fluid surrounding the paddle is included in the pressure integration.

When a paddle is in linearly accelerated motion in a fluid, an added mass of surrounding the paddle m_w will accompany the paddle and the masses of the paddle and the fluid, m_p and m_w, will generate the driving force T_I as a reaction of the inertial force of the paddle and the fluid, or

$$T_I = (m_p + m_w)(\dot{U} - \dot{V}) \simeq m_w(\dot{U} - \dot{V}) \quad (7.1\text{-}6)$$

where the approximation is obtained by assuming that the mass of the paddle itself is smaller than the added mass of the surrounding fluid, i.e., $m_p \ll m_w$. If the paddle is cup-shaped, the added mass m_w can be approximated by the sum of the mass of the fluid inside the cup m_c and the added mass A_{11} associated with a paddle without concavity. Thus, the added mass $(m_w \simeq m_c + A_{11})$ of a cup-shaped body is probably larger than that of a flat plate, making it also effective for the initial dash. The added mass of the fluid, generally a tensor A_{ij}, for typical configurations is given in Tables 7.1-1 and 7.1-2 (Landweber 1956, Saunders 1957; Landweber and Macagno 1957, 1959, 1960).

In the recovery stroke, the paddle is returned to the original position with its drag area reduced.

TABLE 7.1-1. Added masses and moments of inertia of entrained fluid for two-dimensional body. (Saunders 1957)

Form of two-dimensional body	Added mass of entrained liquid	Added moment of inertia of entrained liquid
Rod of circular section	↕ Mode of motion $A_{22} = \rho\pi a^2$	Mode of motion 0
Rod of elliptic section, Major axis	↔ Mode of motion $A_{22} = \rho\pi(b^2\cos^2\alpha + a^2\sin^2\alpha)$	Mode of motion $A_{44} = \frac{1}{8}\rho\pi(a^2 - b^2)^2$
Long flat plate	↕ $A_{22} = \rho\pi a^2$	$A_{44} = \frac{1}{8}\rho\pi a^4$
Rod of square or rectangular section	$A_{22} = c_1\rho\pi a^2$	$A_{44} = c_4\rho\pi a^4$

a/b	c_1
0.1	2.23
0.2	1.98
0.5	1.7
1	1.51
2	1.36
5	1.21
10	1.14

a/b	c_4
0.1	1470
0.2	9.4
0.5	2.4
1	0.234
2	0.15
5	0.15
10	0.147

Thus, Eq. (7.1-6) is actually valid in the accelerating phase of the power stroke of the paddle. In one cycle of the paddling motion composed of the power and recovery strokes, the first part on the right hand side of Eq. (7.1-6), $m_p\dot{U}$, will vanish, so that the contribution of the paddle mass is canceled, because the paddle motion in one cycle does not result in any shift or movement of the center of gravity of the paddle with respect to the body of a creature moving with a constant speed.

The ratio of inertial thrust T_I and the drag force T_D can be approximately given by

$$T_I/T_D = m_w(\dot{U} - \dot{V})/\tfrac{1}{2}\rho S C_D (U - V)^2 \quad (7.1\text{-}7)$$

For a flat plate of elliptic shape, since the area and added mass are given respectively by $S = \pi ab$ and $m_w = A_{33} = \frac{3}{4}\rho\pi a^2 b k_4$ (see Table 7.1-2), the above equation can be rewritten as follows:

$$T_I/T_D \simeq (8/3)(k_4/C_{D,f})\{a(\dot{U} - \dot{V})/(U - V)^2\}$$
$$= (8/3)(k_4/C_{D,f})(\ddot{x}/\dot{x}^2) \quad (7.1\text{-}8)$$

where a and x are respectively the half span of the major axis and the nondimensional distance with respect to the fluid defined by

$$x = \int (U - V)\mathrm{d}t/a \quad (7.1\text{-}9)$$

As the oblateness ratio increases, the ratio $k_4/C_{D,f}$ decreases appreciably, so that the locomotion must rely on the drag force exclusively. However, if a sudden initial dash is required, the oblateness has to be close to 1.

Effects of water surface. When a paddle approaches and pierces the water surface, a water spray is formed at the front side of the paddle, and a cavity filled with air is generated at the rear side, as shown in Fig. 7.1-6. For the surface-piercing state or "ventilating condition" of a rectangular plate (Hoerner 1965), the drag increases by

$$\Delta C_{D,f} = 0.3b/h + 1/\mathrm{Fr}^2 \quad (7.1\text{-}10)$$

where Fr is the Froude number based on the submerged height h, or

$$\mathrm{Fr} = V/\sqrt{gh} \quad (7.1\text{-}11)$$

and b is the width of the plate. On the right-hand side of Eq. (7.1-10), the first and second terms give the spray effect and ventilation effect respectively.

Representative paddling methods. Paddling as a mode of locomotion is often encountered in aquatic insects, as well as some mollusks and a few species of slow-moving fish.

WATER BEETLES. It is interesting to note that many aquatic larvae and some adult insects, typically

TABLE 7.1-2. Added masses and moments of inertia of entrained fluid for three-dimensional body. (Saunders 1957)

Form of three-dimensional body	Added mass of entrained liquid	Added moment of inertia of entrained liquid
Sphere $V_B = \frac{4}{3}\rho\pi a^3$ $2a$	$A_{11} = \frac{1}{2}\rho V_B$ $= \frac{2}{3}\rho\pi a^3$	$A_{44} = 0$
Circular disk moving normal to its plane or rotating about a diameter $2a$	Mode of motion $A_{11} = \frac{8}{3}\rho a^3$	Mode of motion $A_{55} = \frac{16}{45}\rho a^5$
Radius b a a Prolate ellipsoid	Moving broadside $A_{33} = k_2(\frac{4}{3}\rho\pi ab^2)$	$A_{55} = k_5(\frac{4}{15}\rho\pi ab^2)(a^2 + b^2)$

a/b	k_2	k_1	k_5	a/b	k_2	k_1	k_5
1.0	0.5	0.5	0	4.99	0.895	0.059	0.701
1.5	0.621	0.305	0.094	6.01	0.918	0.045	0.764
2.0	0.702	0.209	0.240	6.97	0.933	0.036	0.805
2.51	0.763	0.156	0.367	8.01	0.945	0.029	0.840
2.99	0.803	0.122	0.465	9.02	0.954	0.024	0.865
3.99	0.860	0.082	0.608	9.97	0.960	0.021	0.883
				∞	1.0	0	1.0

$$A_{11} = k_1(\frac{4}{3}\rho\pi ab^2)$$

Form of three-dimensional body	Added mass of entrained liquid	Added moment of inertia of entrained liquid
Planetary ellipsoid or oblate spheroid moving parallel to its polar axis a Radius a b	$\frac{3}{4}\rho\pi a^3\left[\dfrac{e - (\sin^{-1}e)\sqrt{1-e^2}}{\sin^{-1}e - e\sqrt{1-e^2}}\right]$ where $e = \sqrt{1 - \dfrac{b^2}{a^2}}$	
Elliptic disk moving perpendicular to its plane b a	$A_{33} = k_4(\frac{4}{3}\rho\pi a^2 b)$, $k_4 = \dfrac{1}{\displaystyle\int_0^{0.5\pi} \sqrt{\left(\dfrac{a}{b^2}\sin^2\theta + \cos^2\theta\right)}\,d\theta}$	

a/b	1.0	2.0	5.0	10.0
k_4	0.637	0.41	0.19	0.098

represented by the silver water beetle (*Acilius sulcatus*), take a bubble of air down with them for respiration. These bubbles are always attached to the spiracles. The oxygen in the bubbles is used for breathing and is continuously supplied from the surrounding water by diffusion. The entrained rate of any gas is so large in undersaturated conditions that the insects can continue breathing from the air bubble for some time. However, the bubble slowly decreases in size until the insect has to surface either to get a new bubble or to replace the air in the bubble.

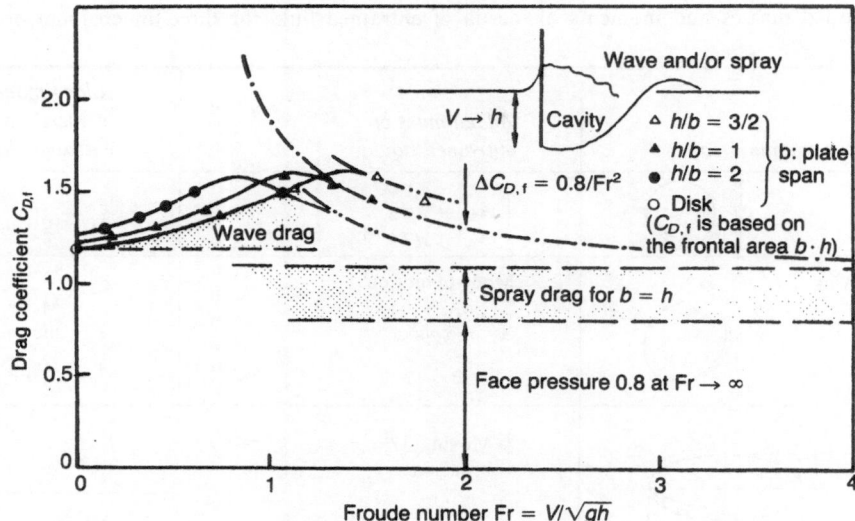

FIG. 7.1-6. Drag coefficient of surface-piercing flat plates. (Hoerner 1965).

The hindlegs of the water beetle not only are broadly flattened but also have many rowing appendages or hairs for effective paddling in the water. These legs and appendages are fully stretched and mechanically arranged to move with the flattened side vertically during the power stroke. In the recovery stroke, the broad sides are turned up 90° and drawn forward by folding their tarsal sections and sliding slowly along the ventral side to reduce unfavorable drag.

Nachtigall (1980a) conducted wind tunnel tests on four species of water beetles and found that at the Reynolds number of $Re \simeq 10^4$, which corresponds to swimming at a high speed of about 0.3 m/s, the drag coefficients based on the frontal area, between $C_{D,f} = 0.38$ and 0.43, were well below that of the sphere, $C_{D,f} = 0.47$, at the same Reynolds number.

BACKSWIMMERS. The backswimmer or water boatman, *Notonecta* of Heteroptera, can swim on its back with the hindlegs serving as oars, in the manner of a water beetle and rests at an angle of 30° to the water surface. As it kicks with its swimming legs, this angle increases to 55°, so that the insect penetrates the water (Chapman 1969).

7.1.2 SERIAL PADDLING

A series of paddling motions can be seen in the swimming of many species of Crustacea. These species usually execute a series of whipping or paddling motions with fringed bristles or webbed legs called swimmerets. In some cases the locomotion resembles the oar-propelled motions of a

rowboat, but the description of this mode will be saved for the following section.

Similar to the whipping motion of cilia and flagella, which was explained in Sect. 5.1.4, paddling is very simple and effective, especially in small animals moving with low speed or at a low Reynolds number. In the mechanical sense, the motion is not different from the whipping motion mentioned earlier. Here, it is related to the pressure drag as well as to the inertial force, rather than to the friction drag as in cilial motion.

The mechanical aspect of a series of paddling motions is such that in the power stroke the respective paddles are driven in the opposite direction to the body motion. In order to send the mass of water backward effectively, the phase of the respective paddles must be adjusted so that the wave generated by the envelope of the series of paddle tips appears to move forward. In the recovery stroke the paddles are moved forward so as to avoid undesirable pressure drag.

Members of the Euphausiacea, Decapoda, and Stomatopoda families have many legs for walking on the sea- or riverbed and for paddling in the water. In the spiny lobster (*Panulirus*), the walking legs have pointed tips for digging into the bottom when walking against the water current. In contrast, the swimming abdominal legs, called pleopods, are formed as shallow caps to increase the hydrodynamic drag and inertial force during the

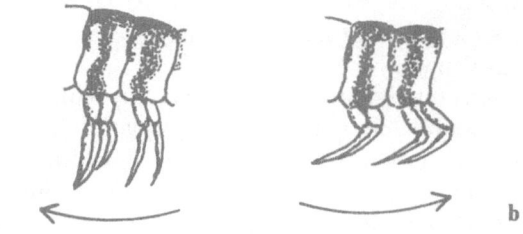

FIG. 7.1-7. **a–c.** Swimming by pleopods of prawn. **a** Power stroke. **b** Recovery stroke.

power stroke. In the case of the prawn (*Peneus*), as shown in Fig. 7.1-7, each of five pairs of pleopods has two hinges and moves in an extended state like a paddle for the power stroke, and in a retracted state for the recovery stroke.

The prawn can also escape from danger with a single quick ventral flexure of its tail fan, which is formed from appendages of the last abdominal somite. The speed of the backward spurt resulting from this motion ranges from 1.2 m/s to 8 m/s for lobsters (Lindberg 1955; Herrick 1985) and generates an acceleration of about 10 G in the common dock shrimp (Daniel and Meyhöfer 1989).

7.1.3 PADDLING OF FEET, LEGS, AND TAIL

Birds. The feet of waterfowl usually have either long toes with a broadened and flattened under-side for walking, or webbed toes for swimming. The former type of foot is convenient for walking on soft and marshy places under shallow water and, for some birds like rails and jacanas, on floating vegetation. The latter type is adapted to produce a driving force by using the drag and inertial force acting on the web during its motion. Some birds, like the grebes, have lobed toes rather than webs. During the driving stroke, the mechanism of force generation is very similar to that of a boat oar, except that the motion is performed in a vertical plane rather than a nearly horizontal one, and the recovery stroke is carried out not in the air but in the water, with the webs narrowed and tightly folded.

It is generally true for any swimming system that the propelling devices should be disposed at the rear portion of the body so as to avoid the adverse effect of the slip stream generated by the devices, and to prevent flow separation by acceleration of the fluid around the stern. This arrangement, however, is not suitable for aquatic species in balancing the center of gravity when they come ashore. Thus, they paddle with their feet kept directly under the body when floating (Fig. 7.1-8a). Then the moment variation is generated to raise the head during the power stroke and lower it during the recovery stroke. This induces a slight change of body attitude that is balanced with the moment variation by a shift of the center of

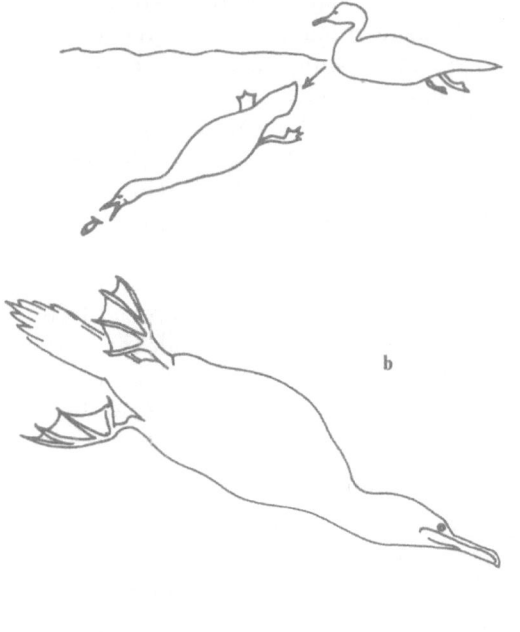

FIG. 7.1-8. **a–b.** Driving modes on and in the water. (From Nelson 1980) **a** Shearwater. **b** Shag.

buoyant force. For this purpose, waterfowls have large rumps covered thickly with plumage that holds in air bubbles for buoyancy as well as for insulation.

In diving, however, they usually paddle with their feet at a rear portion of the body, as shown in Fig. 7.1-8b, and keep the thrust line pointed to the center of gravity so as not to generate any moment during either the driving stroke or the recovery stroke.

In penguins this problem is solved by the positioning of the feet at the rear of the body. However, this arrangement requires the birds to stand upright on land and makes them vulnerable to sudden attack by their enemies. It is thus not surprising that they live in places such as the Antarctic Continent where there are fewer dangerous assailants and where they are able to escape rapidly by "tobogganing."

Frogs. The frog (Anura) is a small amphibian vertebrate characterized by short forelegs and long hindlegs that are adapted for leaping and swimming and have webs between the toes. The frog swims by paddling its flexible hindlegs and webbed feet. Its movements are quite different from the rowing of a boat oar consisting of a stiff, solid blade. The articulated legs and feet of the frog are fully extended in the power stroke, but folded and narrowed in the recovery stroke.

Pinnipedia. Pinnipedia make up an order of mammals that have re-adapted to the marine environment; they include seals, sea lions and walruses. The 32 species of pinnipeds are divided into two super families, Otariidae and Phocidae (Coffey 1977). While pinnipeds have streamlined bodies that are well adapted for aquatic life, they can also stay and move about on land. The fore- and hindlimbs are modified into flat flippers. The front flippers are typically long, and it is their beating that provides the main means of locomotion. The back flippers are shaped and positioned to form a continuous contour, and the tail fits well between them, except in the walrus, which has a web joining the back flippers completely into the tail.

7.1.4 WHIPPING OF LEGS AND TAIL

There are many animals that depend for locomotion on the whiplike motion or whipping of legs and tail. Mechanically this kind of locomotion is not different from the paddling of feet discussed in the preceding section and is also similar to a single

stroke of the snaking of an almost straight (but flexible) plate although we cannot see a complete wavelength.

Otters. Otters, members of the family Mustelidae, are completely adapted to an aquatic life. Their head is flattened, their vibrissae or whiskers are as stiff as in seals, and their ears are small and provided with a membrane to close them when they dive. The five-toed feet are webbed for three-quarters of their lengths and have nonretractile claws (Burton 1980).

Swimming is mainly performed by the paddling of webbed feet. The broad and flattened tail is said to be used for steering in water. However, it can also cooperate with the body in sinuous swimming movements. This motion is closer to whipping than paddling.

The sea otter also swims by paddling webbed hindfeet and whipping a small flattened tail. For precise control of the body motion during flotation on the water surface, the tail is used as a *kai*, a short blade used in oriental boats, the function of which will be described later.

Razor shell or jackknife clam (Ensis). The muscular foot of mollusks, with its primitive, broad "sole," is not only responsible for the temporary attachment of these animals to a hard stratum, but is also their prime means of locomotion. Most bivalves burrow into soft mud or sand and live well below the surface. The razor shells have perfected the art of burrowing. Their way of swimming is to bend an extended foot and whip it repeatedly (Fig. 7.1-9): by which the hydrodynamic drag and inertial forces of added mass acting on the foot

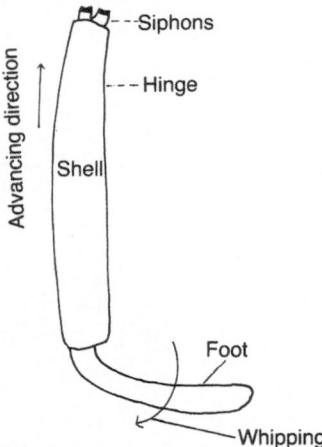

FIG. 7.1-9. Whipping of foot in a razor shell.

propel the body along its longitudinal axis. The motion is whipping rather than the paddling. At the individual stages of the whipping motion the shell can propel its body in the longitudinal direction.

7.2 Jetting

Jet propulsion is widely used by a variety of the larger aquatic animals for normal swimming or for an emergency dash and escape. Unlike microorganisms, the Coelenterata and Mollusca treated here are not so small, and their typical Reynolds numbers in swimming are higher than $Re = 10^3$. The inertial force is thus as important a term in the fluid-dynamic analyses as the viscous force. The most common species of such marine animals are the cephalopods, particularly the squids, octopuses, and cuttlefish, and medusae like jellyfish and scallops. They propel themselves by drawing in and jetting out fluid from a device made of a cavity fitted with a nozzle or nozzles. It is interesting that this type of device, though also found in dragonfly larvae and some fish, is best developed in soft-bodied invertebrates.

7.2.1 MOMENTUM BALANCE

Propulsion system with single nozzle. Let us consider a squirting device like the one illustrated in Fig. 7.2-1. The device can draw in a certain amount of fluid from a nozzle and exhaust it through the same nozzle by contracting the cavity.

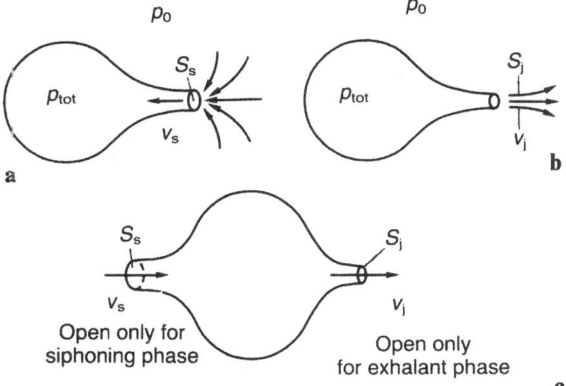

FIG. 7.2-1. **a–c.** Jet propulsive devices. **a** Siphoning phase for single nozzle: $t < t_s$. **b** Exhalent phase for single nozzle: $t_s < t < t_s + t_j$. **c** Siphoning and exhalant phases for dual nozzles.

By assuming, as an ideal case, that (1) the fluid is inviscid and incompressible, and has a density ρ that is constant; (2) the flow in the nozzle is homogeneous with a constant speed or a constant rate of mass; (3) the pressure is balanced with the static pressure p_0 of the surrounding fluid; and (4) the device is at rest at first and the attained speed is very low in comparison with the flow speed, it is possible to obtain the following momentum balance equations for a system consisting of the device and the related fluid: Drawing or "inhalant phase," or "siphoning." At an early initial stage of the siphoning, the fluid is accelerated from rest and sucked into the cavity through a nozzle of cross-sectional area S_s at a mean speed of v_s and a rate of mass $\dot{m}_s$, as shown in Fig. 7.2-1a. Then the device, the mass of which is m, is pulled to the nozzle side, i.e., the direction opposite to the fluid flow, by the instantaneous sucking force F_s given by

$$F_s = \dot{m}_s v_s = \rho S_s v_s^2 \qquad (7.2\text{-}1)$$

or its integration over time, the "impulse"

$$I_s = \int_0^{t'_s} F_s \, dt = \dot{m}_s v_s t'_s = m'_s v_s \qquad (7.2\text{-}2)$$

where

$$\dot{m}_s = \rho S_s v_s \quad \text{and} \quad t'_s = m'_s / \dot{m}_s \qquad (7.2\text{-}3)$$

and where t'_s is some intermediate time far before the cavity is filled with the fluid and m'_s the mass of fluid sucked into the cavity at this time. Then, the equation of the total momentum of the system, which consists of the momentum of the device itself mU_1 and that of the entrained fluid $m'_s v_s$ is given by

$$m'_s v_s + mU_1 = 0 \quad \text{or} \quad U_1 = -(m'_s/m)v_s \qquad (7.2\text{-}4)$$

The above equation states that the velocity of the device, which is propelled in the opposite direction to the sucking velocity, is proportional to the sucking speed v_s times the mass ratio m'_s/m.

However, at the late stage of this siphoning, the fluid is stopped by the inner wall of the cavity, by what is called the "water hammer effect," and the device is decelerated by the force acting on the inner wall if the loss of momentum of the fluid decelerated at the wall is larger than the momentum of the fluid accelerated at the nozzle. The cavity is finally filled with the fluid, the total mass of which is m_s. Then, the momentum equation of the system, after the cavity has been filled with the fluid, is given by

$$(m + m_s)U_2 = 0 \quad \text{or} \quad U_2 = 0 \quad (7.2\text{-}5)$$

where U_2 is the finally attained velocity of the device.

During the above process, the fluid is sucked and accelerated from the outside to the inside, or from the right to the left in Fig. 7.2-1a, and decelerated and stopped in the cavity. Corresponding to the above fluid motion, the device itself is accelerated from the left to the right, and then stopped. The distance of movement to the right-hand side, $-X_1$, is then counterbalanced with the translational distance of the center of gravity of the related fluid X_s, having shifted to the left-hand side with the above siphoning process, as follows:

$$-X_1 = (m_s/m)X_s \quad (7.2\text{-}6)$$

These processes are plotted in Fig. 7.2-2 by solid lines within the time of $t \leq t_s$.

EXHAUSTING OR EXHALANT PHASE, OR EJECTING. In "exhalant phase" the fluid drawn into the cavity is ejected from the nozzle with a relative velocity v_j at a minimum cross-sectional area S_j, called "vena contracta" (Lamb 1930), where the pressure in the jet stream is equal to the pressure outside the jet stream p_0 (Fig. 7.2-1b), and the device is pushed left by the instantaneous jet force F_j, which is directed in the opposite direction to the jet flow and is given by

$$F_j = \dot{m}_j v_j = \rho S_j v_j^2 \quad (7.2\text{-}7)$$

or impulse I_j given by

$$I_j = \int_0^{t_j} F_j \, dt = \dot{m}_j v_j t_j = m_j v_j \quad (7.2\text{-}8)$$

where

$$\dot{m}_j = \rho S_j v_j, \quad t_j = m_j/\dot{m}_j, \quad \text{and} \quad m_j = m_s \quad (7.2\text{-}9)$$

and where the jet exhaust area S_j is different from S_s.

It can be said that (1) the force and impulse are respectively proportional to the square and linear velocities at the terminal end of the nozzle; (2) the impulse is proportional to the related mass of the fluid or the capacity of the cavity; and (3) unlike during the inhalant phase, the jet of inviscid fluid will not be stopped after having been ejected.

The equation for the momentum balance of the system is

$$0 = -m_s v_j + \dot{m}U_3 \quad \text{or} \quad U_3 = (m_s/m)v_j \quad (7.2\text{-}10)$$

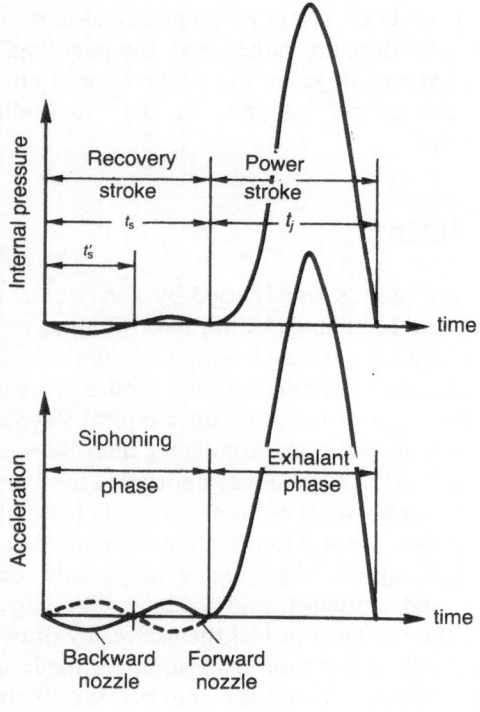

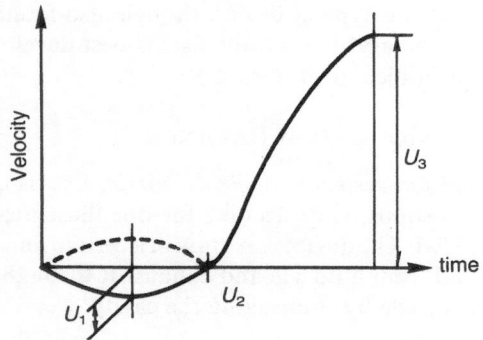

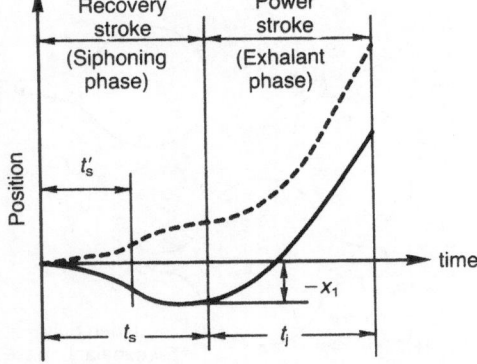

FIG. 7.2-2. A process of jet locomotion from rest.

where U_3 is the finally attained velocity. This velocity is proportional to the product of the jet velocity v_j and the mass ratio m_s/m. It can thus be said that a higher rate of jetting and a larger ratio of mass of the fluid will generate a larger driving force. These processes are also shown in Fig. 7.2-2 by solid lines within the time of $t_s < t \le t_s + t_j$.

By combining the above two phases as a locomotive process, the device can first move right by the distance given by Eq. (7.2-6) in the siphoning phase (the recovery stroke), and then move left with the velocity given by Eq. (7.2-10) in the exhalant phase (the power stroke).

The forces expressed by the momentum change during the siphoning and exhalant phases are actually obtained by the horizontal component of the pressure integration over the internal surface of the cavity. For example, in the exhalant phase, if the cavity is assumed to be so large that the internal pressure p_{tot} can be expressed by Bernoulli's equation:

$$p_{tot} = p_0 + \tfrac{1}{2}\rho v_j^2 \qquad (7.2\text{-}11)$$

then the jet force and impulse in the power stroke are given by

$$\left.\begin{aligned} F_j &= \dot{m}_j v_j = \rho s_j v_j^2 = 2(p_{tot} - p_0)S_j \\ &= 2(p_{tot} - p_0)S(S_j/S) \\ I_j &= 2(p_{tot} - p_0)S_j t_j = 2(p_{tot} - p_0)S(S_j/S)t_j \end{aligned}\right\} \\ (7.2\text{-}12)$$

where S is the cross-sectional area of the exhaust nozzle.

The above equations state that (1) the jet thrust is proportional to the exhaust area of the jet exit S times the increment of internal pressure in the cavity and the area ratio S_j/S (called the coefficient of contraction), and (2) the jet impulse is also proportional to the product of jet duration.

The internal pressure increases by rapid contraction of the cavity, as illustrated by the solid lines in Fig. 7.2-2, and is reduced by its expansion. It must be mentioned, however, that for a given power, if the area of the jet nozzle S increases, the internal pressure must usually decrease.

Let us consider the effect of vena contracta. As shown in Fig. 7.2-3a, at the initial stage of the exhalant phase, the pressure difference between internal pressure and external pressure $(p_{tot} - p_0)$ accelerates the water with the force of $(p_{tot} - p_0)S$, and the device is driven by the reaction to this force. However, as shown in Fig. 7.2-3b, the inter-

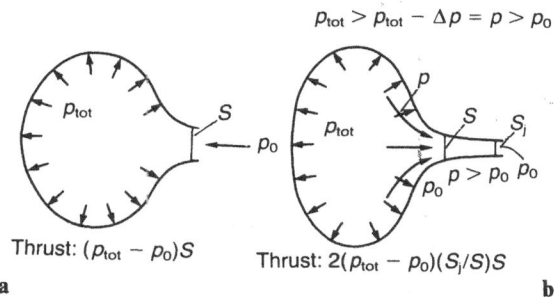

$$p_{tot} > p_{tot} - \Delta p = p > p_0$$

FIG. 7.2-3. **a, b.** Contraction of the jet. **a** Initial stage. **b** During exhalant phase.

nal flow thereafter reduces the pressure around the nozzle and causes flow contraction outside the nozzle. This process generates an increases thrust which is represented by the factor of $2(S_j/S)$ in Eq. (7.2-12). For a circular nozzle, Lamb (1930) gave the coefficient of contraction of $S_j/S = 0.62$ or $2(S_j/S) = 1.24$. This means that the driving force by a circular nozzle attains 1.24 times the static thrust $(p_{tot} - p_0)S$ at the initial stage of the exhalant phase.

Propulsion system with dual nozzles. As shown in Fig. 7.2-1c, if a sucking nozzle is directed left and a jet nozzle is directed right, they are operated alternatively, so that (1) the sucking nozzle is active first, in the siphoning phase during which the jet nozzle is closed; and (2) the jet nozzle is active second, in the exhalant phase during which the sucking nozzle is closed. This process is shown by the dotted lines in Fig. 7.2-2.

The sucking direction and speed of the fluid of the siphoning nozzle are not so important, but sucking forward is slightly better than sucking backward for an earlier start, and slow sucking is better than fast for preventing unfavorable effects such as friction loss.

A fish breathes by drawing in water from its forward directed mouth and discharging it through its rearward-directed opercula slots, whereas the squid swims with a series of cyclic jettings in which power and recovery strokes are performed alternately by exhaling and inhaling through the apertures of a funnel system, both of which are directed backward.

Jetting in translational motion. If the siphoning is performed by a moving squid with a translational speed V, then the water sucked into the cavity obtains the momentum of $m_s V$ which generates a negative thrust called "momentum drag." Thus, unlike Eq. (7.2-8), the jet inpulse I_j in one stroke cycle is given by

$$I_j = m_j v_j - m_s V = m_j(v_j - V) \quad (7.2\text{-}13)$$

The above equation states that the jet speed v_j must óvercome the forward speed V for positive thrust.

7.2.2 ANIMALS USING JET PROPULSION

Soft-bodied animals, which derives support from hydrostatic pressure instead of rigid skeletons, usually lack the muscle power for sustained rapid movement; thus, many of them are sedentary. In squids and jellyfish, however, the jet propulsion technique has been adapted both for a rapid dash and for sustained swimming. On the other hand, larval Anisoptera use jet propulsion for escape and predatory movement.

Jellyfish. Jellyfish (medusae) and the medusoid forms of two classes of Cnidaria (Hydrozoa and Scyphozoa) have a parachutelike or bell-shaped gelatinous mass and can swim by alternate rhythmical extension and contraction of the bell. In the driving stroke, water is expelled from the subumbrella cavity by contraction of the circular subumbrella muscles, so as to slowly propel the animal in the opposite direction. In the recovery stroke, water is drawn into the cavity through a circular rim that is rounded inward. DeMont and Gosline (1988) showed that the bell structure is resilient and works at the resonant frequency of the structure by storing contraction energy in the power stroke and releasing it in the recovery or refilling phase.

Since there is no definite funnel for jetting, the propulsive force is not so large but the efficiency may be none the worse because of the large exhaust area of the jet. Essentially, medusae are carried about by tides and currents.

Squids. Most cephalopods, commonly called cuttlefish, specifically decapod squids, possess fins that are given an undulatory motion for maneuvering at relatively low speed. In certain squids, however, the habitual mode of swimming is by periodic rapid expulsion of the surrounding fluid. The Japanese squid (*Todarodes pacificus*) is known to migrate 2,000 km by swimming continuously for 2.5 months at an average speed of 0.3 m/s or 0.6 knots (Gosline and DeMont 1985).

In the inhaling phase of *Loligo opalescens*, the increment in the diameter of the mantle and the volume of the internal cavity are, compared

TABLE 7.2-1. Swimming performance and respiratory metabolism for the squid, *Illex illecebrosus*, in comparison with the sockeye salmon, *Oncorhynchus, nerka*. (Webber and O'Dor 1986)

Item	I. illecebrosus	O. nerka[a]
Temperature (°C)	15	15
Total length (m)	0.42	0.37
Mass (g)	400	500
Critical speed (ms^{-1})	0.76	1.35
Active metabolism (mlO_2kg^{-1}h^{-1})	1047	480
Standard metabolism (mlO_2kg^{-1}h^{-1})	313	40
Rest metabolism (mlO_2kg^{-1}h^{-1})	202	—
Scope for activity (mlO_2kg^{-1}h^{-1})	734	440
Gross cost of transport (J kg^{-1}m^{-1})[b]	7.6	1.9
Net cost of transport (J kg^{-1}m^{-1})[b]	5.4	1.7

[a] Salmon data from Brett (1965) and Brett and Glass (1973).
[b] Cost of transport calculated after Schmidt-Nielsen (1972).

with the relaxed state, approximately 10% and 22% respectively (Gosline and DeMont 1985). In the exhalant phase, the water jet is expelled under high pressure; the gauge pressure is 30 kPa (Trueman and Packard 1968; Trueman 1980). This is achieved by contracting the mantle to about 75% of its relaxed diameter. Nearly all the water, equivalent to about 60% of the relaxed mantle volume, is expelled through a narrow funnel tube in a powerful jet. A *Loligo* with a mass of 100 g can accelerate from rest to 2 m/s in a single squirt, and maximum speeds of 8 m/s have been reported for other squids (Trueman 1975). The direction of the jet flow can be controlled by adjusting the nozzle direction.

Webber and O'Dor (1986) reported that the jet frequency and the pressure in the mantle chamber increase with the swimming speed. In *Sepia* the internal pressure rises to more than 20 kPa (0.2 atm) for a period of about 250 ms during the power stroke, then drops to 1.5 kPa below ambient pressure during the recovery stroke for 50 ms (Trueman 1980).

The swimming performance and respiratory metabolism for an *Illex illecebrosus* weighing 4N ($\simeq 0.4$ kgf) are given in Table 7.2-1 in comparison with those of a sockeye salmon (*Oncorhynchus nerka*) weighing 5 N. It can be seen that the fish can swim by a fanning motion nearly twice as fast

Advancing direction

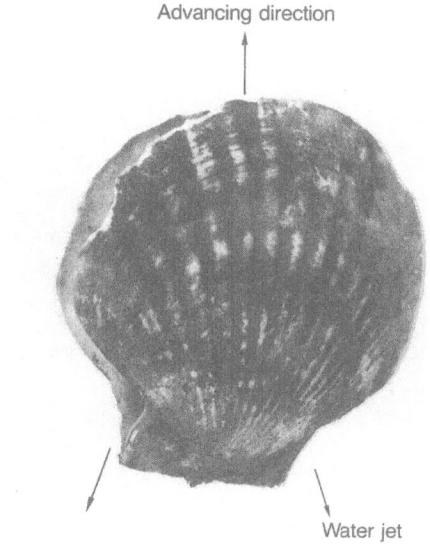

a   Water jet

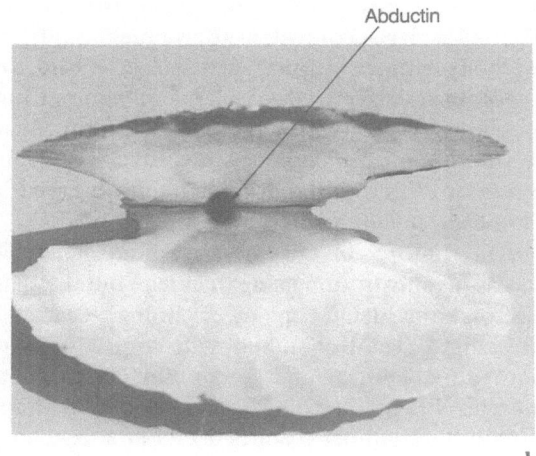

Abductin

b

FIG. 7.2-4. **a, b.** Scallop. **a** Shell. **b** Abductin.

as the squid, but using less than half the energy, as reflected in the cost of transport figures. Thus the Froude efficiency of squid jet propulsion must be low in comparison with that of fish in fanning motion, because the squid must accelerate much less water than the salmon to increase velocity. In the case of *Illex illecebrosus*, the Froude efficiency is $\eta = 0.42$. (Webber and O'Dor 1986).

Nautilus. In *Nautilus*, younger stages of the shell are enclosed and strengthened by the later growing parts, forming a spiral series of chambers. The gas in these chambers remains at effectively the same pressure, whatever the depth (Currey 1977, Ward et al. 1980).

The water jet of *Nautilus* is generated by a hyponome, which is actually a pair of muscular flaps that curl around each other to form a highly flexible funnel. Since the distance between the center of gravity and the center of buoyancy is short, the nautilus repeats the rocking motion with the momentary stimulation of each respiratory jetting.

Scallops. Shells of Ptericonchina such as *Patinopecten*, *Pecten*, and *Amusium* have two valves that are joined together at the hinge by an outer hinge ligament that is flexible but inextensible (Fig. 7.2-4). Just inside this is a block of the elastic protein, abductin, that forms the inner hinge ligament. Its Young's modulus is, as already noted in Table 4.4-3, $(1.3–4.0) \times 10^7$ dyn/cm^2 or $(1.3–4.0) \times 10^6$ Pa (Alexander 1968).

The scallop can swim by opening and closing its valves repeatedly and rapidly like a castanet.

When the shell is closed by the contraction of an adductor, the inner hinge ligament is compressed and a part of the energy is stored in the abductin. This tends to make the valves open again when the adductor muscle relaxes. That is to say, as explained in Sect. 4.4.4, the locomotion is performed with a resonant frequency of this swimming system (DeMont 1990). In the opening or recovery phase, the fluid can be introduced into the shell, through its round margin. In this process, the fluid rounding the rim certainly generates the suction force like the take-off phase in butterflies (Sect. 4.5.3). Usually, in the closing or power phase, the pallial curtain (velum) around the water is initially closed near the margin of the two valves, two holes located on opposite sides of the flat hinge are opened, and the re-entrained water is expelled through them (Dadswell and Weihs 1990). This, as shown in Fig. 7.2-4a, pushes the shell in the direction opposite to the water jet. The motion resembles jumping and is sustained by the repeated opening and shutting of the valves. The shell seems to use this intermittent jet propulsion for escaping from danger; thus, the swimming direction of the ventral side is usually upward and against the shell's weight. In some cases, when the ventral side of the shell is stimulated, by starfish for example, the scallop can evade danger with the dorsal side front by ejecting the ventral jet (Moore and Trueman 1971). Furthermore, the giant scallop, *Placopecten magellanicus*, is known to swim level at a speed of 0.3–0.6 m/s (Dadswell and Weihs 1990).

As stated before, jet propulsion is not always efficient but, in the case of the scallop, the amount of jetted water is equivalent to as much as 50% of the body volume during each jet cycle, and the use of this relatively large volume of water at low pressure (300–600 Pa) suggests that scallops swim in an economical manner as regards energy consumption. Another remarkable habit observed in *Patinopecten* will be presented in Sect. 7.5.2.

Larval Anisoptera. Larval Anisoptera can walk across the substratum using their legs, but they can also make sudden escape by ejecting water from a specialized rectal chamber with contractions of both dorso-ventral and longitudinal abdominal muscles. In the case of *Aeschna cyanea*, the pressure in the chamber reaches 6 kPa at a frequency of 2.2 Hz (Trueman 1980). Then, the legs usually drift backward.

7.3 Sweeping

A number of devices have been developed for propelling a boat by man power. These include the paddle, punt, oar, and *ro* or Oriental scull. The first three devices use the hydrodynamic drag and direct inertial force generated by the added mass of the fluid in the accelerated motion of the devices. Thus, they are convenient for powerful driving, and more specifically, for a dash start. The last is a unique device utilizing the hydrodynamic lift acting on a blade in sweeping motion. It is, therefore, convenient for power saving in long voyages. Living creatures adopted similar devices in their body configuration and skillfully utilized them for propulsion.

7.3.1 MECHANISM OF PROPULSION

In *ro* sculling, as shown in Fig. 7.3-1, the rower stands near the stern of the boat and repeatedly pulls and pushes the loom back and forth. This action imparts to the blade a sinusoidal sweeping or lead-lag motion in a conical plane against hydrodynamic torque Q. Further, the superposition of a sinusoidal twist along the longitudinal axis gives the blade a sinusoidal feathering motion coupled with the lead-lag motion. The rounded surfaces of the underside of the blade and of the supporting dip or receptacle for a short stake extending from the blade help to generate this feathering motion as a universal joint. This sweep-

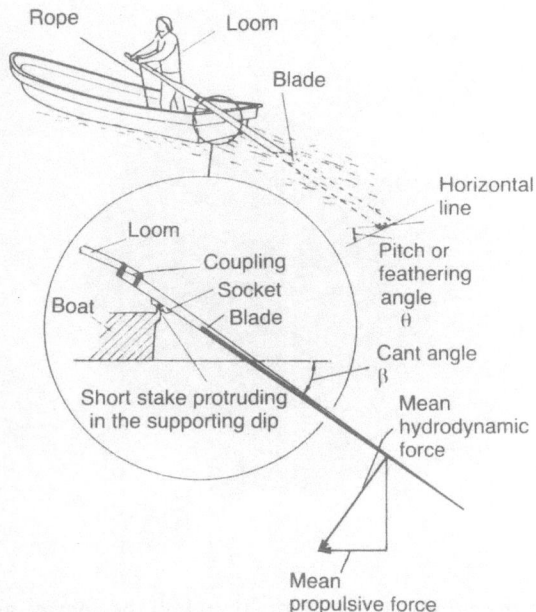

FIG. 7.3-1. Operation of *ro* (From Azuma et al. 1989).

ing creates lift on the blade, as well as drag. The horizontal component of the lift gives a propulsive force T to the boat, but the lift also causes a very large flatwise-bending moment M_B around the pivot. This moment is counterbalanced with the tension acting on a rope that is tied to the floor of the boat, so that the rower must not allow the rope to come loose during the operation. The rower can also steer the boat by pushing the loom to turn right and pulling it to turn left.

The mechanics of the *ro* were analyzed by Azuma et al. (1989). By applying the local circulation method (LCM), they determined the hydrodynamic force and moments acting on the blade. This is performed by analyzing the very complex flow around a blade element in sweeping motion, and by integrating numerically the elemental forces and moments generated by the blade element along the span under the water surface. Under the assumption of constant profile drag, they also obtained analytical expressions of the mean hydrodynamic forces and moments by taking the mean values within one complete stroke, as given in Table 7.3-1. It was found from these analyses that: (1) both the thrust and moment are approximately proportional to the blade pitch angle θ; (2) the thrust is also proportional to the sine of the cant angle β or $\sin\beta$; (3) the thrust and moment are proportional to the lift slope, whereas the torque is proportional to the drag coefficient; (4) all

TABLE 7.3-1. Mean thrust, bending moment, and torque. (Azuma, et al. 1989)

Item	Formula
Mean thrust $\overline{T}$	$\frac{1}{2}\rho a c R^3 \omega^2 \sin\beta [\frac{1}{6}(B^3 - x_0^3)\theta_0(\zeta_{1c}^2 + \zeta_{1s}^2) + \frac{1}{2}(B^2 - x_0^2)\hat{V}\zeta_0(-\theta_{1s}\zeta_{1c} + \theta_{1c}\zeta_{1s})\cos\beta$ $\quad + \frac{1}{2}(B - x_0)\hat{V}^2\{\theta_0(\sin^2\beta + (2\zeta_0^2 + \zeta_{1c}^2 + \zeta_{1s}^2)\cos^2\beta) + 2\zeta_0(\theta_{1c}\zeta_{1c} + \theta_{1s}\zeta_{1s})\}]$
Mean bending moment $\overline{M}_B$	$\frac{1}{2}\rho a c R^4 \omega^2 [\frac{1}{8}(B^4 - x_0^4)\theta_0(\zeta_{1c}^2 + \zeta_{1s}^2) + \frac{1}{3}(B^3 - x_0^3)\hat{V}\zeta_0(-\theta_{1s}\zeta_{1c} + \theta_{1c}\zeta_{1s})\cos\beta\}$ $\quad + \frac{1}{4}(B^3 - x_0^2)\hat{V}^2\{\theta_0(\sin^2\beta + (2\zeta_0^2 + \zeta_{1c}^2 + \zeta_{1s}^2)\cos^2\beta) + 2\zeta_0(\theta_{1c}\zeta_{1c} + \theta_{1s}\zeta_{1s})\}]$
Mean torque $\overline{Q}$	$\frac{1}{2}\rho c R^4 \omega^2 [\frac{1}{8}(1 - x_0^4)\delta_0(\zeta_{1c}^2 + \zeta_{1s}^2) + \frac{1}{4}(1 - x_0^2)\hat{V}^2\{\sin^2\beta + (2\zeta_0^2 + \zeta_{1c}^2 + \zeta_{1s}^2)\cos^2\beta\}$ $\quad \frac{1}{2}\delta_2(1 - x_0^2)\hat{V}^2\sin\beta]$
Lead-lag motion $\zeta(t)$	$\zeta_0 + \sum\limits_{n=1}^{3}\{\zeta_{nc}\cos(n\omega t) + \zeta_{ns}\sin(n\omega t)\}$
Feathering motion $\vartheta(t)$	$\theta_0 + \sum\limits_{n=1}^{5}\{\theta_{nc}\cos(n\omega t) + \theta_{ns}\sin(n\omega t)\}$ where $C_l = a\alpha$ $C_d = \delta_0 + \delta_1\alpha + \delta_2\alpha^2$ $\hat{V} = V/R\omega$

quantities are proportional to the square of the angular velocity ω^2 times cR^3 for force, and cR^4 for moment, where R and c are the blade length and chord respectively; (5) the propulsive efficiency η, as well as the mean thrust and torque, is optimal for stepwise or on-off change of feathering motion θ_0, and departs from its optimal value with a dulling of the feathering motion; (6) as the period of locomotion $2\pi/\omega$ decreases, the thrust and torque increase, but the efficiency is kept almost constant; and (7) there is an optimal amplitude of the feathering angle θ_0 that is a function of the Froude number based on the forward velocity and the length of the hull $V/\sqrt{gl}$, for maximizing the efficiency.

7.3.2 SWEEPING DEVICES IN ANIMALS

Pectoral fins. It is known that fish generally use the pectoral fins at low activity levels and the caudal fin for burst activity. The percoid family, Embiotocidae, and the swellfish family are known groups of fish that employ noncaudal fin propulsion at high sustained swimming speeds. As stated in Sect. 6.1.4, one of the motions of the pair of pectoral fins of such fish may be considered a sweeping motion.

Flattened tails. As stated in Sect. 7.1.4, the flattened tail seen in sea otters and beavers can be used in many ways like a *kai*, which is a short scull operating in the same way as the *ro* but having a larger cant angle β than that of the *ro*. Propulsive force can be produced either by whipping or by sweeping. Because the length of the tail is not so

large, it can be used for precise control of the attitude and position of the body.

Epibatic tails. Many sharks have a heterocercal or epibatic tails (Breder 1926) with a large upper lobe and a small lower lobe (Alexander 1968). This form has probably been adopted for life near the sea bottom. The tail moves from side to side as the shark swims with snake-like motion, and the long lobe, which resembles half of a swept wing, works like a *ro* operating with coupled heaving and feathering motions. This locomotion is a combination of fanning and sweeping. It differs from the operation of a *ro* in that the leading edge is always the upper edge of the lobe, whereas in the *ro* the edges on opposite sides alternately serve as the leading edge and the trailing edge.

Crabs. Crustaceans are notable for their skill in a wide variety of types of locomotion. Their swimming and walking modes using pleopods, legs, and tail fans were discussed earlier in Sect. 7.1.2. Here, however, let us take a look at a species that swims in another way, namely, by sweeping.

Portunid crabs (Portunidae) can swim sideways with the powerful sweeping action of specialized blades on the last pair of thoracic limbs. As shown in Fig. 7.3-2, these blades are made up of four or five paddles, each of which is oval and flat, or in some cases slightly spoon-shaped. The paddles are connected with each other without looseness by three single hinges and one universal hinge at the root. Two single hinge lines are vertical axes for lead-lag motion of an amplitude of about 90°, and one is a horizontal axis for a feathering motion of about 90°. With the additional universal joint at

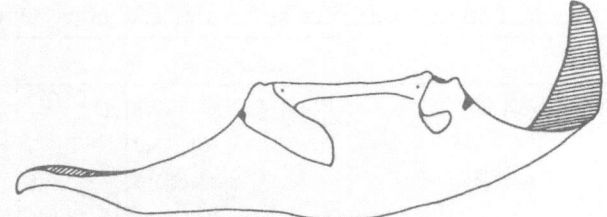

Fig. 7.4-1. Manta utilizing beating.

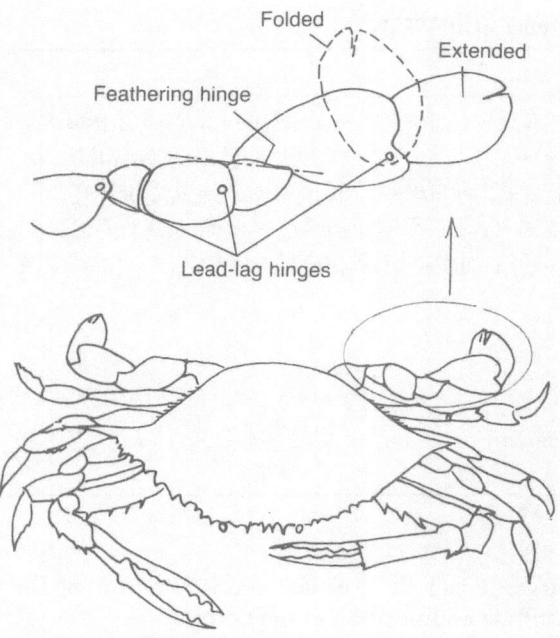

Fig. 7.3-2. Swimming legs of *Portunus*.

the blade root, the whole blade can simulate the motion of a *ro* and produce enough thrust for speeds up to 1 m/s.

7.4 Beating

As we have seen, beating is the only method capable of generating both propulsive and lifting forces near the center of gravity of a flying creature. However, the lifting force is close to zero for swimming creatures, because their specific gravity is nearly equal to that of the surrounding fluid. Aside from swimming birds, therefore, few swimming species need use beating to propel themselves.

Since a propulsive force can be generated in not only the downstroke but also the upstroke, the beating of swimming creatures in water is performed nearly symmetrically for both strokes; thus, their beating looks somewhat curious, as we are used to the asymmetric beating of birds and insects.

7.4.1 BEATING OF FINS AND FORELEGS

As explained in the preceding chapter, the pectoral fins are used in various ways, one of which is

beating, an active method of obtaining propulsion for swimming. Pectoral fins used for beating are located close to the center of gravity to obtain greater stability.

Rays. The body of a ray (Rajiformes) is depressed from top to bottom instead of from side to side, as is usual in the flat fish form. One of the giants of the ocean, the manta or devilfish (*Mobula mobular*) shown in Fig. 7.4-1 has a span of 6–7 m and weighs approximately 1 t. Because of its huge size, it is also called the giant ray. Like giant whales and whale sharks, it feeds on tiny plankton, small crustaceans, and small fish by scooping them into its mouth.

The ray swims by slowly beating the tips of a pair of winglike fins that are enlarged into triangles with pointed tips. Since the reduced frequency of their beating motion in cruising flight appears to be of the order of $k \simeq 10^{-1}$, the combined feathering and flapping mode is no doubt the same as that of the maximum efficiency mentioned in Sect. 4.2.3. Two other rays known to use beating for locomotion are *Rhinoptera bonasus* and *Myliobatis*. The former has wings shaped like an airfoil section, and in swimming, a vertical fin of triangular shape is elevated like the vertical tail wing of an airplane, probably contributing to the lateral stability of the ray. The latter repeatedly leaps clear out of the water, but for what reason remains unknown. It often does a complete somersault, one of the most spectacular sights of the sea.

Mola. Mola (Molidae, a family of Tetraodontiformes), uses its high-aspect-ratio wing consisting of long dorsal and anal fins by flapping them laterally. This locomotion is therefore called "tetraodontiform" (Breder 1926). There is no trunk musculature at all (Harris 1953). The reduced frequency is low, and this fact guarantees a high efficiency in slow cruising motion. The *Mola*'s food consists of soft-bodied oceanic animals, jellyfish, salps, ctenophores, and siphonophores (Burton

1980), all of which are slow swimmers and easy to capture. They have the unusual habit of basking at the surface, lying on their side as though dead.

Squid. Many squid can swim by sending deflection waves along a pair of fins. However, instead of making an undulatory motion of the fins, some species such as Tarodes or *Ommastrephes* beat them in a manner similar to rays.

Flying fish. Hatchetfish (Gasteropelecidae) and butterfly fish (*Pantodon buchholzi*) are also known to flap their wings (pectoral fins) with powerful muscles and fly in the air. Theirs is indeed a beating flight unlike the gliding flight of *Exocoetus* or *Cypselurus* seen in Sect. 3.5.2.

Cetaceans. The humpback whale (*Megaptera novaeangliae*) inhabits most of the world's oceans and has a pair of very long and bumpy flippers, and skin nodules around the head and flippers. The span or total extended width of the flippers is almost equal to the length of the animal, roughly 30 m. In addition to fluke locomotion, the flippers can be used to propel the body by beating them.

Turtles. Unlike tortoises, which although not swimmers are very fond of water and of wallowing in mud, turtles are good swimmers with finlike legs. The forelegs are used as beating wings, and the hindlegs are used as sculling wings. Interestingly, there is a small protuberance at the midspan of each foreleg that probably acts to prevent flow separation during beating, like the alulae of birds.

7.4.2 BIRD FLIGHT IN WATER

Most birds like water and come to the edge of bodies of water to drink, bathe, and gather food. Some species such as storks, herons, flamingoes, and snipes, use their long legs to walk in shallow water in search of food; other species, such as divers, grebes, petrels, cormorants, pelicans, waterfowl, rails, gulls, terns, and loons, move on or under the water surface using their webbed feet. Some birds also fly in water by beating their wings, and these include penguins, auks, water ouzels, some diving ducks, petrels, and cormorants. These birds are fully covered by plumage with a thick layer of underfeathers and are usually equipped with small wings that have a large enough area to enable the bird to maneuver when pursuing prey. By preening, they keep their feathers constantly greased with waxy substances that keep their skin from direct contact with the water and protect it from heat loss. Heat loss is most severe

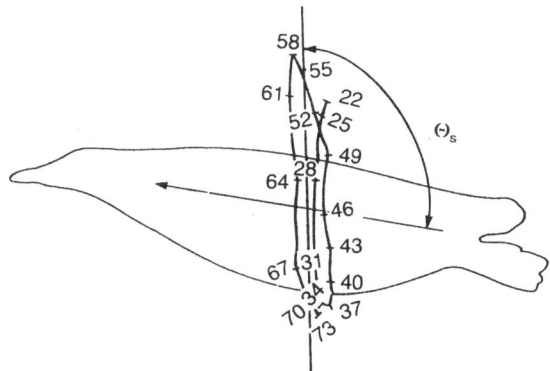

FIG. 7.4-2. Stroke plane of Humboldt penguin. (From Hui 1988 with permission). *Numbers* represent time steps.

at the extremities, like the beak and legs, although these, too, are adapted to save as much heat as possible. The long bones are not pneumatic and the number of air sacs is fewer than in other birds (Meister 1962). It is known that in diving ducks most of the power output is used for overcoming buoyancy (Stephenson et al. 1989).

Penguins. Penguins swim by beating their wings in order to travel to and from foraging sites, and to pursue prey. Thus, the wing of a penguin (Spheniscidae), here also called a "flipper," is quite small and formed mostly of very flattened bones covered with small, almost scalelike, feathers. They are adapted solely for use in swimming. The penguin swims mainly with the powerful beating of its flippers, as shown in Fig. 7.4-2. This mode of beating in cruising is very similar to that of the swift flying in air at high speed, and the flippers are not folded on the upstroke. Then, the effective or outer part of the wing seems to work with almost the same amplitude in the spanwise direction, which is more like the "cyclo-gyro wing" (Sect. 8.2) than the beating wing in which the amplitude increases toward the wing tip.

When traveling long distances, king penguins porpoise by accelerating in the water over about 6–12 m, and then jumping through the air for about 3–4 m. As in the case of porpoises and dolphins discussed in Sect. 6.4.3, they do this to conserve energy for swimming while breathing. The critical speed V_0^* for Humboldt penguins or other penguins of similar size is about 3.0 m/s (Hui 1987a).

Clark and Bemis (1979) measured the speed and wingbeat frequency for four species of penguins and gave the following relation, which is very similar to that given for fish in Sect. 6.3.3;

$$U/l \propto 1.3f \quad \text{or} \quad U/lf \propto 1.3. \quad (7.4\text{-}1)$$

Considering that the ratio of wing chord to body length is typically about one tenth, the reduced frequency is estimated roughly to be $k \simeq 0.2$ or $\sigma \simeq 5$ [see Eq. (6.3-22)].

For the swimming of Humboldt penguins, Hui (1988) reported the following interesting facts: (1) The stroke plane, as seen in Fig. 7.4-2, is tilted forward ($\theta_s > \pi/2$); (2) the feathering angle with respect to the stroke plane is positive for the downstroke and negative for the upstroke; and (3) the reduced frequency based on the tip speed of the beating wing and its chord is less than 0.1. These facts pointed out by Clark and Bemis as well as Hui imply that the penguin needs to generate mostly driving force, and only a slight lift, which, unlike that of airborne birds, is negative, probably to overcome the buoyant force. Therefore, the wingbeat serves to produce an even driving force during both the up- and downstrokes, and the beating is executed economically.

Auks, water ouzels and murres. Auks (*Alca*) and water ouzels are also well adapted for flying in the water by beating their reduced wings, and frequently do so while feeding. The great auk (*Pinguinus impennis*), now extinct, could not fly but was able to move swiftly in the water. Murres glide swiftly downward from their nesting cliffs at a steep angle, then sweep outward in a long curve before leveling off. After diving under the water surface, they fly rapidly through the water and, with repeated swift strokes of their wings, herd small fish into a group. Then, they dart among them, snatching one here and there in their sharp-pointed bills.

Cormorants. The cormorant, which is also a good pursuit swimmer, propels itself with its feet and the beating of half-folded wings. After fishing, it rests on rocks or buoys, and holds its wings open to dry like the shag (*Phalacrocorax aristotelis*).

7.5 Sailing

"Sailing" is a driving method used by many living creatures and man-made craft for travel on the water surface. It involves a complex mixture of hydrodynamic and aerodynamic principles. In sailboats, the sail shape, its trim angle, and the center

of gravity must be varied greatly in response to the wind, boat velocity and surface conditions. However, in general, in the sailing of living creatures the sail shape and the trim angle cannot be changed, and the animal or plant has no way of adjusting its sailing mode for particular wind and surface conditions. It can only drift at the mercy of the wind and water. However, some species are probably able to control the trim angle of the sail and the center of gravity and buoyancy of the immersed part. This makes it possible for them to take a specified course to their destination. Sailing can thus be used not only to disperse the species by utilizing seasonal winds and ocean currents, but also to capture food by towing a net in the water for catching plankton and fish.

7.5.1 Sails and Sailing Mechanics

The most direct way to utilize wind power is sailing. Sailing does not involve thermodynamic cycles or heat exchange, nor does it involve dynamic components such as a rotating or beating mechanisms. A sailboat typically consists of a vertical wing (sail) in the air, joined to an inverted wing (center-board) in the water, all assumed to be stable, buoyant, and deformable for all points of sailing (B. Smith 1980).

Let us consider a sailboat propelled by a unirig. It is assumed, for simplicity, that movable crew weight is sufficient to keep the boat nearly upright; thus, only the horizontal components of aero- and hydrodynamic forces are taken into consideration.

As shown in Fig. 7.5-1, under the assumption of small side slip of the boat, the aerodynamic force acting on a sail is generated by the apparent wind velocity V_A, which is a vectorial summation of the true wind velocity V_W and the negative boat velocity $-V_B$. Two components of the aerodynamic force are the drag parallel to V_A and the lift perpendicular to V_A and are given by a function of angle of attack α_A of the sail as shown by a polar curve in the figure. The angle α_A can be given by the angle of apparent wind velocity β_A minus the sail angle δ_A. The aerodynamic force can also be decomposed into two components, a driving force in a direction approximately parallel to the center line of the boat and a normal force perpendicular to the course. The latter force is balanced with the hydrodynamic force acting on the centerboard immersed in the water, which results from a very small sideslip of the boat.

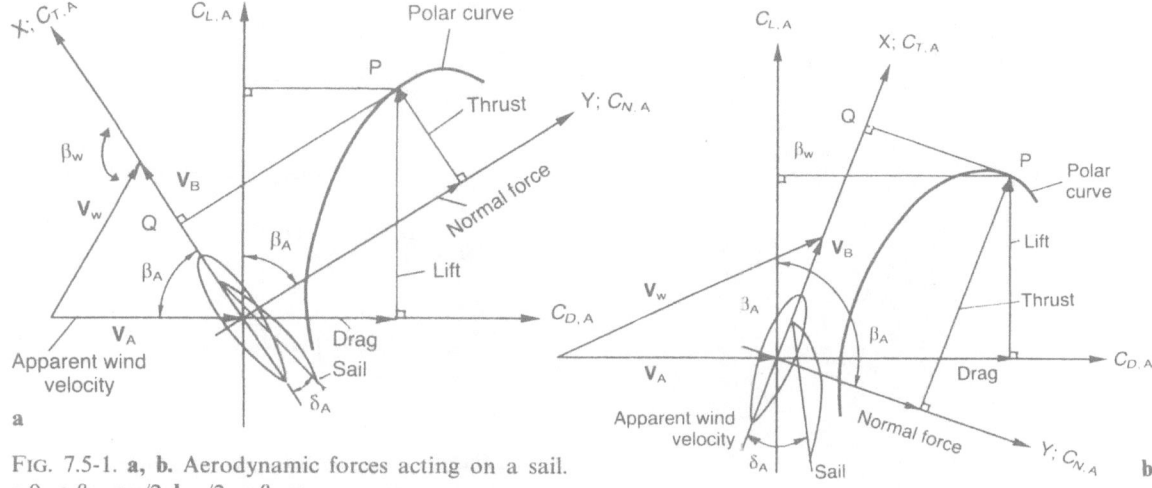

FIG. 7.5-1. **a, b.** Aerodynamic forces acting on a sail.
a $0 < \beta_A < \pi/2$. **b** $\pi/2 < \beta_A < \pi$.

These forces nondimensionalized in coefficient forms are, for example, represented by a point P on the polar curve. Then, the maximum thrust can be given by making the line $\overline{PQ}$ in Fig. 7.5-1 tangential to the polar curve for a given β_A. In Fig. 7.5-1a, when the angle β_A is less than $\pi/2$ or the boat moves against the apparent wind, a larger thrust can be obtained for large aspect ratio sails than small ones. However, in Fig. 7.5-1b, when the angle β_A is more than $\pi/2$ or the boat moves toward the apparent wind, a larger thrust can be obtained for small aspect ratio sails (See Sect. 3.5.5).

Figure 7.5-2 depicts a Japanese fishing boat used on lakes mostly for catching pond smelts. The boat is driven by a low aspect ratio sail that makes use of a tailwind to overcome the drag of a laterally extended fishing net and the boat towing the net.

7.5.2 SAILING ANIMALS

Siphonophores. Siphonophores (Siphonophorae) form floating colonies composed of both polypoid and medusoid forms. Two colonial species, *Physalia*, known as the Portuguese man-of-war, and *Velella*, are shown in Fig. 7.5-3. Each comprises a colony of polyps (gastrozooids) divided into a sail, a gas-filled float (pneumatophore), and different clusters on the undersurface of the float, one made up of feeding polyps, another of reproductive polyps, and a third cluster forming the long stinging tentacles (dactylozooids) (Burton 1980).

Propelled in midocean by the aerodynamic force acting on a saillike projection on the float, they tow their submerged portion like a net, much like the fishing boat shown in Fig. 7.5-2. In view of the small aspect ratio of the sail, it would seem that not only the drag but also the lift is effective. As we saw in the preceding section, a low aspect ratio sail is very effective in a tailwind, with the maximum aerodynamic driving force.

Fig. 7.5-4 shows two types of Physalia which are blown at angles up to about 45° (or $\beta_W \leq 135°$) to the left or right of the downwind direction. The angle is dependent on the wind condition. They have evolved a remarkable dimorphism as regards the curvature of the sail and asymmetrical form of the float. This enables them to disperse their population equally to the north and south in the trade wind zone, and automatically ensures broader dispersal than if they all sailed in the same direction (Mackie, 1974).

Scallops (Patinopecten). In Japanese, the word

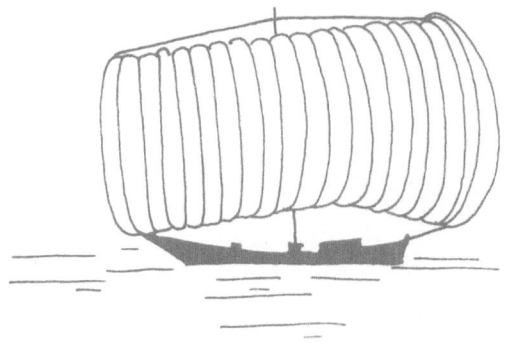

FIG. 7.5-2. One of the Japanese fishing boats.

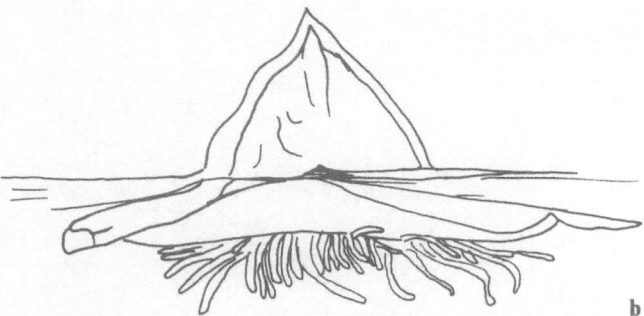

FIG. 7.5-3. **a, b.** Siphonophores. **a** *Physalia* (Portuguese man-of-war). **b** *Velella*.

for *Patinopecten* is *Hotategai*, which means sail-hoisting shellfish. *Patinopecten* is believed by some local fishermen to migrate in groups by sailing on the sea surface. From my experimental work, I have observed that it can slide or windsurf, using its shells. Its flat valve catches the wind, and its concave valve floats on the sea (Fig. 7.5-5). This picture was taken by a camera located at the side of a water basin in which a dead scallop was floated, so as to create "windsurfing" under the coaction of the buoyant force and a driving force in the form of a wind of about 12 m/s from a fan. The buoyant force was obtained by closing a part of the open mouth of the shell with a strip of cellophane tape applied to simulate natural fringes. A live scallop

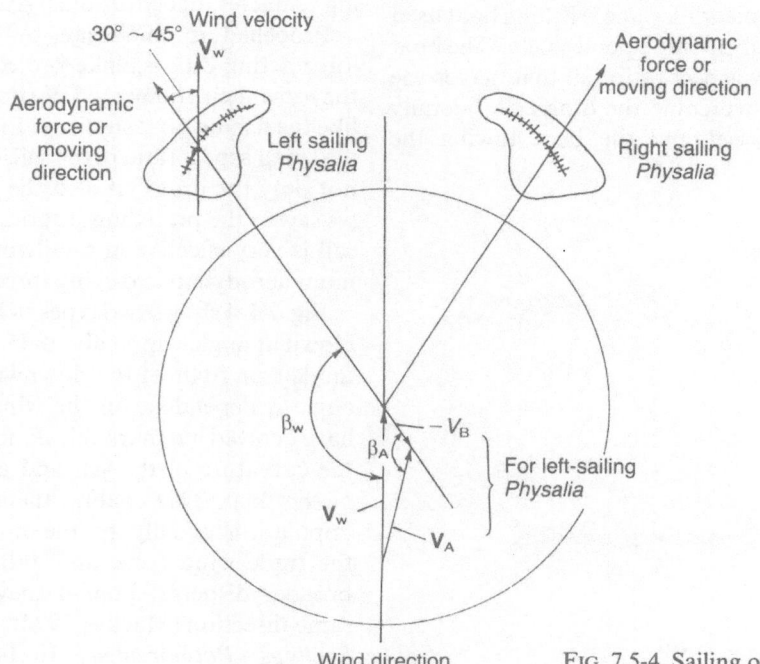

FIG. 7.5-4. Sailing of *Physalia*.

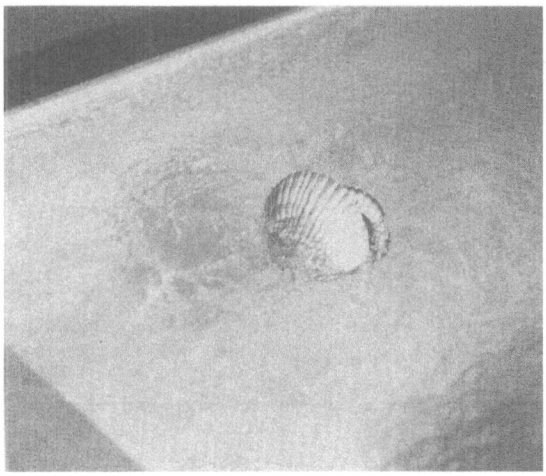

FIG. 7.5-5. Wind-surfing scallop.

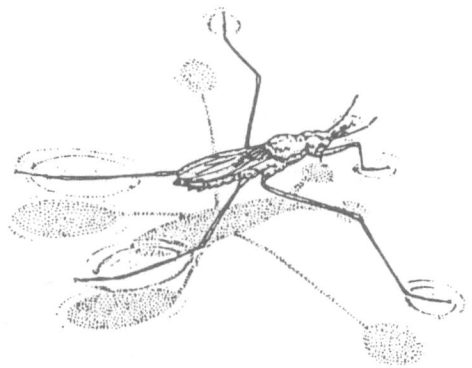

FIG. 7.6-1. A pond skater.

7.6.1 SKATING

Figure 7.6-1 shows a pond skater sculling over the surface of still or slow-flowing water. It is supported by long middle- and hindlegs and has an unwettable hairpile over the ventral surface of its body that helps it float (Burton 1980). The length of the body and legs is no more than a few centimeters. The legs depress the water surface into concave meniscuses. Then the surface tension acting on the legs can support the weight of the pond skator.

As a measure of the maximum lifting force, we can use the ratio of the maximum lifting force to the weight $L_{s,max}/W$, which is called the safety coefficient. The surface tension of pure water decreases almost linearly from 76 to 60 dyne/cm as the temperature increases from 0°C to 100°C. The surface tension of seawater is about 26% of that of pure water. Thus, the safety coefficient of several species is reported to be more than 9 (Baudoin 1980).

The weight of the body is proportional to the cube of the linear dimension, whereas the surface tension is proportional to the linear dimension itself. Therefore, beyond a certain length or weight of the body, the safety coefficient becomes less than one. The maximum body length (or mass) can be estimated from the maximum bulk of the floating insects. Small insects of a suitable design can walk on water, although a large insect of exactly the same design cannot.

However, as a floating insect becomes smaller, the adhesive force acting on it will become larger and prevent its free locomotion on the water

probably accomplishes the same effect by closing its fringes, as explained in Sect. 7.2.2 on jet propulsion. In this way, the attitude of the shell and, thus, of the sail can be made very stable for various wind speeds and mild surface conditions. This mechanism of locomotion can overcome the difficulty of jet propulsive swimming, which is pointed out by O'Dor and Webber (1991) as follows: The biomechanical data on the locomotion of scallops indicate that they require a tremendous long-term power for returning by migration to their spawning area as suggested by Dadswell and Weihs (1990). Although there is no confirmed observation of the live scallops, I believe that the scallops must utilize this potential of windsurfing (or the ability of energy-saved locomotion) for their migratory movement.

7.6 Skating and Wave Riding

Some insects, specifically those that live on water, such as members of the orders Hemiptera and Collembola, can walk on the water surface. These insects have hydrofuge cuticles that prevent them from getting wet. Unlike marine water striders, almost all freshwater striders have some kind of wing growth along their back.

Water striders are rarely found on turbulent water or alga-covered surfaces. They are fond of skating on water only a few centimeters deep, presumably to avoid the danger of being eaten by large fish.

surface. Thus, the minimum bulk of the floating insect is also limited.

Locomotion in slow movement is performed mostly by the pair of long middle-legs. The claw-bearing tarsi are immersed in the water and pushed against the rear wall of the depressed water surface by the rotation of all parts of the leg (tarsus, tibia, femur, and trochanter) (Walker 1983).

Pond skaters can communicate with each other by sending out a series of ripples at a frequency and wave-speed of about 10–30 Hz and 20 cm/s respectively. Transmission of information between two pond-skaters separated by 20 cm takes 1 s, about the same time that it took astronauts on the lunar surface to hear a call from NASA in Houston.

It is interesting to note that tiny nonaquatic insects that fall on the water, and even small floating insects that have had their ability to repel water impaired by, for example, foreign matter sticking to them, cannot easily escape from the water surface because for such small insects, as stated before, the water is very sticky and its surface tension is extremely large in comparison with inertial forces.

7.6.2 WAVE RIDING

The static and dynamic pressure of a surface wave drives floating objects in the direction of wave propagation. This phenomenon is called "wave riding." Surfing is one kind of wave riding. Many living creatures take advantage of wave riding for locomotion and dispersion of the species by utilizing the static pressure.

Effects of Hydrostatic Pressure. Let us consider a two-dimensional sinusoidal wave of amplitude a, wavelength λ, and wave propagation speed C in (X, Y, Z) coordinate frame, as shown in Fig. 7.6-2. The potential function ϕ is given by

$$\phi = -aC \frac{\cosh\{(2\pi/\lambda)(h+Z)\}}{\sinh(2\pi h/\lambda)}$$
$$\cdot \cos\{(2\pi/\lambda)(X - Ct)\} \qquad (7.6\text{-}1)$$

where h is the depth of the mean wave surface, and the wave propagation speed C is given by

$$C = \sqrt{(g\lambda/2\pi)\tan h(2\pi h/\lambda)}. \qquad (7.6\text{-}2a)$$

which is approximated as

$$\left. \begin{array}{ll} C = \sqrt{g\lambda/2\pi} & \text{for deep sea} \\ C = \sqrt{gh} & \text{for shallow water} \end{array} \right\} \quad (7.6\text{-}2b,c)$$

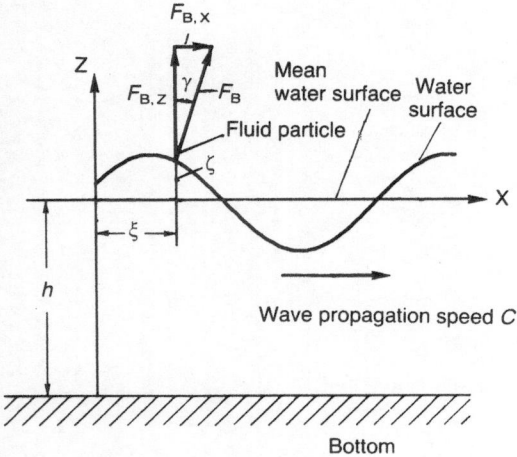

FIG. 7.6-2. Buoyant force acting on a particle near the wave surface.

The ordinates of a fluid particle on the wave of a free surface can then be expressed by

$$\left. \begin{array}{l} \xi = \overline{\xi}_0 + a\coth(2\pi h/\lambda)\cos\{(2\pi/\lambda)(\overline{\xi}_0 - Ct)\} \\ \zeta = a\sin\{(2\pi/\lambda)(\overline{\xi}_0 - Ct)\} \end{array} \right\}$$
$$(7.6\text{-}3)$$

where $\overline{\xi}_0$ is the mean position of the particle.

The pressure p is, then, given by

$$\left. \begin{array}{l} \mathrm{p} = -\rho(\partial\phi/\partial t) - \rho g Z \\ \\ = \rho a \left(\dfrac{2\pi}{\lambda}\right) C^2 \dfrac{\cosh\{(2\pi/\lambda)(h+Z)\}}{\sinh(2\pi h/\lambda)} \\ \\ \cdot \sin\{(2\pi/\lambda)(X - Ct)\} - \rho g Z \end{array} \right\} \quad (7.6\text{-}4)$$

Since the hydrostatic force or buoyant force acting on an immersed body in the fluid, whose immersed part of the volume is V_B, is given by

$$\mathbf{F}_B = \begin{bmatrix} F_{B,X} \\ F_{B,Y} \\ F_{B,Z} \end{bmatrix} = -\int_{V_B} \nabla p \, d\tau$$

$$= \begin{bmatrix} -\int_{V_B} (\partial \mathrm{p}/\partial X) d\tau \\ 0 \\ -\int_{V_B} (\partial \mathrm{p}/\partial Z) d\tau \end{bmatrix} \qquad (7.6\text{-}5)$$

Substituting Eq. (7.6-4) into the above equation yields the buoyant force as given in Table 7.6-1. The exact expressions tell us that: (1) in order to get a large driving force $F_{B,X}$ the immersed body

TABLE 7.6-1. Buoyant force in wave riding

Item	Expression
Horizontal component, $F_{B,x}$	$-\displaystyle\int_V \rho ac^2 (2\pi/\lambda)^2 \frac{\cosh\{(2\pi/\lambda)(h+Z)\}}{\sinh(2\pi h/\lambda)}\cos\{(2\pi/\lambda)(X-Ct)\}d\tau$
	$\simeq \begin{cases} -\int \rho ag(2\pi/\lambda)e^{(2\pi/\lambda)Z}\cos\{(2\pi/\lambda)(X-Ct)\}d\tau & \text{for deep sea} \\ -\int \rho ag(2\pi/\lambda)\cos\{(2\pi/\lambda)(X-Ct)\}d\tau & \text{for shallow water,} \end{cases}$
Vertical component, $F_{B,z}$	$-\displaystyle\int_V \left[-\rho ac^2(2\pi/\lambda)^2 \frac{\sinh\{(2\pi/\lambda)(h+Z)\}}{\sinh(2\pi h/\lambda)}\sin\{(2\pi/\lambda)(X-Ct)\} - \rho g\right]d\tau$
	$\simeq \begin{cases} \iint \rho g[-a(2\pi/\lambda)e^{(2\pi/\lambda)Z}\sin\{(2\pi/\lambda)(X-Ct)\}+1]d\tau & \text{for deep sea} \\ \iint \rho g[-a(2\pi/\lambda)^2(h+Z)\sin\{(2\pi/\lambda)(X-Ct)\}+1]d\tau & \text{for shallow water} \end{cases}$
Inclination angle of buoyant force[a] γ	$\tan^{-1}(F_{B,x}/F_{B,z})$
	$\simeq \begin{cases} \tan^{-1}\left[\dfrac{-a(2\pi/\lambda)e^{(2\pi/\lambda)Z_B}\cos\{(2\pi/\lambda)(X_B-Ct)\}}{-a(2\pi/\lambda)\sin\{(2\pi/\lambda)(X_B-Ct)\}+1}\right] & \text{for deep sea} \\[4mm] \tan^{-1}\left[\dfrac{-a(2\pi/\lambda)\cos\{(2\pi/\lambda)(X_B-Ct)\}}{-a(2\pi/\lambda)^2(h+Z_B)\sin\{(2\pi/\lambda)(X_B-Ct)\}+1}\right] & \text{for shallow water} \end{cases}$

[a] $(X_B, 0, Z_B)$ is the center of buoyant force.

should be small in comparison with the wavelength, or $V_B^{1/3} \ll \lambda$; (2) the sinusoidal horizontal and vertical forces are superposed on a stationary buoyant force given by $\int \rho g d\tau$; (3) the maximum horizontal force and the minimum vertical force are obtained while the body is moving with the velocity of the wave; and (4) the respective forces are proportional to the wave amplitude a.

The approximate expressions further state that: (1) the horizontal force is positive on the front surface of a wave, or $\frac{1}{4}\lambda < X - Ct < \frac{3}{4}\lambda$; (2) the horizontal force is inversely proportional to the wavelength; (3) the force is maximum at the mid-slope of the wave or $Z = \zeta = 0$, and (4) the vertical force is maximum at the valley of the wave and is inversely proportional to the wavelength in the deep sea, but to the square of the wavelength in shallow water. Near the wave surface, or $Z_B \simeq 0$ in the deep sea, the inclination of the buoyant force is approximately normal to the wave surface, because the wave surface given by Eq. (7.6-3) yields

$$\mathrm{d}\zeta/\mathrm{d}\xi = \frac{\mathrm{d}\zeta/\mathrm{d}\bar{\xi}_0}{\mathrm{d}\xi/\mathrm{d}\bar{\xi}_0} = -\gamma(Z_B = 0) \quad \text{for deep sea}$$

(7.6-6)

Aquatic plants and animals in the lower classes, including almost all plankton, floating seaweeds, and jellyfish, drift about at the mercy of the waves and can be driven by an inclined buoyant force.

This driving process is called "free riding." When the driven object floats on the wave surface, the process is called surf riding. If the specific weight of the object is lighter than, or about equal to, that of the water, so that the object floats on the wave surface, the object can move with the water surface, which has a propagation speed C, without sinking into the water with the fluid particles surrounding its body.

In addition to the pressure gradient caused by the surface wave, there can be another pressure gradient at the front of a floating body such as seen with a boat in the water. A pressure rise Δp results from the slowing of the local velocity from U_0 to U, due to the flow divergence in front of the obstacle,

$$\Delta p = \tfrac{1}{2}\rho U_0^2\{1 - (U/U_0)^2\} \qquad (7.6\text{-}7)$$

Actually, dolphins and porpoises are known to get a partial or complete free ride on the waves produced by wind, breakers observed at beaches (as discussed in the next subsection), and the waves generated by the bow of a ship and known as bow waves or side waves (Woodcock 1948; Hayes 1953; Perry et al. 1961). A pressure gradient pointed in the forward direction can also be found at other places, such as ahead of a blunt bow (Fejer and Backus 1960; Hertel 1969), or behind the maximum cross area of an immersed body (Norris and Prescott 1961).

FIG. 7.6-3. A woodcut print by Hokusai Katsushika.

Breaking waves. When a regular wave approaches the beach, its characteristics change with decreasing water depth in such a way that the wave finally runs ashore and breaks up in surf. Breakers on evenly sloping beaches may be of two kinds, plunging and spilling. As the famous woodcut print by Hokusai shows (Fig. 7.6-3), the plunging breaker has a well-rounded and concave front and a convex back. In contrast, the spilling breaker as also shown in Fig. 7.6-3 is concave on both faces and is generated by a wave with a steepness greater than 0.01, particularly if there is an onshore wind. These breaking waves are capable of moving small objects on the beach with the strong pressure generated by the fast flow surrounding the objects.

Small clams. Small clams such as *Chion semigranosus, Latona cuneata,* and *Donax denticulata* live in a sloping area of the beach or wash zone, where the surf breaks. They shift up and down the beach with the tidal cycle. They can burrow into the sand within a few seconds, in less time than the cycle of wave action; thus, they can evade unfavorable wave motion by hiding under the sand. During rising tide, the clams emerge from the sand only for an incoming wave and get pushed up by the wave, while during ebbing tide, the clams emerge from the sand only for a receding wave.

Coconuts. As stated in Sect. 2.3.2 coconuts and seeds of *Crinum* are distributed widely by floating on currents and are tossed onto the seashore by breaking waves. The voyage begins when a seed drops from the treetop into the sea. The wind and currents determine the course taken by the seed, as well as the landing site.

CHAPTER 8

Conclusions

8.1 Concluding Remarks by Chapter

Chapter 1 introduced a general outline of the basic theory required for studying biokinetics, touching on the fluid-dynamic characteristics of air and water to help the reader understand the dynamic behavior of living creatures of various configurations that dwell in different environments and pursue diverse modes of life. By treating the equations of motion representing the physical and kinematical phenomena in a nondimensional form, it is possible to clarify the generalized rules of motion and the modes of locomotion in a unified form for various sizes and even for different species, and thus to characterize the motion of living creatures by a few fundamental nondimensional parameters, such as Reynolds number, reduced frequency, and Froude number.

Chapter 2 discussed the mechanics of dragging, floating, and jumping. It was seen that locomotion, whether through air or water, is performed using drag force and/or buoyant force against the force of gravity. Dragging and floating in air are possible only for tiny creatures which lack a flying device or for small creatures having very thin threads, because for these small creatures both air and water have extremely viscous characteristics and using the inertial effect or the gravity effect is almost ineffective. In contrast, it is easy for all other creatures, even large ones, to use these modes of locomotion in water because their density is nearly equal to that of the water. Species dispersal in air or water over long distances can therefore be performed without consuming locomotory energy.

It was also seen that the highest standing jump an animal can achieve is about 2 m, but somewhat higher running jumps are possible because the kinetic energy of running is converted into potential energy. In addition, falling through air onto the ground or water is not dangerous for small organisms.

In *Chap. 3*, the fluid-dynamic characteristics of a fixed wing, statistical data on bird configuration, and the gliding performance of various plant seeds and animals were introduced. The body configuration and wing structure of large birds were found to be well adapted to gliding flight in air. The observations made here include:

1. Large land birds have a pair of wings of medium aspect ratio with slotted tips and a large and fully extendable tail fan for supporting steady gliding flight in gusty winds and for enabling them to maneuver adroitly while preying in complex enviroments such as among trees and bushes. Land birds can utilize the up currents which occur frequently over land. Sea birds seldom have this opportunity.
2. Large sea birds have a pair of wings of high aspect ratio with pointed tips and a small tail fan to support high-performance gliding flight. They utilize the vertical shear of horizontal wind near the sea surface for performing the continuous gliding flight called dynamic soaring.
3. Other gliding animals, such as flying squirrels and flying reptiles, are not good fliers because of the low aspect ratio of their membranous wings. Their occasional flights are, however, energy saving and sufficient to keep them safe from harm. Flying fish and flying squid have a pair of medium aspect ratio wings and make optimal gliding flight after taking off from the sea surface. They probably fly to evade predators. They usually fly horizontally with decreasing speed.

4. Samara or other winged seeds take advantage of the wind for dispersal through either steady gliding flight or autorotational flight. The wing configuration and center of gravity are well adapted for optimal dispersive flight.

Chapter 4 described the various modes of beating flight and the flight performance of birds and insects. The power that has to be consumed to maintain the profile, induced, and parasite powers in steady horizontal flight is termed the necessary power, and this has to be balanced with the available power generated by the muscles for wing beating. Small flying creatures have sufficient available power for continuous hovering flight, which, like high-speed flight with high beating frequency, requires much necessary power. However, as the body size and body mass increase, the available power becomes insufficient for continuous hovering flight and the bird has to gain air speed for takeoff by running on land or water surfaces, or by falling from a high surface.

For cruising flight, the optimal phase shift between feathering and flapping motions is $\pi/2$ as the wing surface follows the tangential line of the wing orbit. This is performed by a combination of active twist and elastic twist of the outer wing during the beating motion.

Flying creatures are limited to a mass of about 15 kg. The beating frequency of large birds is nearly 1 Hz. As the body size decreases, beating becomes more pronounced than gliding and the beating frequency increases. For small birds and insects, the beating frequency is locked on the resonant frequency of the beating mechanism so as to cancel the unfavorable power need for inertial force at high frequency by introducing an elastic constraint in the mechanism.

For reducing energy consumption, large beating birds either fly in formation or take advantage of the ground effect, while some small birds perform intermittent beating or bounding flight.

Unlike a bird's wing, which has a streamlined airfoil section, an insect's wing has a nonstreamlined airfoil section with rough surfaces. The sharp leading edge and rough surface of the wing provide a good lift-to-drag ratio in the low Reynolds number range. Insects having two pairs of high aspect ratio wings utilize mainly the steady or quasi-steady aerodynamic force generated by beating motion with a low reduced frequency, whereas insects having two pairs or one pair of low aspect ratio wings rely on the unsteady aerodynamic force resulting from the separated flow around the wing, which moves with high reduced frequency and low Reynolds number.

In *Chap. 5*, the snaking motion of elongated organisms was introduced and analyzed. It was seen that almost all microorganisms use either a snaking or spiral motion of a flagellum or bundle of flagella, with a constant amplitude of undulation. Some small animals rely on the motion of cilia arranged on the body surface. These methods of locomotion, as well as the devices themselves, are very simple and well adapted for swimming in relatively viscous fluid. However, the efficiency of motion is very low in comparison with the snaking motions of larger animals with an elongated body or fin that rely on fluid inertial forces.

The snaking motion in large elongated animals is characterized by a lateral undulatory body (or fin) motion with increasing amplitude toward the tail along the longitudinal axis of the body, which has a constant height. Since the wavelength of this snaking motion is usually smaller than the body length, a few complete waves can be observed in the undulatory body motion. The inertial force as well as the drag force related to the body undulation can be utilized to generate propulsive force and thus the efficiency is far larger than that of flagellar undulation with a constant amplitude. By increasing the body or fin height toward the tail, the inertial force is further augmented by the added mass in proportion to the square of the height.

Snaking motion is well adapted to the needs of an animal which wants to maintain a precise position relative to an object in the water. The motion and the mechanism of the devices for producing it are so simple that the undulation can be sent easily either forward or backward and the switching of the wave direction can be quickly altered without any appreciable delay.

Chapter 6 analyzed the various configurations of caudal fins and their use in swimming. Animals that swim in water have highly streamlined shapes and smooth body surfaces exhibiting special hydrodynamic properties that may provide lower friction drag. The propulsive force generated by such a streamlined body in undulatory motion results from the inertial force of an added mass of fluid and thus increases by enlarging either the wave amplitude or the body depth (width in vertical direction) towards the tail. Enlarging the wave amplitude is effective for increasing the fluid acceleration and is used by the slender, elongated creatures discussed in Chap. 5. On the other hand, enlarging the body depth towards the tail increases the added mass and is used by creatures with flat bodies; a specific manifestation is the caudal fin. Locomotion modes are thus strongly related to life behavior in given environmental conditions.

Fish living in rivers have very flexible slender bodies and can move even in shallow and rapid streams by sending a lateral deflection wave with backwardly increasing amplitude along the longitudinal axis of the body and fanning a triangular caudal fin (an optimally designed wing having a highly swept leading edge) with large amplitude. Such a caudal fin can bear the violent fanning of slender fish. Fish and cetaceans inhabiting deep water have widely extended bodies and large laterally extended caudal fins (vertically for fish and horizontally for cetaceans). They are strong swimmers over a broad range of speeds.

In very low speed swimming, as during precise position keeping relative to a prey or other object, locomotion is performed by a pair of pectoral fins (low aspect ratio wings) that exibit various locomotory behaviors. Fish with a crescent or lunate caudal fin (an optimally designed high aspect ratio wing) swim continuously with high cruising speed and dynamically pursue their prey. For these fish, the dynamic lift generated by a pair of pectoral fins as a high aspect ratio wing makes up for the insufficient buoyant force against the body weight. Intermittent swimming called porpoising is an economic way of locomotion for dolphins, porpoises, and penguins.

In *Chap. 7*, swimming that essentially utilizes the inertial force was discussed, including paddling, whipping, and jetting. Paddling and whipping rely on the inertial force and the hydrodynamic drag generated by legs and feet, with or without membranes. The inertial force is generated by these flat or blunt bodies in unsteady motion within the acceleration phase of the power stroke. The hydrodynamic drag results mainly from the inertial force of fluid around the bodies in the separated state. Jetting utilizes the momentum change of water motion between inhalant and exhalant phases. A high speed jet is effective for dashing, but for slow economical movement the jet speed should be low and the duration of the exhalant phase should be long for a given impulse. The mechanism of the above locomotion modes is simple and such modes are sometimes used as a secondary method of locomotion in animals having other well-developed locomotive devices. Good examples are the swimming of water birds using webbed feet and the swimming of fish using pectoral fins.

Other active swimming methods that rely on hydrodynamic lift include sweeping and beating, which are used for active locomotion during feeding, preying, and migrating, and skating, which is utilized only by small insects and fish in special restricted environments, such as at the boundary between air and water or between water and a solid surface. Sailing and wave riding are also used in these environments for passive locomotion of plants and animals. Although no complex mechanisms are used, the movement is nevertheless often very dramatic.

8.2 Wing Types and Body Forms

There are many methods of locomotion for active flying and swimming, and various ways of locomotion using a wing or wings. Figure 8.2-1 shows several modes of wing motion for lifting and propelling living creatures in a fluid as classified by the name of the wing: (a) stationary or fixed wing for gliding, soaring, and sailing, (b) rotary wing for braking or slow descent, (c) wheeling or tumbling wing for the same purpose, (d) beating wing for flight in air and water, (e) paddling plate or rowing blade for positive propulsion in water, (f) fanning wing for either dashing or power-saving propulsion in water, (g) sweeping wing for propulsion in water, and (h) cyclo-gyro wing for lifting and propelling.

The stationary wing must have forward speed to generate lifting force. The rotary wing, however, has a rotational speed about an axis close to one end of, and perpendicular to, the plane of the wing, and thus creates a lifting force along the rotational axis mainly near the other end or tip of the wing. The direction of the lifting force can be controlled by tilting the rotational axis. On the other hand, the wheeling wing, having an angular motion about the spanwise axis through the midchord of the wing, generates lift in forward motion by utilizing the Magnus effect, transverse force generated by a spinning object moving in fluid.

In the beating wing, locomotion is characterized by three motions around the respective hinges

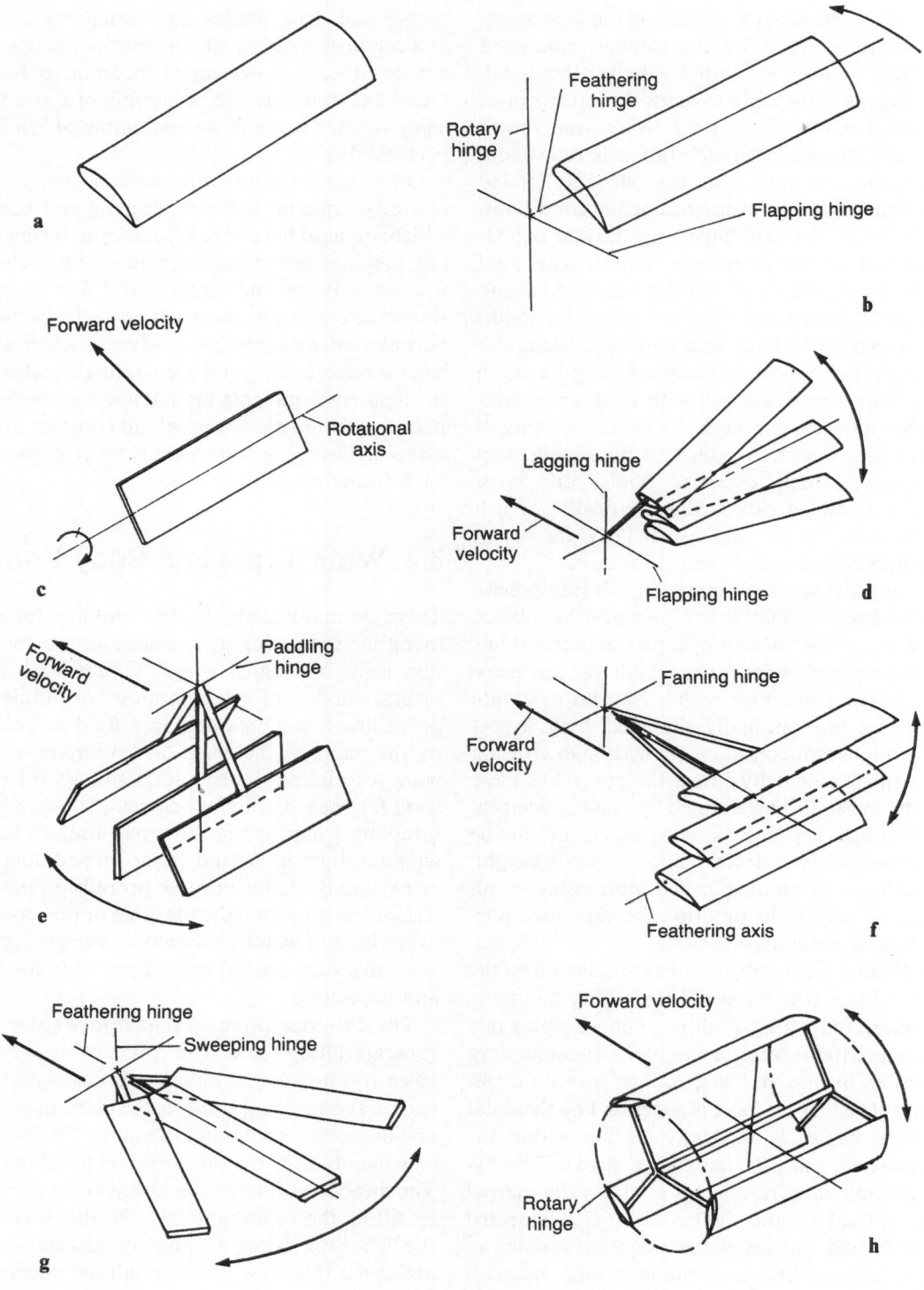

F<small>IG.</small> 8.2-1. **a–h.** Wing types. **a** Stationary wing. **b** Rotary wing. **c** Wheeling or tumbling wing. **d** Beating wing.
e Paddling plate. **f** Fanning wing. **g** Sweeping wing. **h** Cyclo-gyro wing.

or joints located at one end of the wing: flapping (normal to the wing surface), lagging (parallel to the wing surface), and feathering (twisting along the spanwise axis). Both the propulsive and lifting forces are generated near the other end of the wing. When the wing has forward speed, the inner wing is mainly responsible for producing the lifting force. By keeping the joints close to the center of gravity, flight is possible without constant use of another wing for trimming the body. This is the only mode of flight that makes it possible to generate both lifting and propulsive forces near the center of gravity of the creature, and is thus widely used by creatures that fly in the air. The only other device that might be used by flying creatures is the rotary wing but it has not been adopted, except in rotation of the whole body, because of the physiological difficulties experienced during rotation of any part of the body.

In the paddling plate, the winglike plate is moved perpendicularly to the plate surface; thus, the lifting force is not important. The propulsive force relies on the inertial force of the associated fluid as well as the hydrodynamic drag force, the former of which is especially important at low plate speed. The mechanical characteristics of the rowing blade are similar to those of the paddling plate.

The fanning wing has heaving and feathering motions made possible by adopting two or more locomotive hinges arranged parallel to the wingspan. Thus, the wing can produce a lifting force as well as a propulsive force over the whole span of the wing. However, as the wing cannot be arranged close to the center of gravity, the lifting force cannot be utilized effectively. Wide use of the fanning wing for propulsion is therefore limited to aquatic animals.

In the sweeping wing, locomotion is achieved through a coupled lagging and feathering motion of a long slender blade. The leading and trailing edges of the blade are symmetrical around a pivot or universal joint located at one end, and the blade is immersed in the water diagonally. The forward component of the lift generated by the blade is used as the propulsive force. Since the mechanism can be used even near the center of gravity, aquatic animals can use it for propulsion only.

The cyclo-gyro wing is a kind of rotating wing but differs from the ordinary wheeling wing in that its rotational axis is parallel, not perpendicular, to the wing-span. By introducing adequate cyclic feathering motion, the mean lifting force can be directed in any desired direction. If the radius of the cylinder on which the wings are arranged is reduced to zero, the cyclo-gyro wing becomes a wheeling wing.

It is interesting to observe that swimming in water makes possible a much wider variety of body and/or wing motions than can be used for flying in air, since there is little or no need to counterbalance the force of gravity and the locomotion can be concentrated principally on propelling the body forward. The propulsive force can thus be generated by utilizing the fluid dynamic drag, lift, and other inertial force generated by the steady and/or unsteady body or wing motion. That is to say, very small creatures utilize the friction drag predominant in viscous fluid by sending a deflection wave with constant amplitude, whereas large creatures swimming in comparably less viscous fluid utilize either the pressure drag and inertial force by sending a deflection wave with increasing amplitude along the body of almost constant height, or the lift and inertial force by sending a deflection wave with increasing amplitude along the body of enlarged height toward the tail, typically a winglike caudal fin.

8.3 In Summary

All creatures known to man, and man himself, evolved on Earth, and over the eons of the Earth's history, these creatures developed in response to their environments, the atmosphere, the ocean, and the land, resulting in the present states that we see now. They will no doubt continue to evolve in the future. Evolution proceeds through the natural selection of those members of the species best adapted to the changing environmental conditions. Many environmental changes, such as changes in the composition of the atmosphere, are not independent natural phenomena but phenomena caused in large part by the activity of the living creatures themselves, specifically of man. Occasionally, some species have become extinct because of their failure to cope with abrupt changes in their surroundings, while other species were able to adapt to the new environmental conditions and survive.

The activities of plants and animals are strongly influenced by the daily and seasonal variation in sunlight and darkness and the monthly waxing and waning of the moon. The Earth's creatures grow, feed, and breed to these rhythms, in various environments in which they live together with others, either in cooperation or in competition.

Although each respective species has several ways of locomotion corresponding to environmental conditions, its modes of locomotion are limited by the body configuration including its size. In order to survive, each creature has developed a body configuration and a mode of locomotion that seems to be optimal for its style of life in a given habitat and a specified situation. Thus, in this book, the modes of locomotion are classified into various fundamental patterns. Through the description of these patterns and the inclusion of living examples of each, it can be seen that either single or multiple patterns which are well fitted to the creature's body configuration and life style in its environment are selected. Therefore, the reasons behind why locomotion is limited to such patterns become clear.

The application of the variational method in mathematical analysis to solve various optimal problems proposed in each chapter clarifies this point of view further. Solutions of these problems obtained either analytically or numerically showed that optimal modes of locomotion certainly exist, and by comparison with the observed data the relation shown in Fig. 1.1-1 was confirmable.

In the analysis of the fluid-dynamic load acting on wings in steady and unsteady motions, the local circulation method (LCM) was used for a wide range of locomotion modes, such as gliding, beating, fanning, and sweeping. This method was originally developed for performance, load, and noise analyses of rotary wings by myself and my colleagues. The method was fundamentally based on the lifting line theory for wings but was extended to the field of the lifting surface theory for wings in unsteady motion under Weissinger's approximation (Weissinger 1947, De Young 1947, De Young and Harper 1948). The application of the LCM to wings in reciprocating motion instead of rotating motion was realized in the low reduced frequency range by assuming a rigid vortex wake system generated from the wing and moving with the induced velocity. Highly precise results have been obtained in, for example, the calculation of dragonfly flight performance. This method of calculation could be extended further to the analysis of the dynamic behaviour of other flying and swimming creatures and to reconfirm the optimality of their lifestyle, once detailed information on the body and wing configurations and the modes of locomotion of the wings become available.

Instead of giving a simple collection of data and materials, this entire approach allows not only a fundamental analysis of the various modes of locomotion but also a bird's eye view on the locomotion of species from a novel and unique engineering perspective. This is in contrast to previous studies where locomotion used to be treated separately as an individual problem.

References

Abbott IH, von Doenhoff AE (1959) Theory of wing sections. Dover, New York

Abbott BC, Bigland B, Ritchie JM (1952) The physiological cost of negative work. J Physiol (Lond) 117: 380–390

Abe J (1981) Vortices generated by fish locomotion (in Japanese). J Flow Visualization Soc Japan (frontispiece)

Abrahams MV, Colgan PW (1985) Risk of predation, hydrodynamic efficiency and their influence on school structure. Environ Biol Fish 13:195–202

Alexander RMcN (1959) The physical properties of the swimbladder in intact Cypriniformes. J Exp Biol 36:315–332

Alexander RMcN (1968) Animal mechanics. University of Washington Press, Seattle, pp 266–325

Alexander RMcN (1971) Animal mechanics. Sidgwick and Jackson, London

Alexander RMcN (1973) Muscle performance in locomotion and other strenuous activities. In: Bolis L, Maddrell SHP, Schmidt-Nielsen K (eds) Comparative animal physiology. North Holland, Amsterdam, pp 1–22

Alexander RMcN (1977) Swimming. In: Alexander RM, Goldspink G (eds) Mechanics and energetics of animal locomotion. Chapman and Hall, London

Alexander MB, Camp DW (1983) Significant events in low-level flow conditions hazardous to aircraft. NASA TM 82522

Alexander MB, Camp DW (1985) Analysis of low-altitude wind speed and direction shears. J Aircraft 22:705–712

Altringham JD, Johnston IA (1990a) Scaling effects on muscle function: Power output of isolated fish muscle fibres performing oscillatory work. J Exp Biol 151: 453–467

Altringham JD, Johnston IA (1990b) Modeling muscle power output in a swimming fish. J Exp Biol 148: 395–402

Anderson GW (1973) An experimental investigation of a high lift device on the owl wing. Air Force Institute of Technology, Wright-Patterson Air Force Base, Ohio, March AD-769492

Anderson RA (1974) Formation of the bacterial flagellar bundle. In: Wu TYT, Brokaw CJ, Brennen C (eds) Swimming and flying in nature, 1. Plenum Press, New York, pp 45–56

Anderson SO, Weis-Fogh T (1964) Resilin. A rubber-like protein in the insect cuticle. Adv Insect Physiol 2:1–65

Anikouchine WA, Sternberg RW (1973) The world ocean. An introduction to oceanography. Prentice-Hall, New Jersey

Anon (1967) Angler's Encyclopedia. Magazine 67-7, Tokyo

Archer SD, Johnston IA (1989) Kinematics of labriform and subcarangiform swimming in the Antarctic fish Nontothenia neglecta. J Exp Biol 143:195–210

Ashenberg J, Weihs D (1984) Minimum induced drag of wings with curved planform. J Aircraft 21:89–91

Asmussen E (1952) Positive and negative musclar work. Acta Physiol Scand 28:364–382

Au D, Weihs D (1980a) At high speeds dolphins save energy by leaping. Nature 284:548–550

Au D, Weihs D (1980b) Reply on Gordon's comments. Nature 287:759

Augspurger CK (1986) Morphology and dispersal potential of wind-dispersed diaspores of neotropical trees. Am J Bot 73(3):353–363

Azuma A (1979) Flying behaviors in living creatures. Blue Backs B-378 (in Japanese). Kodansha, Tokyo

Azuma A (1980) Swimming behaviors in living creatures. Blue Backs B-412 (in Japanese). Kodansha, Tokyo

Azuma A (1981a) A reasoning on the flight of squid (in Japanese). J Pop Sci: 81–85

Azuma A (1981b) Unsteady aerodynamics observed in locomotion of living creatures (in Japanese). J Japan Aero and Astro Sci 29:91–96

Azuma A (1983) Biomechanical aspects of animal flying and swimming. International Series on Biomechanics, vol 4A, Biomechanics VIII-A. Human Kinetics Publishers, Champaign, Illinois

Azuma A (1986) Splendid locomotions in living creatures (in Japanese). Kyoritsu, Tokyo

Azuma A, Kawachi K (1979) Local momentum theory and its application to the rotary wing. J Aircraft 16:6–14

Azuma A, Yasuda K (1985) The flight of rotary samara (in Japanese). Proceedings of the 23rd Aircraft Symposium. The Japan Society for Aeronautical and Space Science. October 21–22, Nagoya, pp 354–357

Azuma A, Okuno Y (1987) Flight of samara *Alsomitra macrocarpa*. J Theo Biol 129:263–274

Azuma A, Watanabe T (1988) Flight performance of a dragonfly. J Exp Biol 137:221–252

Azuma A, Yasuda K (1989) Flight performance of rotary seeds. J Exp Biol 138:23–54

Azuma A, Nasu K, Hayashi T (1983) An extension of the local momentum theory to the rotors operating in twisted flow field. Vertica 7:45–59

Azuma A, Azuma S, Watanabe I, Furuta T (1985) Flight mechanics of a dragonfly. J Exp Biol 116:79–107

Azuma A, Furuta T, Iuchi M, Watanabe I (1989) Hydrodynamic analysis of the sweeping of a "ro"—an oriental scull. J Ship Research 33(1):47–62

Baba SA (1974) Developmental changes in the pattern of ciliary response and the swimming behavior in some invertebrate larvae. In: Wu TYT, Brokaw CJ, Brennen C (eds) Swimming and flying in nature, vol 1. Plenum, New York pp 317–323

Bagley JA (1956) The pressure distribution on two-dimensional wings near the ground. RAE TN Aero 2472

Bainbridge R (1958) The speed of swimming of fish as related to size and the frequency and amplitude of the tail beat. J Exp Biol 35:109–133

Bainbridge R (1960) Speed and stamina in three fishes. J Exp Biol 37:129–153

Bainbridge R (1961) Problems of fish locomotion: Vertebrate locomotion. Symp Zool Soc [Lond] 5:13–32

Bainbridge R (1963) Caudal fin and body movements in the propulsion of some fish. J Exp Biol 40:23–56

Baker PS, Cooper RJ (1979) The natural flight of the migratory locust *Locusta migratoria* L: Wing movements and gliding. J Comp Physiol 131:79–87, 89–94

Balasubramanian R (1980) Analytical and design techniques for drag reduction studies on wavy surfaces. NASA CR 3225

Batchelor GK (1970) Slender-body theory for particles of arbitrary cross-section in Stokes flow. J Fluid Mech 44:419–440

Baudoin R (1980) The world of water skaters (in Japanese). Anima 84:53–59

Baylor ER, Baylor MB, Blanchard DC, Syzbek LD, Appel C (1977) Virus transfer from surf to wind. Science 198:575–580

Beamish FWH (1978) Swimming capacity. In: Hoar WS, Randall DJ (eds) Fish Physiology, vol VII, Locomotion. Academic, New York

Belyayer VV, Zuyer GV (1969) Hydrodynamic hypothesis of schooling in fishes. J Ichthyology 9:578–584

Bennet-Clark HC (1975) The energetics of the jump of the locust *Schistocerca gregaria*. J Exp Biol 63:53–83

Bennet-Clark HC (1977) Scale effects in jumping animals. In: Pedley TJ (ed) Scale effects in animal locomotion. Academic, New York

Bennet-Clark HC (1980) Aerodynamics of insect jumping. In: Elder HY, Trueman ER (eds) Aspects of animal movement. Cambridge University Press, Cambridge

Bennet-Clark HC, Lucey ECA (1967) The jump of the flea: A study of the energetics and a model of the mechanism. J Exp Biol 47:59–76

Bennett L (1966) Insect aerodynamics: Vertical sustaining force in near-hovering flight. Science 152:1263–1266

Berg HC (1975) How bacteria swim. Sci Am 233:36–44

Berg HC (1976) How spirochetes may swim. J Theor Biol 56:269–273

Berger AJ (1961) Bird study. Dover, New York

Bernstein MH, Thomas SP, Schmidt-Nielsen K (1973) Power input during flight of the fish crow *Corvus ossifragus*. J Exp Biol 58:401–410

Bernstein MH, Curtis MB, Hudson DV (1979) Independence of brain and body temperature in flying American kestrels. Am J Physiol 237:58–62

Bernstein MH, Duran HL, Pinshow B (1984) Extrapulmonary gas exchange enhances brain oxygen in pigeons. Science 226:564–566

Betts CR, Wooton RJ (1988) Wing shape and flight behaviour in the butterflies *Lepidoptera papilionoidea* and *L. hesperioidea*: A preliminary analysis. J Exp Biol 138:271–288

Betz A (1932) Verhalten von Wirbelsystemen. Z Angew Math Mech 3:164–174 (Behavior of vortex systems. NACA TM-713)

Bilanin AJ, Teske ME, Williamson GG (1977) Vortex interactions and decay in aircraft wakes. AIAA J 15:250–260

Bishop DW (1958) Mobility of the sperm flagellum. Nature 182:1638–1640

Bisplinghoff RL, Ashley H, Halfman RL (1957) Aeroelasticity. Addison-Wesley, Reading, Massachusetts

Blake JR (1971) Infinite models for ciliary propulsion. J Fluid Mech 49:209–222

Blake RW (1979) The mechanics of labriform locomotion I. Labriform locomotion in the angelfish *Pterophyllum eimekei*: An analysis of the power stroke. J Exp Biol 82:255–271

Blake RW (1980) The mechanics of labriform locomotion. J Exp Biol 85:337–342

Blake RW (1981a) Influence of pectoral fin shape on thrust and drag in labriform locomotion. J Zool

(London) 194:53–66

Blake RW (1981b) Mechanics of drag-based mechanisms of propulsion in aquatic vertebrates. Vertebrate locomotion. Symp Zool Soc [Lond] 48:29–52

Blake RW (1981c) Mechanics of ostraciiform locomotion. Can J Zool 59:1067–1071

Blake RW (1983a) Mechanics of gliding in birds with special reference to the influence of the ground effect. J Biomech 16:649–654

Blake RW (1983b) Energy of leaping in dolphins and other aquatic animals. J Mar Biol Assoc UK 63:61–70

Blake RW (1983c) Fish locomotion. Cambridge University Press, Cambridge

Blake RW (1983d) Median and paired fin propulsion. In: Webb DW, Weis D (eds) Fish biomechanics. Praeger Special Studies, Praeger Scientific, New York

Blick EF, Walter RR (1968) Turbulent boundary layer characteristics of compliant surfaces. J Aircraft 5:11–16

Blum JJ, Hines M (1979) Biophysics of flagellar motility. Quart Rev Biophysics 12(2):103–180

Boettiger EG, Furshpan E (1952) The mechanics of flight movements in Diptera. Biol Bull Mar Biol Lab, Woods Hole 102:200–211

Bone Q (1975) Muscular and energetic aspects of fish swimming. In: Wu TYT, Brokaw CJ, Brennen C (eds) Swimming and flying in nature, vol 2. Plenum New York, pp 493–528

Borror DJ, Delong DM, Triplehorn CA (1976) An introduction to the study of insects, 4th edn. Holt, Rinehart, and Winston, New York

Breder CM Jr (1924) Respiration as a factor in locomotion of fishes. Am Nat 58:145–155

Breder CM Jr (1926) The locomotion of fishes. Zoologica NY 4:159–297

Brett JR (1963) The energy required for swimming by young sockeye salmon with a comparison of the drag force on a dead fish. Trans R Soc Can (IV) 1:441–457

Brett JR (1965) The relation of size to rate of oxygen consumption and sustained swimming speed of sockeye salmon (*Oncorhynchus nerka*). J Fish Res Bd Can 22:1491–1501

Brett JR, Glass NR (1973) Metabolic rates and critical swimming speeds of sockeye salmon (*Oncorhynchus nerka*) in relation to size and temperature. J Fish Res Bd Can 30:379–387

Brody S (1945) Bioenergetics and growth. Reinhold, New York

Brokaw CJ (1965) Non-sinusoidal bending waves of sperm flagella. J Exp Biol 43:155–169

Brokaw CJ (1972) Flagellar movement: A sliding filament model. Science 178:455–462

Brooker BE (1965) Mastigonemes in a bodonid flagellate. Exp Cell Res 37:300–305

Brooks JD, Lang TG (1967) Simplified method for estimating torpedo drag. In: Greiner L (ed) Under-

water missile propulsion. Compaso, Arlington, Virginia, pp 117–146

Brown CE (1973) Aerodynamics of wake vortices. AIAA J 11:531–536

Brown RHJ (1948) The flight of birds: Cycles of the pigeon. J Exp Biol 25:322–333

Brown RHJ (1953) The flight of birds: Wing function in relation to flight speed. J Exp Biol 30:90–103

Brown RHJ (1963) Jumping arthropods. Times Sci Rev Summer: 6–7

Brown CE, Muir BS (1970) Analysis of ram ventilation of fish gills with application to skipjack tuna (*Katsuonus pelamis*). Ibid 27:1637–1652

Bryson RA, Kutzbach (1968) Air pollution. Assoc Am Geog Comm Coll Geog, Washington DC

Buchthal F, Weis-Fogh T (1956) Contribution of the sarcolemma to the force exerted by resting muscle of insects. Acta Physiol Scand 35:345–364

Buckholz RH (1986) The functional role of wing corrugations in living systems. J Fluids Eng 108:93–97

Buller AHR (1934) Researches on fungi, vol 6. Longmans Green, New York

Burdak VD (1969) The ontogenetic development of the scale cover of the mullet *Mugil saliens* risso. J Zool 47:732–738

Bursnall WJ, Loftin LK (1951) Experimental investigation of the pressure distribution about a yawed circular cylinder in the critical Reynolds number range. NACA TN 2463

Burton M (ed) (1980) The new Larousse encyclopedia of animal life. Hamlyn, London

Burton AJ, Sademan DC (1961) The lift provided by the elytra of the rhinoceros beetle *Oryctes boas*. S Afr J Sci 57:107–109

Bushnell DM, Hefner JN, Ash RL (1977) Compliant wall drag reduction for turbulent boundary layers. Phys Fluids 20:531–548

Byrne DN, Buchmann SL, Spangler HG (1988) Relationship between wing loading, wingbeat frequency, and body mass in homopterous insects. J Exp Biol 135:9–23

Campbell A (1983) Seashore life. Natural Library, Exeter Books, New York

Carr LW, McAlister KW, McCroskey WJ (1977) Analysis of the development of dynamic stall based on oscillating airfoil experiments. NASA TN D-8382

Carta FO (1967) Unsteady normal force on an airfoil in a periodically stalled inlet flow. J Aircraft 4:416–421

Casarella MJ, Parsons M (1970) Cable systems under hydrodynamic loading. Marine Tech Soc J 4:27–44

Chapman RF (1969) The insects: Structure and function. English University Press, London

Cheng HK, Murillo LE (1984) Lunate-tail swimming propulsion as a problem of curved lifting line in unsteady flow: Asymptotic theory. J Fluid Mech 143:327–350

Childress S (1981) Mechanics of swimming and flying. Cambridge University Press, Cambridge

Chopra MG (1974) Hydrodynamics of lunate-tail swimming propulsion. J Fluid Mech 64(2):375–391

Chopra MG, Kambe T (1977) Hydromechanics of lunate-tail swimming propulsion. J Fluid Mech 79(2): 49–69

Chwang AT, Wu TYT (1971) A note on the helical movement of micro-organisms Proc R Soc Lond [Biol] 178:327–346

Chwang AT, Wu TYT (1974) Hydromechanics of flagellar movements. In: Wu TYT, Brokaw CJ, Brennen C (eds) Swimming and flying in nature, vol 1. Plenum New York, pp 13–30

Ciffone DL (1974) Correlation for estimating vortex rotational velocity downstream dependence. J Aircraft 15:716–717

Ciffone DL, Orloff KL (1975) Far-field wake-vortex character-characteristics of wings. J Aircraft 12: 464–470

Clark BD, Bemis W (1979) Kinematics of swimming of penguins at the Detroit zoo. J Zool Lond 188: 411–428

Clements RR, Maull DJ (1973) The rolling up of a trailing vortex sheet. Aero J R Aero Soc 77:46–51

Coakley CJ, Holwill MEJ (1972) Propulsion of micro-organisms by three-dimensional flagellar waves. J Theor Biol 35:525–542

Coffey DJ (1977) Dolphins, whales, and porpoises. An encyclopedia of sea mammals. Collier, New York

Cone CA (1960) A theoretical investigation of vortex sheet deformation behind a highly loaded wing. NASA TN D-657

Cone CD Jr (1962a) The soaring flight of birds. J Sci Am 206:130–140

Cone CD Jr (1962b) The theory of induced lift and minimum induced drag of nonplanar lifting systems. NASA TR R-139

Cone CD Jr (1964) The design of sailplanes for optimum thermal soaring performance. NASA TN D-2052

Cooter RJ, Baker PS (1977) Weis-Fogh clap and fling mechanism in locusta. Nature 269:53–54

Cornelis P (1981) Effect of winglets on performance and handling qualities of general aviation aircraft. J Aircraft 18:587–591

Cox RG (1970) The motion of long slender bodies in a viscous fluid. Part 1. General theory. J Fluid Mech 44:791–810

Currey JD (1977) Problems of scaling in the skeleton. In: Pedley TC (ed) Scale effects in animal locomotion. Academic, London, pp 153–181

Currey JD (1980) Skeletal factors in locomotion. In: Elder HY, Trueman ER (eds) Aspects of animal movement. Cambridge University Press, Cambridge, pp 27–48

Dadswell ML, Weihs D (1990) Size-related hydrodynamic characteristics of the giant scallop, *Placo-*

pectern magellanicus (Bivalvia: Pectinidae). Can J Zool 68:778–785

Dam CP van (1987a) Efficiency characteristics of crescent-shaped wings and caudal fins. Nature 325: 435–437

Dam CP van (1987b) Induced-drag characteristics of crescent-moon-shaped wings. J Aircraft 24:115–119

Dam CP van, Holmes RJ, Pitts C (1981) Effects of winglets on performance and handling qualities of general aviation aircraft. J Aircraft 18:587–591

Dam CP, Vijgen PMHW, Holmes BJ (1991) Aerodynamic characteristics of crescent and elliptic wings at high angles of attack. J Aircraft 28:253–260

Daniel TL, Meyhöfer E (1989) Size limits in escape locomotion of carridean shrimp. J Exp Biol 143: 245–265

De Young J (1947) Theoretical additional span loading characteristics of wings with arbitrary sweep, aspect ratio, and taper ratio. NACA TN 1491

De Young J (1977) Nonplanar wing load-line and slender wing theory. NASA CR-2864

De Young J, Harper CW (1948) Theoretical symmetrical span loading of subsonic speeds for wings having arbitrary planform. NACA Rep. 921

DeMont ME (1990) Tuned oscillation in the swimming scallop, *Pecten maximus*. Can J Zool 68:786–791

DeMont ME, Gosline JM (1988) Mechanics of jet propulsion in the hydromedusan jellyfish *Polyorchics penicillatus*. I, II, III. J Exp Biol 134:313–322, 333– 345, 346–361

Dinkelacker A (1966) Preliminary experiments on the influence of flexible walls on boundary layer turbulence. J Sound Vibration 4:187–214

Du Bois AB, Cavagna GA, Fox RS (1976) Locomotion of bluefish. J Exp Zool 195:223–235

Dudley R, Ellington CP (1990) Mechanics of forward flight in bumblebees, I, II. J Exp Biol 148:19–52, 53–88

Eames MC (1967) Steady-state theory of towing. The Royal Institution of Naval Architects, London, pp 185–206

Edmunds M (1974) Defence in animals. Longman, New York

Edwards RH, Cheng HK (1982) The separation vortex in the Weis-Fogh circulation-generation mechanism. J Fluid Mech 120:463–473

Elder HY, Trueman ER (1980) Aspects of animal movement. Cambridge University Press, Cambridge

Eliraz Y, Ilan D (1977) Performance of the ARAVA aircraft with wing-tip winglets. Israel J Tech 15:35–43

Ellington CP (1975) Non-steady-state aerodynamics of the flight of *Enearsia formosa*. In: Wu TYT, Brokaw CJ, Brennen C (eds) Swimming and flying in nature, 2. Prenum, New York, pp 783–796

Ellington CP (1984a) The aerodynamics of hovering insect flight. Philos Trans R Soc Lond [Biol] 305: 1–181

Ellington CP (1984b) The aerodynamics of flapping animal flight. Am Zool 24:95–105

Encyclopedia Americana (1963) International edn. American Corporation, New York

Encyclopaedia Zoologica (1963) (in Japanese) Toyama I, Abe M, Tokioka T (eds) Hokuryukan, Tokyo

Ennos AR (1987) A comparative study of the flight mechanism of Diptera. J Exp Biol 127:355–372

Ennos AR (1989a) Inertial and aerodynamic torque on the wings of Diptera in flight. J Exp Biol 142:87–95

Ennos AR (1989b) The kinematics and aerodynamics of the free flight of some Diptera. J Exp Biol 142:49–85

Ennos AR, Wooton RJ (1989) Functional wing morphology and aerodynamics of *Panorpa germanica* (Insecta: Mecoptera). J Exp Biol 143:267–284

Farrand J Jr (1988) Eastern Birds. An Audubon handbook. McGraw-Hill, New York

Favier D, Maresca C, Rebont J (1982) Dynamic stall due to fluctuations of velocity and incidence. AIAA J 20:865–871

Fejer AA, Backus RH (1960) Porpoises and bow-riding of ships under way. Nature 188:700–703

Fisher DH, Blick EF (1966) Turbulent damping by flabby skins. J Aircraft 3:163–164

Flechner SG (1979) Effect of winglet on a first-generation jet transport wing. VI-stability characteristics for a full-span model at subsonic speeds. NASA TR 1330

Franzisket L (1965) Beobachtungen und Messungen am Flug der fliegenden Fische. Abt Allg Zool Physiol 70:235–240

Freymuth P (1988) Propulsive vortical signature of plunging and pitching airfoils. AIAA J 26:881–883

Freymuth P, Bank W, Palmer M (1985a) Further experimental evidence of vortex splitting. J Fluid Mech 152:289–299

Freymuth P, Finaish F, Bank W (1985b) Three-dimensional vortex patterns in a starting flow. J Fluid Mech 161:239–248

Freymuth P, Jackson S, Bank W (1989) Toward dynamic separation without dynamic stall. Exp Fluids 7:187–196

Fulford GR, Blake JR (1986) Macro-ciliary transport in the lung. J Theor Biol 121:381–402

Ganguly DN, Mitra B (1962) On the vertebral column of the teleostean fishes of different habits and habitats. *Mugil corsula*, *Pama pama*, *Triacanthus brevirostris*, and *Andamia heteroptera*. Anat Anz 110:289–311

Garrick IE (1936) Propulsion of a flapping and oscillating airfoil. NACA Rep 567

Gawn RWL (1948) Aspects of the locomotion of whales. Nature 161:44

Gero DR (1952) The hydrodynamic aspects of fish propulsion. American Museum Novitates 1601:1–32

Gessow A, Myers GC Jr (1952) Aerodynamics of the helicopter. Macmillan, New York

Gilliard E (1967) Living birds of the world. Doubleday, New York

Goldspink G (1977a) Energy cost of locomotion. In: Alexander RM, Chapman GG (eds), Mechanics and energetics of animal locomotion. Chapman and Hall, London

Goldspink G (1977b) Design of muscles in relation to locomotion. In: Alexander RM, Chapman GG (eds) Mechanics and energetics of animal locomotion. Chapman and Hall, London

Goldspink G (1977c) Mechanics and energetics of muscle in animals of different sizes with particular reference to the muscle fibre composition of vertebrate muscle. In: Pedley TJ (ed) Scale effects in animal locomotion. Academic, London, pp 37–55

Goldstein S (1929) The forces on the solid body moving through viscous fluid. Proc R Soc Lond [Biol] 123:225–235

Gordon CN (1980) Leaping dolphins. Nature 287:759

Gosline JM, DeMont ME (1985) Jet-propelled swimming in squids. Sci Am 252:74–79

Graham RR (1934) The silent flight of owls. J R Aero Soc 38:837–843

Graham JB, Lowell WR, Rubinoff I, Motta J (1987) Surface and subsurface swimming of the sea snake *Pelamis platurus*. J Exp Biol 127:27–44

Gray J (1933a) Studies in animal locomotion: The movement of fish with special reference to the eel. J Exp Biol 10:88–104

Gray J (1933b) Studies in animal locomotion: The relationship between waves of muscular contraction and the propulsive mechanism of the eel. J Exp Biol 10:386–390

Gray J (1933c) Studies in animal locomotion: The propulsive mechanism of the whiting. J Exp Biol 10:391–400

Gray J (1936) Studies in animal locomotion: The propulsive powers of the dolphin. J Exp Biol 13:192–199

Gray J (1939) Studies in animal locomotion: The kinetics of locomotion of *Nerris diversicolor*. J Exp Biol 16:9–17

Gray J (1953) How animals move. Cambridge University Press, Cambridge

Gray J (1955) The movement of sea urchin spermatozoa. J Exp Biol 32:775–801

Gray J (1957) How fishes swim. Sci Am 192:48–54

Gray J (1968) Animal locomotion. Weidenfeld and Nicolson, London

Gray J, Hancock GJ (1955) The propulsion of sea urchin spermatozoa. J Exp Biol 32:802–814

Gray J, Lissaman HW (1964) The locomotion of nematodes. J Exp Biol 41:135–154

Greenewalt CH (1962) Dimensional relationships for flying animals. Smithsonian miscellaneous collections. Smithsonian Institution, Washington, vol 144, no 2, pp 1–46

Greenewalt CH (1975) The flight of birds. Trans Am Philos Soc 65(4):1–67

Gyorgyfalvy D (1967) Possibilities of drag reduction by

use of a flexible skin. J Aircraft 4:186–192

Haines P, Luers J (1983) Aerodynamic penalties of heavy rain on landing airplanes. J Aircraft 20: 111–119

Hainsworth FR (1987) Precision and dynamics of positioning by Canada geese flying in formation. J Exp Biol 128:445–462

Hainsworth FR (1988) Induced drag savings from ground effect and formation flight in brown pelicans. J Exp Biol 135:431–444

Halfman RM, Johnson HC, Haley SM (1951) Evaluation of high-angle-of-attack aerodynamic-derivation data and stall-flutter prediction techniques. NACA TN 2533

Hall-Craggs ECB (1965) An analysis of the jump of the lesser galago (*Galago senegalensis*). J Zool Lond 147:20–29

Ham ND (1968) Aerodynamic loading on a two-dimensional airfoil during dymamic stall. AIAA J 6:1927–1934

Ham ND, Garelick MS (1968) Dynamic stall considerations in helicopter rotors. J Am Heli Soc 13:49–55

Hancock GJ (1953) Self-propulsion of microscopic organisms through liquids. Proc R Soc Lond [Biol] 217:96–121

Hancock GJ (1970) On rolling up of a trailing vortex sheet. Aero J R Aero Soc 74:749–752

Hankin EH (1913) Animal flight. Iliffe, London

Harper DG, Blake RW (1990) Fast-start performance of rainbow trout *Salmo gairdneri* and northern pike *Esox lucius*. J Exp Biol 150:321–342

Harper DG, Blake RW (1991) Prey capture and the fast-start performance of northern pike *Esox lucius*. J Exp Biol 155:175–192

Harris JE (1936) The role of the fins in the equilibrium of the swimming fish. Wind tunnel test on a model of *Mustelus canis* (Mitchell). J Exp Biol 13:476–493

Harris JE (1938) The role of fins in the equilibrium of the swimming fish. The role of the pelvic fins. J Exp Biol 15:32–47

Harris JE (1953) Fin pattern and mode of life in fishes. In: Marshall SA, Orr D (eds) Essays in marine biology. Oliver and Boyd, Edinburgh, pp 17–28

Hartman FA (1961) Locomotor mechanisms of birds. Smithsonian miscellaneous collections. Smithsonian Institution, Washington DC

Hartt AC (1966) Migrations of salmon in the north Pacific Ocean and Bering Sea as determined by seining and tagging, 1959–1960. Int North Pac Fish Comm Bull 19:1–141

Haussling HJ (1979) Boundary-fitted coordinates for accurate numerical solution of multibody flow problems. J Comp Phys 30:107–124

Hayes WD (1953) Wave riding of dolphins. Nature 172:1060

Henderson Y, Haggard HW (1925) The maximum of human power and its fuel. Am J Phys 72:264–282

Herrick FH (1985) The American lobster. A study of its habits and development. Bull US Fish Comm 15: 1–252

Hersh AS, Soderman PT, Hayden RE (1974) Investigation of acoustic effects of leading-edge serrations on airfoils. J Aircraft 11:197–202

Hertel H (1966) Structure-form-movement. Reinholds, New York

Hertel H (1969) Hydrodynamics of swimming and wave-riding dolphins. In: Andersen HT, Harold T (eds) The biology of marine mammals. Academic, New York

Hess JL, Videler JJ (1984) Fast continuous swimming of saith (*Pollachius virens*): A dynamic analysis of bending moments and muscle power. J Exp Biol 109:229–251

Heyson HH (1959) An evaluation of linearized vortex theory as applied to single and multiple rotors hovering in and out of ground effect. NASA TN D-43

Heyson HH (1960) Measurements of the time-averaged and instantaneous induced velocities in the wake of a helicopter hovering at high tip speeds. NASA TN D-393

Heyson HH, Riebe GD, Fulton CL (1977) Theoretical parametric study of the relative advantages of winglets and wing-tip extensions. NASA Technical Paper 1020

Higuchi H (1978) Avian ecology and evolution (in Japanese). Shisaku, Tokyo

Hill AV (1950) The dimensions of animals and their muscular dynamics. Sci Prog 38:209–230

Hill AV, Howarth JV (1959) The reversal of chemical reactions in contracting muscle during an applied stretch. Proc R Soc Lond [Biol] 151:169–193

Hiramoto Y, Baba SA (1978) A qualitative analysis of flagellar movement in echinoderm spermatozoa. J Exp Biol 76:85–104

Hoerner SF (1965) Fluid dynamic drag. Hoerner, New Jersey

Hoerner SF, Borst HV (1975) Fluid-dynamic lift. Hoerner, New Jersey

Holst E von, Küchemann D (1942) Biological and aerodynamical problems of animal flight. J Roy Aero Soc 46:39–56

Holwill MEJ (1966) Physical aspects of flagellar movement. Physiol Rev 46:696–785

Holwill MEJ (1974) The role of body oscillation in the propulsion of microorganisms. In: Wu TYT, Brokaw CJ, Brennen C (eds) Swimming and flying in nature, vol 1. Plenum, New York, pp 133–141

Holwill MEJ (1977) Low Reynolds numer undulatory propulsion in organisms of different sizes. In: Pedley TJ (ed) Scale effects in animal locomotion. Academic, London

Holwill MEJ, Burge RE (1963) A hydrodynamic study of the motility of flagellated bacteria. Arch Biochem Biophys 101:249–260

Horridge CA (1956) The flight of very small insects. Nature 178:1334–1335

Horton-Smith C (1938) The flight of birds. Witherby, London

Houston AI (1986) The optimal flight velocity for a bird exploiting patches of food. J Theor Biol 119: 345–362

Hoyt JW (1975) Hydrodynamic drag reduction due to fish slimes. In: Wu TYT, Brokaw CJ, Brennen C (eds) Swimming and flying in nature, vol 2. Plenum, New York

Hubbs CL (1933) Observations on the flight of fishes, with a statistical study of the flight of the cypselurinae and remarks on the evolution of the flight of fishes. Mich Acad Sci Arts Letts 17:575–611

Hui CA (1987a) The porpoising of penguins: An energy-conserving behavior for respiratory ventilation? Can J Zool 65:209–211

Hui CA (1987b) Power and speed of swimming dolphins. J Mamm 68(1):126–132

Hui CA (1988) Penguin Swimming. I. Hydrodynamics. Physiol Zool 61(4):333–343

Hummel D (1978) Recent aerodynamic constrictions to problems of bird flight. 11th ICAS Congress, Lisbon, Sept 10–16, A1–05 pp 115–129

Hunter JR, Zweifel JR (1971) Swimming speed, tail beat frequency, tail beat amplitude, and size in mackerel *Trachurus symmetricus* and other fishes. Fish Bull US 69:253–266

Idrac P (1932) Experimentelle Untersuchungen über den Segelflug. Ordenbourg, Munich

Isaacs JD (1969). The nature of oceanic life. Sci Am special edn., pp 189–201

Iwago M (1981) A message from the sea (in Japanese). Asahi Newspapers, Tokyo

Izumi K, Kuwahara K (1983) Unsteady flow field, lift and drag measurements of impulsively started elliptic cylinder and circular arc airfoil. AIAA 16th Fluid and Plasma Dynamics Conference July 12–14, Danvers, Massachusetts

James EC (1975) Lifting line theory for an unsteady wing as a singular perturbation problem. J Fluid Mech 70:735–771

Jarosch R (1967) Studien zur Bewegungsmechanik der Bakterien und Spirochäten des Hochmoores. Oster Botan Z 114:255–306

Jensen M (1956) Biology and physics of locust flight: The aerodynamics of locust flight. Philos Trans R Soc Lond [Biol] 239:511–552

Johannessen CL, Harder JA (1960) Sustained swimming speeds of dolphins. Science 132:1550–1551

Johnson W (1969) The effect of dynamic stall on the response and airloading of helicopter rotor blades. J Am Heli Soc April: 68–79

Johnson TP, Johnston IA (1991) Power output of fish fibres performing oscillatory work: Effects of acute and seasonal temperature change. J Exp Biol 157: 409–423

Johnson JL Jr, McLemore HC, White R, Jordan FL Jr (1979) Full scale wind tunnel investigation of an ayres S2R-800 thrush agricultural airplane. SAE Paper 79-0618

Jones RT (1946) Properties of low aspect-ratio pointed wings at speeds below and above the speed of sound. NACA Rep 835

Jones RT (1950) The spanwise distribution of lift for minimum induced drag of wings having a given lift and a given bending moment. NACA TN 2249

Jones RT (1980) Wing flapping with minimum energy. Aero J 84 (834):214–217

Kanwisher JW, Ridgway SH (1983) The physiological ecology of whales and porpoises. Sci Am 248:102–111

Kanwisher J, Sundnes G (1966) Thermal regulation in cetaceans. In: Norris KS (ed) Whales, dolphins, and porpoises. University of California Press, Berkeley

Kármán T von (1954) Aerodynamics. Selected topics in the light of their historical development. Cornell University Press, Ithaca, New York

Karpouzian G, Spedding G, Cheng HK (1990) Lunate-tail swimming propulsion: Performance analysis. J Fluid Mech 210:329–351

Katz J, Weihs D (1979) Large amplitude unsteady motion of a flexible slender propulsor. J Fluid Mech 90:713–723

Kawachi K, Azuma A (1988) Propulsive mechanism by means of the caudal fin in fishes (in Japanese). Twenty-Sixth Aircraft Symposium, October, 19–21, Sendai

Kawachi K, Inada Y, Azuma A (1989) A study on the gliding flight of flying fish by means of the optimal control theory (in Japanese). 20th Annual Forum of Japan Society Space and Aeronautical Science, April 4–5, Tokyo

Kimura H (1943) Flying characteristics of Zanonia's seeds (in Japanese). Aero Magazine 4:60–63

Kingsolver TG (1985) Butterfly engineering. Sci Am 253:90–97

Kinoshita H, Murakami A (1967) Control of ciliary motion. Physiol Rev 47:53–82

Kirkpatrick SJ (1990) Short communication. The moment of inertia of bird wing. J Exp Biol 151:489–494

Knite M, Hefner RA (1941) Analysis of ground effect on the lifting airscrew. NACA TN 835

Kokshaysky NV (1977) Some scale-dependent problems in aerial animal locomotion. In: Pedley TJ (ed) Scale effects in animal locomotion. Academic, London

Kramer MO (1957) Boundary-layer stabilization by distributed damping. J Aero Sci 24:459–460

Kroeger RA et al. (1972) Low speed aerodynamics for ultra quiet flight. AF Flight Dynamics Laboratory Wright-Patterson AFB, Ohio, AFFDL-TR-71-75

Küchemann D (1953) The distribution of lift over the surface of swept wings. Aero Quart 4:261–278

Küchemann D (1978) The aerodynamic design of aircraft. Pergamon, Oxford

Kuethe DO (1988) Fluid mechanical valving of air flow in bird lungs. J Exp Biol 136:1–12

Kuribayashi S (1981) Insects in flight (in Japanese). Heybonsha, Tokyo

Lamb H (1930) Hydrodynamics (5th edn.). Cambridge University Press, Cambridge

Lamer JE (1976) Prediction of vortex flow characteristics of wings at subsonic and supersonic speeds. J Aircraft 13:490–494

Lan CE (1979) The unsteady quasi-vortex-lattice method with applications to animal propulsion. J Fluid Mech 93:747–765

Landahl MT (1961) On the stability of laminar incompressible boundary layer over a flexible surface. J Fluid Mech 13:609–632

Landgrebe AJ (1972) The wake geometry of a hovering helicopter rotor and its influence on rotor performance. J Am Heli Soc 17:3–15

Landweber L (1956) On a generalization of Taylor's virtual mass relation for Rankin bodies. Quart Appl Math 14:51–56

Landweber L (1961) Motion of immersed and floating bodies. In: Streeter VL (ed) Handbook of fluid dynamics. McGraw-Hill, New York

Landweber L, Macagno NC (1957) Added mass of two-dimensional forms oscillating in a free surface. J Ship Res 1:20–30

Landweber L, Macagno M (1959) Added mass of three parameter family of two-dimensional force oscillating in a free surface. J Ship Res 2:36–48

Landweber L, Macagno M (1960) Added mass of a rigid prolate spheroid oscillating horizontally in a free surface. J Ship Res 3:30–36

Lang TG (1966) Hydrodynamic analysis of cetacean performance. In: Norris KS (ed) Whales, dolphins, and porpoises. University of California Press, Berkeley

Lang TG (1975) Speed, power, and drag measurements of dolphins and porpoises. In: Wu TYT, Brokaw CJ, Brennen C (eds) Swimming and flying in nature, vol 2. Plenum, New York, pp 553–572

Lang TG, Daybell DA (1963) Porpoise performance tests in a seawater tank. Naval Ordinance Test Stat Tech Rep 3063

Lang TG, Pryor K (1966) Hydrodynamic performance of porpoise (Stennella attenuata). Science 152:531–533

Langmuir I (1938) Surface motion of water induced by wind. Science 87:119–123

Large E (1981) The optimal planform, size, and mass of a wing. Aero J 85:(842) 103–110

Lasiewski RC, Dawson WR (1967) A reexamination of the relation between standard metabolic rate and body weight in birds. Condor 69:13–23

Lewis T, Taylor LR (1967) Introduction to experimental ecology. Academic, London, p 167

Liebeck RH (1978) Design of subsonic airfoils for high lift. J Aircraft 15:547–561

Liefson E (1960) Atlas of bacterial flagellation. Academic, New York

Lighthill MJ (1960a) Mathematics and aeronautics. J R Aero Soc 64:373–394

Lighthill MJ (1960b) Note on the swimming of slender fish. J Fluid Mech 9:305–317

Lighthill MJ (1969) Hydrodynamics of aquatic animal propulsion. Ann Rev Fluid Mech 1:413–446

Lighthill MJ (1970) Aquatic animal propulsion of high hydromechanical efficiency. J Fluid Mech 44:265–301

Lighthill MJ (1971) Large-amplitude elongated-body theory of fish locomotion. Proc R Soc Lond [Biol] 179:125–138

Lighthill MJ (1973) On the Weis-Fogh mechanism of lift generation. J Fluid Mech 60:1–17

Lighthill J (1975a) Aerodynamic aspects of animal flight. In: Wu TYT, Brokaw CJ, Brennen C (eds) Swimming and flying in nature, vol 1. Plenum, New York, pp 423–491

Lighthill J (1975b) Mathematical biofluiddynamics. Society for Industrial and Applied Mathematics, Philadelphia

Lighthill J (1976) Flageller hydrodynamics. The John von Newman lecture (1975). SIAM Review 18:161–230

Lighthill J (1977) Introduction to the scaling of aerial locomotion. In: Pedley TJ (ed) Scale effects in animal locomotion. Academic, London

Lighthill J (1979) A simple fluid-flow model of ground effect on hovering. J Fluid Mech 93:781–797

Lighthill J, Blake R (1990) Biofluiddynamics of balistiform and gymnotiform locomotion: Biological background and analysis by elogated-body theory. J Fluid Mech 212:183–207

Lilienthal O, Lilienthal G (1911) Bird flight as the basis of aviation. Green, New York

Lin JC, Walsh MJ (1984) Turbulent roughness drag due to surface waviness at low roughness Reynolds numbers. J Aircraft 21:978–979

Lindberg PG (1955) Growth, population dynamics, and fluid behavior in the spiny lobster Panulirus interruptus. Publ Zool University of California, Berkeley 59:157–248

Lindsey CC (1978) Form, function, and locomotion habits in fish. In: Hoar WS, Randall DJ (eds) Fish Physiology: Locomotion. Academic, New York

Lissmann HW (1963) Electric location by fishes. In: Eisner T, Wilson EO (eds) (1975) Animal behavior. Freeman, San Francisco, pp 180–189

Lissaman PB, Harris GL (1969) Turbulent friction in compliant surface. AIAA J 7:1625–1627

Lissaman PB, Shollenberger CA (1970) Formation flight of birds. Science 168:1003–1005

Lobert G (1981) Spanwise lift distribution of forward-

and aft-swept wings in comparison to the optimum distribution form. J Aircraft 18:496

Losos JB, Papenfuss TJ, Macey JR (1989) Correlates of sprinting, jumping and parachuting performance in the butterfly lizard *Leiolepis belliani*. J Zool [Lond] 217:559–568

Lowy J, Spencer M (1968) Structure and function of bacterial flagella. Symp Soc Exp Biol 22:215–236

Luers JK (1983) Wing contamination: Treat to safe flight. Astronautics and Aeronautics 21:54–59

Luers J, Haines P (1983) Heavy rain influence on airplane accidents. J Aircraft 20:187–190

Mackie GO (1974) Locomotion, flotation, and dispersal. In: Muscatine L, Lenhoff HM (eds) Reviews and new perspectives, coelenterate biology. Academic, New York pp 313–357

Magnan A (1930) Les caractéristiques géométriques et physiques des Poissons. Ann Sci Nat 13:197–198

Magnuson JJ (1970) Hydrostatic equilibrium of *Euthynus affinis*, a pelagic teleost without a gas bladder. Copeia 1:56–85

Magnuson JJ (1973) Comparative study of adaptations for continuous swimming and hydrostatic equilibrium of Scombridae and Xiphoid fishes. US Fish Wild Sew Fish Bull 71:337–356

Magnuson JJ (1978) Locomotion by scombrid fishes: Hydromechanics, morphology, and behavior. In: Hoar WS, Randall DJ (eds) Fish physiology: Locomotion. Academic, New York, pp 239–313

Magnuson JJ, Prescott JH (1966) Courtship, feeding and miscellaneous behavior of Pacific bonito (*Sarda chiliensis*). Animal Behav 14:54–67

Marey EJ (1894) Le mouvement. Librairie de l'Academie de Medecine, Paris

Margaria R (1968) Positive and negative work performances and their efficiencies in human locomotion. Int Z Angew Physiol 25:339–351

Markness MLR, Markness RD, McDonald DA (1957) The collagen and elastin content of the arterial wall in the dog. P R Soc Lond [Biol] 146:541–551

Marshall NB (1965) The life of fishes. Weidenfeld and Nicolson, London

Maruyama K (1984) Elastic protein of muscles (in Japanese). Science 54:747–753

Maskew B (1982) Prediction of subsonic aerodynamic characteristics: A case for low order panel methods. J Aircraft 19:157–163

Massey BS (1979) Mechanics of fluids, 4th edn. Van Nostrand Reinhold, New York

Maxworthy T (1979) Experiments on the Weis-Fogh mechanism of lift generation by insects in hovering flight: Dynamics of the "flying." J Fluid Mech 93:47–63

Maxworthy T (1981) The fluid dynamics of insect flight. Ann Rev Fluid Mech 13:329–350

McCutchen CW (1977) The spinning rotation of ash and tulip tree samaras. Science 197:691–692

McMahon TA, Bonner JT (1983) On size and life. Scientific American Library, New York

McMasters JH (1984) Reflections of a paleoaerodynamicist (invited paper) AIAA-84-2167. AIAA 2nd Applied Aerodynamics Conference, August, Seattle

McMasters JH (1989) The flight of the bumblebee and related myths of entomological engineering. Am Sci 77:164–169

McMasters JH, Henderson ML (1979) Low speed single-element airfoil synthesis. NASA CP2085, Part 1

Meister W (1962) Histological structure of the long bones of penguins. Anat Rec 143:377–388

Mertens R (1960) Fallschirmspringer und Gleitflieger unter den Amphibien und Reptilien. In: Schimidt H (ed) Der Flug der Tiere. Kramer, Frankfurt am Main

Miley S (1974) On the design of airfoils for low Reynolds numbers. AIAA 74-1017

Miller N, Tang JC, Perlmutter AA (1968) Theoretical and experimental investigation of the instantaneous induced velocity field in the wake of a lifting rotor. USAAVLABS Tech Rep, pp 67–68

Miyan JA, Ewing AW (1988) Further observations on dipteran flight: Details of the mechanism. J Exp Biol 136:229–241

Montoya LC, Flechner SG, Jacobs PF (1977) Effect of winglets on a first-generation jet transport wing: Pressure and spanwise load distributions for a semi-span model at high subsonic speeds. NASA TN D-8474

Moore JD, Trueman ER (1971) Swimming of the scallop, *Chlamys operculans* (L.). J Exp Mar Biol Ecol 6:179–185

Moreau RE (1961) Problems of Mediterranean-Sahara migration. IBIS 103a:373–427, 580–623

Moriya T (1959) On the aerodynamic theory of an arbitrary wing section. Selected scientific and technical papers. The Moriya Memorial Committee, Tokyo

Myall JO, Berger SA (1970) Recent progress in the analytical study of sails. Proceedings of the 2nd AIAA Symposium on Sailing, AIAA, Los Angeles

Nachtigall W (1974) Insects in flight. McGraw-Hill, New York

Nachtigall W (1976) Wing movements and the generation of aerodynamic forces by some medium-sized insects. In: Rainey RC (ed) Insect Flight. Blackwell, Oxford, pp 31–49

Nachtigall W (1980a) Mechanics of swimming in water beetles. In: Elder HY, Trueman ER (eds) Aspects of animal movement. Cambridge University Press, Cambridge, pp 107–124

Nachtigall W (1980b) Some aspects of the kinematics of wing beat movements in insects. In: Elder HY, Trueman ER (eds) Aspects of animal movement. Cambridge University Press, Cambridge, pp 169–175

Nagai M, Maeda G, Irabu K (1982) Study of swimming

motions of various swimmers and a mechanical fish. ASME, Advances in Bioengineering, Winter Annual Meeting, Nov 14–19, Phoenix, pp 56–58

Narsall JR (1956) The lateral musculature and the swimming of fish. Proc Zool Soc [Lond] 126:127–143

Nelson B (1980) Seabirds. Their biology and ecology. Hamlyn, London

Newman JN (1977) Marine hydrodynamics. MIT, Cambridge, Massachusetts

Newman BG, Low HT (1981) Two-dimensional flow at right angles to a flexible membrane. Aero Quart 32:243–269

Newman BG, Savage SB, Schouella D (1977) Model tests on a wing section of an Aeschna dragonfly. In: Pedley TJ (ed) Scale effects in animal locomotion. Academic, London

Nishiki S (1983) Spiders flying in sky (in Japanese). Wildlife 57:15–16

Nishimura S (1976) Current and dispersion of living creatures over long distances. Study on bio-oceanography (in Japanese). Kaiyo Shuppan, Tokyo

Noll RB, Ham ND (1982) Effects of dynamic stall on SWECS. Trans ASME 104:96–101

Nonweiler T (1963) Qualitative solution of the stability equation for a boundary layer in contact with various forms of a flexible surface. Aero Res Coun Rep 622

Norberg RA (1972) The pterostigma of insect wings is an inertial regulator of wing pitch. J Comp Physiol 81:9–22

Norberg RA (1973) Autorotation, self-stability and structure of single-winged fruits and seeds (samara) with comparative remarks on animal flight. Biol Rev 48:561–596

Norberg RA (1975) Hovering flight of the dragonfly Aeschna juncea L., kinematics and aerodynamics. In: Wu TYT, Brokaw CJ, Brennen C (eds) Swimming and flying in nature, vol 2. Plenum, New York, pp 763–778

Norris KS, Prescott JH (1961) Observations on Pacific cetaceans of Californian and Mexican waters. Univ Calif Publ Zool 63:291–402

O'Dor RK, Webber DM (1991) Invertebrate athletes: Trade-offs between transport efficiency and power density in cephalopod evolution. J Exp Biol 160:93–112

Oehme H (1977) On the aerodynamics of separated primaries in the avian wing. In: Pedley TJ (ed) Scale effects in animal locomotion. Academic, New York

Onda Y, Azuma A, Yasuda K (1986) The wake structure trailed from samara in autorotation (in Japanese). Special issue of 14th Symposium on Flow Visualization, July 16–18, Tokyo Flow Visualization Society in Japan, vol 6 (22) pp 311–314

Orszag SA (1977) Prediction of compliant wall drag reduction. NASA CR-2911

Orszag SA (1979) Prediction of compliant wall drag reduction. NASA CR-3071

Osborne MFM (1951) Aerodynamics of flapping flight with application to insects. J Exp Biol 28:221–245

Oseen CW (1913) Über de Gültigkeitsbereich der stokesschen Widerstandsformel. Arkiv Mat Astr Phys 9:1–15

Ovchinnikov VV (1966) Turbulence in the boundary layer as a method for reducing the reistance of certain fish in movement. Biofizika 11:213–215

Pankhurst RC (1964) Dimensional analysis and scale factors. Chapman and Hall, London

Pao HP (1970) Dynamical stability of a towed thin flexible cylinder. J Hydronautics 4:144–150

Parrott GC (1970) Aerodynamics of gliding flight of the black vulture Coragyps stratus. J Exp Biol 53:363–374

Parry DA (1949) The swimming of whales and a discussion of Gray's paradox. J Exp Biol 26:24–34

Partridge BL (1982) The structure and function of fish schools. Sci Am 246:90–99

Paturi FR (1974) Nature, mother of invention, the engineering of plant life. Penguin, London

Pennycuick CJ (1968a) A wind tunnel study of gliding flight in the pigeon Columba livia. J Exp Biol 49:509–526

Pennycuick CJ (1968b) Power requirements for horizontal flight in the pigeon Columba livia. J Exp Biol 49:527–555

Pennycuick CJ (1969) The mechanics of bird migration. IBIS 3:525–556

Pennycuick CJ (1972) Animal flight. Edward Arnold, London

Pennycuick CJ (1975) Mechanics of flight. In: Farner DS, King JR. Parkes KC (eds) Avian Biology, vol 5. Academic, New York

Pennycuick CJ (1990) Predicting wingbeat frequency and wavelength of birds. J Exp Biol 150:171–185

Pennycuick CJ, Lock A (1976) Elastic energy storage in primary feather shafts. J Exp Biol 64:677–689

Pennycuick CJ, Obrecht HH III, Fuller MR (1988) Empirical estimates of body drag of large waterfowl and raptors. J Exp Biol 135:253–264

Pennycuick CJ, Fuller MR, McAllister L (1989) Climbing performance of harris hawks (Parabuteo unicinctus) with added load: Implications for muscle mechanics and for radiotracking. J Exp Biol 142:17–29

Perkins CD, Hage RE (1965) Airplane performance, stability and control, 10th edn. John Wiley, New York

Perrins C, Cameron AD (1976) Birds: Their life, their ways, their world. Harry N. Abrams, New York

Perrins CM, Middleton ALA (1985) The encyclopedia of birds. Facts on File, New York

Perry P, Acosta AJ, Kiceniuk T (1961) Simulated wave-riding dolphins. Nature 192:148–149

Polhamus EC (1966) A concept of the vortex lift of sharp-edge delta wings based on a leading-edge-suction analogy. NASA TN D-3767

Polhamus EC (1968) Application of the leading-edge-suction analogy of vortex lift to the drag-due-to-lift of sharp-edge delta wings. NASA TN D-4739

Pope A, Harper JJ (1966) Low-speed wind tunnel

testing. John Wiley, New York

Prampero PE di, Cortili G, Celentano F, Cerretelli P (1971) Physiological aspects of rowing. J Applied Phys 31:853–857

Prandtl L, Betz A (1927) Vier abhandlungen zur Hydrodynamik und Aerodynamik. E AVA 1:50

Pressnell M (1987) Winning wings. Aeromodeller 52:492–495

Prezant RS, Chalermwat K (1984) Floatation of the bivalve *Corbicula fluminea* as a means of dispersal. Science 225:1491–1493

Pringle JWS (1948) The gyroscopic mechanism of the halters of Diptera. Phil Trans R Soc Lond [Biol] 233:47–384

Pringle JWS (1957) Insect flight. Cambridge University Press, Cambridge

Pringle JWS (1975) Insect flight. Oxford Biology Reader 52. Oxford University. Press, Oxford

Pringle JWS (1976) The muscles and sense organs involved in insect flight. In: Rainey RC (eds) Insect flight. Blackwell, Oxford, pp 3–15

Pyatetskiy VY (1970) Kinematic swimming characteristics of some fast marine fish (in Russian). Bionika 4:11–20

Rainey AG (1956) Preliminary study of some factors which affect the stall-flutter characteristics of thin wings. NACA TN 3622

Rainey RC (1969) Effects of atmospheric conditions on insect movement. Quart J R Met Soc 95:424–434

Raj P, Iversen JD (1978) Inviscid interaction of trailing vortex sheets approximated by point vortices. J Aircraft 15:857–859

Raspet A (1950) Performance measurements of a soaring bird. Aero Eng Rev 9:14–17

Raspet A (1960) Biophysics of bird flight. Soaring 24:12–20

Rayleigh L (1983) The soaring of birds. Nature 27:534–535

Rayner J (1977) The intermittent flight of birds. In: Pedley TJ (ed) Scale effects in animal locomotion. Academic, New York

Rayner JMV (1979a) A new approach to animal flight mechanics. J Exp Biol 80:17–54

Rayner JMV (1979b) A vortex theory of animal flight: The vortex wake of a hovering animal. J Fluid Mech 91:697–730

Rayner JMV (1979c) A vortex theory of animal flight: The forward flight of birds. J Fluid Mech 91:731–763

Rayner JMV (1985a) Vorticity and propulsion mechanics in swimming and flying vertebrates. Principles of construction in fossil and present reptiles. Konzepte SFB 230:89–117

Rayner JMV (1985b) Bounding and undulating flight in birds. J Theor Biol 117:47–77

Reavis MA, Luttges MW (1988) Aerodynamic forces produced by a dragonfly. AIAA-88-0330, AIAA 26th Aerospace Science Meeting, January 11–14, Reno, Nevada

Rees CJC (1975a) Form and function in corrugated insect wings. Nature 256:200–203

Rees CJC (1975b) Aerodynamic properties of an insect wing section and a smooth aerofoil compared. Nature 258:141–142

Reiss E (1984) A nonlinear structural concept for drag-reducing compliant walls. AIAA J 22:399–402

Reynolds AJ (1965) The swimming of minute organisms. J Fluid Mech 23:241–260

Richardson EG (1936) The physical aspects of fish locomotion. J Exp Biol 13:63–74

Ritchile D (1950) The frequency of beat of sperm tails. Science 111:172–173

Rosen MW, Cornford NE (1970) Fluid friction of the slime of aquatic animals. Naval Undersea Research and Development Center, Tech Report 193:1–45

Rosen MW, Cornford NE (1971) Fluid friction of fish slimes. Nature 234:49–51

Rossow VJ (1977) Convective merging of vortex cores in lift-generated wakes. J Aircraft 14:283–290

Rubner M (1883) Über den Einfluss der Körpergrösse auf Stoff-und Kraftwechsel. Z Biol 19:535–562

Rüppell G (1977) Bird flight. Van Nostrand Reinhold, New York

Rüppell G (1989) Kinematic analysis of symmetrical flight manœuvres of odonata. J Exp Biol 144:13–42

Saffman PG (1967) The self-propulsion of a deformable body in a perfect fluid. J Fluid Mech 28:385–389

Sarpkaya T (1974) On the performance of hydrofoils in dilute polyox solutions. International Conference on Drag Reduction, September, Cambridge

Satir P (1974) The present status of the sliding microtube model of ciliary motion. In: Sleigh MA (ed) Cilia and flagella. Academic, London

Sato M (1980) Analyses of wing motion observed in the flight of living creatures (in Japanese). Thesis, University of Tokyo

Sato M, Azuma A, Saito S (1979) Analysis of wing motion in living creatures (in Japanese). 17th Airc Symp Japan Soc Aero and Space Sci, November 14–17, Tokyo

Sannders HE (1957) Hydrodynamics in ship design, vol 2. The Society of Naval Architects and Marine Engineers, Jersey City, NY, chap. 62

Savage SB, Newman BG, Wong D T-M (1979) The role of vortices and unsteady effects during the hovering flight of dragonflies. J Exp Biol 83:59–77

Schlichting H (1942) Analysis of flying in formation. Mitt D Akademie Lufo, Trans TMB 239 and Doct ZWB UM2066, 1944, Trans TMB 240

Schmidt-Nielsen K (1971) How birds breathe. Sci Am 225:72–79

Schmidt-Nielsen K (1972) Locomotion: Energy cost of swimming, flying and running. Science 177:222–228

Schmidt-Nielsen K (1975) Recent advances in avian respiration. In: Peaker M (ed) Symp Zool Soc Lond 35:33–47

Schmidt-Nielsen K (1977) Problems of scaling: Locomotion and physical correlates. In: Pedley TJ (ed) Scale effects in animal locomotion. Academic, London, pp 1–21

Schmitz FW (1967) Aerodynamik des Flugmodells. Trans N70-39001, Nat Tech Info Service, Springfield, pp 45–32

Schwind RG, Allen HJ (1973) The effects of leading-edge separations on reducing flow unsteadiness about airfoils—An experimental and analytical investigation. NASA CR-2344

Scorer RS (1958) Natural aerodynamics. Pergamon, London

Shapiro AH (1961) Basic equations of fluid flow. In: Streeter VL (ed) Handbook of fluid dynamics. McGraw-Hill, New York

Shepard MP, Hartt AC, Yonemori T (1968) Salmon of the North Pacific Ocean: Chum salmon in offshore waters. Int North Pac Fish Comm Bull 25:1–69

Shimada K, Yoshida T, Asakura S (1974) Cinemicrographic analysis of the movement of flagellated bacteria. In: Wu TYT, Brokaw CJ, Brennen C (eds) Swimming and flying in nature, vol 1. Plenum, New York, pp 31–43

Sinnarwalla AM, Sundaram TR (1978) Lift and drag effects due to polymer injections on a symmetric hydrofoils. J Hydro 12:71–77

Skews BW (1990) Autorotation of rectangular plates. J Fluid Mech 217:33–40

Sleigh MA, Blake JR (1977) Methods of ciliary propulsion and their size limitations. In: Pedley TJ (ed) Scale effects in animal locomotion. Academic, London

Sleigh MA, Barlow DI (1980) Metachronism and control of locomotion in animals with many propulsive structures. In: Elder HY, Trueman ER (eds) Aspects of animal movement. Cambridge University Press, Cambridge, pp 49–70

Smith B (1980) New approaches to sailing. Astronautics and Aeronautics 18:36–47

Smith DS (1965) The flight muscles of insects. In: Eisner T, Wilson EO (eds) (1977) The insects. Freeman, San Francisco

Smith EW (1985) About the exhibits by Elizabeth and Max Hall. Museum of Comparative Zoology, Harvard University, Cambridge

Smith RL, Blick EF (1969) Skin friction of compliant surfaces with foamed materials substrate. J Hydro 3:100–102

Somps C, Luttges M (1985) Dragonfly flight: Novel uses of unsteady separated flows. Science 228:1326–1329

Spedding GR (1986) The wake of a jackdaw (*Corvus monedula*) in slow flight. J Exp Biol 125:287–307

Spedding GR (1987a) The wake of a kestrel (*Falco tinnunculus*) in gliding flight. J Exp Biol 125:287–307

Spedding GR (1987b) The wake of a kestrel (*Falco tinnunculus*) in flapping flight. J Exp Biol 127:59–78

Spedding GR, Rayner JMV, Pennycuick CJ (1984) Momentum and energy in the wake of a pigeon (*Columba livia*) in slow flight. J Exp Biol 111:81–102

Spillman JJ (1978) The use of wing tip sails to reduce vortex drag. Aero J 618:387–395

Standard Atmosphere (1955) Table and data for altitudes to 65,800 foot. NACA Rep 1235

Stephenson R, Lovvorn JR, Heieis MRA, Jones DR, Blake RW (1989) A hydromechanical estimate of the power requirements of diving and surface swimming in lesser scaup (*Aythya affinis*). J Exp Biol 147:507–519

Stepniewski WZ (1979) Rotary-wing aerodynamics: Basic theories of rotor aerodynamics (with application to helicopters). NASA Contract Rep 3082

Stepniewski WZ, Keys CN (1984) Rotary-wing aerodynamics, vols I and II. Dover, New York

Storer JH (1948) The flight of birds. Bull 28. Cranbrook Institute of Science, Bloomfield Hills, Michigan

Sunada S, Azuma A (1988) Flow visualization of vortices generated by a triangular wing in beating motion (in Japanese). 16th flow visualization symposium. J Flow Visual 8:335–338

Sunada S, Kawachi K, Azuma A (1989) Vortices generated by butterfly's wing at take-off phase. 5th International Symposium on Flow Visualization, Aug 21–25, Prague

Sunada S, Kawachi K, Watanabe I, Azuma A (1989) Visualization of vortex generated from two rotating plates (in Japanese). J Flow Visual 9(Suppl):31–34

Sundnes G (1963) Swimming speed of fish as a factor in gear research. Fiskeridir Skr, Ser Havunders 13:126–132

Tanaka K (1976) The Galapagos. Wildlife of Kojo Tanaka, I (in Japanese). Shakai Shiso, Tokyo

Tani I (1943) On the design of aerofils in which the transition of the boundary layer is delayed (in Japanese). Report of the Aeronautical Research Institute, Tokyo Imperial University, vol XIX, no. 250

Tani I (1988) Drag reduction by riblet viewed as roughness problem. Proc Jan Aca 64:21–24

Tani I (1989) Drag reduction by sand grain roughness. 2nd IUTAM symposium on structure of turbulence and drag reduction, July 25–28, Zurich

Taylor G (1951) Analysis of the swimming of microscopic organisms. Proc R Soc Lond [A] 209:447–461

Taylor G (1952a) The action of waving cylindrical tails in propelling microscopic organisms. Proc R Soc Lond [A] 211:225–239

Taylor G (1952b) Analysis of the swimming of long and narrow animals. Proc R Soc Lond [A] 24:158–183

Taylor G (1953) The action of waving cylindrical tails in propelling microscopic organisms. Proc R Soc Lond [A] 211:225–239

Terres JK (1968) How birds fly. Hawthorn, New York

Theilheimer F (1943) The influence of sweep on the spanwise lift distribution of wings. J Aero Sci 10:101–104

Theodorsen T, Garrick IE (1933) General potential

theory of arbitrary wing sections. NACA Rep no. 452

Thom A, Swart P (1940) The forces on an airfoil at very low speeds. J R Aero Soc 44:761–770

Thomas J (ed) (1984) Swimming and buoyancy in fish. Animal physiology, unit 13. Open University Press, Milton Keynes

Thompson DW (1942) On growth and form, vols 1 and 2. Cambridge University Press, Cambridge

Tillet JPK (1970) Axial and transverse Stokes flow past slender axsymmetric bodies. J Fluid Mech 44:401–418

Tomotika S, Nagamiya T, Takenouti Y (1933) The lift on a flat plate place near a plane wall with special reference to the effect of the ground upon the lift of a monoplane aerofoil (report). Aeronautical Research Institute, Tokyo

Trueman ER (1975) The locomotion of soft-bodied animals. Arnold, London

Trueman ER (1980) Swimming by jet propulsion. In: Elder HY, Trueman ER (eds) Aspects of animal movement. Cambridge University Press, Cambridge, pp 93–105

Trueman ER, Packard A (1968) Motor performance of some cephalopods. J Exp Biol 49:495–507

Tuck EO (1968) A note on a swimming problem. J Fluid Mech 31:305–308

Tucker VA (1968) Respiratory exchange and evaporative water loss in the flying budgerigar. J Exp Biol 48:67–87

Tucker VA (1970) Energetic cost of locomotion in animals. Comp Biochem Physiol 34:841–6

Tucker VA (1971) Flight energetics in birds. Am Zool 11:115–124

Tucker VA (1972) Metabolism during flight in the laughing gull *Larus atricilla*. Am J Physiol 222:237–245

Tucker VA (1973) Bird metabolism during flight: Evaluation of a theory. J Exp Biol 58:689–709

Tucker VA (1974) Energetics of natural avian flight. In: Paynter RA Jr (ed) Avian energetics. Natural Ornithologists Club, Cambridge, Massachusetts

Tucker VA (1975) Flight energetics. In: Peaker M (ed) Avian physiology. 35th Symposium of the Zoological Society of London. Springer, New York, pp 49–63

Tucker VA (1977) Scaling and avian flight. In: Pedley YJ (ed) Scale effects in animal locomotion. Academic, London

Tucker VA (1987) Gliding birds: The effect of variable wing span. J Exp Biol 133:33–58

Tucker VA, Parrott GC (1970) Aerodynamics of gliding flight in a falcon and other birds. J Exp Biol 52:345–367

Uchida H (1983) A scrappy bird, the Japanese wagtail (in Japanese). Anima, Magazine of Natural History 121:34–39

Victory M (1943) Flutter at high incidence. British ARC R and M 2048

Videler JJ, Hess F (1984) Fast continuous swimming of two pelagic predators, saithe (*Pollachius virens*) and mackerel (*Scomber scombrus*): A kinematic analysis. J Exp Biol 109:209–228

Vinogradov IN (1951) The aerodynamics of soaring bird flight (in Russian) Dosarm, Moscow

Vlymen WJ (1974) Swimming energetics of the larval anchovy *Engraulis mordax*. Fish Bull 72:885–899

Vogel S (1966) Flight in *Drosophila*: Flight performance of tethered flies. J Exp Biol 44:567–578

Vogel S (1967a) Flight in *Drosophila*: Variation in stroke parameters and wing contour. J Exp Biol 46:383–392

Vogel S (1967b) Flight in *Drosophila*: Aerodynamic characteristics of flying and wing models. J Exp Biol 46:431–443

Vugts HF, Wingerden van WKRE (1976) Meteorological aspects of aeronautic behaviour of spiders. OIKOS 27, Copenhagen, pp 433–444

Walker J (1983) Loading into the ways of water striders, the insects that walk (and run) on water. Sci Am 249:160–169

Walsh MJ (1982) Turbulent boundary layer drag reduction using riblets. AIAA paper, 82-0169

Walsh MJ, Lindemann AM (1984) Optimization and application of riblets for turbulent drag reduction. AIAA paper, 84-0347

Walsh MJ, Weinstein LM (1978) Drag and heat transfer on surfaces with small longitudinal fins. AIAA paper 78-1161, AIAA 11th Fluid and Plasma Dynamics Conference, Seattle, Washington

Walsh MJ, Sellers WL III, McGinley CB (1989) Riblet drag at flight conditions. J Aircraft 26:570–575

Walters V (1962) Body form and swimming performance in scombroid fishes. Am Zool 2:3–149

Ward PG, Greenwald L, Greenwald OE (1980) The buoyancy of the chambered nautilus. Sci Am 243(4):162–175

Wardle CS (1975) Limit of fish swimming speed. Nature 255:725–727

Wardle CS (1977) Effects of size on the swimming speeds of fish. In: Pedley TJ (ed) Scale effects in animal locomotion. Academic, New York

Warner F, Satir P (1974) The structural basis of ciliary bend formation. Radial spoke positional changes accompanying microtuble sliding. J Cell Biol 63:35–63

Washburn J, Washburn L (1984) Active aerial dispersal of minute wingless arthropods: Exploitation of boundary-layer velocity gradients. Science 223:1088–1089

Watanabe Y, Kimura M (1976) The movement of a fish: A biomechanical study on motional function. 8th International Congress on Cybernetics, September 6–11, Namur, Belgium

Watanabe I, Azuma A, Watanabe T (1986) Wake vortex of flying dragonflies. 4th International Symposium on Flow Visualization, August 26–29, Paris

Watson RD, Balasubramanian R (1984) Wall mass transfer and pressure gradient effects on turbulent skin friction. AILL J 22:143–145

Webb PW (1971a) The swimming energetics of trout: Thrust and power output at cruising speeds. J Exp Biol 55:489–520

Webb PW (1971b) The swimming energetics of trout: Oxygen consumption and swimming efficiency. J Exp Biol 55:521–540

Webb PW (1973a) Kinematics of pectoral fin propulsion in *Cymatogaster aggregata*. J Exp Biol 59:697–710

Webb PW (1973b) Effects of partial caudal-fin amputation on the kinematics and metabolic rate of underyearling sockeye salmon (*Oncorhynchus nerka*) at steady swimming speed. J Exp Biol 59:565–581

Webb PW (1974) Efficiency of pectoral-fin propulsion of Cymatogaster aggregata. In: Wu TYT, Brokaw CJ, Brennen C (eds) Swimming and flying in nature, vol 2. Plenum, New York, pp 573–584

Webb PW (1975) Hydrodynamics and energetics of fish propulsion (bulletin). Fisheries Research Board of Canada 190:1–159

Webber DM, O'Dor RK (1986) Monitoring the metabolic rate and activity of free-swimming squid with telemetered jet pressure. J Exp Biol 126:205–224

Weihs D (1972) A hydrodynamical analysis of fish turning manoeuvres. Proc R Soc Lond [Biol] 182:59–72

Weihs D (1973a) The mechanism of rapid starting of slender fish. Biorheology 10:343–350

Weihs D (1973b) Mechanically efficient swimming techniques for fish with negative buoyancy. J Marine Res 31:194–209

Weihs D (1974) Energetic advantages of burst swimming of fish. J Theor Biol 48:215–229

Weihs D (1977) Effects of size on sustained swimming speeds of aquatic organisms. In: Pedley TJ (ed) Scale effects in animal locomotion. Academic, New York, pp 333–338

Weis-Fogh T (1956) Biology and physics of locust flight II. Flight performance of the desert locust (*Schistocerca gregaria*). Philos Trans R Soc Lond [Biol] 239:459–510

Weis-Fogh T (1961) Molecular interpretation of the elasticity of resilin, a rubber-like protein. J Mol Biol 3:648–667

Weis-Fogh T (1964) Biology and physics of locust flight. VIII. Lift and metabolic rate of flying locusts. J Exp Biol 41:257–271

Weis-Fogh T (1965) Elasticity and wing movements in insects. Proc XIIth Int Congr Ent pp 186–188

Weis-Fogh T (1967) Metabolism and weight economy in migrating animals, particularly birds and insects. In: Beament JWL, Treherne JE (eds) Insects and physiology. Oliver and Boyd, Edinburgh, pp 143–159

Weis-Fogh T (1972) Energetics of hovering flight in hummingbirds and in Drosophila. J Exp Biol 56:79–104

Weis-Fogh T (1973) Quick estimates of flight fitness in hovering animals, including novel mechanisms for lift production. J Exp Biol 59:169–230

Weis-Fogh T (1975a) Flapping flight and power in birds and insects, coventional and novel mechanisms. In: Wu TYT, Brokaw CJ, Brennen C (eds) Swimming and flying in nature, vol 2. Plenum, New York, pp 729–762

Weis-Fogh T (1975b) Unusual mechanisms for the generation of lift in flying animals. Sci Am 233:80–87

Weis-Fogh T (1977) Dimensional analysis of hovering flight. In: Pedley TJ (ed) Scale effects in animal locomotion. Academic, London

Weis-Fogh T, Alexander R McN (1977) The sustained power output from striated muscles. In: Pedley TJ (ed) Scale effects in animal locomotion. Academic, London, pp 511–525

Weis-Fogh T, Jensen M (1956) Biology and physics of locust flight: Basic principles in insect flight. A critical review. Philos Trans R Soc Lond [Biol] 667:415–458

Weissinger J (1947) The lift distribution of swept-back wings. NACA TM 1120, translated from ZWB Forschungsbericht, Nr. 1553, Berlin-Adelershof, 1942

Welty C (1955) Birds as flying machines. In: Griffin DR (ed) (1974) Animal engineering. Freeman, San Francisco

Westwater FL (1935) The rolling up of the surface of discontinuity behind an airfoil of finite span. British ARC R and M 1692

Whipp BJ, Wasserman K (1969) Efficiency of muscular work. J Appl Physiol 26:644–648

Whitcomb RT (1976) A design approach and selected wind tunnel results at high subsonic speeds for wing-tip mounted winglets. NASA TN D-8260

Whitehead H (1985) Why whales leap. Sci Am 252:70–75

Whitfield P, Orr R (1978) The hunters. Simon and Schuster, New York

Wigley WCS (1953) Water forces on submerged bodies in motion. Trans INA 95:268–279

Wilkie DR (1959) The work output of animals: Flight by birds and by manpower. Nature 183:1515–1516

Wilkie DR (1977) Metabolism and body size. In: Pedley TJ (ed) Scale effects in animal locomotion. Academic, London, pp 23–36

Williams GM (1974) Viscous modelling of wing-generated trailing vortices. Aero Quart 25:143–154

Williamson K (1969) Weather systems and bird movements. Quart J R Met Soc 95:414–423

Withers SP, Timko PL (1977) The significance of ground effect to the aerodynamic cost of flight and energetics of the black skimmer (*Rhyncops nigra*). J Exp Biol 70:13–26

Woodcock AH (1948) The swimming of dolphins. Nature 161:602

World Atlas of Birds (1974) Crescent Books, New York

World of wildlife. (1970) Life in the ocean. Orbis, London

Wortmann FX (1961) Progress in the design of low drag airfoils. In: Lachmann GV (ed) Boundary layer and flow control. Pergamon, London, pp 748–770

Wortmann FX (1972) Airfoils with high lift/drag ratio at a Reynolds number of about one million. Motorless Flight Research, NASA CR-2315

Wu JC, Hu-Chen H (1984) Unsteady aerodynamics of articulate lifting bodies. AIAA-84-2184, AIAA 2nd Applied Aerodynamic Conference, August 21–23, Seattle, Washington

Wu TYT (1961) Swimming of a waving plate. J Fluid Mech 10:321–344

Wu TYT (1971a) Hydromechanics of swimming of fishes and cetaceans. In: Yih CS (ed) Advances in applied mechanics, vol II. Academic, New York

Wu TYT (1971b) Hydrodynamics of swimming propulsion: Swimming of a two-dimensional flexible plate at variable forward speeds in an inviscid fluid. J Fluid Mech 46 2:337–355

Wu TYT (1971c) Hydrodynamics of swimming propulsion: Some optimum shape problems. J Fluid Mech 46:521–544

Wu TYT (1971d) Hydromechanics of swimming propulsion: Swimming and optimum movements of slender fish with side fins. J Fluid Mech 46:545–568

Wu TYT (1977) Flagella and cilia hydromechanics. In: Skalak R, Schultz AB (eds) Biomechanics Symposium, presented at the 1977 Joint Applied Mechanics Fluids Engineering and Bioengineering Conference. Am Soc Mech Eng, New York

Wu AK, Miele A (1980) Sequential conjugate gradient restoration algorithm for optimal control problems with non-differential constraints and general boundary conditions: Optimal control applications and methods. John Wiley, New York, pp 69–88, 119–130

Wu TYT, Newman JN (1972) Unsteady flow around a slender fish-like body. Proceedings of the International Symposium on Directional Stability and Control of Bodies Moving in Water. Institution of Mechanical Engineers, London

Yamana M, Nakaguchi H (1968) Design philosophy and engineering of aircraft (in Japanese). Yokendo, Tokyo

Yasuda K, Azuma A (1992) The autorotation boundary in Samaras. SEB Annual Meeting, April 5–10, University of Lancaster, UK

Yates GT (1983) Hydromechanics of body and caudal fin propulsion. In: Webb PW, Weihs D (eds) Fish biomechanics. Pragaeyer, New York

Yates GT (1986) Optimum pitching axes in flapping wing propulsion. J Theor Biol 120:255–276

Yih CS (1961) Ideal-fluid flow. In: Streeter VL (ed) Handbook of fluid dynamics. McGraw-Hill, New York

Young AD (1939) The calculation of the total and skin friction drags of bodies of revolution at zero incidence. British ARC R and M, no. 1874

Zbrozek JK (1950) Ground effect on the lifting rotor. Aero Res Counc Rep 2347. HMSO, London

Zimmer H (1979) The significance of wing end configuration in airfoil design for civil aviation aircraft. NASA TM-75711

Zimmer H (1983) The aerodynamic optimization of wings in the subsonic speed range and the influence of the design of the wing tips. Dissertation, University of Stuttgart

Zorgniotti AW, Hotchkiss RS, Wall LC (1958) High-speed cinephotomicrography of human spermatozoa. Med Rad Phot 34:44–49

Further Reading

Alexander RM, Goldspink G (eds) (1977) Mechanics and energetics of animal locomotion. Chapman and Hall, London

Anderson HT, Harold T (eds) (1969) The biology of marine mammals. Academic, New York

Beament JWL, Treherne JE (eds) (1967) Insect and physiology. Oliver and Boyd, Edinburgh

Bolis L, Maddrell SHP, Schmidt-Neilsen K (eds) (1973) Comparative animal physiology. North Holland, Amsterdam

Eisner T, Wilson EO (eds) (1975) Animal behavior. Freeman, San Francisco

Eisner T, Wilson EO (eds) (1977) The insects. Freeman, San Francisco

Elder HY, Truman ER (eds) (1980) Aspects of animal movement. Cambridge University Press, Cambridge

Farner DS, King JR, Parkes KC (eds) (1975) Avian biology, vol. V. Academic, New York

Greiner L (ed) (1967) Underwater missile propulsion. Compaso, Arlington, Virginia

Griffin DR (ed) (1974) Animal engineering. Freeman, San Francisco

Hoar WS, Randall DJ (eds) (1978) Fish physiology. Academic, New York

Kuethe AM, Chow CY (1986) Foundations of Aerodynamics—Bases of Aerodynamic Design, Fourth ed. John Wileys & Sons, New York

Lachmann GV (ed) (1961) Boundary layer and flow control. Pergamon, London

Marshall SA, Orr D (eds) (1953) Essays in marine biology. Oliver and Boyd, Edinburgh

Muscatine L, Lenhoff HM (eds) (1974) Reviews and new perspectives, coelenterate biology. Academic, New York

Norris KS (ed) (1966) Whales, dolphins, and porpoises. University of California Press, Berkeley

Paynter RA Jr (ed) (1974) Avian energetics. Natural Ornithologists Club, Cambridge, Massachusetts

Pedley TJ (ed) (1977) Scale effects in animal locomotion. Academic, New York

Rainey RC (ed) (1976) Insect flight. Blackwell, Oxford

Sleigh MA (ed) (1974) Cilia and flagella. Academic, London

Streeter VL (ed) (1961) Handbook of fluid dynamics. McGraw-Hill, New York

Webb PW, Weihs D (eds) (1983) Fish biomechanics. Praeger, New York

Wu TYT, Brokaw CJ, Brennen C (eds) (1974) Swimming and flying in nature, 1 and 2. Press, New York

Yih CS (ed) (1971) Advances in applied mechanics. Academic, New York

Index